Microscopy of Semiconducting Materials, 1983

Microscopy of Semiconducting Materials, 1983

Proceedings of the Institute of Physics Conference held in
St Catherine's College, Oxford, 21–23 March 1983

Edited by A G Cullis, S M Davidson and G R Booker

S
M   M
Ⅲ

Conference Series Number 67

The Institute of Physics
Bristol and London

CODEN IPHSAC 67 1–520 (1983)

British Library Cataloguing in Publication Data

Microscopy of semiconducting materials, 1983.
  —(Conference series/Institute of Physics, ISSN
  0305-2346; no. 67)
  1. Semiconductors—Research—Congresses
  2. Electron microscopy—Congresses
  I. Cullis, A G   II. Booker, G R
  III Davidson, S M   IV. Series
  537.6'22   QC611.4

ISBN 0-85498-158-6
ISSN 0305-2346

Conference Co-Chairmen
    A G Cullis and S M Davidson
Local Organiser
    G R Booker
Honorary Editors
    A G Cullis, S M Davidson and G R Booker

Published by The Institute of Physics, Techno House, Redcliffe Way, Bristol BS1 6NX, and 47 Belgrave Square, London SW1X 8QX, England.

Printed in Great Britain by J W Arrowsmith Ltd, Bristol.

Preface

This volume contains invited and contributed papers presented at the conference on Microscopy of Semiconducting Materials which took place on 21–23 March 1983 in St Catherine's College, Oxford. The conference was the third in the series devoted to advances in microscopical studies of semiconductors and was organised under the auspices of the Electron Microscopy and Analysis Group of The Institute of Physics with co-sponsorship by the Materials Section of the Royal Microscopical Society. Over 220 scientists and technologists from the UK and ten other countries attended the three-day meeting.

As was clear at previous conferences in the series, research advances in both fundamental and applied semiconductor microscopy are taking place extremely rapidly. Special sessions on the properties of dislocations, high-resolution microscopy and transient annealing phenomena highlighted the progress in these three particularly dynamic fields. The semiconductor general sessions covered a wider range of studies of as-grown or, often, ion-implanted and heat-treated material. Electronic device assessment by transmission and scanning electron microscopy and by acoustical and light optical techniques was featured particularly strongly. An individual session was also devoted to developments in x-ray imaging and diffraction work. Overall, the conference gave a comprehensive picture of the state-of-the-art in the different topic areas.

Each camera-ready paper submitted for publication in this proceedings volume was scrutinised by at least one referee and modified accordingly. We are most grateful for the detailed consideration given to the various papers by members of the referee panel, which was constituted as follows:

P D Augustus, C B Carter, C Claeys, G Fontaine, D B Holt, D C Joy, S S Lau, K Löhnert, S Mahajan, J L Merz, W C Nixon, H Oppolzer, J R Patel, F A Ponce, R C Pond, D J Smith, D J Stirland, H P Strunk, H K Wickramasinghe and E Wolfgang.

The conference also derived particular benefit from financial sponsorship and assistance provided by a number of commercial firms and laboratories. Contributions made by the following are gratefully acknowledged:

British Telecom Research Laboratories
Centronic Ltd
Ferranti plc
GEC Hirst Research Centre
Hitachi Scientific Instruments
JEOL (UK) Ltd
Link Systems Ltd
Lintech Instruments Ltd
Motorola Ltd
Philips Electronic and Associated Industries Ltd
Plessey Research (Caswell) Ltd

Standard Telecommunication Laboratories Ltd
VG Microscopes Ltd

Finally, we are very pleased to thank colleagues in our own laboratories (and in particular Mr N G Chew at RSRE) for help provided throughout the organisational period of this conference. Special thanks are due to Mrs D M Handley for expert, rapid secretarial work carried out at all stages in this same period.

June 1983

A G Cullis
S M Davidson
G R Booker

Contents

Section 3: Transient annealing phenomena

† Invited

Contents

† Invited

Contents

Section 7: X-ray techniques

Section 8: Non-conventional microscopy

Section 9: Device assessment by scanning microscopy

†Invited

          †Invited

Inst. Phys. Conf. Ser. No. 67: Section 1
Paper presented at Microsc. Semicond. Mater. Conf., Oxford, 21–23 March 1983

1

Effect of doping on mechanical properties, recrystallisation and diffusion in semiconductors

P B Hirsch

Department of Metallurgy and Science of Materials, University of Oxford, Parks Road, Oxford OX1 3PH, UK

Abstract   The effects of doping on diffusion, mechanical properties and grain boundary mobility in semiconductors are attributed to deep levels associated with defects and their migration. The paper reviews the theory of the effect of doping on dislocation velocity, a method of assessing the doping effect from indentation rosettes, and the doping effect on grain boundary migration.

1. Introduction

It is well known that doping semiconductors with electronically active impurities affects diffusion (Fairfield and Masters 1967, Jain and Overstraeten 1975, Tannenbaum 1961), dislocation velocity, and mechanical properties (Alexander and Haasen 1968, George and Champier 1979, Haasen 1979). More recently enhanced grain growth has been observed in doped polycrystalline silicon (Wada and Nishimatsu 1978, Mei, Rivier, Kwart and Dutton 1981, Smith, Tan and Fontaine 1982). All these phenomena have as a common feature the movement of a lattice defect, i.e. vacancy, interstitial, dislocation in the bulk crystal, grain boundary dislocation, or point defect in a grain boundary. The structure of such defects involves strained or broken bonds, and these in turn are associated with electron energy levels in the band gap; this leads to the possibility of charged as well as neutral defects, and the relative concentration of these in thermo-dynamic equilibrium with the electron gas depends on the position of the Fermi level, i.e. on the doping level. The movement of the defects involves the breaking of bonds, and the saddle point of the process at which the bonds are broken will be associated with deep levels in the band gap. The saddle point energy and therefore the activation energy of migration of the defect is likely to depend on the charged state of the defect.

The effects of doping on diffusion in and mechanical properties of semiconductors therefore have similar origins, i.e. they are due to electronic levels in the band gap associated with defects and their migration. This paper includes a review a) of the present state of the theory of the effect of doping on the dislocation velocity in semiconductors, b) of a method of obtaining information about the effect from indentation experiments, and c) of the effect of doping on grain boundary migration.

2. Theory of effect of doping on dislocation velocity

After deformation at relatively low temperatures the dislocations in Si and Ge tend to lie along crystallographic directions, and the velocities

are therefore expected to be controlled by the stress necessary to overcome
the intrinsic lattice (i.e. Peierls) resistance. The movement of the
dislocations occurs by the generation and motion of kinks on the dislocation
line (fig. 1). The theory for this mechanism has been developed for two
limiting regimes (Hirth and Lothe 1968). At relatively high temperatures
and low stresses the drift model applies in which the velocity is proportional to the concentration of kinks ($\exp - F_k/kT$, F_k = kink formation
energy), the velocity of the kinks ($\alpha \exp - W_m/kT$, W_m is activation energy
for migration of kinks along the dislocation line), and the driving force,
τ^n (τ is resolved shear stress and n a constant, usually $n \sim 1 \div 2$). In 1978
a model for the doping effect on dislocation velocity was proposed (Hirsch
1979), within the framework of the drift model, in which doping affects the
concentration of charged kinks, and the activation energy of migration of
charged kinks. The basic dislocation types in the diamond lattice are 60°
and screw dislocations; the dislocations are generally found to be dissociate
into partials, and the basic partial dislocation types which constitute the
60° and screw dislocations are 30° and 90° partials. Kinks in these partials
may be either reconstructed or have dangling bonds (Hirsch 1980, 1981a, Jones
1980). Reconstruction of the core of the 30° and 90° partials involves
either bonding between pairs of adjacent atoms along the core (for the 30°
partial), or between pairs of atoms across the core (for the 90° partial).
In both cases any particular atom has the choice of bonding with one of two
neighbours in equivalent positions, leading to different relative displacements. Where the bonding changes from one set of neighbours to the other
along the dislocation core, an antiphase defect (APD) occurs, associated
with a dangling bond (Hirsch 1980). (This has been called a soliton by
Heggie and Jones (1982)). An APD reacts with a reconstructed kink to form
a dangling bond kink, and APDs can be formed by the reverse reaction.

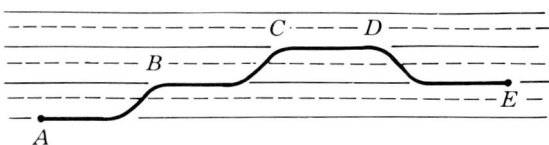

Fig. 1 Movement of dislocations by generation of motion of kinks.

Fig. 2 shows schematically a set of electron energy levels which might be
expected from reconstructed dislocations and kinks. The acceptor and donor
levels for the reconstructed dislocations (E_{da}, E_{dd}) and kinks ($_1E_{Ka}$, $_1E_{Kd}$)
are likely to be shallow levels; the dangling bond levels associated with
dangling bond kinks ($_2E_{Ka}$, $_2E_{Kd}$) and the saddle point configuration for the
two types of kink (where bonds are broken, $_{1,2}E_{Ksa}$, $_{1,2}E_{Ksd}$), are expected
to be deeper levels.

A reconstructed kink has a relatively low energy, but the activation energy
for migration will be relatively large because the saddle point configuratior
involves two broken bonds. A dangling bond kink is likely to have a higher

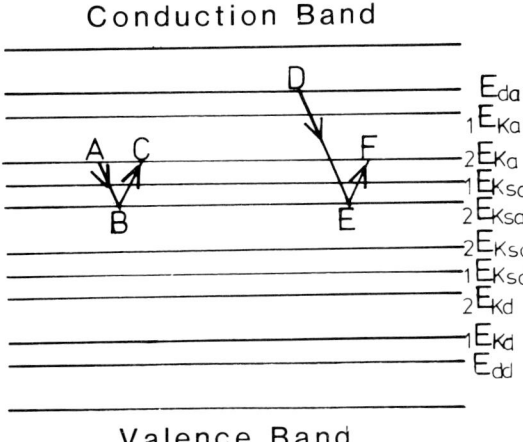

Conduction Band

Fig. 2 Schematic set of acceptor and donor energy levels for reconstructed dislocations (E_{da}, E_{dd}), reconstructed kinks ($_1E_{Ka}$, E_{Kd}), dangling bond kinks ($_2E_{Ka}$, $_2E_{Kd}$), saddle point configurations ($_{1,2}E_{Ksa}$, $_{1,2}E_{Ksd}$). An electron on a charged (dangling bond) kink moves from A to B to C as the kink moves one atomic distance (see fig. 3). For a double kink nucleated at a negative charge on a dislocation, an electron moves from D to the saddle point energy level E, to the kink level F.

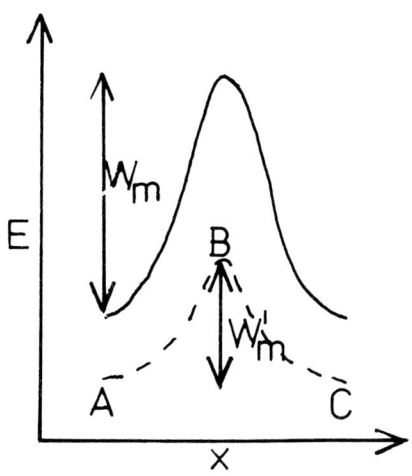

Fig. 3 Total energy (E) versus displacement (x) of a charged (dashed line) and uncharged (full line) kink along the dislocation line.

Fig. 4 Form of indentation rosette on a {100} silicon surface (Hu 1973).

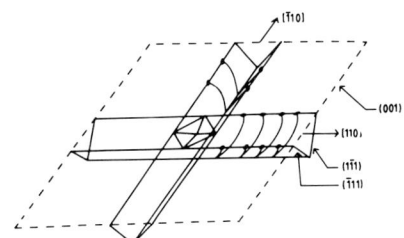

energy, but the migration energy should be smaller because the movement of such a kink involves only the switching of bonds without creating additional dangling bonds; dangling bond kinks may be stabilised by the presence of charge.

If we neglect the concentration of positively charged kinks, the concentration of negatively charged kinks, $_-c_k$, relative to that of neutral kinks, $_oc_k$, is given by Fermi-Dirac statistics as

$$_-c_k/_oc_k = \exp\left(E_F - {_i}E_{Ka} - eV\right)/kT \tag{1}$$

where E_F is the Fermi energy, and eV is the electrostatic energy of the charged dislocation, $i = 1,2$ depending on the type of kink; there is a similar expression for positively charged kinks (Hirsch 1979). Doping changes the concentration of charged kinks in thermodynamic equilibrium with the electron gas, through E_F. The concentration of neutral kinks, $_oc_k$, is constant at a given temperature, independent of the doping level, provided the interaction between charged and neutral kinks is neglected and the concentrations of all types of kinks are small. The velocity of the dislocations controlled by negatively charged kinks (assuming $_+c_k$ to be negligible) is given by

$$\frac{v_-}{v_o} = \exp\left(E_F - {_i}E_{Ka} - eV - \Delta W_m\right)/kT \tag{2}$$

where $\Delta W_m = W'_m - W_m$ is the difference between the migration energy for charged (W'_m) and uncharged (W_m) kinks). Fig. 3 shows the energy as a function of displacement for a charged and uncharged kink along the dislocation line, the energy barrier being due to the second Peierls potential. Assuming that the Born-Oppenheimer approximation holds, the electron will follow the motion of the ions and move from the energy level $_iE_{Ka}$ to $_iE_{Ksa}$ and back to $_iE_{Ka}$ during the activated process (see fig. 2). Thus

$$\Delta W_m = {_i}E_{Ksa} - {_i}E_{Ka} + \Delta W'_m \tag{3}$$

where $\Delta W'_m$ is the contribution to ΔW_m from changes in positions of the ions due to the charge. Thus the energy term in (2) becomes

$$\Delta U = \left(E_F - {_i}E_{Ksa} - eV - \Delta W'_m\right) \tag{4}$$

as if, as in Jones' theory (1980), equilibrium with the electron gas was established during activation at the saddle point. The present treatment avoids this difficulty inherent in the Jones theory (Hirsch 1981b), in that equilibrium with the electron gas is already established in the ground state of the kink. However, as suggested by Jones (1980), the electron energy level important in the doping effect is that corresponding to the saddle point configuration. This energy level is not expected to be the same as that found

from DLTS, Hall conductivity or optical data obtained from specimens containing static dislocation distributions. Furthermore ΔU contains a term
$\Delta W'_m$, which is not known, and although eV is expected to be small, it may
not be negligible (Heggie and Jones 1982).

The discussion so far leaves open the question as to whether the mobile
kinks are of the reconstructed ($_1 E_{Ka}$, $_1 E_{Ksa}$) or dangling bond type ($_2 E_{Ka}$,
$_2 E_{Ksa}$). The activation energy for kink migration is likely to be smaller
for dangling bond kinks, because motion occurs by dangling bond exchange,
rather than by the formation of two dangling bonds which occurs when reconstructed kinks move. Thus, it seems likely that charged dangling bond
kinks are the most mobile. Heggie and Jones (1982) suggest that these
would not be stable, that they would tend to dissociate into reconstructed
kinks and mobile antiphase defects (APDs), and that a kink would move only
if an APD would be temporarily bound to it. This theory is however
effectively identical to the treatment given above since both depend
essentially on the concentration of charged dangling bond kinks.

Direct measurements of migration energy are available only for intrinsic
Si (Hirsch, Ourmazd and Pirouz 1981, Louchet 1981). The high migration
energy (~ 1.2 ev), and the small kink formation energy inferred from this
value and the activation energy for dislocation motion, suggest that the
kinks controlling dislocation motion in intrinsic Si are reconstructed
(Heggie and Jones 1982). Analysis of the experimental data by Schröter,
assuming that the dislocation velocity v can be approximated by $v = v_o +
v^+ + v^-$, gives for 60° dislocations an acceptor level $E_a = 0.67 \pm 0.04$ ev,
with $\partial \bar{E}_a / \partial T = (- 1.6 \pm 0.5) \times 10^{-4}$ ev K^{-1}, and a donor level $E_d = 0.28 \pm
0.17$ ev with $\partial E_d / \partial T = (- 2.0 \pm 2.1) \times 10^{-4}$ ev K^{-1}, all measured relative
to the valence band (see Hirsch 1981b). Heggie and Jones (1982) have
reanalysed the data, making different assumptions about the temperature
dependence of the gap and making some allowance for eV, and quote $E_a =
0.55$ ev and $E_d = 0.34$ ev respectively. The results imply that for
intrinsic Si, i.e. with E_F between E_a and E_d, neutral kinks may be
controlling the dislocation velocity. While these may be reconstructed,
the velocity of dislocations in strongly doped Si may be controlled by
charged dangling bond kinks. This introduces another uncertainty in the
interpretation of the apparent energy levels deduced from the experiments,
because not only is the $\Delta W'_m$ term in the activation energy not known, but
there may be another term due to the possibility that the velocity controlling kinks in intrinsic Si may be of the reconstructed kind, while those in
strongly doped material are of the dangling bond kind. Thus, while the
important electron energy levels are likely to be those corresponding to
the saddle point configuration, as suggested by Jones (1980), derivation
of these levels from apparent values inferred from the experiments, which
include the ionic terms, is subject to uncertainties.

For Ge the early results of Patel and Chaudhuri (1966) suggested the
existence of an acceptor level close to the valence band. Recent results
by Pirouz and Freeland (unpublished) seem to confirm this, but the data
have not yet been analysed in detail.

At higher stress and lower temperatures, the dislocation velocity is controlled by the nucleation of double kinks (Hirth and Lothe 1968). For the
experimental conditions under which the doping effect on dislocation
velocity has been determined the dislocation velocity v is independent of
the length of the dislocation segment, and is given by

$$v = 2h \sqrt{J \, v_K} \qquad (5)$$

where h is the distance between neighbouring Peierls valleys, J is the rate of nucleation of double kinks per unit length of dislocation lines, and v_K is the kink velocity (Hirth and Lothe 1968). Hirsch (1981b) has suggested a) that preferential nucleation may occur at charged points along the dislocation line, in thermodynamical equilibrium with the electron gas, or b) that equilibrium is established with the electron gas in a metastable position of the double kink before the saddle point position is reached. In the former case the electrons (or holes) move from their dislocation levels, E_{da}, to the saddle point level (for simplicity this is shown as $_2E_{Ksa}$ in fig. 2), and beyond the saddle point to a charged kink level $_2E_{Ka}$. Several alternative reaction paths are possible. Preferential nucleation may occur at a single charge, resulting in two kinks, one charged, the other not. The former is likely to be more mobile than the latter. Under these conditions (for a singly negatively charged double kink)

$$_-J \; \alpha \; f \, n_T \; \exp \, (- F_{DK}/kT) \qquad \exp(- W_m'/kT) \, \exp(E_{da} - \, _2E_{Ka})/kT \qquad (6)$$

where F_{DK} is the formation energy of a double kink consisting of two dangling bond kinks, f is the occupation probability of the original charged dislocation level, E_{da}, n_T is total concentration of possible sites. With $fn_T = n_o f/(1-f) = n_o \exp-(E_{da} - E_F + eV)/kT$, where n_o is concentration of uncharged sites

$$_-J \; \alpha \; n_o \; \exp(- F_{DK}/kT) \; \exp(-(W_m + \Delta W_m)/kT) \exp(E_F - \, _2E_{Ka} - eV)/kT$$

and from (3)

$$= J_o \; \exp \, (E_F - \, _2E_{Ksa} - eV - \Delta W_m')/kT \qquad (7)$$

where J_o is the nucleation rate for a similar but uncharged double kink. After nucleation equilibrium is established with the electron gas, and the kink velocity controlled by negatively charged kinks will be

$$_-v_K = \, _ov_K \; \exp \, (E_F - \, _2E_{Ka} - eV)/kT \; \exp \, (_2E_{Ka} - \, _2E_{Ksa} - \Delta W_m')/kT \qquad (8)$$

the first exponential being the fraction of negatively charged kinks, and the second due to the change in activation energy for migration associated with the charge. Thus, it follows from (5), (7), (8) that

$$v_- = v_o \; \exp \, (- \Delta U/kT) \qquad (9)$$

where ΔU is given by (4) as in the drift model.

Alternatively equilibrium is established with the electron gas before the saddle point is reached. For example, when $E_F - \, _2E_{Ka}$ is large, the great majority of kinks would be negatively charged. This can be achieved by nucleation at two closely spaced charges along the dislocation line. Thus

$$J = \alpha \, f^2 \; \exp \, (- F_{DK}/kT) \; \exp \, (- W_m'/kT) \; \exp \, 2(E_{da} - \, _2E_{Ka})/kT$$

$$\alpha \; \exp \, (- F_{DK}/kT) \; \exp \, (- W_m'/kT) \; \exp \, 2(E_F - \, _2E_{Ka} - eV)/kT$$

$$= J_o \; \exp \, (- \Delta W_m/kT) \; \exp \, 2(E_F - \, _2E_{Ka} - eV)/kT \qquad (10)$$

In this case $_-v_K = \, _ov_K \; \exp \, (_2E_{Ka} - \, _2E_{Ksa} - \Delta W_m')/kT \qquad (11)$

and from (3), (5), (10), (11) equation (9) is recovered, with a ΔU which

differs from (4) in that eV includes the Coulomb interaction energy between
two point charges a distance r apart, which is approximately $0.15b/r$ ev
(per electronic charge) for Si, where b is the Burgers vector. For typical
values of the critical double kink width (Haasen 1975), this electrostatic
term is < 0.05 ev, which is small compared to the energy reductions due to
the charged kink mechanisms, although it will affect the critical double
kink width. Another possibility is that following nucleation at a single
charge of a metastable double kink, a second charge arrives before the
saddle point is reached. This should again give an expression similar to
(9).

Heggie and Jones (1982) have suggested that preferential nucleation may occur
at an antiphase defect (APD). Then J depends on the concentration of APDs
which are in equilibrium with the electron gas. In their paper the predicted dependence on doping when $v \alpha J$ (which applies when the velocity is
proportional to the segment length) is correlated with that observed experimentally, obtained however under conditions when $v \alpha J^{1/2}$ (i.e. eqn. (5)).
However, Heggie and Jones (1983) point out that in their model both
nucleation and migration of a kink depend on the presence of an APD. It
follows from (5) that the dependence on E_F is again the same as on the
drift model.

In conclusion, several alternative paths to the saddle point configuration
have been suggested, all of which predict similar dependencies of v on E_F.
Calculations of the energies of charged dangling bond kinks, APDs and
reconstructed kinks will be necessary before an assessment can be made of
the relative importance of the mechanisms depending on preferential
nucleation at charged points of a dislocation and at APDs.

3. Effect of doping on indentation rosettes

The effect of doping on dislocation velocity will be reflected in certain
mechanical properties. Recently indentation experiments, using a Vickers
indenter, have been carried out on variously doped Si specimens (Roberts,
Pirouz and Hirsch 1983). After indentation of (100) faces at 400°C, the
specimens were annealed for 30 minutes at that temperature. Specimens were
then etched at room temperature in a reagent consisting of one part of
0.25M potassium dichromate and four parts of 40% HF (Secco d'Aragona 1972).
The rosettes were examined optically.

The results showed that the indentation diagonal, and thus the microhardness,
are effectively independent of doping. However the extent of the rosettes
varies systematically with doping. In this geometry, the rosette consists
of a central plastic region, and extending outwards from it, in the form
of a cross two arms of dense etch pits along [110] directions (fig. 4).
These arms are formed by half loops punched out along these two directions,
with the dislocations expected to lie on the (111) planes intersecting in
the [110] directions (Hu 1973). Fig. 5 shows the size of the rosettes
(total side-to-side span) as a function of load for differently doped
silicon specimens; error bars for each load are also shown. For all but
the smallest load, the diameters of the rosettes for the n and p type
specimens with doping concentrations of 2×10^{18} cm^{-3} are considerably
larger than those with low doping levels, with those for the n type being
larger than those for the p type. This trend is qualitatively that found
for dislocation velocity as a function of doping in Si (e.g. George and
Champier 1979). In addition the rosette for n type float zone Si is larger
than that for Czochralski Si, showing the same trend as that found for

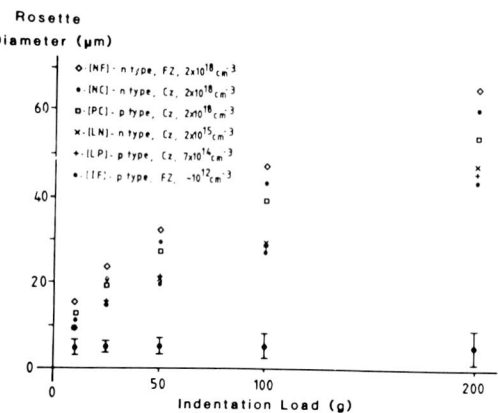

Fig. 5 Size of rosettes (total side-to-side span) as a function of load for differently doping silicon specimens. Error bars are shown for each load.

dislocation velocity (Harada and Sumino 1982), and which is attributed to oxygen.

The results have been interpreted on the basis that the dislocations punched out from the central plastic zone move until the resultant stress on any dislocation due to the indentation stress field and the interactions with the other dislocations along the arm is insufficient to overcome an effective lattice friction stress controlling the dislocation motion; the indentation stress field is assumed to be of the type proposed by Yoffe (1982), which consists of a compressive radial stress and a tensile tangential stress, both decreasing as x^{-3}, where x is the distance from the indentation centre. Table 1 gives the mean effective friction stresses (averaged over a set of values obtained for each specimen from rosette sizes for different loads) for Si with various doping concentrations.

Table 1

Values of effective lattice friction stress in Si
at 400°C, as a function of doping

Si type and doping (cm^{-3}) (Fz ≡ float zone; Cz ≡ Czochralski)	Effective lattice friction stress (MPa)
Fz, p, 10^{12}	102.7
Cz, p, 7×10^{14}	98.7
Cz, n, 2×10^{15}	98.6
Cz, p, 2×10^{18}	83.0
Cz, n, 2×10^{18}	77.5
Fz, n, 2×10^{18}	73.1

These values of the effective lattice friction stress are reasonably consistent with values of stress at which plastic deformation has been found by Pirouz (unpublished) to begin in compression experiments in this temperature range.

4. Effect of doping on grain boundary migration in polysilicon

Recent experiments on polysilicon have revealed an effect of doping on

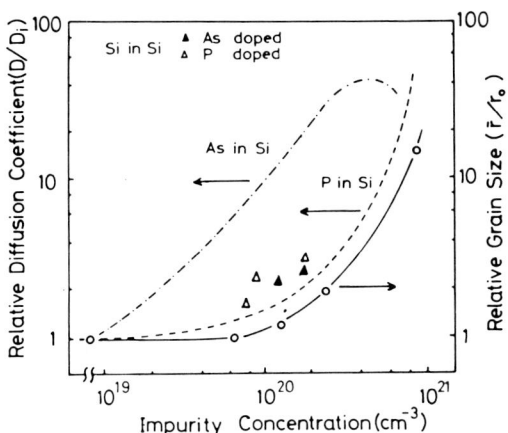

Fig. 6 Grain size of P doped polysilicon relative to that of intrinsic material as a function of concentration of P; comparison with effect of doping by As and P on diffusion coefficient of As, P, and Si in Si (Wada and Nishimatsu 1978).

grain boundary migration (Wada and Nishimatsu 1978, Mei, Rivier, Kwart and Dutton 1981). Fig. 6 reproduces fig. 6 of Wada and Nishimatsu (1978) in which the grain size in P doped poly-silicon (solid line), relative to that in intrinsic Si, is plotted as a function of concentration of P. The trend in the enhanced grain boundary migration rate is compared with that of diffusion of As, P, and of self diffusion in Si as a function of doping concentration. The experiments of Mei, Rivier, Kwart and Dutton (1981) show a similar effect on grain boundary mobility due to doping by As, but no effect is found with B. It seems likely that the enhancement of the grain boundary mobility is again due to an electronic effect. Both the above sets of authors correlate the effect with diffusion controlled mechanisms, for which the electronic effect also operates. However, Smith, Tan and Fontaine (1982) point out that grain boundary migration is likely to take place by movement, i.e. glide and climb, of grain boundary dislocations, which in certain configurations have dangling bonds associated with their cores. The kinetics will depend on a) kink generation and migration, and b) jog generation and migration. Both processes involve breaking of bonds in the saddle point configuration, and b) depends on grain boundary diffusion. Smith, Tan and Fontaine point out that B enhances self diffusion in Si, but does not affect grain boundary migration, and the effect is therefore likely to be controlled by mechanisms involving electron energy states specific to defects and their migration in the grain boundaries themselves. But no analysis in terms of effective electron energy levels has yet been published.

5. Discussion and conclusion

In conclusion the doping effects on diffusion, dislocation mobility and grain boundary migration are all basically due to similar mechanisms, i.e. due to the dependence of the concentration of charged defects on the Fermi energy, which is controlled by the doping level, and to the dependence of the mobility of defects on their charged state. These processes are a consequence of thermodynamic equilibrium. Deviations from thermodynamic equilibrium by injection of carriers e.g. by light or electron irradiation also leads to important effects which are generally due to different mechanisms. Thus the photoplastic effect in Si (Küsters and Alexander 1983) and the electron beam induced glide in GaAs (Maeda and Takeuchi 1983) have been attributed to non-radiative recombination at dislocations, which can enhance the mobility of kinks e.g. by a multiphonon process. Similar mechanisms have been invoked to explain stimulated point defect migration (Kimerling 1978).

References

Alexander H and Haasen P 1968 Solid State Phys. 22 28
Fairfield J M and Masters B J 1967 J. Appl. Phys. 38 3148
George A and Champier G 1979 Phys. Stat. Sol. 53a 529
Haasen P 1975 Phys. Stat. Sol. 28a 145
Haasen P 1979 J. de Physique Colloque C6 40 C6-111
Harada H and Sumino K 1982 J. Appl. Phys. 53 4838
Heggie M and Jones R 1982 J. de Physique Colloque C1 43 C1-45
Hirsch P B 1979 J. de Physique Colloque C6 40 C6-117
Hirsch P B 1980 J. Microscopy 118 3
Hirsch P B 1981a Proc. Materials Research Soc. Symp. on Defects in
  Semiconductors ed J Narayan and T Y Tan (New York: North Holland)
  pp 257-271
Hirsch P B 1981b J. de Physique Colloque C3 42 C3-149
Hirsch P B, Ourmazd A and Pirouz P 1981 Inst. Phys. Conf. Ser. 60 29
Hirth J P and Lothe J 1968 Theory of Dislocations (New York: McGraw Hill)
  pp 484-497
Hu S M 1973 J. Appl. Phys. 46 1470
Jain R K and Van Overstraeten R J 1975 J. Electrochem. Soc. 122 552
Jones R 1980 Phil. Mag. 42B 213
Kimerling L C 1978 Sol. State Electron. 21 1391
Küsters K H and Alexander H 1983 Proc. Intern. Conf. Defects in
  Semiconductors Amsterdam, Physica 116B 594
Louchet F 1981 Inst. Phys. Conf. Ser. 60 35
Maeda K and Takeuchi S 1983 J. de Physique Colloque Properties and Structure
  of Dislocations in Semiconductors, Aussois 1983, in press
Mei L, Rivier M, Kwart Y and Dutton R W 1981 Proc. 4th Int. Symp. on Silicon
  Materials Science and Technology ed H R Huff, R J Kriegler and Y Takeishi
  (Electrochemical Soc.) pp 1007-1019
Patel J R and Chaudhuri A R 1966 Phys. Rev. 143 601
Roberts S G, Pirouz P and Hirsch P B 1983 J. de Physique Colloque Properties
  and Structure of Dislocations in Semiconductors, Aussois 1983, in press
Secco d'Arragona F 1972 J. Electrochem. Soc. 119 948
Smith D A, Tan T Y and Fontaine C 1982 IBM Research Report RC9361 (No
  41211)
Tannenbaum E 1961 Sol. State Electron. 2 123
Wada Y and Nishimatsu S 1978 J. Electrochem. Soc. 125 1499
Yoffe E H 1982 Phil. Mag. 46A 617

Inst. Phys. Conf. Ser. No. 67: Section 1
Paper presented at Microsc. Semicond. Mater. Conf., Oxford, 21–23 March 1983

11

The motion of charged dislocations in $A^{II}B^{VI}$ semiconductors

Yu A Ossipyan

Institute of Solid State Physics of the USSR Academy of
Sciences, 142432, Chernogolovka, USSR

Abstract The moving dislocations electrical charge value
has been determined experimentally in $A^{II}B^{VI}$ semiconductors. It is of dynamic nature, is involved in disloca-
tions-point centers interaction and essentially affects
the flow stress value.

The discovery of the photoplastic effect (PhPE) in cadmium sulphide (Ossipyan, Savchenko 1968) has made it apparent
that dislocations in $A^{II}B^{VI}$ semiconductors interact readily,
as they move, with the electron subsystem of the crystal.
    Fig.1 shows a stress-strain diagram of a ZnSe crystal.
The photoplastic effect is seen on the diagram. At the moment
of illuminating by visible light of the deforming crystal its
flow stress value increases by a factor of 2. The initial
value is recovered after the light is switched off. The spectral dependence of the PhPE is shown in Fig.2. It exhibits a
maximum in the self-absorption edge region and a smooth decrease towards the long-wave region.

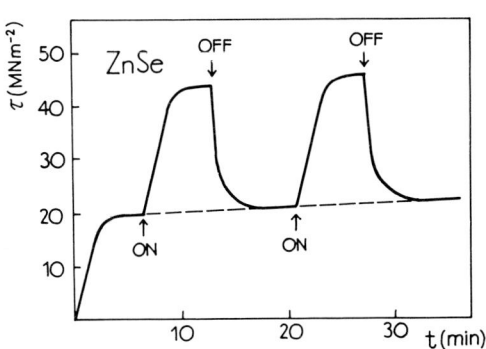

Fig.1.

The investigations
of the PhPE in different
$A^{II}B^{VI}$ crystals evidence
for the fact that the
PhPE mechanism involves
a change of the condi-
tions of the electric in-
teraction between a mo-
ving dislocation and
point centers.
    This is confirmed
by our subsequent disco-
veries of such phenomena
as infrared quenching of
the PhPE (IRQPhPE)
(Ossipyan et al. 1971),
impurity photoplastic ef-
fect (IPhPE) (Ossipyan et
al. 1974) deformation luminescence (DL) and injection plastic
effect (Ossipyan, Petrenko 1973).
    Fig.3 shows the IRQ PhPE for CdS. Fig.4 shows its spectral
dependence. The quenching maximum is seen to fall within
the region about 1.3 eV. For CdS this corresponding to the

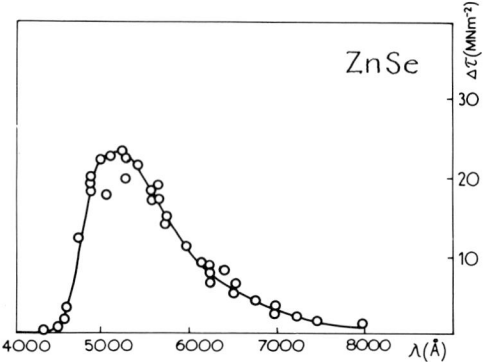

Fig.2.

R-center local level absorption (Bube, 1967). The subsequent investigations of plastic deformation of $A^{II}B^{VI}$ semiconductors have, however, shown, that the electric charge of the dislocations themselves plays rather an important role in the deformation process. We have observed the phenomena, originating from the motion of charged dislocations which bring the charge to the crystal surface (Ossipyan, Petrenko 1975). Figs.5 to 7 illustrate the experimentally observed effects caused by a change of the electrical conditions on the crystal boundary during its plastic deformation. The short-circuit effect (ShSE), Fig.5. can be presented as follows. If in the course of the plastic deformation the opposite faces of the crystal are shorted through the external circuit, then the flow stress value

Fig.3.

can markedly decrease. This effect can be explained in the following way. The electric charge accumulated at the surface during the preceeding plastic deformation interferes with the motion of charged dislocations and their emergence on the surface. That is, hardening is produced. Said factor is eliminated if this charge is removed by means of the external circuit. The flow stress is then decreased.

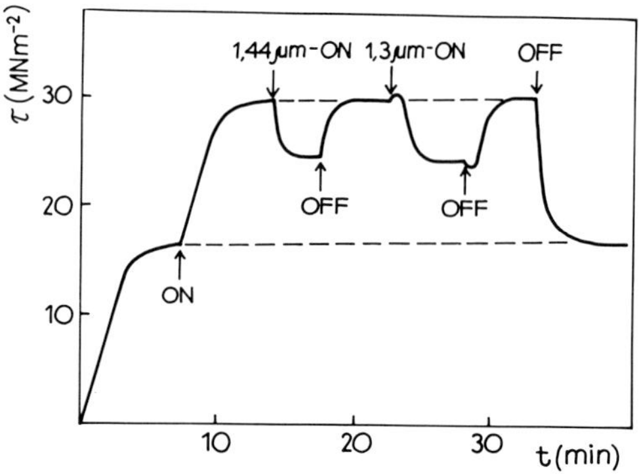

The dislocation current (DC) in the external circuit (Fig.6) can easily be measured in the course of the plastic deformation when charged dislocations are moving in the crystal volume. The DC measurement method for $A^{II}B^{VI}$ crystals is much more sensitive than the other methods of mechanical measurements.

The electroplastic effect consists in the fact that the external electric field applied to the deforming crystal either increase, depending on its strength and sign, or decrease the flow stress value (Fig.7). Its mechanism is related to the influence of the external field on charged dislocations and is apparent from the above explanations.

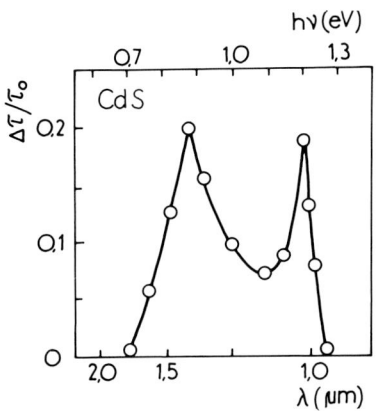

Fig.4.

The aforementioned facts show convincingly that the electric charge of dislocations plays an important role in the processes of plastic deformation of semiconductors. Therefore, we are confronted by the task to determine quantitatively the value of the charge related to different dislocation types and to study the effect of various factors (temperature, strain rate, illumination, electric field etc) exerted on this value.

With this aim in view, we have developed and employed several experimental methods.

<u>The dislocation-current method.</u> In the general case, the dislocation current density is given as

$$j_D = \sum_i q_i < \rho_i \overline{v}_i > \qquad (1)$$

where q_i , is the linear density of the dislocation charge; ρ_i and v_i are the concentration and the velocity of the dislocation.

Summation is performed over all the types of charged dislocations.

Here, noteworthly is an essential difference between the $A^{II}B^{VI}$ semiconductors and alkali halide crystals (AHC) which, too, exhibited certain electrical phenomena. In the inversion crystals like AHC the dislocations of opposite mechanical sign, moving in opposite directions, are physically equivalent, the identical electric charge included. As seen from eqn.(1) this gives, at a uniform deformation, $<j_D>=0$.

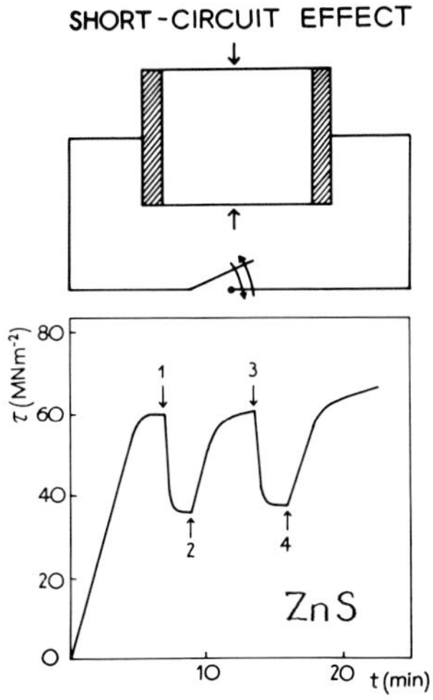

SHORT-CIRCUIT EFFECT

Fig.5.

The experimental observation of the dislocation currents in AHC is only possible when spatial or temporal fluctuations occur in the dislocation fluxes ($\rho_i v_i$) of identical mechanical sign (Whitworth 1975).

As already mentioned, the plastic deformation of $A^{II}B^{VI}$ compounds exhibited an absolutely different situation. We observed the electric currents whose value and direction were constant and which were proportional to the strain rate. The measurement of the dislocation current value on mutually perpendicular sample faced enables one to determine the direction of the dislocation current vector j_D . This direction

DISLOCATION CURRENT

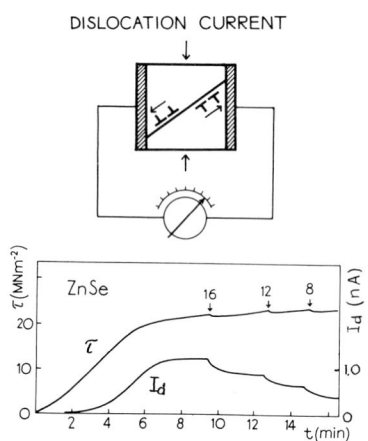

Fig.6.

was coincident with that of 60°
dislocations when dislocations moved both in the basal plane of hexagonal (CdSe, CdS, ZnO) and in
the [111] plane of cubic crystals
(CdTe, ZnSe, ZnS, ZnTe), i.e. it
lied in the glide plane and was
normal to 60°-dislocation lines.
So, it is natural to assume that
it is these dislocations that transfer the charge.

It turns out that in all the
$A^{II}B^{VI}$ compounds the mobilities of
α and β -60° dislocations are different. Therefore, in this case,
the main contribution to the dislocation current, is made by one dislocation type. This enables one
to determine the linear density of
the dislocation charge q by measuring the dislocation current I_D and
the rate of the plastic flow $\dot{\ell}_n$.

In the simplest case when
one-type dislocations move in parallel with the current contracts,
with a distance dx travelled by
ine dislocations, the charge, passing through the external circuit,
is given as

$$dQ = \frac{q \cdot L_D \cdot dx}{H} = \frac{q \cdot D \cdot dx}{H} \qquad (2)$$

and the reduction of the sample
length in the compression direction is

$$d\ell_n = \frac{b_{\bar{z}} \cdot dS}{S_1} = \frac{b_{\bar{z}} \cdot dx}{H} \qquad (3)$$

Here, H is the contact-to-contact
distance along the glide plane,
$b_{\bar{z}}$ - is the Burgers vector projection into the compression axis,
dS - is the area swept off by the
dislocation as it covers dx ,
$S_1 = D \cdot H$ is the glide plane area,
D is the contacts width.

ODD ELECTROPLASTIC EFFECT

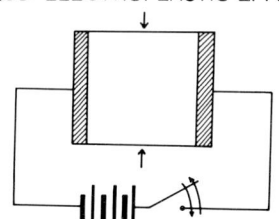

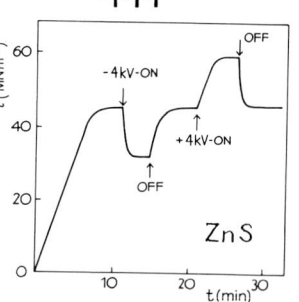

Fig.7.

For the motion of any arbitrary number of dislocations
per unit time we have

$$Q = \frac{q \cdot D \cdot \Sigma dx_i}{H} ; \quad \ell_n = \frac{b_{\bar{z}} \Sigma dx_i}{H} \qquad (4)$$

$$\frac{Q}{\ell_n} = \frac{I_d}{\dot{\ell}_n} = q \cdot \frac{D}{b_{\bar{z}}} ; \quad q = \frac{I_d \cdot b_{\bar{z}}}{\dot{\ell}_n \cdot D} \qquad (5)$$

As seen from eqn. (5), the ratio of the dislocation current value, I_d, to the plastic flow rate, $\dot{\ell}_n$, is dependent neither on the dislocation density nor on their velocity. This enables one to derive the linear density of the dislocation charge from the measurements of I_d and $\dot{\ell}_n$.

Let us consider a more general case when a dislocation loop broadens in the glide plane. If we assume that the mobility of any one type of the loop components surpasses that of the other types, for instance, $V_\alpha > V_\beta, V_{scr}$ then we obtain the equation to determine the linear charge density of the most mobile dislocation type

$$\frac{dQ}{d\ell_n} \cong q \cdot \frac{D}{bz}$$

It is, however, to be noted that as the plastic deformation is being stored in the crystal, the low-mobility loop components can accumulate (L_β/L_α, increases) and the $dQ/d\ell_n$ ratio will decrease in time. Under these conditions eqn. (5) cannot be employed to determine q. Therefore, we consider that eqn. (5) can only be employed when, in experiment, $I_d/\dot{\ell}_n$ is independent of the degree of deformation. There are two more restrictions imposed on the application of the dislocation currents method.

In the stationary case when the current passing through the electroscope I_m is constant, the potential difference arises at the contacts $U = I_m \cdot R_{in}$, where R_{in} is the input resistance of the electroscope. The potential difference gives rise to the conduction current (electron) in the sample $I_{con} = U/R_s$ where R_s is the resistance of the sample. Since

$$I_m = I_d - I_{con}. \text{then} \quad I_d = I_m \left(1 + \frac{R_{in}}{R_s}\right) \qquad (6)$$

As seen from (6) the dislocation current $I_d = I_m$ only when $R_{in} \ll R_s$.

It is obligatory that this relation be fulfilled in the measurements. Another restriction is that the dislocation charge can be screened by free carriers and charged point defects. If the screening is complete, neutral objects (dislocation + screening charge) move in the sample volume, and no dislocation currents arise. The additional investigations of the electronic properties of the crystals investigated have shown that no such screening occurs at dislocation velocities from 10^{-3} to 1 cm/sec.

The dislocation-band method (Zaretskii et al. 1977) enables one to determine the charge of not only the most mobile dislocations. As evidenced by experiment, the principal dislocation sources, at least at the initial stages of plastic deformation, are crystal surfaces.

Having removed by polishing certain stress concentrators from the surface, one can load the crystal up to a stress τ, when $\tau_m < \tau < \tau_s$. Here τ_m are the starting stresses for the dislocation motion in the volume, τ_s is the operating stress of the surface dislocation sources. If we now prick or scratch the surface, this will produce a glide band consisting of dislocations whose type is determined by the surface type. The band will be gliding into the crystal interior. Figs.8 and 9 illustrate this method. Here we measure ΔQ (the charge after it has passed through the external circuit) and ΔL (the deformation of the sample caused by the surface-initiated glide plane passing through it). As before,

$$q = \alpha \frac{\Delta Q}{\Delta L} \cdot b_z$$

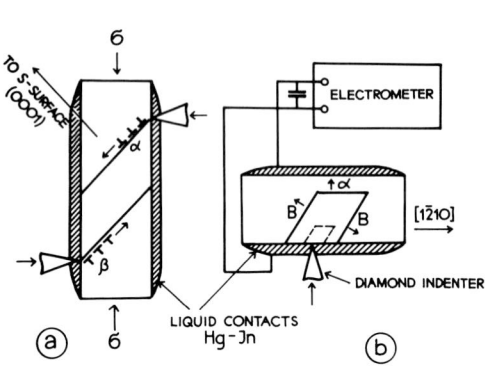

Fig.8.

where α is a geometrical multiplier dependent on the dislocation-loop type. Fig. 9 illustrates the measurements of ΔQ and ΔL while α – and β –glide bands are passing through the CdS sample. Rather small values of ΔL (from 0.01 to 1 m) were registered by a movable-electrode tube.

The method of the compensating field application. This method is based on the conception that the action of the elastic stresses exerted on the dislocation by loading is compensated by the external electric field, i.e. electro-plastic effect (Fig.7).

The force operating on the unit length of the charged dislocation is

$$F = q \cdot E + b\tau \qquad (7)$$

If, in experiment, the force, F, remains invariable under the applied external field E (this can be judged

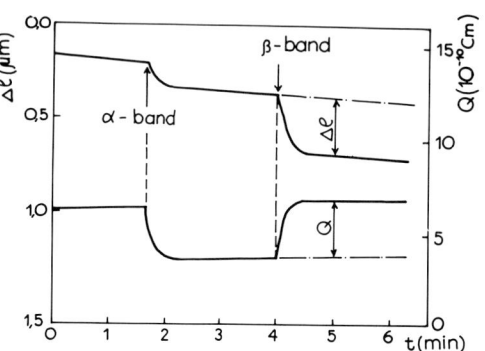

Fig.9.

upon by the invariability of the strain rate $\dot{\ell}_n$) then

$$q = \alpha \cdot b \cdot \frac{\Delta\tau}{E} \qquad (8)$$

Here, $\Delta\tau$ is a change of the deforming stress needed to retain the value of $\dot{\ell}_n$ invariable under the applied E.

The application of the three aforementioned methods to determine the dislocation charge densities in $A^{II}B^{VI}$ crystals yielded similar results. These results are tabulated in Table 1. They pertain to the measurements in darkness at room temperature and are averaged over a great number of ingots.

The tabulated values of q pertain to the dislocations with an edge component. The charge density in the screw dislocations is so small that we have failed to observe it by said methods. In n-type materials all the dislocations were

Table 1.
Types of mobile dislocations and their charge

Material	Structure and type of conductivity	Dislocation type	Mean charge in the dark (q.b/e)	Maximum charge under illumination (q.b/e)
ZnO	n, wurtcite	60° perfect Zn(g)- β	-0.54	
		screw (g)	uncharged	uncharged
ZnS	n,sphalerite	edge partial S(g)	-0.80	-1.15
		screw (g)	uncharged	uncharged
ZnSe	n,sph.	60° perfect Se(g)- α	-0.50	-0.92
		screw (g)	uncharged	uncharged
ZnTe	p,sph.	60° perfect Te(g)- α	+0.42	+0.78
		screw (g)	uncharged	uncharged
CdS	n,wur.	60° perfect Cd(g)- β	-0.36	-0.54
		60° perfect S(g)- α	-0.7	
		screw(g)	uncharged	uncharged
CdSe	n,wur.	60° perfect Cd(g)- β	-0.17	-0.35
		screw(g)	uncharged	uncharged
CdTe	n,sph.	60° perfect Te(g)- α	-0.12	
		screw(g)	uncharged	uncharged

negatively charged, and in p-ZnTe positively charged. Noteworthy is the fact that polar dislocations (α and β) in CdS have charge of the same sign though its value is different. The effect of illumination on the dislocation charge was detected in ZnS (Ossipyan, Petrenko 1976), CdS and CdSe, ZnSe (Petrenko, Whitworth 1980) and ZnTe. Always, the charge increased under the illumination. The spectral dependence of the charge value is shown in Fig.10. A special technique involving the switch-over of the deforming rate was employed to study the dependence of the q value on the deforming rate and on the temperature. In this case, the conditions were created such that the dislocation density might be assumed constant during the measurements (Ossipyan, Petrenko

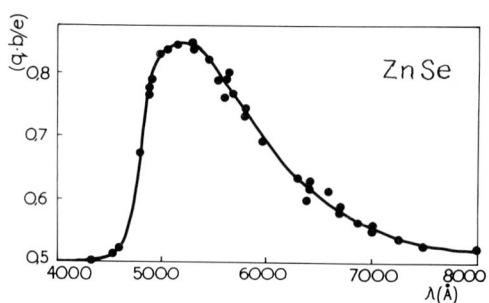

Fig.10.

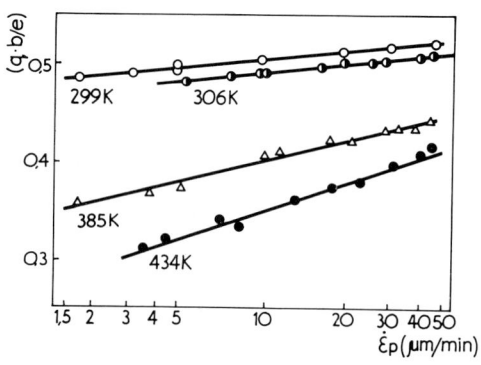

Fig.11.

1978) that is, the deforming rate was proportional to the dislocation velocity. Here, it turns out that q grows with increasing velocity in proportion to the logarithm of the deforming rate, and the proportionality factor increases with temperature (Fig.11, ZnSe). We have employed this technique,

varying, in turn, the loading ading rate, the temperature and the dislo-

cation charge (by light or by free carriers in injection). The analysis of the obtained curves show (Ossipyan, Petrenko 1979) that in all the cases for the investigated $A^{II}B^{VI}$ compounds the relation

$$\dot{\varepsilon}_n = \dot{\varepsilon}_o \exp\left\{-\frac{E_o - \gamma\tau + aq}{T}\right\} \quad (9)$$

is fulfilled.

The coefficient γ (activation volume) does not change under the illumination and remains constant within the accuracy 10 to 15% with a change of the degree of plastic deformation,

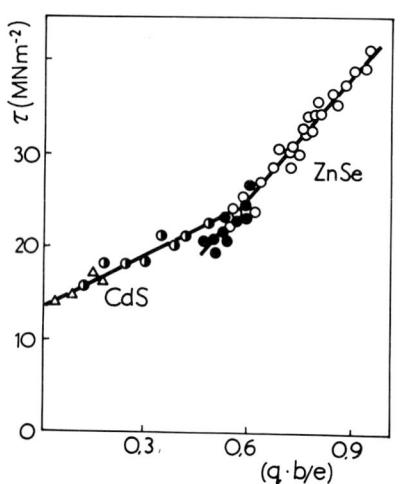

Fig.12.

It equals (20 ± 2) b^3 for ZnSe, $(25 \pm 4)b^3$ for ZnS and $(50 \pm 10)b^3$ for CdS.

This gives grounds to assume that in $A^{II}B^{VI}$ compounds the dislocation motion is governed by the Peirls mechanism. This is also evidenced by the micrographs at which are seen long even dislocation segments, lying in the directions of the Peierls relief-valleys. Fig.12 illustrates the experimental dependence of the critical resolved shear stress in CdS and ZnSe on the dislocation charge value. (Samples of various boules in dark and by illumination). I

It is seen that for CdS it is essential, and for ZnSe very essential. This suggests the con-

clusion that in $A^{II}B^{VI}$ the plastic deformation processes cannot be described correctly without allowance for the value of
the dislocation charge and its variations.

This is confirmed by the data of fig.13 presented for
all the $A^{II}B^{VI}$ crystals investigated. The correlation is apparent between the mean charge values and the yield stress.

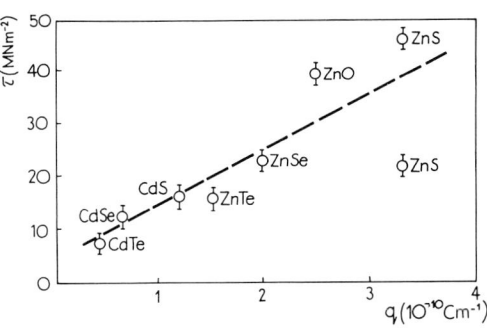

Fig.13.

In conclusion, we
shall dwell on the nature
of such a strong disloca-
tion charge.

The results of our in-
vestigations (including
the aforementioned) show
that neither the value nor,
in particular, the sign of
the dislocation charge van
be explained on the base
of the conceptions about
the ionic character of the
interatomic bonds in these
crystals.

The charge value is
of a dynamic origin and
changes as dislocations move across the crystal in which there are very many electrically charged point centers. Due to
the interaction between the moving dislocations and the local
centers part of their charge is captured by the dislocations.
Simultaneously, part of the local centers are ionized and some quantity of electrons are "popped out" into the conduction
band wherein they recombine (deformation luminescence).

So, the crystal with moving dislocations exhibit a non-
equilibrium electron distribution between the bands, local
centers and dislocations. This differs strongly from that of
the case of "resting" dislocations. Some of these processes
are shown schematically in fig.14, illustrating the band
structure changes around
the dislocations that move
with a velocity V_D . Here
R_l , is the radius at which
the point center charge can
be captured by the dislocation, R_{scr} is the screening
cylinder radius, E_f –is the
Fermi level, E_D –the dislocation level, E_t –is the
position of the local centers level away from the
dislocations. The arrows
indicate the electron transitions from the centers
to the dislocation and to

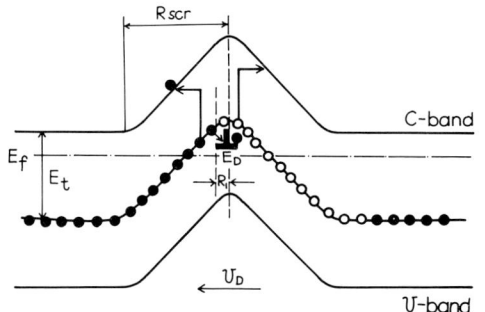

Fig.14.

the conduction band. Black circles are the centers filled
up with electrons, open circles are empty centers.

Basing on this scheme, we can explain the processes of
the electric charge redistribution between dislocations,

local centers, electron and hole bands, taking place as the dislocations move in the course of plastic deformation. We can also explain the effect on these processes of illumination, of carrier injection, temperature, deforming rate etc. This is the way to treat convincingly the above stated phenomena.

In conclusion, the author wishes to express his gratitude to V.Petrenko, I.Savchenko, M.Shichsaidov and A.Zaretskii for their cooperation in obtaining the main results stated in this lecture.

REFERENCES

Bube, R.H., 1967, Physics and Chemistry of II-VI Compounds, edited by M.Aven and J.S.Prener (Amsterdam: North Holland), p.659.

Kirichenko, L.G., Petrenko, V.F., and Uimin, G.V., 1978, Zh. eksp.teor.Fiz., 74, 742 (Soviet Phys. JETP, 47, 389).

Ossipyan, Yu.A., Petrenko, V.F., 1973, Pis'ma Zh.eksp.teor. Fiz., 17, 555.

Ossipyan, Yu.A., Petrenko, V.F., 1975, Zh.eksp.teor.Fiz., 69, 1362 (Soviet Phys. JETP, 42, 695).

Ossipyan, Yu.A., Petrenko, V.S., 1976, Dokl.Akad.Nauk SSSR, 226, 803 (Soviet Phys.Dokl., 21, 87).

Ossipyan, Yu.A., Petrenko, V.F., 1978, Zh.eksp.teor.Fiz., 75, 296 (Soviet Phys. JETP, 48, 147).

Ossipyan, Yu.A., Petrenko, V.F., 1979, J. de Phys. C-6,40,161.

Ossipyan, Yu.A., Petrenko, V.F., Savchenko, I.B., 1971, Pis'ma Zh.eksp.teor.Fiz., 13, 622-624.

Ossipyan, Yu.A., Petrenko, V.F., Shikhsaidov, N.Sh., 1974, Pis'ma Zh.eksp.teor.Fiz., 20, 363-364.

Ossipyan, Yu.A., Savchenko, I.B., 1968, Pis'ma Zh.eksp.teor. Fiz., 7, 130 (Soviet Phys. JETP Lett. 7, 100).

Petrenko, V.F., Whitworth, R.W., 1980, Phil.Mag.A. 41, 681.

Whitworth, R.W., 1975, Adv.Phys., 24, 203.

Zaretskii, A.V., Ossipyan, Yu.A., Petrenko, V.F., and Strukova, G.K., 1977, 1977, Fizika tverd.Tela, 19, 418 (Soviet Phys.St., 19, 240).

Zaretskii, A.V., Ossipyan, Yu.A., and Petrenko, V.F., 1978, Fizika tverd.Tela, 20, 1442 (Soviet Phys., solid St., 20, 829).

Inst. Phys. Conf. Ser. No. 67: Section 1
Paper presented at Microsc. Semicond. Mater. Conf., Oxford, 21–23 March 1983

21

Lattice images of undissociated 60° dislocations in silicon

J L Hutchison, G R Anstis and P Pirouz

Department of Metallurgy & Science of Materials,
University of Oxford, Parks Road, Oxford, OX1 3PH, England.

Abstract   60° dislocations which are undissociated have been found in
thin regions of an ion-thinned, deformed silicon specimen. Computer image matching is used in an attempt to characterise their core structure.
Possible reasons for their formation are discussed.

1. Introduction

Plastic deformation of Si produces dislocations which are generally dissociated into partials; deformation at relatively low temperatures causes the
dislocations to lie along the<110> directions in the {111} glide planes.
Screw dislocations with Burgers vector $\frac{1}{2}[110]$ are dissociated into two 30°
partials, whilst the 60° dislocations form pairs of 30° and 90° partials.
Lattice imaging has been used successfully to identify the 30° core configuration produced by deformation at 420°C as the 'glide' type (Anstis et al,
1981); we now report some observations of undissociated 60° dislocations
encountered in the same specimen, which do not appear to have arisen from
the deformation process. Computer image matching is used in an attempt to
elucidate the core structure.

2. Experimental

A single crystal of Si was deformed by compression under high stress along
[312] at 420°C following predeformation at 750°C. This produced hexagonal
loops of dislocations lying on the (1$\bar{1}$1) glide plane and consisting of screw
and two different 60° dislocations. The screw dislocations lay along the
[110] direction and were dissociated into 30° partials. Sections were cut
normal to the [110] direction and thinned for electron microscopy by Ar$^+$ ion
bombardment. Weak beam experiments (Ourmazd, unpublished) on foils parallel to the (1$\bar{1}$1) planes had shown that practically all of the dislocations
present were dissociated; lattice imaging was then used to characterise the
30° cores, images being recorded using a JEOL 200CX electron microscope equipped with a LaB$_6$ cathode. Seven beams (000, 111 and 002 reflections) were
used to form images under carefully controlled conditions. These images revealed a number of undissociated dislocations close to the foil edge, where
the thickness was less than about 60Å. Fig. 1 shows an example of one of
these undissociated dislocations, situated close to a 30° partial and stacking fault. The extra half plane in the undissociated dislocation is (1$\bar{1}$1),
and careful measurement of the projected Burgers vector suggests that the
dislocation is a 60° type. In the following sections we describe attempts
to identify the core structure of such dislocations by computer image -
matching, and we discuss possible reasons for their occurrence in this particular specimen.

© *1983 The Institute of Physics*

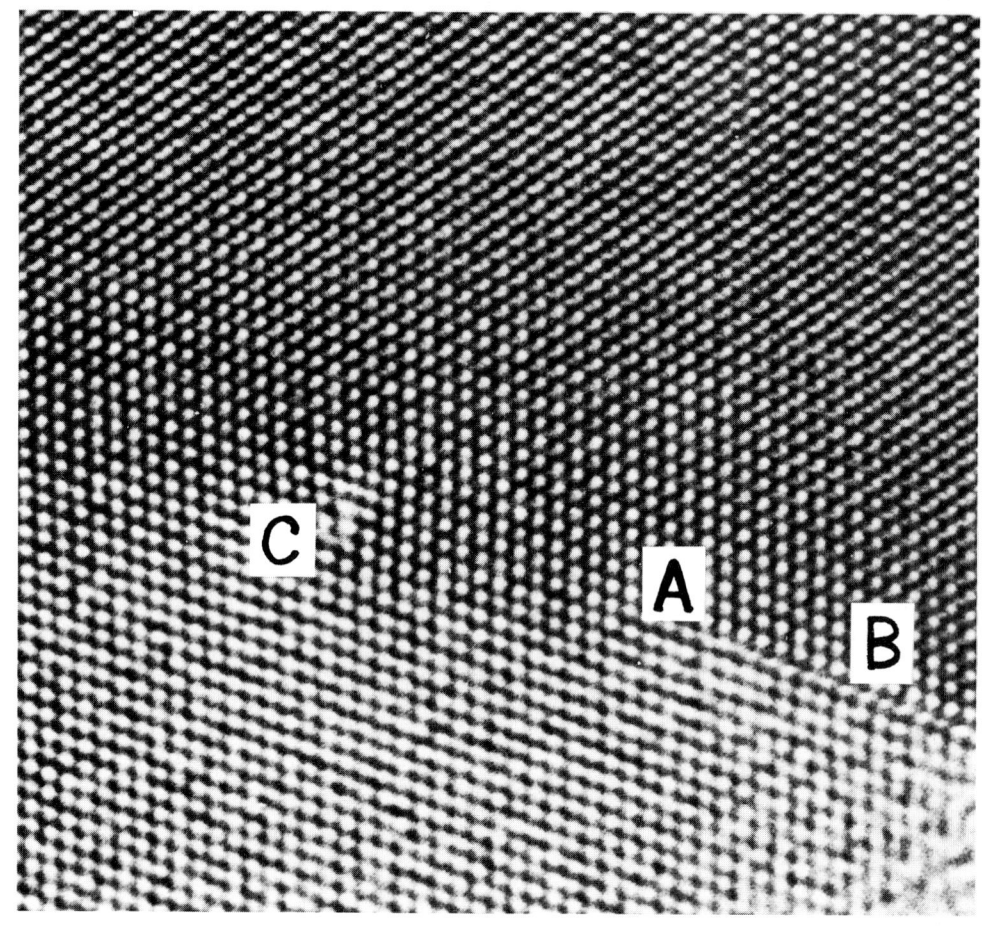

Fig. 1.   [110] lattice image of silicon, showing 30° partial (A), stacking
fault (B) and undissociated 60° dislocation (C).

3. Image Simulation

Simulation of image contrast of a perfect 60° dislocation was performed for
four different configurations of atoms at the core. These are based on the
glide and (interstitial) shuffle models (Hirth and Lothe, 1968). Positions
were first calculated using the theory of isotropic elasticity, the origin
for the calculation being at an arbitrary point which is not at atomic site.
'Relaxation' of the configuration was carried out by requiring bond lengths
to be similar to those in a perfect lattice. The shuffle model differs from
the glide in having an extra column of atoms. These configurations are
shown in Fig. 2.

Simulation of high resolution images was carried out with the multi-slice
method using a PDP 11/23 computer interfaced to a double 512 x 512 x 8 - bit
digitial video framestore. The programmes make use of a dedicated hardware
unit for calculating two dimensional Fourier transforms (Boyes et al, 1982)
which enables them to be used interactively (Goringe, Muggridge and Anstis,
to be published).

For the multi-slice programme, a unit cell of dimensions 28.2Å x 28.2Å x
3.8Å was chosen. Calculations based on 64 x 64 arrays include only those
Bragg reflections out to 220. Since only thin regions of the foil are being
examined this limitation is not a serious limitation. Parameters for the
calculations are: foil thickness, 38Å; Cs = 1.2mm; objective aperture ad-
mitting (111) and (002) reflections; objective lens defocus, - 550Å (i.e.
underfocus). For these values of thickness and defocus atomic pairs appear
dark. Computed images for the different configurations are shown in Fig. 3
which is at the same scale as the experimental image in Fig. 4.

The calculations suggest that there are discernable differences between images of a glide and a shuffle type dislocation. However variations in experimental image detail, possibly due to amorphous surface layers, makes unambiguous characterisation of the core configuration difficult.

4. Origin of the dislocations

The specimen was compressed along the [312] direction which is very favourable for single slip. The primary glide system in this case is $(1\bar{1}1)$ $[110]$
and nearly all the hexagonal dislocation loops produced during the low temperature - high stress deformation stage lie on the $(1\bar{1}1)$ plane. However,
glide on the secondary $(1\bar{1}1)$ $[011]$ and tertiary $(1\bar{1}1)$ $[10\bar{1}]$ slip systems cannot be completely avoided.

The specimens for lattice imaging were cut parallel to the (110) planes and
all the dissociated dislocations observed were 30°-partials obtained from
the dissociation of the screw dislocations according to the following reaction:

$$\frac{1}{2}[110] \rightarrow \frac{1}{6}[21\bar{1}] + \frac{1}{6}[12\bar{1}]$$

Also, as expected, the glide plane for the two partials was parallel to $(1\bar{1}1)$
(see Fig. 1). As this figure shows, the undissociated 60° dislocations also
have the same glide plane. These perfect dislocations cannot be constricted
parts of the dissociated dislocations because constriction would give rise
to undissociated screw dislocations, with no extra half plane, which would
hardly be visible in lattice imaging when viewed end-on.

As the Burgers vector of the 60° dislocations cannot be $\frac{1}{2}[110]$, they must

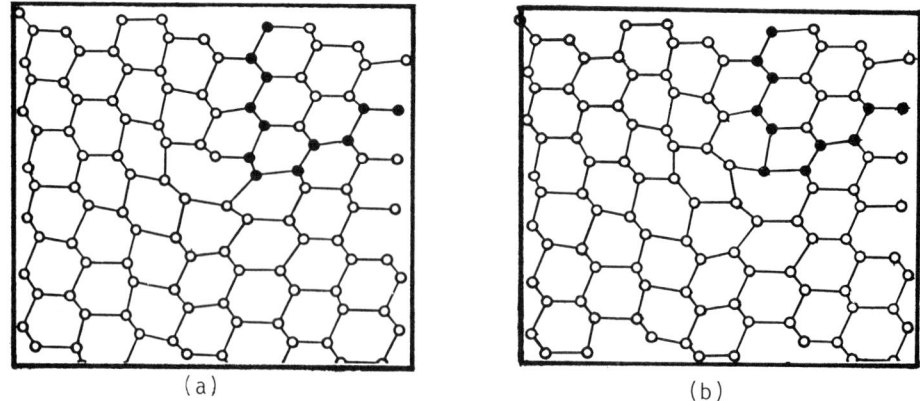

Fig. 2.   Models for 60° core: (a) glide, (b) shuffle.   Extra atomic planes are shaded.

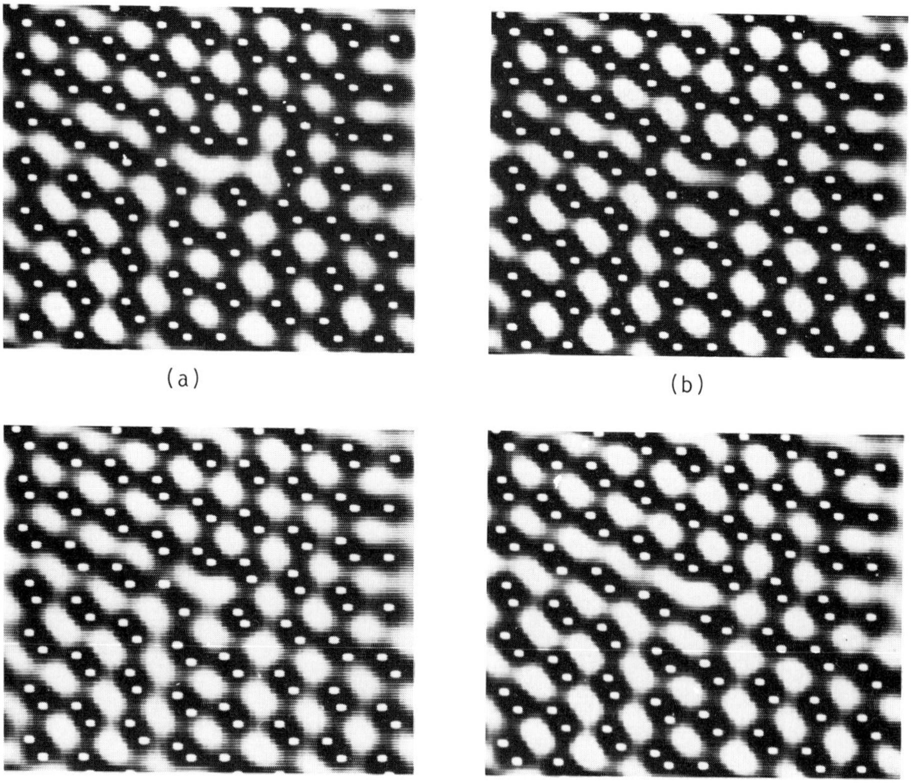

(a)

(b)

(c)

(d)

Fig. 3.   Computed images for 60° core: (a) glide, (b) 'relaxed' glide, (c) shuffle, (d) 'relaxed' shuffle.   Atomic positions are shown as white dots.

Fig. 4. [110] lattice image of a 60° dislocation core.

belong either to the secondary or the tertiary slip system. From the lattice images, it is not possible to determine the Burgers vector of these
dislocations. If they were produced by deformation, one would expect them
to belong to the $(1\bar{1}1)$ [011] slip system which has a Schmid factor three
times greater than that of the tertiary system. However, if they were in
fact produced by deformation, the question remains as to why they are not
dissociated!

The possibility that one of the partials rotates to an edge orientation and
then recombines with a 30° partials to form a perfect 60° dislocations is
also ruled out for a number of reasons. In thin crystals, there is a tendency for dislocations to rotate due to image forces. However this is always
a rotation towards the screw orientation. Further, such a rotation should
occur for both 30°-partials.

It must be remembered that a number of 60° dislocations were observed in
the specimen. However such observations were limited to the very thin part
($\leq 100\text{Å}$) of only one of a number of specimens thinned for lattice imaging.
It is thus possible that in this particular specimen, the thinning process
caused some stress and the 60° dislocations were generated to relieve the
strain. We note also that weak-beam experiments (in somewhat thicker foils)
failed to reveal any dislocations of this type.

References

Anstis G R, Hirsch P B, Humphreys C J, Hutchison J L and Ourmazd A, 1981
    Inst. Phys. Conf. Ser. No. 60 23
Boyes E D, Muggridge B J and Goringe M J, 1982 J. Microscopy 127 321
Hirth J P and Lothe J 1968 Theory of Dislocations (New York: McGraw-Hill)

Acknowledgements

We wish to thank Professor H Alexander for providing the specimen,
Dr A Ourmazd for preparing it for high resolution work and Dr M J Goringe
and Mr B J Muggridge for use of their computer programmes.

Inst. Phys. Conf. Ser. No. 67: Section 1
Paper presented at Microsc. Semicond. Mater. Conf., Oxford, 21–23 March 1983

The mechanism of dislocation climb in GaAs

G Feuillet and D Cherns

Department of Metallurgy & Science of Materials, University of Oxford,
Parks Road, Oxford, OX1 3PH, England.

Abstract   The weak-beam technique of electron microscopy is used to
investigate the climb of dissociated $a/2<1\bar{1}0>$ dislocations in GaAs under
electron irradiation in the high voltage electron microscope. In room
temperature irradiations, climb of both α and β dislocations proceeds by
nucleation on the individual partials of interstitial loops whose
Burgers vector may differ from that of the original dislocation. Irradiations at elevated temperatures indicate climb of the total dislocation
with double jog generation; the climb mechanism is discussed.

1. Introduction

Dislocation climb has been found to play an important role in the degradation of light-emitting diodes and double heterostructure lasers. During
the operation of such devices, the point defects that produce climb migrate
to the dislocation core by recombination-enhanced diffusion (Lang and
Kimerling, 1974). Dislocation climb under conditions of a high supersaturation of point defects introduced by electron-irradiation in the HVEM,
has been studied by transmission electron microscopy in Cu-Al alloys (Cherns
et al, 1980) and in silicon (Ourmazd et al, 1981). In all cases climb
proceeds by the nucleation on the individual partials of small prismatic
dislocation loops, assumed to be of interstitial type owing to the dislocation 'bias'. The character of the loops is such as to minimize the
total energy of the parent partial and loop configuration and to maximize
the edge component of the nucleated loops. Interaction with the opposite
partial can lead to complex configurations where the Burgers vector of the
nucleated loops differs from that of the original dislocation. Subsequent
studies of quenched Cu-13%Al (Decamps et al, 1983) showed broadly similar
climb mechanisms. In this case climb proceeds by absorption of vacancies.
It was found that the inside-outside contrast of the nucleated loops in the
quenching experiments was opposite from that in the electron irradiation
case thus enabling climb by vacancy and interstitial mechanisms to be distinguished.

We report here on the climb of dissociated dislocations in GaAs under
electron irradiation in the HVEM. In the case of GaAs, edge dislocations
have different core structures depending on the sign of their Burgers
vector: the two different types, termed α and β dislocations, have their
extra half planes terminating on the Ga and As faces of a (111) crystal
respectively. There is, moreover, evidence that α and β dislocations in
degraded lasers climb at different rates (Hutchinson and Dobson, 1980).
In this paper, we compare the climb mechanism for α and β dislocations.
The climb of screw dislocations is also discussed.

2. Experimental

GaAs specimens in <111> orientation (n type, $N_d-N_a= 2 \times 10^{16}$ at /cm3) were deformed by bending at high temperature. Thin <$\overline{1}11$ > specimens were prepared
by chemical polishing in a solution of chlorine in methanol. The foils
contained dislocations of Burgers vector a/2<110> type lying in the (111)
glide plane which were analysed using the weak-beam method of electron
microscopy in a JEOL 100B electron microscope operating at 100kV. These
dislocations were found to be dissociated over most of their length, their
Shockley partials bounding an intrinsic stacking fault. Care was taken to
identify α and β dislocations by noting the assymmetry of the weak-beam
image (Feuillet, 1983). The dislocations were subsequently irradiated
by 1MeV electrons in the Oxford AE1 EM7 HVEM to a dose of 1 to 3×10^{20} e/cm2
either at room or elevated temperature (≈450°C). The post-irradiation
dislocation configurations were analysed again using weak-beam methods in
the JEOL 100B, using reflections of the $2\overline{2}0$-type or of the $1\overline{1}1$ and $1\overline{3}1$-
type in ($\underline{g}$, 3$\underline{g}$+) or ($\underline{g}$, 5$\underline{g}$+) conditions.

3. Results

Fig. 1a shows a β type dissociated edge-dislocation prior to irradiation,
imaged in $\underline{g}.\underline{b}_t$ = 2 conditions. Its Burgers vector is $\underline{b}_t$ = BA = $\underline{b}_1$ + $\underline{b}_2$,
$\underline{b}_1$ and $\underline{b}_2$ referring respectively to the top and bottom partials, $\underline{b}_1$ = δA,
$\underline{b}_2$ = Bδ. Fig. 1b to e include micrographs obtained after irradiation at
room temperature of the same disloc ation, taken in $2\overline{2}0$-type reflections.
The partial dislcoations are now heavily decorated with new features, like
those denoted X, Y and Z. The contrast of the new features depended
markedly on the sign of g indicating dislocations in dipole configuration.
Contrast analysis suggested X and Y in the vicinity of δA and Bδ respectively were of δC/Cδ type. For example X and Y̅ are not visible in the
g = $2\overline{2}0$ ($|\underline{g}.\underline{b}|$ = 0) but are visible in g = $02\overline{2}$, $\overline{2}02$ ($|\underline{g} . \underline{b}|$ = 1). Simi-
larly, Z close to the partial Bδ appears to be of Aδ/δA type since it is
invisible for $\underline{g}$ = $\overline{2}02$ and visible for the other two $\overline{2}20$-type reflections.
The sense of the dipole contrast (strong or weak depending on the sign of
$\underline{g}$, e.g. X in figs 1c,d is consistent with that observed in electron-
irradiated Cu-13%Al (Cherns et al, 1980) suggesting that the point defects
which condense on β-dislocations are of interstitial type.

The observations in Fig. 1 are consistent with the mechanism shown in Fig.
2. Suppose loops BA and CA nucleate on the δA partial and BA and BC loops
nucleate on the Bδ partial. These loops interact with their parent partial
as illustrated and, following dissociation of the segments in the upper
(111) plane, lead to the formation of dipoles Bδ/δB and δC/Cδ on the Aδ
partial, and of dipoles Aδ/δA and δC/Cδ on the Bδ partial. The climb forces
acting on the parts of the loops not lying in the original glide plane (the
'loop-jogs') tend to twist the loops CA in direction Bδ and the loops BC
in direction Aδ, as indicated by the dotted arrows. The plan view of Fig.
2 is in good agreement with the experimental results.

Fig.3 shows micrographs of an α edge dislocation irradiated at room temperature.
Features, like X and Y which are out of contrast for g = $2\overline{2}0$
and in contrast for the two other $2\overline{2}0$-type reflections are dipoles of
Cδ/δC type. Inside-outside contrast analysis carried out on these dipoles
again revealed the interstitial character of the nucleated loops.

The room temperature irradiation of dislocations in orientations other
than pure edge also showed the nucleation of new loops on the individual
partials. Observations on α and β dislocations were again similar. In

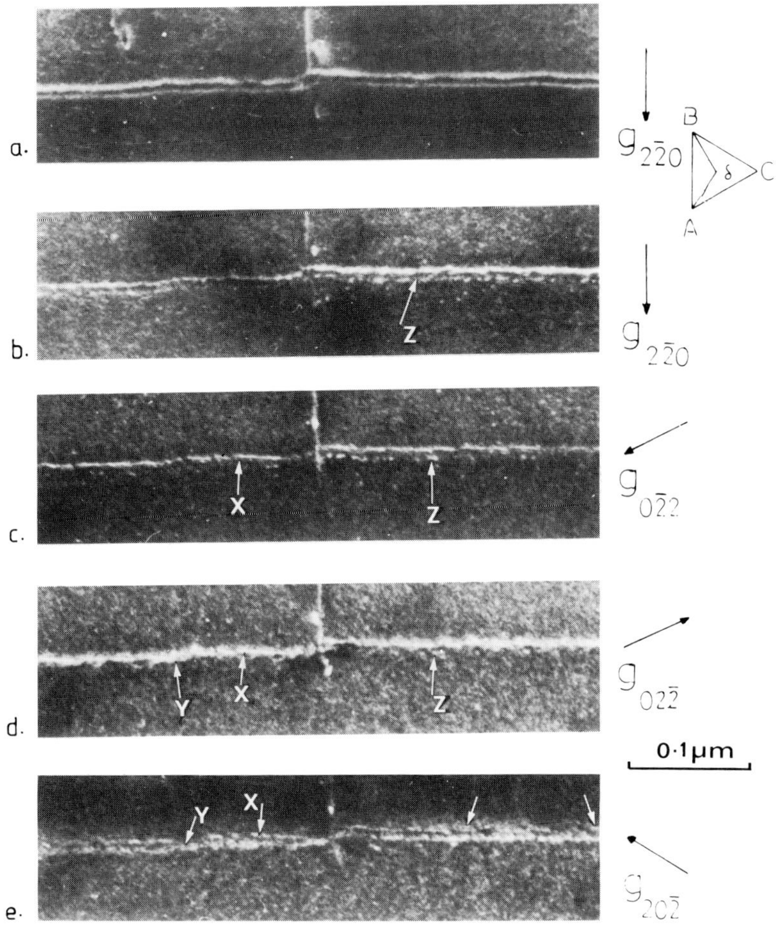

Fig. 1.   Edge dislocation of β-type irradiated by 1MeV electrons at room
temperature (a) before irradiation, (b) - (c) after irradiation.

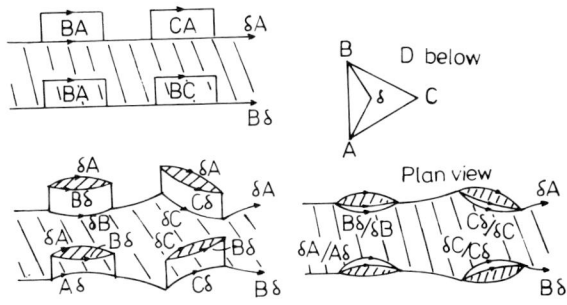

Fig. 2.   Explanation of contrast in Fig. 1 (see text).

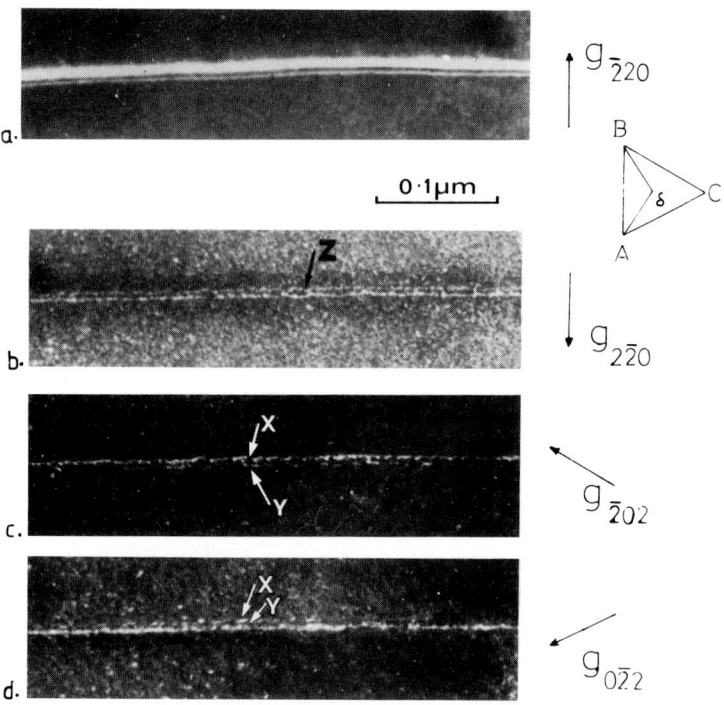

Fig. 3. Edge dislocation of α type irradiated by 1MeV electrons. The top
and bottom partials have Burgers vectors δ̲B̲ and A̲δ̲ . (a) before irradiation, (b) - (d) after irradiation.

contrast irradiations at elevated temperatures seldom showed additional
loops on the dislocations. However the final configurations of these dislocations indicated that climb had occurred leading to the generation of
new double jobs. For example, Fig. 4 shows a near-screw dislocation both
before (a) and after (b, c) irradiation at elevated temperature, viewed in
$\underline{g}.\underline{b}_t = 2$ conditions. This dislocation has a total Burgers vector $\underline{b}_t = \underline{AB}$
and partial Burgers vectors $\underline{b}_1 = \underline{A\delta}$ (top partial), $\underline{b}_2 = \underline{\delta B}$ (bottom partial).
The dislocation is uniformly dissociated prior to irradiation (Fig. 4a)
suggesting the absence of jogs. In the post-irradiation pictures, (fig.
4b, 4c) the pure screw segment, on the right-hand side of the pictures,
has a regular zig-zag configuration, whereas for orientations slightly off-
pure screw, the configuration is zigzagged but less symmetrical. It is
clear that, after irradiation, the dislocation is dissociated along the
large uncusped segments oriented towards 30° orientation and may be constricted elsewhere. Observations in different reflections showed no new
loops along the dislocation.

We may explain the observations in Fig. 4 by a mechanism due to Cherns et
al (1982) (their Fig. 15) shown in Fig. 5. Loops with Burgers vectors, $\underline{AC}$
and $\underline{BC}$ are nucleated on the partials $\underline{A\delta}$ and $\underline{\delta B}$ below and above plane δ
respectively. These loops are energetically favourable (Cherns et al, 1980).
Following annihiliation of $\underline{C\delta}$ and $\underline{\delta C}$ on plane δ (Fig. 5b) cross slip of $\underline{BC}$

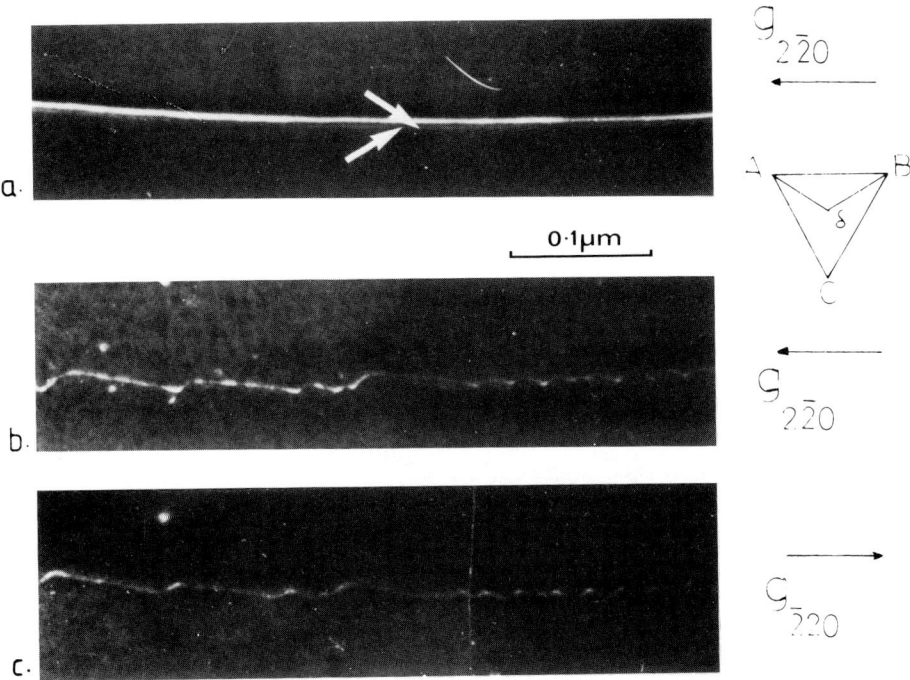

Fig. 4. Screw dislocation irradiated by lMeV electrons at a temperature ∿450° (a) before irradiation, (b), (c) after irradiation. The Burgers vectors of the partials are indicated in (a).

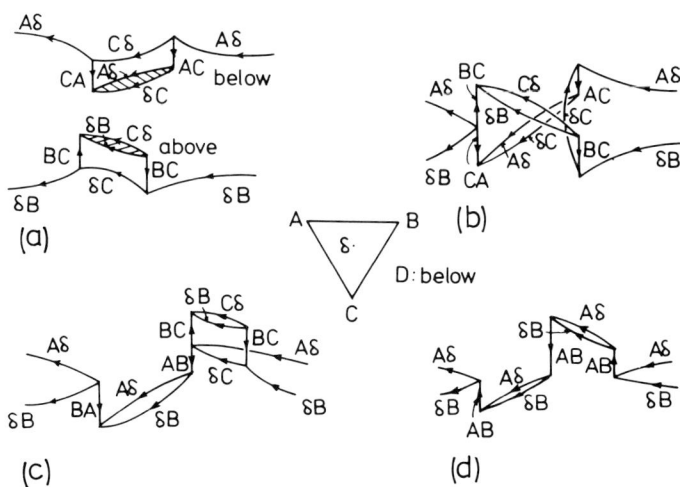

Fig. 5. A mechanism for double job formation on a near screw dislocation (following Cherns et al (1980)).

on α converts CA (arrowed) to AB (Fig. 5c). Cross slip of the remaining CA segment on β subsequently converts the whole dislocation to AB. Absorption of further interstitials causes the new jogs AB to move in a direction perpendicular to AB such that the dislocation attains a zig-zag form in good agreement with the experimental results.

The need for the cross-slip mechanism in Fig. 5 may be avoided, if we assume that only loops of AB type are generated on the original partials. This is at variance with the room temperature observations. However if we assume that the partial separation in GaAs decreases with increasing temperature the increasing elastic interaction between a loop nucleus on the one partial and the opposite partial should eventually favour the nucleation of AB loops. The question of whether the partial separation in GaAs varies with temperature requires further work. A further possibility that loop 'debris' i.e. dipoles of Cδ/δC, is removed from the climbing dislocation by a 'pinching off' process as in Si (Ourmazd et al, 1981) appears unlikely as no pinched-off loops were observed in GaAs.

4. Conclusions

Our results show that climb of both α and β dislocations in GaAs under room temperature electron irradiation results in the nucleation of new interstitial loops on the individual partials. In contrast with the results of Cherns et al (1980) on Cu-13%Al, the new loops are often in near screw orientation, e.g. loops AC and BC in Fig. 2, and thus have a small climb force, i.e. they are not efficient for accommodating interstitials. One factor may be the greater importance of core energy terms in the semiconductor compared to the alloy case where the initial loop nucleation is involved. Observations on GaAs irradiated at elevated temperature indicate climb of the total dislocation. The results may be explained by considering loop interaction with the opposite partial, with the removal of 'climb debris' by a cross-slip mechanism.

References

D Cherns, P B Hirsch and H Saka. Proc. Roy. Soc. A371 (1980) 213.
B Decamps, D Cherns and M Condat. Phil. Mag. (1983) in press.
G Feuillet. M.Sc. thesis, Oxford (1983).
P W Hutchinson and P S Dobson. Phil. Mag. A41 (1980) 601.
D V Lang and L C Kimerling Phys. Rev. Lett. 33 (1974) 489.
A Ourmazd,D Cherns and P B Hirsch. Inst. Phys. Conf. Ser. No. 60 (1981) 39.

Inst. Phys. Conf. Ser. No. 67: Section 1
Paper presented at Microsc. Semicond. Mater. Conf., Oxford, 21–23 March 1983

33

Quantitative TEM analysis of the dislocation structure at the lower yield point of silicon

A Oueldennaoua, J P Michel and A George

Laboratoire de Physique du Solide, ENSMIM, Parc de Saurupt, 54042 NANCY
Cedex, France. (Laboratoire Associé au CNRS N°155)

Abstract TEM investigations of silicon single crystals deformed at the
lower yield point at a shear strain rate of $2.10^{-5}s^{-1}$ between 823 K and
1073 K are reported. Local shear stresses and mobile dislocation
densities were tentatively estimated. The relevance of these microscopic
data is discussed, using the known dislocation velocities and Orowan's
formula.

1. Introduction

These observations aim at a better understanding of the dislocation processes controlling the flow stress in the yield region of silicon. The shear
strain rate $\dot{\gamma}$ is related to the velocity v of dislocations of Burgers
vector b through Orowan's formula $\dot{\gamma} = \rho_m bv$, provided that the density of
mobile dislocations ρ_m is known. In many cases, activation parameters
extracted from macroscopic experiments (strain rate jumps or stress relaxa-
tions) are directly attributed to the dislocation velocity, under the
assumption that ρ_m is a low varying function of straining conditions.
Silicon offers the possibility of a more careful analysis since direct
measurements of dislocation velocities are available (George and Champier
1979). From compression tests supplemented with stress relaxation tests,
activation volumes and enthalpies were measured (Omri 1981) and the lower
yield point appeared to be controlled by a thermally activated process with
$\Delta G \sim$ 2-2.5 eV, in fair agreement with the activation energy for single
dislocation glide. However, it could not be decided from those experiments
how the mobile dislocation density and the effective stress vary as functions of temperature and strain rate. Systematic TEM investigations of the
dislocation structure have been undertaken to answer these questions.
Preliminary results are reported in this paper.

2. Experimental

Dislocation-free, float-zoned silicon crystals (n type, $\rho \gtrsim 50$ Ω.cm, WASO
quality from Wacker) were used. Specimens ($4\text{x}4\text{x}14$ mm^3) were compressed
along the [123] axis (primary slip plane (1̄11)[101̄], Schmid factor : 0.47)
under an atmosphere of 10 % H_2, 90 % N_2. Each crystal was first pre-strai-
ned at 1323 K and $\dot{\gamma} = 2.10^{-5}s^{-1}$, in stage I up to $\gamma \simeq 6.10^{-2}$, then cooled
under load to the testing temperature between 823 K to 1073 K (which was
the investigated range in dislocation velocity measurements) and immediately
deformed up to the lower yield point. After cooling down under load and
chemical thinning, foils were observed in a JEM 200 CX electron microscope
operating at 200 kV.

3. Results

3.1 Stress, strain curves and yield stresses

Results of the compression experiments (Omri 1981) will be reported in details elsewhere. Fig.1a presents typical $\tau(\gamma)$ curves and Fig.1b the temperature dependence of the lower yield stress $\tau_{\ell y}$ at $\dot{\gamma} = 2.10^{-5}s^{-1}$.

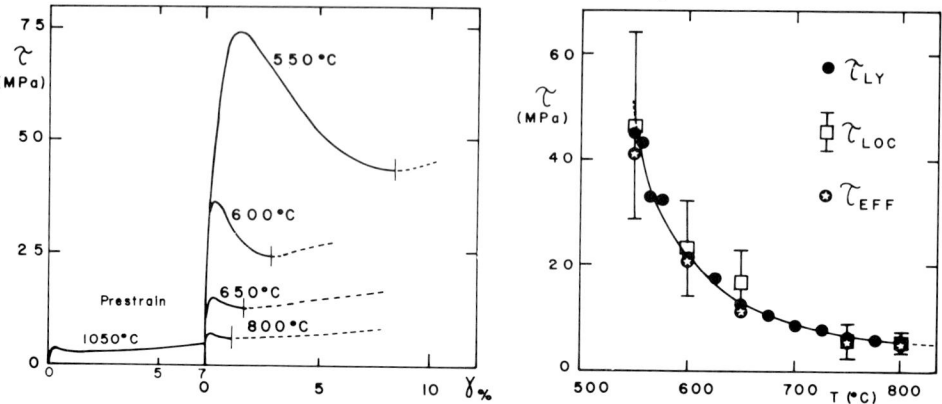

Fig.1a : Stress strain curves at $\dot{\gamma}=2\times10^{-5}s^{-1}$. Pre-strain and second deformation

Fig.1b : Temperature dependence of the lower yield stress $\tau_{\ell y}$ at $\dot{\gamma}=2.10^{-5}s^{-1}$ after pre-strain τ_{loc} local stresses with standard deviation, τ_{eff} calculated effective stress.

3.2 TEM observations

After pre-strain, most of the dislocations form bundles of dipoles with a strong edge character, isolated loops having no special crystallographic orientations, as shown in Fig.2. Observations of foils cut parallel to the (110) plane, perpendicular to the primary slip plane show that dislocations mainly belong to this primary plane, except for some superjogs (S).

At the lower yield point of the second deformation, dislocations are still in the primary slip plane and more than 90 % have the primary Burgers vectors 1/2 [10$\bar{1}$], but their arrangement strongly depends on temperature.

Fig. 2a : Dislocation structure after pre-strain at T = 1323 K, $\gamma \simeq 6 \times 10^{-2}$. Foil parallel to (1$\bar{1}$1)

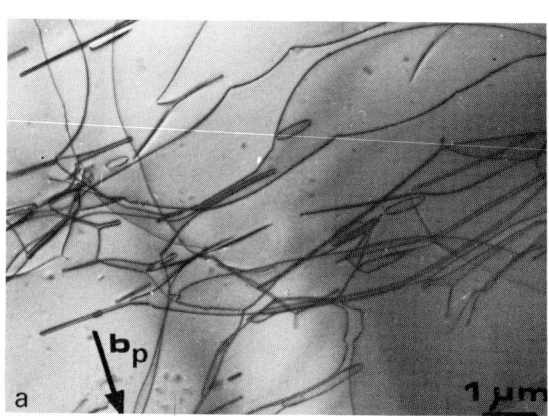

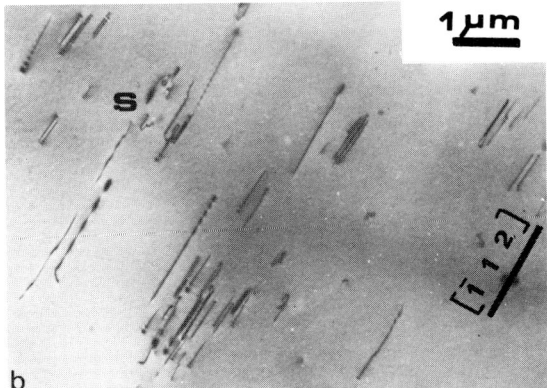

Fig. 2b : Dislocation struc-
ture after pre-strain at
T = 1323 K, $\gamma \simeq 6 \times 10^{-2}$.
Foil normal to ($1\bar{1}1$).

b

At 823 K the distribution is homogeneous and dislocation lines are parallel
to the <110> directions (Fig.3). Screw dislocations are at least as fre-
quent as each of the two types of 60° dislocations. Few dipoles and loops
can be seen. Foils near the (110) plane confirm the homogeneous distribution and the predominance of three families of line orientations.

At 873 K, some qualitative
change is already noticeable.
Fig.4 shows an increased number of edge type dipoles, in
addition to <110> oriented
dislocations. At 923 K and
higher temperatures, the <110>
orientations are no longer
observed and the microstructure evolves towards that
observed after pre-strain.
The distribution is less
homogeneous and the proportion of edge dipoles increases with temperature (Fig.5).

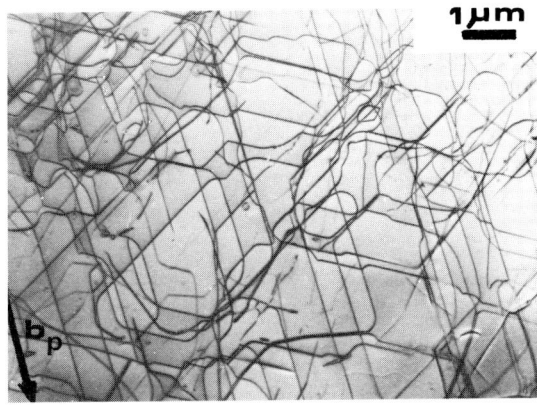

a

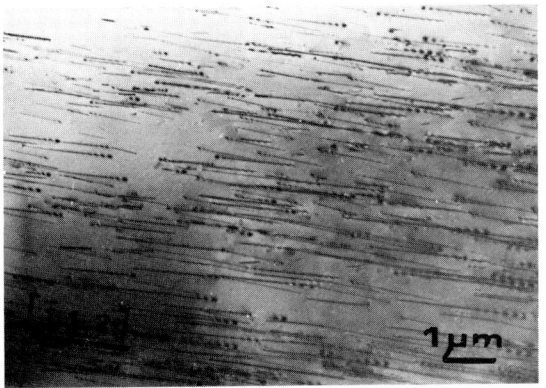

Fig.3 : Dislocation structure at T = 823 K, lower
yield point. Foils parallel
(a) and nearly normal
(b) to ($1\bar{1}1$)

b

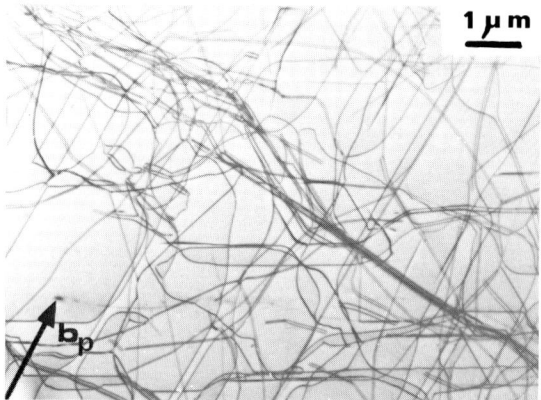

Fig.4 : Dislocation structure
at T = 873 K, lower yield
point. Foil parallel to (1$\bar{1}$1)

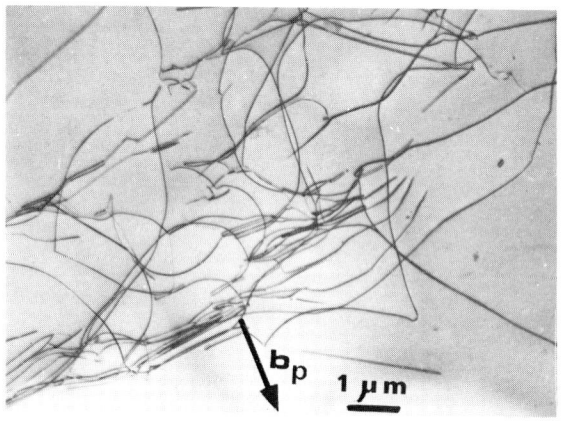

Fig.5 : Dislocation structure at T = 1023 K, lower
yield point. Foil parallel
to (1$\bar{1}$1)

Densities ρ_T and ρ_m of all dislocations and of mobile ones were estimated.
A dislocation is supposed to be mobile if the distance with the next parallel ones is greater than a critical value   d   corresponding to an interac-
tion stress equal to $\tau_{\ell y}$ ($\simeq \tau_{local}$, see below). d was estimated to be $\sim$ 0.1
µm at 823 K and 0.6 µm at 1073 K. When d approaches the foil thickness, it
is probable than this leads to an overestimation of ρ_m, since an apparently
free dislocation could have been pinned in the bulk specimen (by other dislocations lost in the thinning process). The length of the lines was measured on the (1$\bar{1}$1) micrographs using a pen connected to a computer. What-
ever the temperature, the investigated surface and measured length were at
least 700 µm^2 and 10^4 µm, respectively. The foil thickness was determined
by counting equal thickness fringes. It varied from 7.10^2 to 10^3 nm and is
the main cause of uncertainty.

The total dislocation density after pre-strain was 2.6 $\pm$ 0.4x10^8 cm^{-2}, less
than 3.9 $\pm$ 0.5x10^7 dislocations per cm^2 being possibly mobile. The variations of ρ_m and ρ_T with straining conditions are shown in Fig.6. ρ_T varies
only slightly from the value after pre-strain, while ρ_m decreases when
temperature increases. The proportion of mobile dislocations over the total number varies from 70 % at 823 K to (less than) 15 % at 1073 K.

Local stresses can be measured from the radius of curvature at bends of
isolated dislocations. The radii of curvature R were measured by compari-

son with standard circles and histograms of the distribution frequency of
R were established at each temperature. Fig.7 shows some of them. One maximum appears clearly, particularly at low temperatures. The local stress
τ_{loc} is calculated using the formula $\tau_{loc} = \mu b/2\bar{R}$, where $\bar{R}$ is the arithmetic mean radius, μ is the shear modulus, considered equal to $[1/2 \ C_{44}$
$(C_{11}-C_{12})]^{1/2}$ with the elastic constants measured by Burenkov and
Nikanorov (1974). The numerical factor $1/2$ is consistent with a dislocation density of $\sim 5.10^8 cm^{-2}$. The values of τ_{loc} are plotted on Fig.1b. The
limits of confidence are the standard deviation of R. At any temperature
$\tau_{loc} \simeq \tau_{\ell y}$, and standard deviation $\Delta\tau$ increases relatively to the mean
value as temperature increases : ~ 25 % at 823 K and ~ 40 % at 1073 K.

4. Discussion

The estimation of local stresses allows us to calculate the mean dislocation velocity from the data of George and Champier (1979). Local stresses
are described by $\tau = \tau_{loc} \pm \Delta\tau \sin k \ x$ and the average velocity $\bar{v}$ of a
dislocation in that stress field is calculated following Alexander and
Haasen (1968), assuming that the stress dependence of single dislocation
velocity —here written for convenience $v = v_o(T)(\tau/\tau_o)^m$ with m=1.2- is
still obeyed at the scale of TEM. In the calculations, we used $v = (v_s +$
$2 \ v_{60°})/3$ at $T \leq 873$ K and $v = v_{60°}$ at $T > 873$ K. (The analysis is facilitated by the fact that the velocities of screw and 60° dislocations in
these ranges of τ and T are not very different. More questionable is the

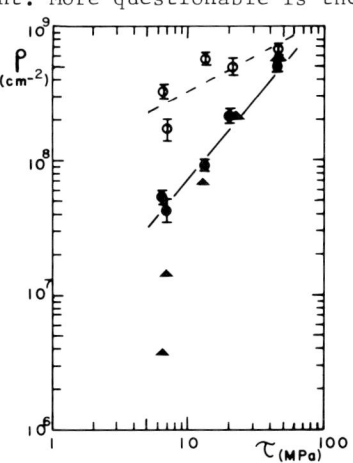

Fig.6 : Dislocation density vs lower
yield stress. Total density :○
mobile dislocation density
measured :●   calculated :▲

Fig.7 : Histograms of the radii of
curvature of dislocations at the
lower yield point, $\dot{\gamma}=2x10^{-5}s^{-1}$.

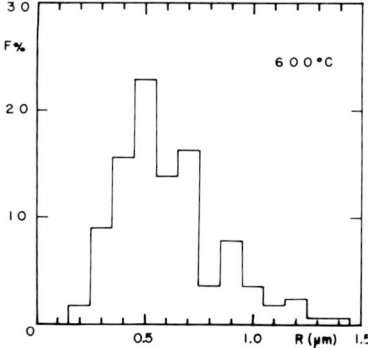

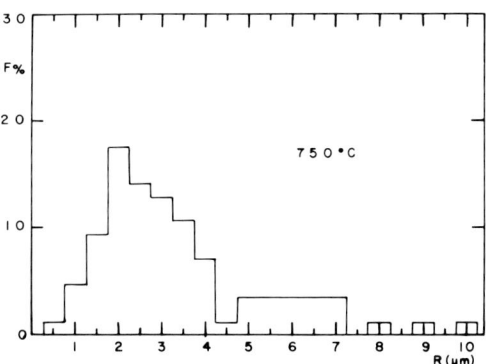

use of velocities measured on <110> segments for curved dislocations at
T > 873 K). This leads to the derivation of an effective stress τ_{eff}, such
as :

$$\bar{v} = v_0(T) \; (\tau_{eff}/\tau_0)^m$$

It results that τ_{eff} is rather close to τ_{loc} and $\tau_{\ell y}$ (Fig.1b).

It is possible to estimate, through Orowan's formula, the actual mobile
dislocation density from the imposed strain rate $\dot{\gamma}$ (Fig.6). A close agreement with the measured values of ρ_m is obtained at low temperatures, but
there is an overestimation of ρ_m on the basis of the micrographs at higher
temperatures and low stresses. As mentioned above, a possible explanation
is that dislocations trapped in dipoles in bulk specimens may appear to be
free in thin foils. This is supported by the observation that "isolated"
dislocations may exhibit reversed curvatures (Fig.5). It is also possible
that the use of the standard deviation $\Delta\tau$ to limit the fluctuations of the
local stress leads to too high values of τ_{eff}.

Further work is under progress to study the influence of the strain rate
and to improve the statistical treatment of the microscopic data.

Acknowledgments

The authors would like to thank Mr. Poirson for his help in operating the
electron microscope.

References

Alexander H and Haasen P 1968 Solid State Physics 22 (Academic Press
    New-York) pp 98-9.
Burenkov Yu A and Nikanorov S P 1974 Sov.Phys.Solid State 16 963.
George A and Champier G 1979 Phys.Stat.Sol.(a) 53 529.
Omri M 1981 Thèse 3°Cycle Nancy ; Omri M, Tête C, Michel J P and George A
    to be published.

Inst. Phys. Conf. Ser. No. 67 © *1983: Section 1*
Paper presented at Microsc. Semicond. Mater. Conf., Oxford, 21–23 March 1983

The dislocation content of some near-coincidence grain boundaries in polycrystalline silicon

Y S Oei, F W Schapink and S Radelaar

Laboratory of Metallurgy, Delft University of Technology, Rotterdamseweg
137, 2628 AL Delft, The Netherlands

Abstract Employing TEM we have investigated the structure of some near-
coincidence grain boundaries in polycrystalline Si. In particular the
dislocation structure of a near $\Sigma = 9$ curved boundary has been analysed.
The dislocation network in this boundary could be described using three
sets of dislocations with DSC Burgers vectors. No evidence for the
existence of partial grain boundary dislocations has been found. From
the occurrence of α-fringes there is clear evidence for the existence of
a rigid translation along the boundary. On the other hand, a near $\Sigma = 13b$
boundary was found to contain partial grain boundary dislocations
separating domains with different states of translation.

1. Introduction

In recent years increasing interest has developed in correlating the
detailed structure of grain boundaries in polycrystalline semiconductors
with their electrical properties. As a consequence, various studies of the
structure of grain boundaries in semiconductors have been performed
(Bacmann et al. 1981, Föll and Ast 1979, Fontaine and Smith 1982, Papon
et al. 1982, Pond 1982). Much of this work has been carried out either on
well-defined boundaries in bicrystals or on first-order ($\Sigma = 3$) twin
boundaries in polycrystalline material. In this paper we describe the
results of a TEM study on some near-coincidence grain boundaries occurring
in commercial polycrystalline Si, some preliminary results of which have
been reported previously (Oei et al. 1982).

2. Experimental details and results

Thin foils of commercial polycrystalline Si (Silso, Heliotronic) were
prepared by ion beam thinning of thin sections cut from a large ingot with
an average grain size of ∿ 1 mm. Foils prepared in this way exhibited
rather irregular surfaces thus limiting the useful areas for TEM work.
Nevertheless several grain boundaries could be investigated over substantial
areas using a Philips EM-400T electron microscope operating at 120 kV.

In this section the results of a TEM study on a near $\Sigma = 9$ coincidence
boundary will be reported. First we describe the grain boundary
crystallography and subsequently the dislocation network will be analysed
from appropriate contrast experiments.

Fig. 1a shows a curved near $\Sigma = 9$ boundary seen in the edge-on position;
the normals of this boundary range from P to R in the stereographic
projection (fig. 1b) and all boundary normals were found to be located in

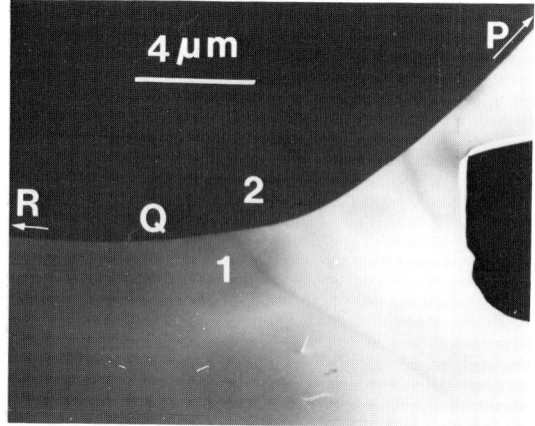

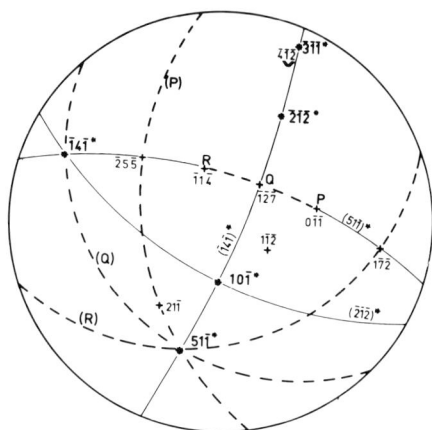

Fig. 1a. The curved near Σ = 9 boundary in an edge-on position.

Fig. 1b. Stereographic projection of the bicrystal, with the centre corresponding to the zero setting of the goniometer stage.

the (51$\bar{1}$)* plane[+]. The deviation from the exact Σ = 9 misorientation (rotation axis/angle: [10$\bar{1}$]*/38.94°) was determined from the shift in Kikuchi line pairs and the rotation of a number of g-vectors from the exact Σ = 9 orientation. Thus, for a [21$\bar{1}$] incident beam direction, diffraction vectors ($\bar{1}$53)*, (113)*, (4$\bar{2}$6)*, (066)* and (3$\bar{3}$3)* were used for this purpose. The deviation from Σ = 9 can be described as an extra rotation θ = (0.43 + 0.05)° of the crystal 2 in a right handed sense approximately around the [$\bar{4}$12] axis (see fig. 1b). The possible Burgers vectors for dislocations in the grain boundary plane can be derived from the shortest translation vectors of the orthorhombic DSC lattice (Warrington and Grimmer 1974), given by the columns of the matrix:

$$DSC = \frac{a}{18} \begin{pmatrix} -1 & 4 & -6 \\ 4 & 2 & -3 \\ -1 & 4 & 3 \end{pmatrix} \qquad R = \begin{pmatrix} 8 & 4 & -1 \\ -4 & 7 & -4 \\ -1 & 4 & 8 \end{pmatrix}$$

where a is the lattice parameter. Also given above is the rotation matrix R transforming coordinates from crystal 1 into crystal 2 (neglecting the contribution from the small deviation from Σ = 9).

The dislocation network existing in this Σ = 9 boundary consists of three sets of dislocations A, B and C, where B and C are mutually parallel, as is shown in fig. 2. For the purpose of analysing the Burgers vectors, contrast experiments have been performed with various common diffracting vectors, as well as individual reflections from either crystal 1 or 2. Table 1 summarizes the results of these observations and an example of the extinction of dislocations A and B is given in fig. 3, both in weak-beam and strong-beam conditions. In these micrographs, taken with the (3$\bar{1}$$\bar{1}$)*

[+] Unless stated otherwise, all planes and directions are given with reference to crystal 1 as in fig. 1b. Planes and directions common to both crystals are indicated by an asterisk.

Table 1: Absolute values of g.b for the dislocations assigned as set A, B and C. The image behaviour is taken from region Q and is tabulated as strong (S), visible (V), weak (W) and invisible (I).

a. crystal 1

set	b	(311)*	(202)*	(113)*	(111)	(1̄11)	(220)	(02̄2)	(022)	(31̄1)
A	$\frac{a}{18}$ [1̄41̄]	0	0	0	0.11	0.33	0.33	0.56	0.33	0.11
B	$\frac{a}{6}$ [1̄1̄2]	0	1	1	0.33	0	0	0.33	1	0.67
C	$\frac{a}{6}$ [211̄]	1	1	0	0.33	0	1	0.67	0	1.22
A		I	I	I	W	V	V	V	W	V
B		I	V	V	S	W	I	V	S	V
C		V	V	V	S	I	S	V	I	S

b. crystal 2 (Only in this table the indices relate to crystal 2)

set	b	(3̄11)*	(202̄)*	(1̄13)*	(1̄11)	(220)	(113)	(311)
A	$\frac{a}{18}$ [1̄41̄]	0	0	0	0.22	0.56	0.44	0.44
B	$\frac{a}{18}$ [21̄7̄]	0	1	1	0.56	0.11	1.11	0.11
C	$\frac{a}{18}$ [71̄2̄]	1	1	0	0.44	0.89	0.11	1.11
A		I	I	I	V	V	V	W
B		I	V	V	V	V	V	I
C		V	V	W	V	V	I	S

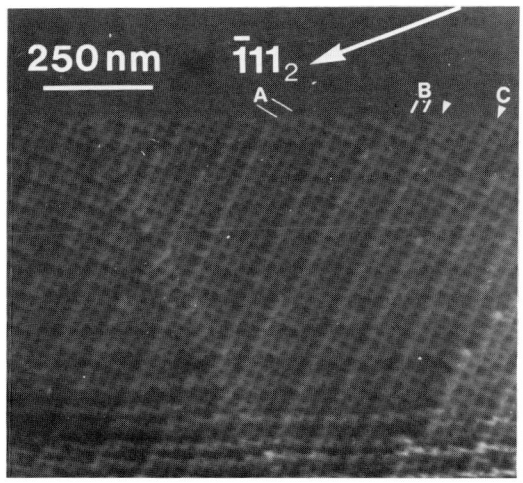

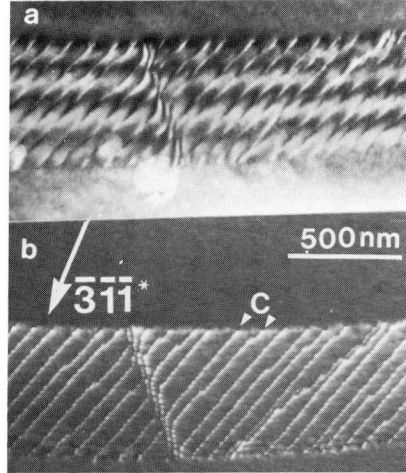

Fig. 2. DF micrograph of the dislocation network in region Q, consisting of the sets A, B and C.

Fig. 3. DF micrographs of the same region as fig. 2, taken with the (3̄11)* diffraction vector.
a) strong two-beam condition
b) weak-beam (∿ g/3g) condition.

reflection, only the dislocation set C is clearly visible. In addition
there is strong fringe contrast visible in fig. 3a due to a rigid relative
translation of both crystals at the boundary (α-fringe contrast) (Papon et
al. 1982). From the observations summarized in table 1 and taking the DSC
vectors into account (see above), it can be deduced that the simplest
possible set of Burgers vectors which can be assigned to the dislocations
of the network are given by $(a/18)[\bar{1}4\bar{1}]$, $(a/6)[11\bar{2}]$ and $(a/6)[2\bar{1}\bar{1}]$, for
sets A, B and C respectively. Table 2 contains the line directions and
spacings of the dislocation network in different boundary areas, marked P,
Q and R in fig. 1b. It is seen that the A set consists of pure screw
dislocations in region Q. Since in this area the network is the most
regular, analysis will be restricted to this region. According to fig. 2
the ratio of the number of B and C dislocations is approximately equal to
3. This observation suggests a model for the dislocation network consisting
of dislocation set A and a set of dislocations in the C direction with an
effective Burgers vector of $3\underline{b}_B + \underline{b}_C = (a/6)[5\bar{2}\bar{7}]$ and with an effective
dislocation line spacing equal to the spacing of the original C set. The
rotation axis $\underline{u}$, associated with a network consisting of two sets of
dislocations with Burgers vectors $\underline{b}_1$ and $\underline{b}_2$, is given by $\underline{u} = \underline{b}_1 \times \underline{b}_2$
(Hirth and Lothe 1968). In the present case, using
$\underline{b}_1 = (a/18)[\bar{1}4\bar{1}]$ and $\underline{b}_2 = (a/6)[5\bar{2}\bar{7}]$, $\underline{u}$ becomes $[5\bar{2}3]$, $7.6°$ away from the
experimentally observed $[4\bar{1}2]$ axis. The calculated dislocation spacings are
26 nm for set A and 123 nm for set C, which are in reasonable agreement
with the corresponding values of table 2 (24 nm and 102 nm). The calculated
line direction for set A deviates ∿ $13°$ from the observed direction in
fig. 2, whereas the two directions coincide for set C.

3. Discussion

The observations regarding the existence of a dislocation network reported
in the previous section permit a reasonably consistent analysis of the
network in a near Σ = 9 coincidence boundary in Si. Starting from the
shortest possible DSC vectors for this boundary, it has been shown that the
observed network can be analysed in terms of three sets of dislocations
with Burgers vectors $(a/6)[11\bar{2}]$, $(a/6)[2\bar{1}\bar{1}]$ and $(a/18)[\bar{1}4\bar{1}]$. The latter
Burgers vector is associated with the A-set, which has screw character in
a large part of the curved boundary. The line direction of the sets B and
C, which are parallel to each other, is $[51\bar{1}]^*$ which is a
common line direction for all boundary orientations, as shown in fig. 1b.
The ratio of the number of B to C dislocations depends to some extent on
their location in the boundary plane, as is shown in table 2. It has
been found that a ratio of 3 gives a reasonable correlation between
the experimentally observed dislocation network parameters and the
calculated parameters based on a model of two sets of dislocations in the
network. It is thus concluded that the network can be analysed in terms of
grain boundary dislocations with Burgers vectors equal to a DSC vector and
no splitting into partial dislocations has to be assumed.

A special contrast feature was observed in this boundary using a non-common
111 diffraction vector from crystal 1. As is shown in fig. 4, there is a
succession of darker and lighter areas in the boundary plane, bounded by B
(or C) dislocations. It should be stressed that this type of contrast has
only been observed for a non-common 111 diffraction vector, and is
therefore not associated with the black-white contrast of rigidly
translated domains as observed in some coincidence boundaries in
semiconductors (Bacmann et al. 1981). Presumably this effect is caused by

Table 2. The observed spacings (D) and determined line directions (L) for the dislocations assigned A, B and C.

set	b	P D[nm]	P L	Q D[nm]	Q L	R D[nm]	R L
A	$\frac{a}{18}$ [$\bar{1}$4$\bar{1}$]	50	[$\bar{2}$5$\bar{5}$]	24	[$\bar{1}$4$\bar{1}$]*	–	[$\bar{1}$72]
B	$\frac{a}{6}$ [1$\bar{1}$$\bar{2}$]	25	[51$\bar{1}$]*	26	[51$\bar{1}$]*	∿ 30	[51$\bar{1}$]*
C	$\frac{a}{6}$ [21$\bar{1}$]	78	[51$\bar{1}$]*	102	[51$\bar{1}$]*	71	[51$\bar{1}$]*

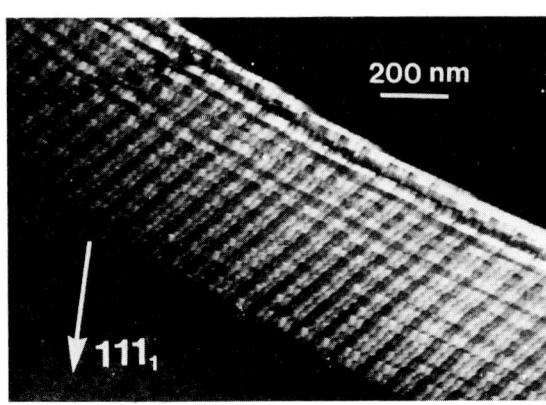

200 nm

111₁

Fig. 4. DF micrograph of a region near Q taken with the non-common 111 diffraction vector. Note the light-dark contrast parallel to the dislocation sets B and C.

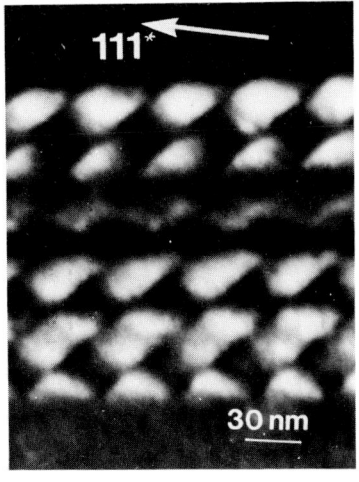

111*

30 nm

Fig. 5. Black-white contrast in a DF micrograph from a near Σ = 13b boundary, taken with the 111* common diffraction vector.

steps associated with the <112> dislocations in the boundary, since the 111 diffraction vector has the smallest extinction distance, which is further reduced when weak-beam diffraction conditions are applied as in fig. 4.

Apart from the network clear evidence has also been obtained for the existence of a rigid translation along the boundary, as demonstrated by the α-fringes in fig. 3a. Although in general the interpretation of such fringes in a boundary deviating from the exact coincidence misorientation is difficult, this does not apply to the conditions of fig. 3a. As can be seen from the stereographic projection in fig. 1b, the $(3\bar{1}\bar{1})^*$ common diffraction vector used to obtain fig. 3 is nearly parallel to the rotation axis of the misorientation from $\Sigma = 9$, hence images taken with this $\underline{g}$-vector should be virtually unaffected by this deviation from $\Sigma = 9$. Thus fig. 3a provides strong evidence for the existence of a rigid relative translation along the boundary plane, the magnitude of which has not yet been established.

Finally we should like to draw attention to a $\Sigma = 13b$ boundary occurring in the same specimen as the $\Sigma = 9$ boundary described here. In this boundary there is clear evidence for the existence of partial grain boundary dislocations, as is evident from the micrograph shown in fig. 5, taken with a common 111 diffraction vector. This micrograph clearly shows a succession of dark and light areas, indicating domains with different states of translation at the boundary bounded by partial grain-boundary dislocations. No detailed analysis of this boundary has as yet been made.

Acknowledgement
This work is part of the research program of the Foundation for Fundamental Research on Matter (FOM-Utrecht) and has been made possible by financial support from the Netherlands Organization for the Advancement of Pure Research (ZWO-The Hague).

References
Bacmann J J, Silvestre G, Petit M and Bollmann W 1981 Phil. Mag. A <u>43</u> 189
Föll H and Ast D 1979 Phil. Mag. A <u>40</u> 589
Fontaine C and Smith D A 1982 Grain Boundaries in Semiconductors, Proc. of the MRS Annual Meeting, Boston (New York: North Holland) p 39
Hirth J P and Lothe J 1968 Theory of Dislocations (New York: Mc Graw Hill) p 652
Papon A M, Petit M, Silvestre G and Bacmann J J 1982 Grain Boundaries in Semiconductors, Proc. of the MRS Annual Meeting, Boston (New York: North Holland) p 27
Pond R C 1982 J. de Physique 43 C1-51
Oei Y S, Schapink F W and Radelaar S 1982 J. de Physique <u>43</u> C1-21
Warrington D H and Grimmer H 1974 Phil. Mag. <u>30</u> 461

Inst. Phys. Conf. Ser. No. 67: Section 1
Paper presented at Microsc. Semicond. Mater. Conf., Oxford, 21–23 March 1983

A theoretical interpretation of dislocation glide in silicon

M Heggie and R Jones

Department of Physics, University of Exeter, Stocker Road, Exeter EX4 4QL,
Devon, England

Abstract   It is likely that glide partials, and hence kinks upon them,
are reconstructed, but soliton-like excitations of reconstruction (antiphase defects) give rise to dangling bonds . Hirsch's theory of doping
dependent dislocation velocity in silicon implies that reconstructed
kinks are strongly bound to solitons, whereas it will be shown that an
equally good description of dislocation behaviour arises by assuming
that their binding energy is small. Theoretical estimates of
formation energies of the soliton in different charge states point to
the instability of the neutral, paramagnetic state. This explains the
lack of e.p.r. activity in the presence of enhanced recombination at
dislocations . Invoking the presence of vacancy-soliton complexes in
low-temperature-deformed silicon leads to an explanation of some of the
features of the e.p.r. spectrum of this material – particularly the
annealing behaviour, which occurs by emission of vacancies from the
complexes.

1. Introduction

There has been much experimental work featuring dislocation motion in
silicon and germanium, including its doping dependence (e.g. George,
Escaravage, Champier and Schröter 1972, Patel, Testardi and Freeland 1976,
Kulkarni and Williams 1976 and recently Hirsch, Ourmazd and Pirouz 1981
and Louchet 1981). Amongst the peculiarities noted in measuring dislocation velocity, the recent observation that those dislocations in swirl-
free float-zone silicon subjected to the highest climb forces glide
fastest, ceteris paribus, seems most strange (Alexander, Kisielowski-
Kemmerich, Weber 1982). Naturally there have been parallel experiments
conducted by Russian groups (e.g. Erofeev and Nikitenko 1971, Bondarenko,
Erofeev, Nikitenko 1973, Nikitenko, Farber, Bondarenko 1982) which have
also uncovered a host of unexplained phenomena (e.g. the asymmetry of
dislocation glide and the existence of glide plane débris in the wake of
moving dislocations).

An attempt to explain the whole experience in this field would be
impossible here, but there is one area which appears amenable to analysis
and that is the high stress (> 10 MPa in silicon) and high temperature
(500°C - 600°C in silicon) régime of dislocation motion. Under these
conditions dislocation velocity is limited by lattice friction and is
relatively unaffected by pinning, so mobility measurements should give an
indication of the intrinsic behaviour of dislocations. Hirsch (1979) has
managed to digest the results of two separate groups (Patel et al 1976,

Erofeev and Nikitenko 1971) using his theory of the doping effect and he points out (Hirsch 1980) that Schröter has successfully applied it to the results of George et al (1972). However, there are many models of double kink nucleation which give identical results to Hirsch for the doping dependence but each has different physical implications. For reconstructed dislocations the key defects are reconstructed kinks and solitons (essentially dangling bonds) and the following four possible models depend on different interactions between them:

(a) Reconstructed kinks move and double kinks nucleate by the 'strained bond' model of dislocation motion, ignoring any role played by solitons (Jones 1980). In this model the doping dependence arises from changes of the kink migration energy, W, with the chemical potential, . However, Hirsch (1980) objects that it would be difficult for the saddle points to be in equilibrium with the electron gas and calculations by us (Heggie 1982, Heggie and Jones 1983) using a parametrised tight-binding Hamiltonian indicate a saddle point electronic structure with levels above mid-gap, contrary to experiment (Hirsch 1979). It is possible that the first objection can be over-come by allowing that equilibrium between the electron gas and shallow levels of the reconstructed kink occurs prior to migration.

(b) Hirsch's theory implies that double kink nucleation gives rise to two kinks with dangling bonds ("Hirsch kinks"). The doping dependence of motion is caused by changes both in Hirsch kink formation energy and migration energy. The same problem of equilibrium with the electron gas as occurred in (a) arises for double kink nucleation in this theory and it may be that this problem may be circumvented by the same postulate of shallow levels (Hirsch 1980b).

(c) Double kink nucleation occurs at a soliton (by a mechanism proposed later) and <u>strong</u> binding between kinks and solitons attracts another soliton, forming two Hirsch kinks, before the double kink critical width is reached.

(d) The last model considered allows that the binding energy between kinks and solitions is negligible (Heggie 1982, Heggie and Jones 1982). It argues that soliton migration is much easier than kink migration and that, as far as double kink nucleation and migration are concerned, solitons act as catalysts altering the effective activation energies for these processes. If a double kink is nucleated at a soliton, then at any subsequent time there is an equal probability that the soliton will be associated with any dislocation site, including either side of the double kink.

The models (a) and (d) are similar in effect, since they lead to changes only in migration energies with doping, causing the kink mean free path, L, which depends on the kink formation energy, to be independent of doping. Models (c) and (d) would give rise to doping-dependent $\bar{L}$. Jones (1983) has discussed (a), (b) and (c); we shall explore the possibilities of (d) in this paper.

2. Properties of Static Dislocations in Silicon

The question of the structure of 90° and 30° glide partials, i.e. the reconstructed vs. the unreconstructed models, is difficult to settle theoretically in a conclusive way (Marklund 1983). This is because of the

difficulties associated with total energy calculations: the simplest
approach - that of valence force potentials - suffers from inadequate
treatment of dangling bonds and more or less arbitrary behaviour for
large deviations from crystalline bond lengths and angles. However,
Marklund's latest results (1983) indicate the likelihood of strong reconstruction. Providing electronic bands associated with deformed silicon
are attributed to 90° and 30° partials, then the smallness of their bandwidth (0.25 eV from DLTS - Kimerling et al 1980) tends to rule out
unreconstructed or weakly reconstructed dislocations. This being so, the
recombination activity of strongly reconstructed dislocations must be
attributed to dangling bonds arising when the reconstruction changes sense
(Jones 1981). Figure 1 depicts these **soliton**-like excitations of reconstruction (otherwise known as antiphase defects) in a glide plane diagram
of a dissociated 60° dislocation (c.f. **solitons** in trans-polyacetylene :
Su, Schrieffer and Heeger 1980).

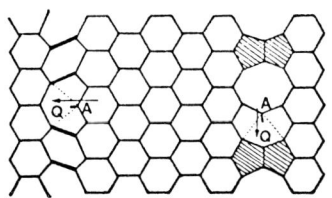

Figure 1 Glide plane diagram of
solitons on the 90°
(left) and 30° (right)
partials.

These defects give rise to half-filled levels around E_v + 0.3 eV that are
strongly coupled to the lattice, particularly movement of atom A in the
direction Q. This coupling may well lead to Anderson's negative-
effective-U behaviour (Heggie and Jones 1982), similar to the experimentally-
proven behaviour of the silicon vacancy (Baraff, Kane and Schlüter 1980,
Newton, Chatterjee, Harris and Watkins 1982), in which an energy lowering
lattice distortion caused by charging overcomes the intra orbital electron-
electron repulsion, U. Because of this behaviour the neutral, paramagnetic
soliton becomes unstable with respect to charged, diamagnetic ones for
all values of the chemical potential, μ, leading to the e.s.r. inactivity
observed for high temperature deformed silicon (Weber and Alexander 1979).

The last reference shows that the position in low temperature deformed
silicon is different and we suggest that a combination of a soliton and
a vacancy might be responsible for some of the e.s.r. signals (Si-K1,K2)
observed in this material. Strain-induced point defects are known to
arise from low temperature deformation (Kimerling and Patel 1979) and
our model of a vacancy-soliton complex (figure 2) is similar to the
segment of 30° shuffle partial proposed by Weber and Alexander (1979).

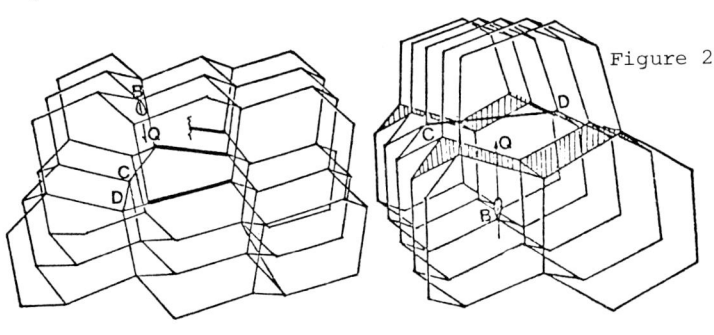

Figure 2 Nearly axial
perspective views
of 90° and 30°
partials each
containing a
vacancy-soliton
complex

We would expect this defect to exhibit positive-U behaviour, because the interaction of electrons trapped on the dangling band on B with the high electron density in the middle of the facing bond CD would be strongly repulsive. Annealing of these complexes, which occurs between $700^{\circ}C$ and $800^{\circ}C$, proceeds by emission of vacancies or interstitial absorption, leaving e.s.r. inactive recombination centres (i.e. solitons).

We cannot calculate the energy of formation of solitons, Fs, theoretically with any confidence, but since we argue the dislocation reconstruction is strong it causes a gap, Erg $\gtrless$ Eg, giving Fs $\gtrless$ Erg/π (Takayama, Lin-Lin, Mak 1980) $\gtrless$ 0.35 eV. However, other authors (Tan, Föll and Krakow 1981) are of the opinion that the energy of a dangling bond in silicon is of the order of 1 eV. The entropy of formation of solitons is also difficult to estimate, but if the high entropy of the vacancy in silicon can be attributed to a 20% softening of the back bonds (as suggested by Allan and Lannoo 1980) and if this softening is due to the presence of dangling bonds, then the soliton formation entropy may be similar to that of the vacancy. This means that, not only would there be an energy barrier to soliton-anti-soliton annihilation, but also a significant entropy barrier (cf. the vacancy and self-interstitial-Gösele, Frank and Seeger 1983). This difficulty in annihilation will mean that solitons observed in low temperature (room temperature) "electrical" experiments must be "frozen in" after introduction during deformation.

We can imagine two ways in which solitons are introduced:

(a) by the breaking of a reconstruction band with thermal agitation

(b) capture of a vacancy by a reconstructed dislocation, giving rise to a vacancy-soliton complex and an anti-soliton.

The process (b) is likely on the grounds of dangling bond (db) arithmetic (an ideal vacancy has 4 db's, an ideal vacancy-soliton complex has 3 db's and an antisoliton has 1 db), but pair-wise coupling of db's may reduce the favourability of this reaction. For germanium Bondarenko and co-workers (1973) conclude that their dislocation velocity data may be explained by (i) facilitation of double kink nucleation caused by point defects and (ii) trapping of a mobile kink by a point defect. The reaction (b) above could explain (i) and the reaction vacancy-soliton vacancy + soliton might account for (ii).

2. Glide of Dislocations

For stresses between 1 and 30 kg mm^{-2} the Hirth-Lothe theory (Hirth and Lothe 1968) of dislocation glide limited by double kink nucleation appears well established now, following the work of Hirsch et al.(1981) and Louchet (1981) which verifies the length dependence of dislocation motion for short segments (e.g. < 4000 Å at $600^{\circ}C$). In this theory the rate of double kink nucleation per unit length of dislocation (J) is determined by the attempt frequency ($\nu_D \sim 10^{13}$ Hz), the applied stress (σ), the temperature (T), dislocation burgers vector (b), kink height (h) and step distance (a), the double kink formation energy ($F_{dk} \approx 2F_k$) and the kink migration energy (W) according to:

$$J = \frac{\nu \sigma bh}{kT} \exp -(F_{dk} + W)/kT$$

This ignores the correction E_σ^* (Jones 1983) to F_{dk} caused by the applied stress. The mean free path of a double kink, $2\bar{L}$, is estimated as $2\bar{L} = 2a \exp(F_k/kT)$ in steady state, giving a dislocation velocity

$$v = 2h\bar{L}J = \frac{2v\sigma abh^2}{kT} \exp -(F_k + W)/kT$$

For short segments of length L kinks cannot traverse the distance $\bar{L}$, so

$$v = hLJ = \frac{v\sigma Lbh^2}{kT} \exp -(F_{dk} + W)/kT$$

Including the stress correction in the exponent, Jones (1983) has estimated $F_{dk} \simeq 1.05$ eV and $W = 1.2$ eV for Louchet's work at 600°C and $F_{dk} = 0.95$ eV, $W = 1.35$ eV for the experiments of Hirsch and co-workers at 420°C.

One additional comment on Hirth-Lothe theory must be made, and that is that F_k and W are free energies of kink formation and activation, i.e. since $F = H - TS$ they cause a temperature-independent term, S, to appear in the exponents above. This gives rise to an "apparent" discrepancy between theory and experiment of $\exp(S/k)$, with $S \approx 3k$ calculated from Louchet's measurements (Jones 1983).

The doping dependence of dislocation velocity is well known now - a review of theories and experiments can be found in Hirsch (1980a). The most suitable theory for this phenomenon is due to Hirsch and it allows that the overall dislocation velocity, V, is the sum of contributions from the differently charged kink saddle points (of energies E_{sp}), i.e.

$$v = v_o \sum_{c=+,o,-} \exp(-E_{sp}^c/kT)$$

where c denotes the charge state and

$$E_{sp}^+ = E_{sp}^o + \mu - E_d - eV \text{ and } E_{sp}^- = E_{sp}^o - \mu + E_a + eV$$

is the chemical potential, V the potential of the dislocation line due to charging and E_d, E_a are the donor, acceptor levels of the saddle point, respectively. This prescription, using standard approximations for μ and ignoring eV, was applied by Schröter to the experiments of George et al. to give $E_d = 0.28$ eV and $E_a = 0.67$ eV, whereas we have taken account of eV, used a different formula for the variation of E_g with T and obtained $E_d = 0.34$ eV, $E_a = 0.55$ eV (Heggie 1982). The temperature dependence of these levels reveals the entropy of formation of differently charged saddle points, but cannot be determined accurately by these fittings. It should be noted that recent work by Imai and Sumino (1983) disagrees with the measurements of George et al - they find no decrease in the activation energy of dislocation motion for silicon with $\sim 10^{19}$ cm^{-3} B compared to intrinsic. This may mean that E_d is in fact lower than that extracted from George's experiments.

The electronic properties of the saddle point having been characterised, the question arises: What is the path to the saddle point and what is the saddle point structure? According to the original theory of Hirsch invoking single kink drift (Hirsch 1979), the answer must be the saddle point of migration of kinks with dangling bonds ("Hirsch kinks"). Constraining this theory to reconstructed glide partials, we find these Hirsch kinks are complexes of reconstructed kinks with solitons and an assumption of his theory then is that the kink and soliton in such a complex are strongly bound to each other (the complex moving as a whole). Since migration energy changes can be included (Hirsch 1980), there can be little objection to this mechanism in the low stress, high temperature régime (providing obstacles are absent), because it requires only one, plausible mechanism for such motion (transfer of a dangling bond). In the high stress régime double kink nucleation is controlling, and, according to the latest version of Hirsch's theory (Hirsch 1983), nucleation of a double kink (comprising two charged Hirsch kinks) will occur at a point on the dislocation line where there are two charge carriers trapped suitably close together by the shallow levels of the reconstructed dislocation. The questions we pose about this theory are:

(a) Can the shallow dislocation bands be sufficiently localised for this close trapping?

(b) Would not the first step involved in creating two Hirsch kinks be rate-limiting, giving an activation volume of order b^3 contrary to experiment (Louchet 1981)? (The first step involves breaking a reconstruction bond to give two dangling bonds and would probably be more difficult than subsequent dangling bond transfers associated with Hirsch kink migration.)

(c) How can a measured kink formation energy as low as 0.5 eV at 600°C be reconciled with kinks containing dangling bonds?

(d) What is the cause of EBIC contrast for straight dislocations?

Let us now explore the possibility (a) of the introduction, which was suggested by what is known intuitively to chemists: reactions involving free radicals ("dangling" bonds) involve less activation than do reactions without them. Starting with the axiom that breaking a good bond is very difficult, we devised a mechanism of double kink nucleation at solitons which depends on only one process, namely dangling bond transfer. The proposed mechanism is illustrated in glide plane diagrams in figure 1 (where the 90° partial is emphasised by thick lines for the reconstruction bonds) and in figure 2 (where the 30° partial is emphasised by shading in the characteristic pentagonal cells).

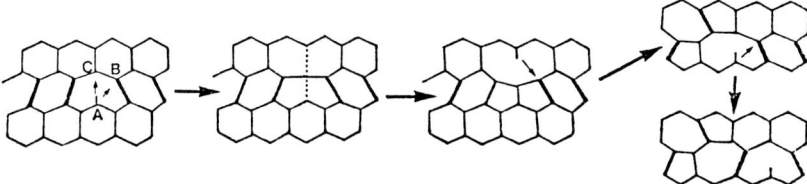

Figure 1   Double kink nucleation on a 90° partial.

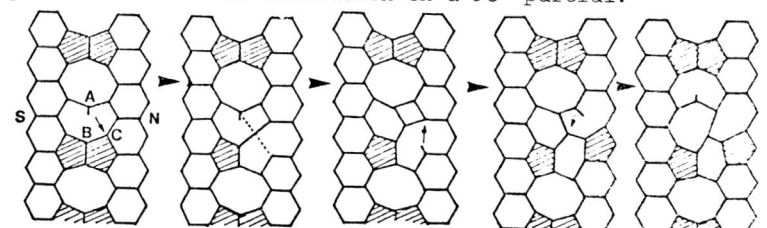

Figure 2   Double kink nucleation on a 30° partial

"Attack" (in a chemical sense) of atom C by the dangling bond ("free radical") on atom A leads to double kink nucleation, whereas attack of atom

B by A leads to soliton migration along the partial. The fact that atom
B is closer to A than is C suggests that soliton motion is easier than
kink-soliton migration and that solitons are not well bound to kinks.
However, we can hypothesise that the binding energy is large and then we
would find that solitons and anti-solitons would annihilate simultaneously
with their kink partners (i.e. annihilation of Hirsch kinks). This would
lead to a situation where dislocation velocity would be limited by the rate
of creation of new soliton pairs, violating our axiom. It would also
present the same problem (c) above as there is with Hirsch's theory. On
the other hand, we can take the kink-soliton binding energy to be small,
thus separating the dynamics of soliton formation and annihilation from
those of kinks. Taking the binding energy to be negligible and assuming,
as seems likely, that soliton migration is faster than kink-soliton migration, a particularly simple treatment of the problem is allowed in which
the presence of solitons modifies the migration energies only.

We argue that kink migration or double kink nucleation at a dislocation
site only occurs when a soliton exists on that site. Because the soliton
migration energy appears to be less than the kink migration energy (by
inspection of figures 1 and 2) and because we postulate a negligible
binding energy between kinks and solitons, then, averaged over the time
of a kink nucleation or migration event the probability of occupation of
any site on the dislocation (including kinks) will be equal to that of
any other. This probability, P_s, is obviously the fractional occupation
of possible dangling bond sites by solitons, which will be $\exp(-F_s/kT)$
if solitons are in thermal equilibrium and each have formation energies,
F_s. Writing the formation energy of the kink-soliton complex as F_{ks}
(strictly we define F_{ks} to be half the formation energy of a kink-
soliton and kink-antisoliton pair) and, similarly, F_{ks}^{sp}, for the saddle
point, the kink-soliton migration energy, W_{ks}, will be given by

$$W_{ks} = F_{ks}^{sp} - F_{ks}.$$

Thus the expressions below for the kink velocity, V_k, and double kink
nucleation rate, J, at a dislocation site comprise the probability of
existence of a soliton at that site multiplied by either the probability
of a successful movement of a kink-soliton complex in the case of V_k
or the probability of successfully nucleating a reconstructed, double kink
with a soliton (of energy $F_{dk} + F_s$) from a soliton for J.

$$v_k = \frac{\nu_0 bha^2}{kT} P_s \exp(-W_{ks}/kT)$$

$$J = \frac{\nu_0 bh}{kT} P_s \exp -(F_{dk} + W_{ks})/kT$$

Once again the stress correction, E_σ^*, in the activation energy of J has
been omitted and it should be noted that F_{dk} is taken to be the energy of
a reconstructed double kink with a soliton minus the energy of a soliton
(a consequence of weak binding between kinks and solitons). Note also
that both V_k and J will depend on the chemical potential, μ, through P_s
(since $F_s = f(\exp \mu/kT)$) and that the different charge states of the
soliton are already in equilibrium with the electron gas, before any
migration steps. Combining the equations for V_k and J with the
expression for P_s when solitons are in thermal equilibrium and the
expression

$$v = 2h\sqrt{Jv_k}$$

for dislocation velocity we obtain

$$v = \frac{2\nu\sigma bh^2 a}{kT} \exp -(\tfrac{1}{2}F_{dk} + W)/kT$$

where $\tfrac{1}{2}F_{dk}$ is the formation energy of half a reconstructed double kink
($\approx$ 0.5 eV from experiment) and W is the effective migration energy of a
reconstructed kink through a kink-soliton saddle point (i.e. $W = F_s + W_{ks}$).
Since F_{dk} does not depend on μ, neither will $\bar{L}$ in this model. According
to this theory F_{dk} is the formation energy of a reconstructed double kink
and the value of 1 eV at 600°C inferred from experiment may not be
incompatible with the theoretical calculation due to Jones (1980), who
used a valence force potential and arrived at $F_{dk} \approx$ 1.5 eV (corresponding
to O K). It seems that a larger discrepancy would be introduced if F_{dk}
included two dangling bonds, as in the theories (b) and (c) of the
introduction.

The levels E_d and E_a extracted from doping dependent velocity experiments
are attributed to a kink-soliton saddle point structure in this model.
We have calculated one-electron levels for such a structure (Heggie 1982,
Heggie and Jones 1983) using a parametrised, tight-binding Hamiltonian
and find that such a saddle point structure gives rise to a half-filled
level, E_o, between E_v + 0.24 eV and E_v + 0.30 eV. We also found that
this level is much less sensitive to lattice distortions than is the
soliton, making positive-U behaviour most likely. In this case one
expects the half-filled level to simulate a donor level at E_o separated
by U, the Hubbard intra-orbital electron-electron energy, from an
acceptor level, $E_a = E_o$ + U. Both fittings to the results of George et
al. (giving E_d = 0.28 eV or 0.34 eV) compare favourably with these
theoretical predictions (comparing E_o with E_d) and our fitting, gives a
value of U (0.21 eV) comparable with that calculated for the vacancy
(0.25 eV - Baraff et al. 1980).

3.   Conclusions

The existence of solitons (independent of kinks) on reconstructed glide
partials in silicon can account both for the electronic properties
of deformed silicon and the behaviour of moving dislocations particularly
in the presence of electrically-active impurities. Experiments to
determine the kink mean free path, $\bar{L}$, as a function of doping would be
useful to discriminate between the different models of double kink
nucleation.

4.   Acknowledgments

M.H. thanks the SERC for support and we have benefitted from discussions
with many experimentalists mentioned, as well as T. King of Exeter.

References

Alexander H, Kisielowski-Kemmerich C, Weber E R 1982
    "Defects in Semiconductors" ed CAS Ammerlaan (North Holland:Amsterdam) 583
Allan G, Lannoo M, 1980 Inst. Phys. Conf. Ser. No.59 : 199
Baraff G A, Kane E O, Schlüter M, 1980 Phys. Rev B21 3563 and 5662
Bondarenko I E, Erofeev V N, Nikitenko V I 1973 Sov.Phys. JETP 37 1109
Erofeev V N, Nikitenko V I, 1971 Sov. Phys. Solid State 13 116
George A, Escaravage C, Champier G, Schröter W, 1972 Phys.Stat. Sol(a)53 483
Gösele U, Frank W, Seeger A, 1983 Sol. Stat. Comms. 45 31
Heggie M I, 1982 PhD Thesis, University of Exeter
Heggie M I, Jones R, 1982 J. de Physique 43 CI-45
Heggie M I, Jones R, 1983 Phil. Mag. to be published
Hirsch P B, 1979 J. de Physique 40 C6-117
Hirsch P B, 1980a J. Microsc.118 3
Hirsch P B, 1980b J. de Physique 42 C3-149
Hirsch P B, 1983 private communication
Hirsch P B, Ourmazd A, Pirouz P, 1981 Inst. Phys. Conf. Ser. No.60:29
Hirth J P, Lothe J 1968 "Theory of Dislocations" McGraw Hill
Imai M, Sumino K, 1983 Aussois conference (7th-11th Mar.)
    to be published in J. de Physique
Jones R, 1980 Phil. Mag. 42 213
Jones R, 1981 Inst. Phys. Conf. Ser. No. 60 : 45
Jones R, 1983 Aussois conference (7th-11th Mar.) to be published in
    J. de Physique
Kimerling L C, Patel J R, 1979 Appl. Phys. Lett. 34 73
Kimerling L C, Patel J R, Benton J L, Freeland P E, 1980 Inst.Phys. Conf.Ser.
    No. 59:401
Kulkarni S B, Williams W S, 1976 J. Appl. Phys. 47 4318
Louchet F, 1981 Inst. Phys. Conf. Ser. No.60:35
Marklund S 1983 Aussois conference (7th-11th Mar.) to be published in
    J. de Physique
Newton J L, Chatterjee A P, Harris R D, Watkins G D, 1982
    "Defects in Semiconductors" ed CAS Ammerlaan (North Holland : Amsterdam)219
Nikitenko V I, Farber B. Ya., Bondarenko I E, 1982 Sov. Phys. JETP 55 891
Patel J R, Testardi L R, Freeland P E, 1976 Phys. Rev.B 13 3548
Su W P, Schrieffer J R, Heeger A J 1980 Phys. Rev. B 22 2099
Takayama H, Lin-Lin Y R, Mak K 1980 Phys. Rev. B 21 2388
Tan T Y, Föll H, Krakow W, 1981 Inst. Phys. Conf. Ser. No.60:1
Weber E, Alexander H 1979 J. de Physique 40 C6-101

Inst. Phys. Conf. Ser. No. 67: Section 1
Paper presented at Microsc. Semicond. Mater. Conf., Oxford, 21–23 March 1983

Valence force-field geometry and electronic states of the 90° partial dislocation in silicon

K W Lodge, A Lapiccirella* and S L Altmann

Department of Metallurgy & Science of Materials,
University of Oxford, Parks Road, Oxford OX1 3PH, United Kingdom
*Istituto Teoria Struttura Elettronica, CNR, CP10,
00016 Monte Rotondo Scalo, Italy

Abstract   Calculations on the structure of the 90° partial dislocation in Si suggest that the reconstructed structure is the more stable. Two electronic bands are found, one well below the top of the valence band and the other in the gap, reaching the bottom of the conduction band. It is shown that the lower band can be lifted by a spurious band alternation associated with the elastic approximation.

1. Introduction

It is now generally accepted (Hirsch 1979, 1980, Jones 1979) that in silicon 60° dislocations, dissociated into 90° and 30° partials separated by a stacking fault, play an important part in determining the properties of the deformed materials. The 90° partial dislocation in silicon has already been studied by Marklund (1979) from the point of view of its geometry and its electronic structure. The geometry was determined from elasticity theory and the energy levels by a tight-binding approach, based on the parameters of Pandey and Phillips (1976). On manually changing atomic positions in order to study the effect of reconstruction on the atomic structure, Marklund concluded at this stage that reconstruction was likely to be favoured. On improving the force field, however, by the use of the Keating potential both with and without anharmonic terms, Marklund (1980) no longer found evidence of reconstruction. We tackle the same problem again in this paper but with substantial changes in approach. First, we use a sophisticatd valence force field, of the Lifson–Warshel type, which is well adapted for the calculation of defect structures (Altmann, Lapiccirella, Lodge and Tomassini 1982). Secondly, we have used the Extended Hückel Theory (EHT) for the calculation of the electronic states. In the present case, this has the following advantage over the tight-binding approach of Pandey and Phillips. In this method, second nearest-neighbour matrix elements have to be scaled and whereas this is easy in the perfect crystal, the problem that arises in defect structures is that one has to decide subjectively whether a given long interatomic distance is first or second-nearest neighbour type. In the EHT all such matrix elements are calculated from formulae that are given as part of the theory.

2. Core Structure

In order to allow for the calculation of electronic states we use the large unit cell model employed by Marklund (1979), which requires a unit cell with two opposing partials, in order to cancel the strain field. We use for this purpose clusters of some 500 silicon atoms, determine their initial positions by anisotropic elasticity theory and then we minimize the total potential energy by using a Lifson–Warshel field that has been fully described elsewhere (Altmann, Lapiccirella, Lodge and Tomassini 1982, Altmann, Lapiccirella and Lodge 1983). The parameters of this field have mainly been obtained from a fitting to the silicon phonon spectrum, but information arising from the value of the stacking fault energy has also been used. The structures obtained with the valence force field with unreconstructed and reconstructed topologies are shown in Fig.1

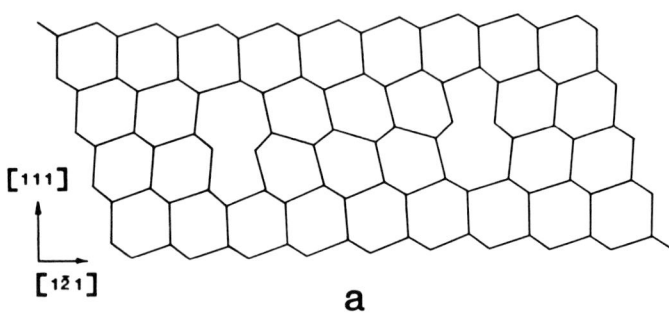

[111]

[12̄1]

a

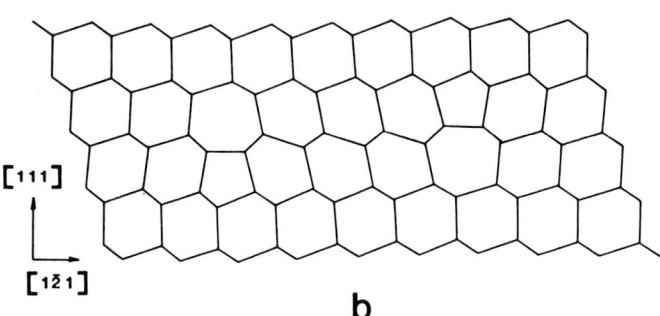

[111]

[12̄1]

b

Fig.1. Projected views of the 500 silicon atoms cluster in the (101̄)projection, computed with the Lifson–Warshel valence force field. (a) Unreconstructed topology. (b) Reconstructed topology.

From the minimization of the energy of the two structures shown in Fig.1 it follows that the reconstructed partial is more stable by 2.66eV with respect to the unreconstructed structure. Admittedly, there are some uncertainties in the unreconstructed structures because of inaccuracies in the treatment of the dangling bond. Their effect, however, can be estimated at a fraction of an eV and therefore it is most unlikely that a better treatment of the dangling bonds could affect the balance between the two structures.

A point that will prove to be most important is the large difference

between the unreconstructed structure in Fig.1a, obtained from the valence
force field, and the calculation of the same cluster by anisotropic
elasticity theory. Detailed comparison of bond lengths shows that elastic
theory introduces a spurious alternation in bond lengths. Thus some of
the long bonds are stretched by about 0.1Å and some of the short bonds
compressed by a similar amount. Indeed, deviations of this type have been
found as large as 0.17Å. This bond length alternation is produced by the
shear component of the edge dislocation strain field and our results
indicate that elastic strains can be incorrect by more than a factor of
two within a distance of about 10Å from the 90° partial in silicon.

3. Electronic Structure

From the cluster calculated as described above a large unit cell is formed
in the manner used by Marklund (1979), which is the cell used in the
calculation of the electronic energy levels. In order to obtain the
parameters to be used in the EHT calculations it was found necessary to
calculate afresh the bands for the perfect silicon crystal so as to ensure
that states near the band gap are accurately obtained. The fitting of the
bands thus obtained not only satisfied this condition but compared well
also with other calculations, such as those of Wang and Klein (1981) or
Pandey and Phillips (1976).

The dislocation bands calculated for the unreconstructed and reconstructed
topologies are shown in Fig.2, where, for comparison, we also show the
bands obtained from a cell with geometry derived from the elastic
approximation.

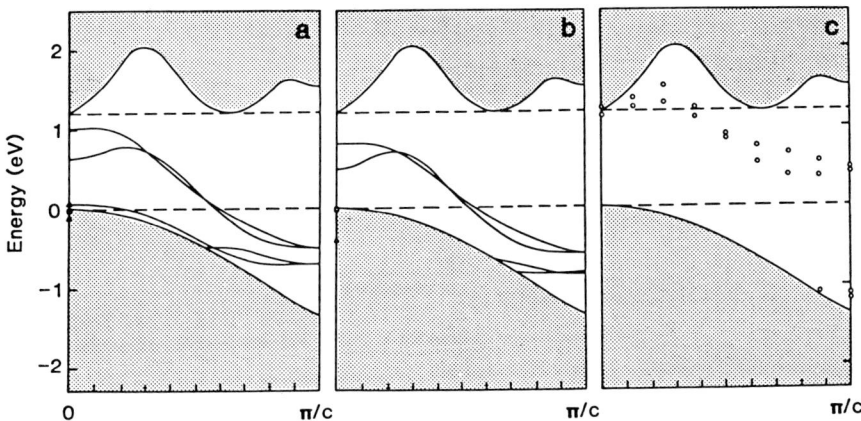

Fig.2. Electronic band structure along $[10\bar{1}]$. The shaded areas give
the extent of the projected valence and conduction bands of the
perfect crystal. (a) elastically generated geometry, (b)
unreconstructed force field geometry, (c) reconstructed force field
geometry, for which the calculated levels are plotted but not
connected into bands.

It must first be noticed that the splitting of the dislocation bands shown in the figure is due to the excessive proximity of the partials in the model used and has no practical meaning. Since it is small, of the order of 0.1eV, it does not affect our description. Also, it is of the same order of magnitude as found e.g. in the pseudopotential calculations of Northrup, Cohen, Chelikowsky, Spence and Olsen (1981).

The bands calculated with the elastic geometry show some resemblance to those of Marklund in the sense that there is a full band near, and partly above, the top of the valence band, and a partly occupied second band, reaching up to about 0.8eV above the top of the valence band. It appears from our results, however, that the fact that the lower band is lifted slightly over the valence band is entirely due to the spurious bond alternation that exists in the elastic structure. It can be seen in Fig.2(b), indeed, that for the improved valence force field geometry, this band is not wider than 0.5eV and that its top lies about 0.2eV below the top of the valence band. As one would expect, reconstruction separates the two bands even further, the lower band dipping to over 1eV below the top of the valence band. There is, on the other hand now, an entirely unoccupied band in the gap, of width ca. 1.2eV reaching up to and overlapping with the bottom of the conduction band. The experimental analysis of Schröter, Scheibe and Schoen (1980) suggests the existence of a half filled band in the gap or, possibly, two bands separated by about 0.3eV, of which the lower one would be full and raised into the gap. Our results would indicate that such bands should be explained by a mechanism other than the one discussed in this paper.

Acknowledgements

This work was made possible by a bilateral agreement between the British Council and the Consiglio Nazionale delle Ricerche (Ufficio Relazioni Internazionali) and the help of these institutions is gratefully acknowledged. One of us (AL) wishes to acknowledge the award of a Visiting Fellowship by the SERC. The authors are grateful for useful discussions with Professor Sir Peter Hirsch.

References

Altmann S L, Lapiccirella A, Lodge K W and Tomassini N, 1982 J.Phys. C, 15, 5581.
Altmann S L, Lapiccirella A and Lodge K W, 1983, Int.J.Quantum Chem., in press.
Hirsch P B, 1979, J.de Physique 40, C-6   27.
Hirsch P B, 1980, J.Microsc. 118 3.
Jones R, 1979, J.de Physique 40 C-6 33.
Marklund S, 1979, Phys.Stat.Sol. (b) 92 83.
Marklund S, 1980, Phys.Stat.Sol.(b) 100 77.
Northrup J E, Cohen M L, Chelikowsky V R, Spence J and Olsen A, 1981, Phys.Rev.B 24 4623.
Pandey K C and Phillips J C, 1976, Phys.Rev.B 13 750.
Schröter W, Scheibe E and Schoen H, 1980, J.Microsc. 118 23.
Wang C S and Klein B M, 1981, Phys.Rev.B 24 3393.

Inst. Phys. Conf. Ser. No. 67: Section 1
Paper presented at Microsc. Semicond. Mater. Conf., Oxford, 21–23 March 1983

Interfacial defects in diamond-structure materials

R C Pond

The University of Liverpool, Department of Metallurgy and Materials Science
P.O. Box 147, Liverpool L69 3BX.

Abstract   The crystallography of grain boundary structures and defects
in materials having the diamond-structure is discussed using a new
methodology. It is shown that significant differences are anticipated
compared to grain boundaries in symmorphic materials.

1. Introduction

Current theories of the structure of interfaces, such as the 'O' lattice
theory (Bollmann 1970), emphasize the crystalline nature of the adjacent
crystals. The methodology used in these theories is to determine the dislocation content of an interface and, therefore, it is only the translation
symmetry of the adjacent crystals that is considered. Some concepts, such
as the coincidence site lattice (csl) and dsc lattice, which have been
developed in these theories, have proved to be very valuable in our understanding of interfacial structure. However, the point symmetry of the
adjacent crystals has not been included in these treatments, and this is
clearly a limitation. A different methodology which overcomes this limitation has been developed recently (Pond and Bollmann, 1979, Pond and
Vlachavas 1983). The purpose of the present paper is to show that valuable
concepts like the dsc and coincidence site lattices can be readily obtained
by using the new approach, and also to illustrate the influence of point
symmetry on interfacial structure. Grain boundaries in materials having
the diamond structure have been chosen to illustrate this result in view of
their importance in semiconductor applications.

2. Crystallographic Methodology

An interfacial structure of interest is imagined to have been created by
the following sequence of events.

i). The lattice of one crystal, designated white and having a space group
$\Phi^*(\lambda)$, is orientated appropriately with respect to a second lattice which
is designated black with space group $\Phi^*(\mu)$. The relative position of the
black and white lattices is chosen so that the space group of the resulting
pattern, called a dichromatic pattern, has maximum symmetry, and this is
designated $⊔⊔^*(p)$.

ii). In order to be able to treat non-holosymmetric and/or non-symmorphic
crystals, the black and white lattices are replaced by black and white
lattice-complexes which have space groups $\Phi(\mu)$ and $\Phi(\lambda)$ respectively.
Lattice-complexes are arrays of points which have the spatial symmetry of
the corresponding crystals (which may not be the same as that of the

crystals' lattices) and have the property that each point has identical
surroundings except possibly for orientation. The pattern created by the
two superposed lattice-complexes is called the dichromatic complex, and
the space group of its most symmetrical form is designated ⊔⊥*(c).

iii). An unrelaxed bicrystal is imagined to be created from a dichromatic
complex by first choosing the orientation of the interfacial plane, and
subsequently locating white bases at white lattice sites on one side of
this plane and black bases at black lattice sites on the other. The space
group of the maximum symmetry form of this bicrystal is designated ⊔⊥*(b).
(Note that these space groups can include translation symmetry in two or
fewer dimensions and chat any point or translation symmetry operations
present must leave the interface plane invariant.)

iv). The final step is to allow the bicrystal to relax; this may involve
relative displacement of the adjacent crystals, migration of the interface,
local relaxation of individual atoms, or removal or addition of material.
Such relaxation may modify the symmetry of the unrelaxed bicrystal, and the
spacegroup of this relaxed bicrystal is designated ⊔⊥(b).

Thus, the creation of a bicrystal is imagined to be a four stage process
carried out in such a way that the space group of the black/white composite
at a particular stage is always either isomorphic to the space group of
the previous stage or reduced to a subgroup of it. Now whenever the
symmetry (translational or point) of a configuration is reduced by some
modification variant configurations arise. The suppression of symmetry
operations in this way is referred to as dissymmetrisation, and the set of
variants so produced are crystallographically equivalent and inter-related
by the symmetry operations suppressed. It follows that variant configurations can arise at each stage of dissymmetrisation in the four stage process
of bicrystal manufacture, as illustrated schematically below. The variants
produced at each stage of dissymmetrisation are designated orientational,
complex, morphological and relaxational in order to emphasize the different
physical consequences of each type of variant (Pond, 1983, Pond and
Vlachavas, 1983). For example, the set of morphological variants
corresponds to the set of symmetrically related bicrystals which can be
created from a given dichromatic complex.

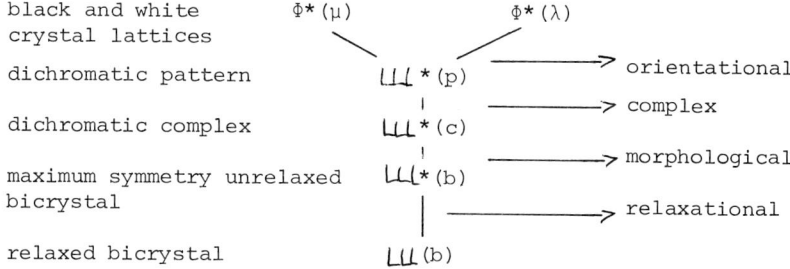

3. Dichromatic Patterns and Complexes For Grain Boundaries

The holosymmetric space group of a dichromatic pattern, ⊔⊥*(p), may be
comprised of two types of symmetry operations, designated ordinary and
antisymmetry operations. The former type relate white points to white and,
simultaneously, black points to black, and the latter relate white points
to black. The presence of antisymmetry operations depends on $\underline{R}'$, the
matrix representing the transformation of coordinates from the white co-

ordinate system into the black. $\underline{R}'$ may represent a proper or improper rotation in the case of grain boundaries, and the prime indicates an operation which relates black points to white. Antisymmetry operations will be present in $\sqcup\!\sqcup^*(p)$ if any of the equivalent descriptions of $\underline{R}'$ has the form of a symmetrising operation of index 2 (Pond and Vlachavas, 1983). Using the coordinate system of the white crystal the alternative descriptions are given by $\underline{D}^*(\lambda)_i\underline{R}'$, where $\underline{D}^*(\lambda)$ are the (symmorphic) point symmetry operations of the group $\Phi^*(\lambda)$. Let any one of this set of equivalent descriptions which has the form of a symmetrising operation be designated $\underline{R}'_s$.

The set of ordinary operations present in $\sqcup\!\sqcup^*(p)$ is given by $\Phi^*(\lambda) \cap \Phi^*(\mu)$; this set is obtained from the solutions of

$$\left[\underline{D}^*(\lambda)_i \big| \underline{\tau}(\lambda)_j\right] = \underline{R}'^{-1}\left[\underline{D}^*(\mu)_k \big| \underline{\tau}(\mu)_l\right]\underline{R}' \tag{1}$$

where $\underline{D}^*(\mu)_k$ are the (symmorphic) point symmetry operations of the black lattice, and $\underline{\tau}(\lambda)_i$ and $\underline{\tau}(\mu)_l$ are the translation vectors of the white and black lattices respectively. Let the solutions to equation (1) be designated $\left[\underline{D}(p)_m \big| \underline{\tau}(p)_n\right]$; the translations $\underline{\tau}(p)_n$ correspond to the set of translation vectors of the csl which we designate T (csl), and the operations $\underline{D}(p)_m$ are all symmorphic. It follows that the group $\sqcup\!\sqcup^*(p)$ is given by

$$\sqcup\!\sqcup^*(p) = T(csl) \left\{\underline{D}(p)_m \cup \underline{D}(p)_m \underline{R}'_s\right\} \tag{2}$$

The space group of a dichromatic complex may also be comprised of ordinary and antisymmetry operations. The ordinary operations will be given by solutions to

$$\left[\underline{D}(\lambda)_p \big| \underline{\alpha}(\lambda)_p + \underline{\tau}(\lambda)_q\right] = \left[\underline{R}'\big|\underline{t}\right]^{-1}\left[\underline{D}(\mu)_r \big| \underline{\alpha}(\mu)_r + \underline{\tau}(\mu)_u\right]\left[\underline{R}'\big|\underline{t}\right] \tag{3}$$

where $\underline{D}(\lambda)_p$ and $D(\mu)_r$ are the symmetry operations of the crystal groups $\Phi(\lambda)$ and $\Phi(\mu)$ respectively, $\underline{\alpha}(\lambda)_p$ and $\underline{\alpha}(\mu)_r$ are the supplementary displacements associated with non-symmorphic operations, and $\underline{t}$ is the displacement of the white crystal away from the previously chosen origin. Let the set of operations which satisfy equation (3) be designated $\{\left[\underline{D}(c)_t\big|\underline{\alpha}(c)_t +\underline{\tau}(c)_v\right]\}$. Antisymmetry operations will also be present if one of the set of equivalent descriptions of the misorientation, $\left[\underline{D}(\lambda)_j\big|\underline{\alpha}(\lambda)_j\right]\left[\underline{R}\big|\underline{t}\right]$ has the form of a symmetry operation designated $\left[\underline{R}_s'\big|\underline{\alpha}'(c)_s\right]$. We note that the set of translation vectors is identical to that for the dichromatic pattern, T(csl), but the set of point symmetry operations may include non-symmorphic operations. It follows that $\sqcup\!\sqcup^*(c)$ is given by

$$\sqcup\!\sqcup^*(c) = T(csl)\left\{\left[\underline{D}(c)_n\big|\underline{\alpha}(c)_n\right] \cup \left[\underline{D}(c)_n\big|\underline{\alpha}(c)_n\right]\left[\underline{R}_s'\big|\underline{\alpha}'(c)_s\right]\right\} \tag{4}$$

$\sqcup\!\sqcup^*(b)$ can be obtained from $\sqcup\!\sqcup^*(c)$ by 'cross-sectioning' on the appropriate plane and at the appropriate location. Thus, the maximum symmetry of an interfacial structure can be established from a knowledge of crystallographic parameters alone. On the other hand, relaxation processes leading to favourable structures depend on physical parameters such as the nature of the chemical bonding in addition to crystallographic ones, and these cannot be predicted at present.

4. Interfacial Defects

Bollmann (1970) showed that the Burgers vectors of perfect grain boundary dislocations correspond to translation vectors of the dsc lattice. This

set of vectors, designated T(dsc) is given by T(dsc) = T(λ) ∪ T(μ),where
T(λ) and T(μ) are the sets $\underline{\tau}$(λ) and $\underline{\tau}$(μ) respectively. Displacements of
one lattice-complex with respect to the other by dsc vectors recreate the
identical dichromatic complex, and hence identical interfacial structures
can be obtained before and after displacement. However, the relative displacement may cause the origin of the complex to be shifted in space; such
shifts imply the existence of an interfacial step at the core of the interfacial defect.

In addition to perfect interfacial defects, various types of imperfect
defects can exist. In particular, defects can exist separating domains of
interfacial structure which are related by symmetry and which consequently
have degenerate energies. For example, if relaxation processes dissymme-
trise ⊔⊔*(b) the domains will correspond to relaxational variants, and
the character of the defects will depend on the nature of the relaxation.
Unlike perfect defects, where the character is determined only by T(λ) and
T(μ) as described above, the nature of imperfect defects depends on the
total symmetry of the adjacent crystals. For example, if ⊔⊔*(c) is dissymmetrised with respect to ⊔⊔*(p), complex variants arise and imperfect
defects can exist which separate equivalent regions of interface created
from the variant complexes. This can occur if one or both of the crystals
is non-holosymmetric or non-symmorphic. Defects in GaAs/Ge and NiSi$_2$/Si
interfaces, as have been reported by Hetherington et al. (1983), are of
this type and have been discussed elsewhere (Pond, 1983). In the present
paper, where we are concerned with grain boundaries between crystals having
the diamond-structure, ⊔⊔*(c) need not be dissymmetrised with respect to
⊔⊔*(p), and hence this type of defect does not arise. However, ⊔⊔*(c)
may be non-symmorphic, and this has interesting consequences. In order to
illustrate the differences between grain boundaries in diamond-structure
materials (Φ*(λ) = Fd3m) and those in isomorphic symmorphic crystals
(Φ*(λ) = Fm3m) we discuss briefly the case of Σ=5 csl related crystals.

Fig.1(a) shows the dichromatic pattern for two fcc lattices rotated by 53.1°
about [001],and where the origin has been taken as a point with symmetry
m3m; ⊔⊔*(p) for this pattern is I4/mm'm', and in the case of crystals with
symmetry Fm3m, ⊔⊔*(c) is identical to ⊔⊔*(p). Figs. 1(b) and (c) depict
two maximum symmetry dichromatic complexes for diamond-structure crystals
having the same relative rotation but choosing a different origin. ⊔⊔*(c)
is I4$_1$/am'd' for (b), and I4$_1$/ac'd' for (c),and (c) can be regarded as
having been obtained from (b) by a relative displacement of ½($\underline{b}_1$+$\underline{b}_2$), where
$\underline{b}_1$ and $\underline{b}_2$ are primitive dsc vectors. Since these complexes have symmetry
isomorphic to ⊔⊔*(p) no complex variants exist. Fig.1(d) shows a complex
obtained from (b) by a relative displacement of ½$\underline{b}_1$ + ¼[001] and in this
case ⊔⊔(c) is I42'm'; it follows that two equivalent dissymmetrised
complexes exist related by the operations present in ⊔⊔*(p) which are not
also present in ⊔⊔(c). These two complexes are also related by a relative
displacement of ½($\underline{b}_1$+$\underline{b}_2$),although this shifts the origin of the complex
into the adjacent (004) plane.

Notable consequences arise from the fact that ⊔⊔*(c) is non-symmorphic;
firstly, for a given interfacial orientation ⊔⊔*(b) may be a lower
symmetry group for Fd3m crystals than for Fm3m; for example twist boundaries
on (001) can have symmetry isomorphic to p42'2' in the latter case but only
p2'2'2 in the former. On the other hand, equivalent interfacial structures
can be created on planes separated by ¼[001] in the diamond case. Another
consequence concerns the nature of imperfect dislocations, and for illustration we consider the instance of dislocations with Burgers vectors equal to

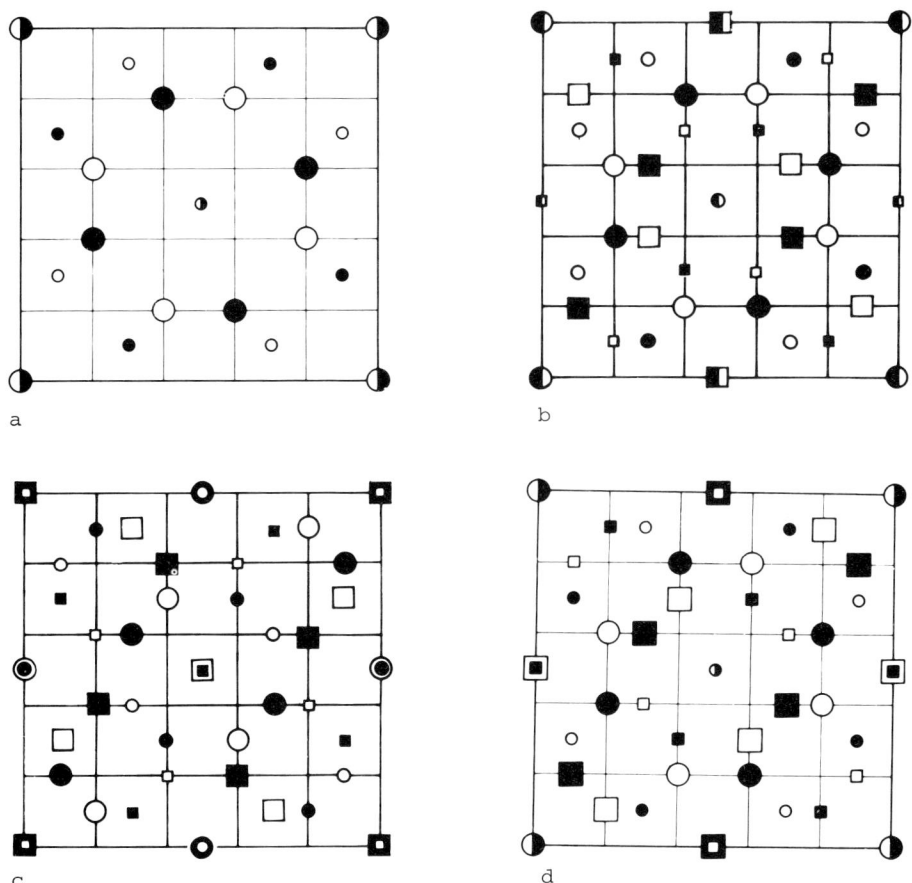

Fig.1 (a) Dichromatic pattern for fcc lattices rotated by 53.1° about
[001]; large and small circles represent the ...ABAB... stacking along
[001]. The origin is taken to be a point with symmetry m3m, and
⊔⊥*(p) = I4/mm'm'. (b) Dichromatic complex for diamond-structure crystals
having the same relative orientation as (a); large and small squares
correspond to sites at height ¼ and ¾ along [001]. The origin is taken as
a point with symmetry 4̄3m, and ⊔⊥*(c) = I4₁/am'd'. (c) Dichromatic
complex obtained from (b) by a relative translation of ½(b₁+b₂) = 1/10<210>.
The space group is I4₁/ac'd', and we note that the space group for the
dichromatic pattern after the same displacement is I4/mc'm'. (d) Dichromatic complex obtained from (b) by a displacement of ¼<111> = ½b₁+¼[001]
modulo a dsc translation vector, and the symmetry group is ⊔⊥(c) = I4̄2'm'.
We note that this complex can be regarded alternatively as having been
obtained by a rotation through the same origin by 36.9° about [001] in the
opposite sense to (b) - it follows that the configuration of circle sites
is identical in (a), (b) and (d).

$\frac{1}{2}(\underline{b}_1+\underline{b}_2)$ lying on the (OO1) plane. Such defects could separate energetically degenerate domains, and in the Fm3m case these would be relaxational
variants, and there would be no step associated with the defect. In the
Fd3m case the domains, which can be imagined as having been created from
the complex shown in Fig.1(d), and its equivalent, need not have been
dissymmetrised by relaxation, and there would be a step of height $\frac{1}{4}\overline{[001]}$
associated with the defect.

5. Discussion and Conclusions

It is suggested that the methodology used in this work is valuable since,
in the study of a given interface, it demonstrates clearly that knowledge
of the crystallographic parameters enables only the prediction of $\underline{\mu}\underline{l}*(b)$,
and that physical arguments must be used in order to predict relaxed inter-
facial structures (and hence $\underline{\mu}\underline{l}(b)$). Moreover, the origin of the different types of variant configurations is readily seen. We note that the
methodology emphasizes the significance of all the symmetry operations
initially present in $\Phi*(\lambda)$ and $\Phi*(\mu)$. Any one of these operations which is
not present in the group of a relaxed interface, $\underline{\mu}\underline{l}(b)$, must relate this
structure to a crystallographically equivalent one. It has also been shown
that the Burgers vectors of perfect interfacial dislocations depend only on
the translation symmetry of the adjacent crystals and their relative
orientation, but that the nature of imperfect defects separating degenerate
domains is profoundly affected by any additional symmetry present. In cases
where $\underline{\mu}\underline{l}*(c)$ is dissymmetrised with respect to $\underline{\mu}\underline{l}*(p)$, defects can exist
which effect the transition from an interfacial structure created from one
dichromatic complex to an equivalent structure created from a complex
variant; defects observed in GaAs/Ge and $NiSi_2$/Si interfaces have been
cited as examples. However, in the case of grain boundaries between holosymmetric non-symmorphic crystals, such as those having the diamond-
structure, $\underline{\mu}\underline{l}*(c)$ may be isomorphic to $\underline{\mu}\underline{l}*(p)$ so that no variants of the
maximum symmetry complex exist. On the other hand, depending on $\underline{R}'$, $\underline{\mu}\underline{l}*(c)$
may itself be non-symmorphic for such crystals and this can lead to inter-
esting interfacial effects as described in the text.

Acknowledgements

The author wishes to thank Dr. J-J. Bacmann for drawing his attention to
interesting and important aspects of diamond-crystallography.

References

Bollman W 1970 Crystal Defects and Crystalline Interfaces, Springer-
    Verlag, Berlin.
Hetherington CJD, Cherns D and Humphreys CJ 1983 this volume.
Pond RC 1983 Phil. Mag. 47A L49.
Pond RC and Bollmann W 1979 Phil. Trans. Roy. Soc. Lond. A292 449.
Pond RC and Vlachavas DS 1983 Proc. Roy. Soc. A386 95.

Inst. Phys. Conf. Ser. No. 67: Section 2
Paper presented at Microsc. Semicond. Mater. Conf., Oxford, 21–23 March 1983

Thermally induced micro-defects in CZ silicon: a high resolution electron microscopy study

F A Ponce*, S Hahn°, T Yamashita*, M Scott* and J R Carruthers*

* Hewlett-Packard Laboratories, Palo Alto, California 94304, U.S.A.
° Siltec Corporation, Mountain View, California 94043, U.S.A.

Abstract   The structure of thermally-induced microdefects in CZ silicon crystals has been studied using high resolution transmission electron microscopy. The observed structures for prolonged thermal treatments at 750, 950 and 1175°C are presented.

1. Introduction

Dislocation-free silicon crystals grown by the Czochralski (CZ) technique are widely used as starting material in integrated circuit fabrication. CZ crystals have small concentrations of oxygen impurities (of the order of a few tens of parts per million) which are incorporated during the crystal growth process. It has been observed that microdefects (MDs) appear when these crystals are submitted to certain thermal treatments, giving rise to the so-called swirl patterns (de Kock 1970) shown in figure 1. These MDs have been associated with the precipitation of oxygen in the silicon lattice, and their structure has been extensively studied using etch-pitting techniques and transmission electron microscopy (TEM) (Tan and Tice 1976, Maher et al 1976, Matsushita 1982). Some questions remain regarding the detailed structure and origin of these MDs. In this article, we present preliminary results of a broad study of thermally induced MDs in CZ silicon using lattice imaging techniques.

2. Experimental Aspects

The material used in this work was grown in the ⟨100⟩ direction and was doped with boron (38-63 Ω-cm). The crystal was not subjected to any prior thermal treatment such as the thermal donor annealing at 650°C. The ingot diameter was 10 cm and the seed portion was selected and sliced into (100) wafers and (110) slugs which were normal and parallel to the direction of growth, respectively. The interstitial oxygen and substitutional carbon content were measured by Fourier transform infrared spectroscopy. The conversion factor for the determination of interstitial oxygen contents was 9.63 from ASTM F121-79. The interstitial oxygen content of the as-grown material was found to be 34 parts per million $(1.7 \times 10^{18}$ cm$^{-3})$ in the axial center of the crystal. The substitutional carbon content was below the detectable limit of the technique (⩽0.5 ppm). X-ray topography indicated the absence of defects and swirl patterns in the as-grown material. Different heat treatment sequences were performed on sets of (100) wafers and (110) slugs. The thermal history of the material was carefully controlled. Slow heating and cooling rates (no more than 60°C/minute) were used in order to avoid the introduction of defects not representative of the annealing temperature under study.

TABLE I. Annealing sequences on as-grown material.

	T(°C)	Atm	time(hrs)	[O]init	[O]final
A	750	O_2	100 hrs	34.0 ppm	26.2 ppm
B	750	O_2	100 hrs	34.0 ppm	
	950	O_2	60 hrs		7.2 ppm
C	750	O_2	100 hrs	34.0 ppm	
	1,175	N_2	64 hrs		12.6 ppm

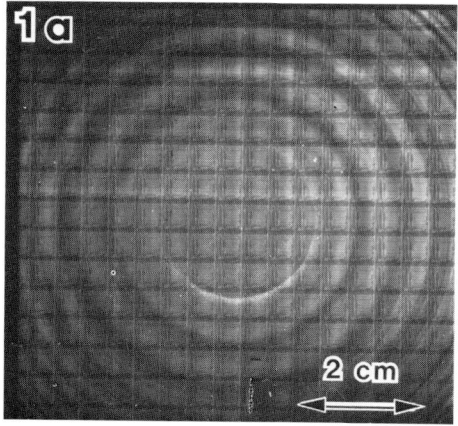

Figure 1. Lang X-ray topographs showing Swirl patterns in CZ silicon after device processing. Sections (a) normal, (b) parallel to the growth direction. MoKα, ($\bar{2}$20) reflection.

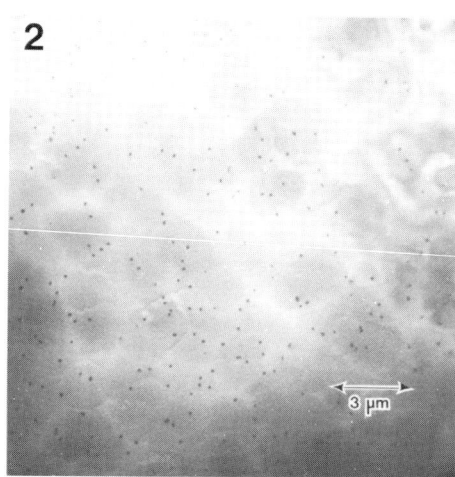

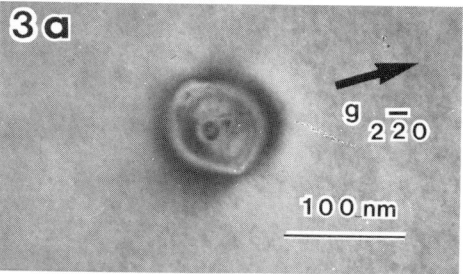

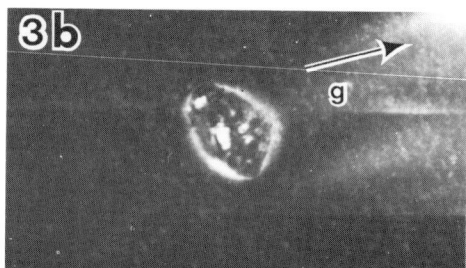

Figure 2. Bright field micrograph showing microdefect distribution following 750°C anneal.

Figure 3. (a) Bright field and (b) weak-beam (g,3g) migrographs of a microdefect in Figure 2; g=(220).

In this paper we discuss the defect structures that result from prolonged heat treatments in three temperature regimes as shown in table 1. After each annealing sequence, the interstitial oxygen and substitutional carbon content was measured by FTIR spectroscopy. X-ray topography and TEM was used to obtain information about the structure. TEM specimens were obtained from the (110) sections. A JEOL 200CX microscope equipped with a high resolution top entry stage was used. The point resolution of the microscope is about 2.5 Å.

3. 750°C anneal

Heat treatments near 700°C produce the highest nucleation rate for the precipitation of interstitial oxygen (Patel 1981). A group of (100) wafers and (110) slugs was annealed at 750°C in dry O_2 for 100 hours to study the nucleation process. The interstitial oxygen content dropped from 34.0 ppm to 26.2 ppm during this anneal. Figure 2 shows a low magnification bright field (BF) TEM micrograph showing the precipitate density resulting from the 750°C anneal. The precipitate density is about 5×10^{12} cm^{-3}, and the average size of the MD is about 1200 Å in diameter. The density of MDs varies with longitudinal position in the crystal, with periodicities roughly equal to the growth striations in X-ray topographs ($\approx$ 100 μm). Figure 3a is a BF micrograph taken under two-beam condition showing the strain field associated with a representative precipitate. These MDs typically have the appearance of small loops. A weak-beam image is shown in Figure 3b which highlights the areas of largest strain. The image indicates that the strain is highly non-uniform within the MD.

The two predominant structure types associated with these MDs are observed in the lattice image in Figure 4. An extrinsic stacking fault (ESF) along a {111} plane is observed in the top portion of the figure. The extrinsic nature of the fault can be directly determined from the image. The MD in the lower portion of the figure lies on a {100} plane and shows evidence of some degree of precipitation with amorphous structure.

4. 950°C anneal

An effect of heat treatment near 950°C which follow an anneal at 750°C is to produce a large drop in interstitial oxygen content. Following the 750°C anneal for 100 hours described above, some material was submitted to a 950°C anneal in dry O_2 for 64 hours. During this process, the interstitial oxygen content dropped from 34 ppm to 7.2 ppm. The observed precipitate density was 3×10^{12} cm^{-3}.

The defect structure following this anneal shows larger precipitates and lower strain fields than the anneal without 950°C treatment. The precipitate structures were similar to the two types described previously. A precipitate of the {100} type is shown in figure 5. The region in the center of the MD appears to be amorphous. These amorphous regions are larger than the ones in the single 750°C anneal. In addition, large ESFs have been observed with typical lengths of a few microns. The extrinsic nature of the stacking faults is directly observed in figure 6.

5. 1170°C anneal

The specimens that have undergone a 1170°C anneal for 64 hours, following the nucleation anneal at 750°C for 100 hrs, appear to be devoid of MDs, except for the presence of large ESFs. Upon close examination, the

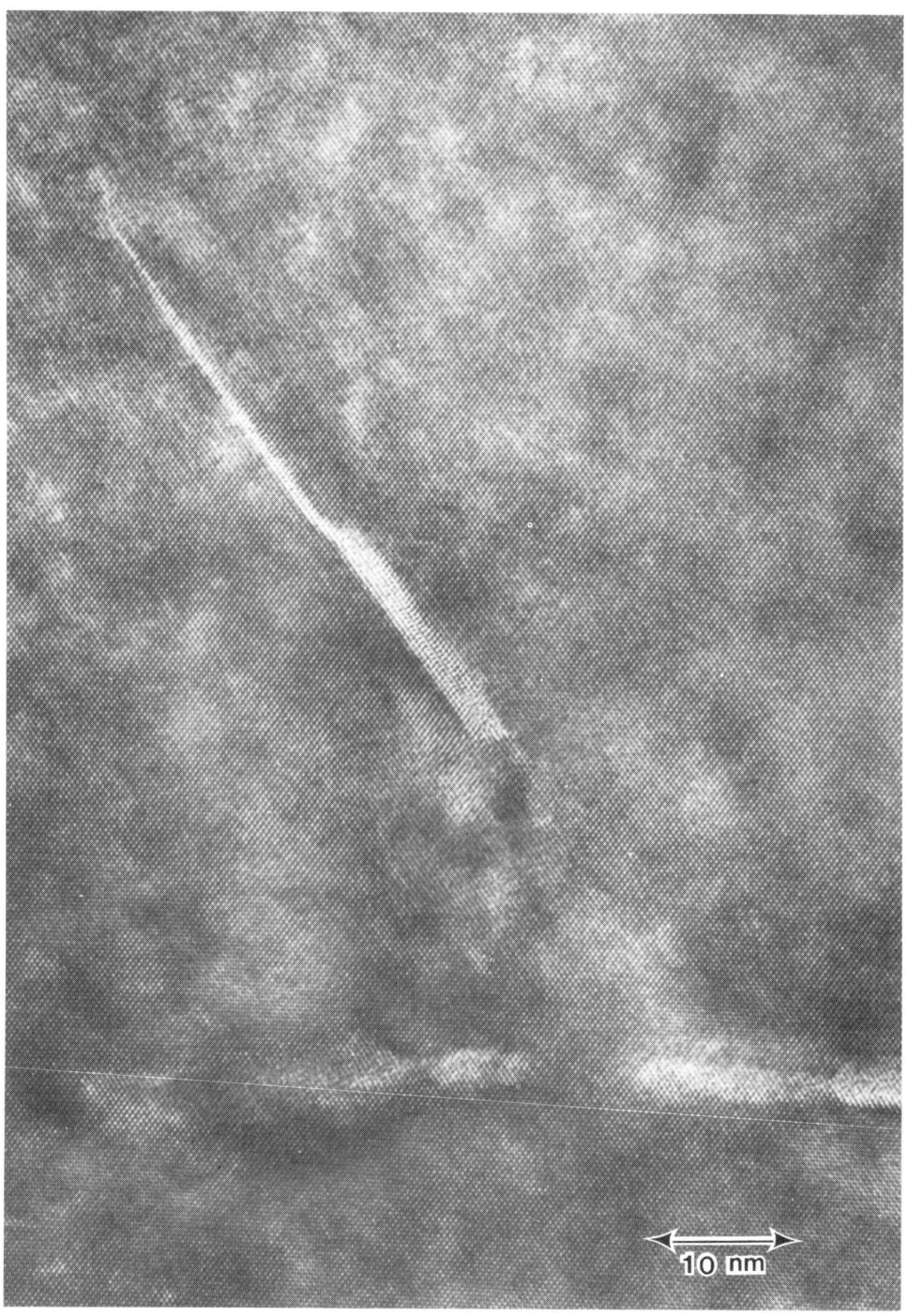

Figure 4. Lattice image of microdefect after 750°C anneal. {111} and {100} defect types are observed.

presence of very small polyhedral particles is observed. Two such particles are shown in figure 7 and 8. They are bound by {111} and {100} silicon planes. The structure of the interior of these precipitates can only be observed in regions of the specimen which have a thickness comparable to the dimensions of the particle (typically 15 nm in diameter). Such is the case in figure 8 which shows what appears to be an amorphous structure in the interior of the particle. These particles always have the same morphology and typical diameters of 15 nm. The proposed geometry for these particles is the truncated octahedra in figure 9, where a [110] section is also shown.

6. Discussion

The aim of these experiments was to obtain information about the state of oxygen precipitation in CZ silicon as close to steady state as possible, hence the choice of long-time anneals. However, it is not clear that equilibrium was reached in each case. The 750°C nucleation heat treatment was common to all three heat treatments discussed here, and the density of microprecipitates was similar in all cases, which indicates that nucleation did occur at the 750°C annealing step. We have briefly studied cases where the nucleation step was omitted, and have observed much lower defect densities. The 750°C anneal indicates that oxygen tends to precipitate predominantly into {100} planes. The observation of ESFs in the vicinity of {100} defects (Figure 4) suggests that precipitation of oxygen promotes the production of silicon self-interstitials which condense into interstitial faults (ESFs). The 950°C anneal gave the largest drop in interstitial oxygen content as measured by FTIR. The size of the MDs observed in the BF images (size of the strain field) was about the same as in the single 750°C anneal. However, larger regions with apparent amorphous structure were observed. The 1170°C annealing treatment produced small faceted polyhedral MDs and large ESFs. The interior of the polyhedra is amorphous and the shape appears to be determined by the silicon matrix.

The strain field associated with the MDs is larger for low temperature anneals and virtually disappears following an anneal at 1170°C. We have not observed prismatic dislocation loops generated by the precipitates in any of the heat treatments discussed in this paper. Further work is being conducted in order to understand the early stages of nucleation and precipitation (i.e. short annealing times).

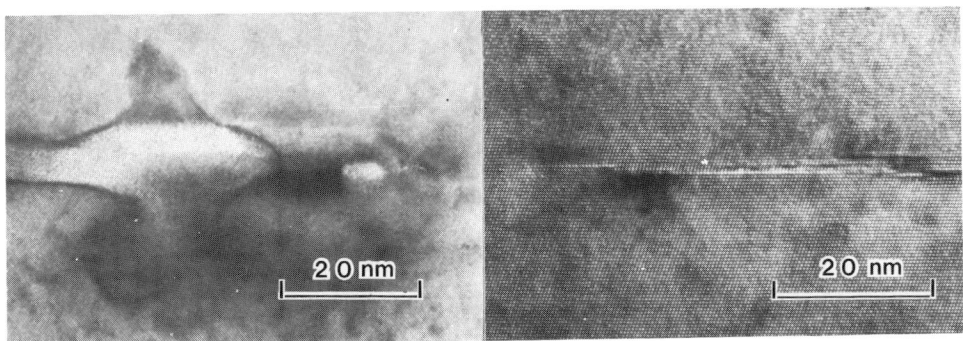

Figure 5. Lattice image of micro-defect following 950°C anneal.

Figure 6. Lattice image of extrinsic stacking fault after 950°C anneal.

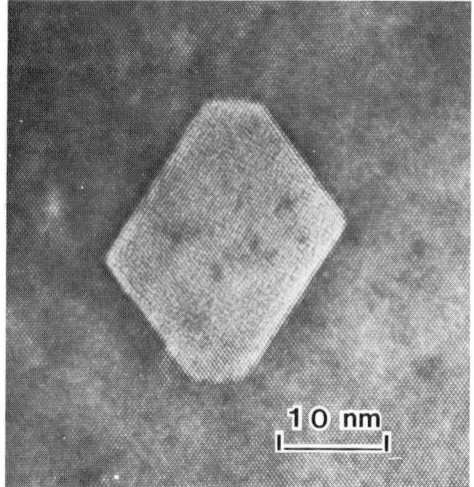

Figure 7. Polyhedral precipitate after 1170°C anneal. Facets follow {111} and {200} silicon planes.

Figure 8. Polyhedral precipitate in thin region of specimen, the interior appears amorphous.

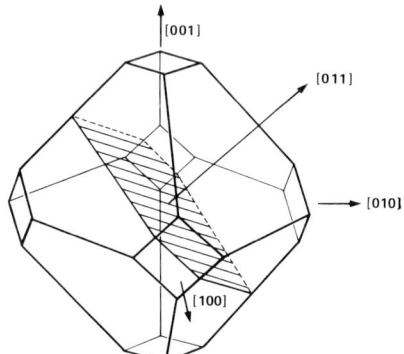

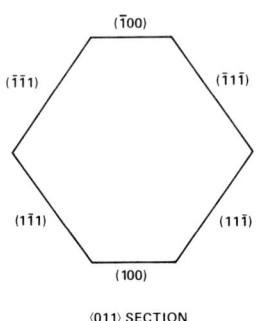

⟨011⟩ SECTION

Figure 9. Geometry of polyhedral precipitate observed after 1170°C anneal. [110] projection shown at right hand side.

Acknowledgments

The authors are very grateful to the late Nick Paredes, Ms. Irene James and to Greg Anderson for technical assistance. Encouragement and support from Robert Burmeister is also gratefully acknowledged.

References

de Kock A J R 1970, Appl. Phys. Lett. 16 100
Maher D M, Staudinger A and Patel J R 1976, J. Appl. Phys. 47 3813
Matsushita Y 1982, J. Crystal Growth 56 516
Patel JR 1981, in Semiconductor Silicon 1981, Huff H R, Kriegler R J and
    Takeishi Y, eds (The Electrochemical Society, Pennington, 1981), 189
Tan T Y and Tice W K 1976, Phil. Mag. 34 615
Tan T Y and Tice W K 1976, Phil. Mag. 34 615

Defects induced by annealing or irradiation in CZ silicon or germanium

J Desseaux-Thibault, A Bourret and J M Penisson

Centre d'Etudes Nucléaires de Grenoble, Département de Recherche Fondamen-
tale - 85X - 38041 Grenoble Cedex, France

Abstract   A comparison is made between defects obtained by annealing
and irradiation. The defects are observed by HREM. Irradiation induced
<110> line defects obtained with 1 MeV electrons are easy to obtain in
germanium. They look similar to rod-like defects in Si produced after
a long-anneal at 650°C. Black dots obtained in both treatments are
compared with the defects produced during ion bombardment in Si.

1. Introduction

Because of their influence on the electrical properties of electronic
devices the defects introduced during processing have been and are still
the subject of many studies. The formation, migration and segregation of
intrinsic defects or impurities have been widely discussed in terms of
vacancies (mono-multi), interstitials and impurity-vacancy or -interstitial
complexes (Watkins et al 1978, Frank et al 1981). The existence of many
of these defects has been deduced from EPR measurements (see the review
given by Corbett et al 1981a). However the structure of the macroscopic
defects produced during oxidation treatments or during irradiation is -
up to now - unknown even if some models exist (Corbett et al 1981b).

High resolution electron microscopy (HREM) can answer a certain number of
structural questions: the aim of this article is to clarify the structure
of the so-called "rod-like" defects and "platelet" defects as well as the
311 stacking faults produced in Si or Ge by annealing (see the review given
by Patel 1981) or by electron irradiation (Matthews, Ashby 1973, Salisbury,
Loretto 1979). Additional defects called "black dots" appearing during
these treatments are compared with the ones obtained by heavy ion bombardment of silicon (M O Ruault et al 1982).

2. Experimental

Heat treatment of dislocation free Si: 1) 650°C (5 days) annealing has
been performed after a pretreatment at 450°C (1 day). 2) 870°C (5 days)
annealing has been also carried out in the same conditions.

The initial p type silicon contained 1.1×10^{18} oxygen atoms cm^{-3} and about
10^{16} carbon cm^{-3}.

Electron irradiation of germanium at 350°C and p type silicon (200°C - 700°C)
was carried out with 1 MeV electrons with a flux rate of 10^{19} $cm^{-2}s^{-1}$.

Ion bombardments were performed on silicon crystals with 200 keV Bi ions at a dose of 6×10^{11} cm^{-2} at room temperature.

HREM observations have been made on a JEM 200 CX at 200 KV with a top entry goniometer stage (Cs = 1.05 mm). All samples are observed along the [011] axis.

3. Thermally Induced Precipitation

650°C: One of the oxygen precipitates is in a crystalline form forming the so-called "rod-like" defects or better "ribbon-like" defects. They lie along the 011 directions. The ribbon plane is mainly a 100 plane of the matrix. The density is 3×10^7 cm^{-2}. Mean size is 7 x 0.8 nm^2. The largest defect is shown in Fig. 1a. The new crystalline phase is clearly obvious on Fig. 1a. It is a precipitate of coesite a high pressure stable monoclinic SiO$_2$ phase with its $\bar{a}$ axis parallel to the [0$\bar{1}$1] matrix axis. The coesite phase has been identified using an image simulation method and nanodiffraction on an optical bench. The (100) planes of Si around the precipitate are distorted by a [100] α extension with $|b| = 0.1$ nm. No extra plane exists. It is to be noticed that the two interfaces parallel to the (100) Si planes do not have the same appearance. Many dislocation dipoles exist aligned with the rod-like defects. The cross-section of one of them is shown in Fig. 1b. The distance between the two dislocations varies and many dislocations have been found to be decorated. Many "black dots" of amorphous structure also exist in the crystal (density $\geq 10^{15}$ cm^{-3}, d $\sim$ 2 nm).

870°C: The defects produced at this temperature are square platelets parallel to the 100 planes of the matrix. The structure of these defects is amorphous (Fig. 1c). These defects are similar to those observed as "black dots" at 650°C but as their size is larger their internal structure is well distinguishable. The platelet density is 1.3×10^{14} cm^{-3} and their average size is 10 x 1.5 nm^2.

4. Electron Irradiation Induced Defects

In germanium: Irradiation in a 1 MeV electron microscope has induced many "rod-like" defects with a density of 10^{10} cm^{-2}. They lie along the 011 directions. Dislocation dipoles are extremely rare. The defects were produced in regions whose thickness was about 0.5 μm. Thus a chemical thinning (HF/HNO$_3$ - 10/90 at -20°C) has been used to remove the denuded zone. The crystal was subsequently observed by HREM. The rod-like defects are similar to those obtained during annealing. The same equivalent Burgers vector exists b = [100]α $|b|$ $\sim$ 0.1 nm. The ribbons have two main planes which are more distinct than in the annealing case: the 100 plane (Fig. 2a) and the 311 plane (Fig. 2b).

At one end on the precipitate (Fig. 2b) the ribbon plane changes from 311 to the 100 plane as in Fig. 1a.

The thickness of the ribbons (0.6 nm) is smaller than for SiO$_2$ precipitate. The difference between the two 100 interfaces of precipitates in Ge should also be noticed. All intermediate size and shape configurations between the defects shown in Fig. 1a, b have been detected. The smallest precipitate (Fig. 3a) and an intermediate one (Fig. 3b) are shown. These kinds of defects have been also obtained in annealed Si. Black dots are also observed but they disappear after a few minutes under the electron beam at 200 kV.

In silicon: After the same electron irradiation time in the temperature range 200-700°C no detectable defects were observed (independently of the oxygen content).

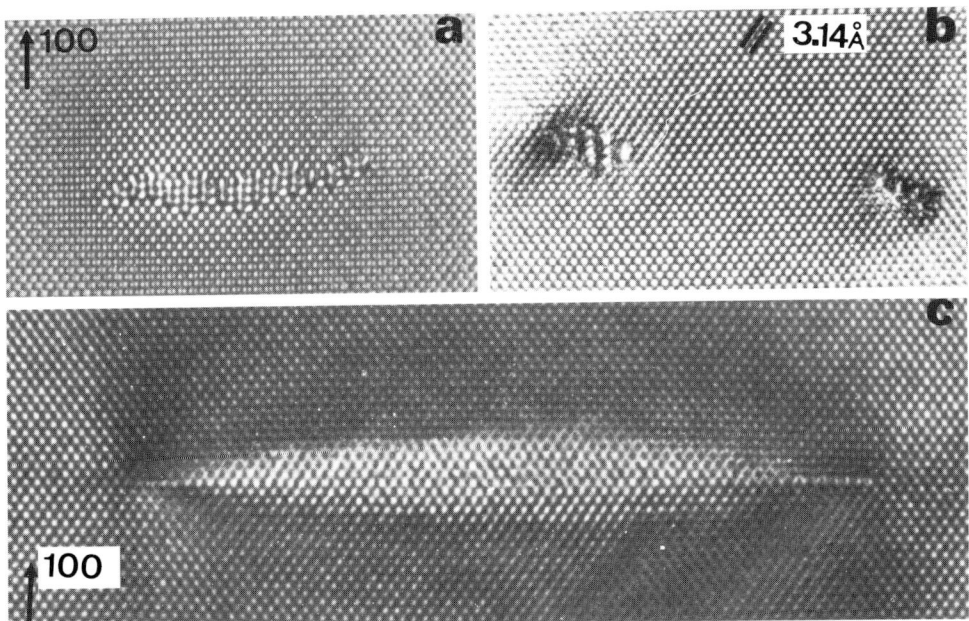

Fig. la,b 650°C annealed silicon a) "ribbon-like" defect cross-section: the new crystalline phase is well recognized; b) dislocation dipole cross-section: both dislocations are also decorated; c) 870°C annealed silicon: the "black dot" is an amorphous pocket parallel to the 100 matrix plane.

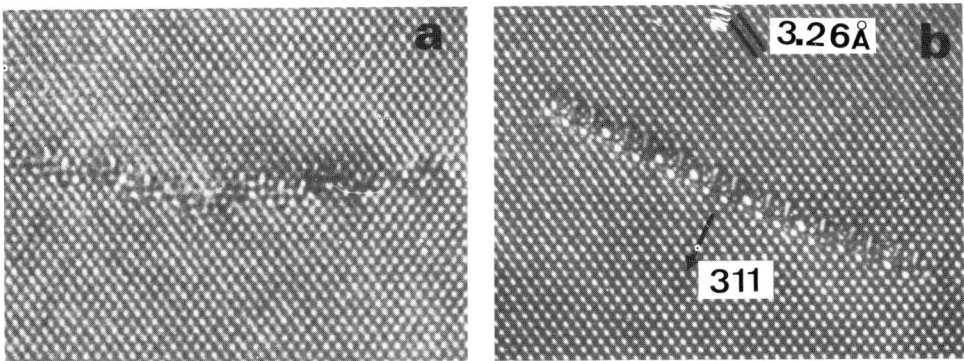

Fig. 2 Defects produced by 1 MeV electron-irradiation at 350°C in germanium a) "ribbon-like" defect; b) "311" defect.

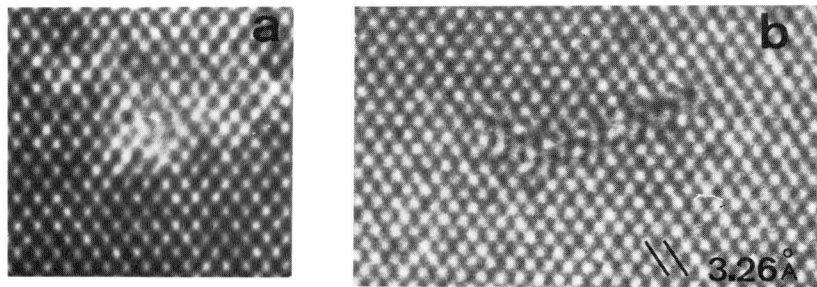

Fig. 3 a) The smallest ē irradiation defect in Ge; b) irradiation induced defect of intermediate size between (3a) and (2a,b).   The same kinds of defects have been also obtained in Si after 650°C heat treatment.

5. Ion Bombardment Induced Defects

The defects produced by Bi bombardment are tracks aligned along the ion bombardment direction.   The greatest length is 15 nm (Fig. 4a).   They are strings of amorphous regions (Werner et al 1982).   No strain field is detectable around the defects and the contrast looks similar to that of a thickness variation (Fig. 4b).

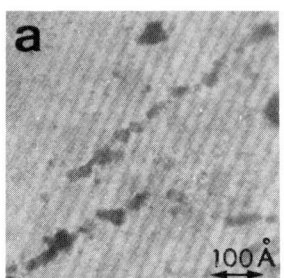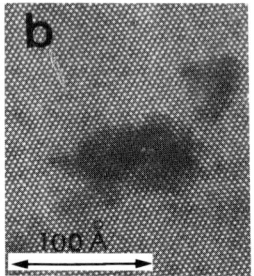

Fig. 4   a) ion bombardment tracks; b) HREM image of the damaged region.

6. Discussion

The following points emerge from the previous observations.

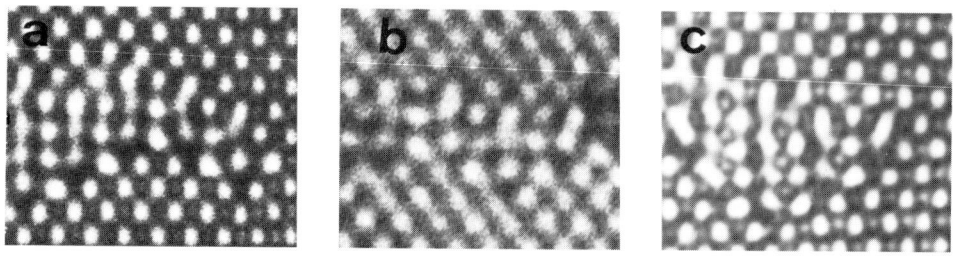

Fig. 5 Comparison between the structure of parts of the precipitates observed in 650°C annealed Si (a); irradiated Ge (b) and in the [100] Si dislocations (c).

6.1   The precipitates observed after Si annealing at 650°C (Fig. 5a) and Ge irradiation (Fig. 5b) are completely identical to the precipitates observed in the [100] dislocation core in Si or Ge (Fig. 5c). They are identified as coesite in silicon and are isomorphous in germanium.

The oxygen has been detected by STEM in the dislocation core in germanium (Bourret et al 1982) using electron energy loss spectroscopy measurements. After a careful examination of the coesite structure, a faceted Si/SiO_2 interface with alternate 311 and 100 matrix planes gives a good match to the growth of coesite. The so-called "311 stacking faults" seem to be precipitates grown on this (311) plane.

6.2   The easy growth of the coesite precipitate along the 011 direction might suggest that the nucleus could be the multivacancy oxygen chain proposed by Corbett (1981a).

6.3   Silicon contains 5×10^{22} atoms cm^{-3}, whereas coesite has 3×10^{22} Si atoms cm^{-3} and 6×10^{22} oxygen atoms cm^{-3}. Similarly the "Ge coesite" contains 2.7×10^{22} Ge atoms cm^{-3} and 5.4×10^{22} oxygen atoms cm^{-3} as compared to 4.4×10^{22} Ge atoms cm^{-3} in the pure germanium. During annealing, at the most, two Si interstitials have been ejected to compensate the density change between the matrix and coesite. The number of interstitials is certainly less because of the remaining strain field around the precipitate. Furthermore due to the number of Si interstitials found in the dislocation dipoles the growth of these dipoles must be produced by the pipe diffusion of ejected interstitials from coesite but also by interstitials coming from the black dots. These interstitials precipitate in the dipole along the rod-like line. During irradiation the production of defect pairs enhances the oxygen segregation by the formation of vacancy - oxygen complexes. The global demand is 5.4 oxygen atoms per 1.7 Ge vacancies (in the case of total strain relaxation). The interstitial diffusion mechanism is not involved. This might be confirmed by the nonexistence of dislocation dipoles along the rod-like lines.

6.4   The oxygen concentration deduced from the crystalline coesite in 650°C annealed Si is 10% of the total precipitated oxygen deduced from IR analysis, before and after treatments. The remaining part is in the "black dots" of the amorphous small platelets. The oxygen concentration as deduced from a rough estimate of the ribbon-like defects after Ge irradiation is surprisingly high ($>10^{18}$ cm^{-3}). However the two interfaces are so close together that the stoichiometry of the precipitate is certainly GeO_x ($x < 2$). Chemical analysis is in progress to clarify this point.

6.5   The irradiation induced precipitates in Ge do not grow in the thinnest parts of the specimen due to the wellknown sink efficiency of the surface. On the other hand it was impossible to obtain the same e irradiation defects in 1 μm thick Si crystals at 650°C. The greater efficiency of the surface in the silicon case than in the germanium one, might be invoked. The oxide layer on the Si surface is one of the obvious consequences of this phenomenon. Moreover the impurity or the dopant role is not yet clear and might not be negligible. However, Aseev et al 1980 were able to produce many "rod-like" defects in silicon and have shown that surface conditions are important to control their production, which is certainly also greatly influenced by irradiation temperature and impurity content.

6.6   The ion bombardment of silicon crystals always induced amorphous regions. No strain field is detectable contrary to the black dots produced during annealing.   The contrast of the amorphous pocket given by HREM is very similar in both cases;   but in the first case some amorphous oxide might be present, whereas the ion bombardment induces amorphisation in Si.

Acknowledgements

The authors are grateful to J. Pelissier who is responsible for the 1 MeV electron microscope (CEN-G), to G. Sylvestre who provided the Ge crystals and to M.O. Ruault who provided the ion bombarded Si samples.   C. Bouvier is also acknowledged for technical assistance.

References

Aseev A L and Astakhov V M 1980 High Voltage Electron Microscopy 4 244
Bourret A and Colliex C 1982 Ultramicroscopy 9 183
Corbett J, Kleinhenz R and Wilsey N 1981a Defects in Semiconductors p 1
    ed Narayan,Tan - North Holland
Corbett J, Karins J and Tan T 1981b Nuclear Inst. Meth. 182/183 457
Frank W, Seeger A and Gosele U 1981 Defects in Semiconductors p 31 ed
    Narayan,Tan - North Holland
Ferreira-Lima C and Howie A 1976 Phil. Mag. 34 1057
Matthews M and Ashby S 1973 Phil. Mag. 27 1313
Patel J R 1981 Semiconductor Silicon Proc. IV Int. Symp. Silicon Materials
    Science and Technology,   Electrochem. Soc. Princeton NJ
Ruault M O, Chaumont J and Bernas H 1982 7th Conf. on Applications of
    Accelerators in Research and Industry,Dexton
Salisbury I and Loretto M 1979 Phil. Mag. A39 317
Stavola M, Patel J R, Kimerling L C and Freeland P E 1983 Appl. Phys. Lett.
    42 73
Watkins G, Troxel J and Chatterjee A 1978 Defects and Irradiation Effects
    in Semiconductors Inst. Phys. Conf. Ser 46
Werner P and Pasemann M 1982 Ultramicroscopy 7 267

Inst. Phys. Conf. Ser. No. 67: Section 2
Paper presented at Microsc. Semicond. Mater. Conf., Oxford, 21–23 March 1983

High resolution electron microscopy studies of native oxide on silicon

J H Mazur, R Gronsky and J Washburn

Materials Science and Molecular Research Division, Lawrence Berkeley
Laboratory, Berkeley, CA 94720

Abstract   High resolution electron microscopy (HREM) of cross-sectional
specimens has shown that the thickness of the native oxide on silicon
is 20 ± 3A, independent of surface orientation. This result has confirmed the value 21 ± 4A determined by ellipsometry assuming a
stoichiometric SiO_2 native oxide. Previous reports of a nonstoichiometric transition layer between Si and SiO_2 containing an excess of
10^{15} cm^{-2} Si atoms have also been alternatively explained by the
observed morphological features of the Si-SiO_2 interface.

1. Introduction

Knowledge of the structure and chemistry of very thin (<100Å) silicon
oxide films on silicon is important both for an understanding of the
initial stages of the oxidation process and for the optimization of
processing steps used in VLSI MOS device and solar cell technology
(Plummer (1982), Godfrey and Green (1979)). In this paper attention
is directed toward the native oxides grown on silicon in air at room
temperature. Earlier ellipsometry studies of native oxide growth kinetics
(Archer (1957), Lukas̆ (1972), Raider et al. (1975)) indicated that
within a few hours after removal of the oxide from the Si surface about
10A of a new native oxide was formed. It is therefore expected that
in many technological processes involving deposition of materials on
silicon a thin layer of native oxide will always be present at the interface (D'Anterroches (1983)). These oxides can obscure understanding of
interfacial phenomena as for example the initial stages of high temperature oxidation. In the present study high resolution electron microscopy was used for direct measurements of the native oxide thickness.
These measurements were compared to thickness measurement by ellipsometry.
In addition the morphology of the Si-SiO_2 interfaces of the native oxide
was compared to that of the interfaces developed during high temperature
oxidation.

2. Experimental Procedures

The native oxide was grown in air for 29 days on HF etched, p-type,
B doped 1.5-17Ωcm resistivity Si wafers having three different orientations: exact (100), 2 ° off (100) and 3° off (111). In addition observations were made on a 200Å-thick oxide grown on a (100) Si surface at
900°C in dry oxygen.

Cross-sections of the specimens were prepared by gluing two pieces of
wafers face to face with epoxy, then cutting the sandwich with a diamond

saw perpendicular to a {110} trace. After mechanical grinding and double-sided polishing to less than 100 μm, the section was glued to a support grid and ion milled to perforation. In this method of specimen preparation the edge containing the native oxide film was protected against milling by a layer of glue which was finally evaporated in the microscope column under a highly-focused electron beam. This method of preparation avoided the high temperature associated with deposition of a more conventional surface protective coating. The highest temperature to which the specimens were exposed was that due to ion beam heating, and was less than 150°C, (Ahn, 1979).

All high resolution electron microscopy observations were made in a JEOL JEM 200CX electron microscope with ultrahigh resolution pole piece.

The ellipsometry measurements were carried out on a Geartner ellipsometer having the He-Ne laser light source (λ= 6328Å).

3.   Experimental Results and Discussion

3.1 Thickness measurement by HREM: The thickness of the oxide films on silicon was measured directly from the high resolution micrographs of the cross-sections imaged along a <110> crystallographic direction. The {111} lattice fringes having a spacing of 3.14Å served as an internal standard of distance.

Observed cross-sections of native oxides on (100), 2° off (100) and 3° off (111) Si surfaces are shown in Figs. 1,2,3 respectively.

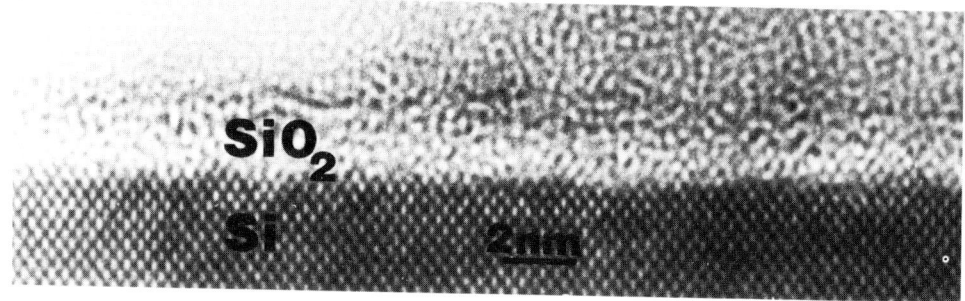

Fig. 1. Native oxide on (100) Si surface; thickness 20 ± 3Å.

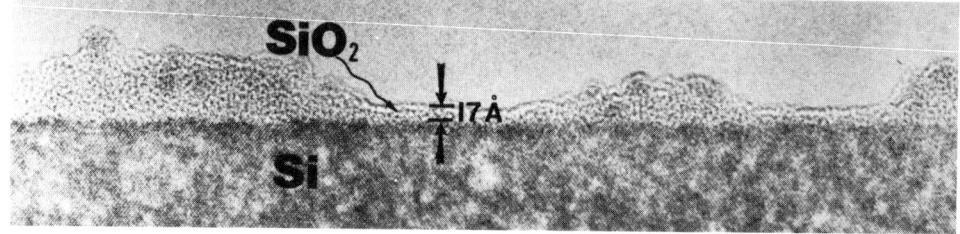

Fig. 2. Native oxide on 2° off (100) Si surface; thickness 17 ± 3Å.

The thickness of oxides as measured between the Si-SiO$_2$ and SiO$_2$-vacuum interfaces (fragments of the epoxy glue were still present on some surface areas) is the same for all Si surfaces and is within 20 $\pm$ 3Å. The oxide is uniform over the entire length of the observed interfaces; its observed mottled contrast is typical of that for amorphous materials.

Fig. 3. Native oxide on 3° off (111) Si surface; thickness 20 $\pm$ 3Å.

3.2 Measurement of native oxide thickness by ellipsometry: In the ellipsometry experiments, the relative phase change (Δ) and relative attenuation (tanψ) are measured for light reflected from the semiconductor surface which is covered by the oxide film. These measurements allow determination of film thickness and the refractive indices of the substrate and film. However for films of very small thicknesses, the refractive index values for both the substrate and the thin film must be assumed as these two parameters cannot be determined directly from the experiment. In addition, it is sufficient to measure the relative phase change to determine the thickness of a very thin film (Twu (1979)).

The measured values of Δ for oxides on (100), 2° off (100), and 3° off (111) Si surfaces, averaged over four zones, were 172.8°, 172.7° and 172.8° respectively. The corresponding thickness d of the oxides was calculated as a function of refractive index (Fig. 4) from equation (1) derived by Twu (1979),

(1) $d = \lambda(\Delta - \Delta_0) \{n_s^2(n_s^2-1)(n^2-1) \cos\theta \sin^2\theta/n^2(n_s^2-1)(n_s^2\cos^2\theta - \sin^2\theta)\}/720$

where λ= 6328Å is wavelength of light, θ = 70° is incidence angle of light, n_s = 3.89 is the refractive index of silicon substrate at 6328Å, Δ_0 = 178.87° is the relative phase change for bare silicon, n is the thin film refractive index, and Δ is the measured relative phase change for substrate with film. The curve "a" in Fig. 4 was obtained for the average value of Δ = 172.8°.

Fig. 4. Variation of the oxide thickness d with the refractive index n of the oxide film calculated for a) Δ = 172.8°, b) Δ = 173.8°, c) Δ = 171.8°.

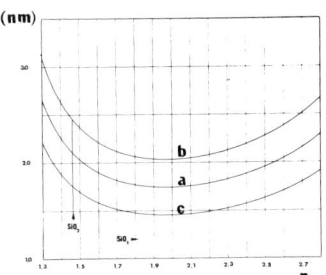

The curves b and c represent the error of the calculated thickness d corresponding to a variance of $\sigma = 1°$ of the   values measured in four zones. Assuming that the native oxide is stoichiometric SiO_2, (n = 1.46), the calculated oxide thickness, $(21 \pm 4\text{Å})$ is in good agreement with the HREM thickness measurements $20 \pm 3\text{Å}$. However, as can be seen from Fig. 4 the calculated thickness for a whole series of compositions with corresponding refractive index ranging from 1.46 (SiO_2) to 2.5 (SiO_{2-x}) would still give agreement within the experimental error of the HREM measurements.

Interest in the chemical composition of the interfacial region has stimulated ESCA studies of the $Si-SiO_2$ interface (Raider et al. (1975), Flitsch and Raider (1975), Raider and Flitsch (1976)). These investigators suggested a model of a nonstoichiometric region of the interface less than 15Å wide containing about 10^{15} cm^{-2} nonoxidized Si-Si bonds/ cm^2. The transition region was narrower and less steeply graded in very thin oxides formed on (100) Si then on (111) Si substrates, but remained independent of oxidation conditions, oxidant or substrate doping.

These Si rich transition layers were assumed to account for discrepancies between ellipsometry and ESCA measurements of ultrathin oxide thickness. Further ellipsometry studies of thermally grown oxides on Si (Corbes and Taft (1976)), Aspnes and Theteen (1980)) were also interpreted using a three layer model with an intermediate layer containing nonstoichiometric, Si-rich oxide. The present work suggests however that there is no need to make such assumptions. Even with a fully stoichiometric oxide (Ishizaka and Iwata (1980)), the apparent nonstoichiometric transition layer can be explained by the observed morphology of the $Si-SiO_2$ interface.

3.3 Morphology of the $Si-SiO_2$ interface:

The morphology of the $Si-SiO_2$ interface depends critically on the silicon surface orientation. For an exact (100) Si surface with native oxide as shown in Fig. 1 the interface can be characterized by a hill-and-valley structure with an asperity of 2-4Å and correlation length of about 20Å. These values are similar to those observed for the 200Å thick oxide grown at 900°C in dry O_2 shown in Fig. 5, as well as those reported earlier for oxides grown under different oxidation conditions (Krivanek and Mazur (1980), Krivanek et al. (1978)).

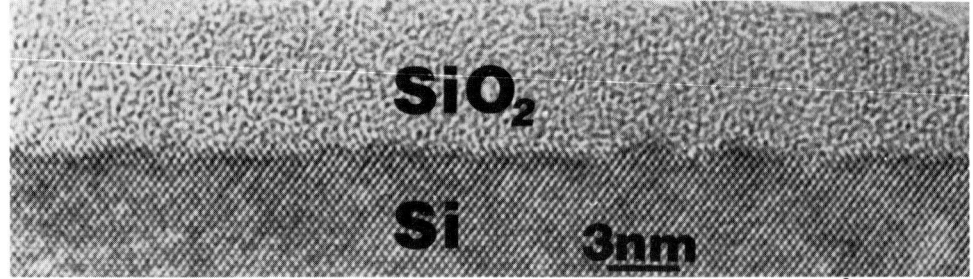

Fig. 5. High resolution image of the $Si-SiO_2$ interface of a 200Å thick oxide grown on (100) Si at 900°C in dry O_2.

The native oxide interface on the 2° off (100) Si substrate, Fig. 6,
appears to be rougher although qualitatively similar to the (100) interface. By comparison the structure of the 3° off (111) interface is best
described as a terrace-ledge configuration, with atomically smooth
terraces about 60Å wide and connecting ledges of 3.14Å height. A similar
structure has also recently been observed for oxides grown at 1000°C
in dry oxygen (Mazur (1983)).

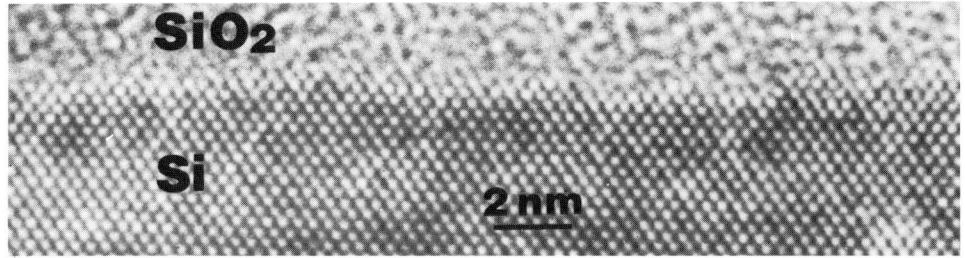

Fig. 6. High resolution image of 20Å thick native oxide on 2° of (100)
Si surface.

Significantly, these morphologies produce an average excess of about
10^{15} cm^{-2} Si atoms in the layers of oxide immediately adjacent to the
silicon. An example of the morphology of the (100) Si-SiO$_2$ interface
is shown schematically in Fig. 7, where the atomic positions on the
side of the interface are identified. Note that the small irregularities
marked "a", "b" and "c" can result in the observed interface with a
degree of roughness that corresponds to that estimated from mobility
measurements in MOS inversion layers at high gate fields (Chang and
Sullivan (1973)). Computer simulations are in progress to determine
whether the model is consistent with the observed details of phase
contrast changes at the Si-SiO$_2$ interface.

Fig. 7. Possible arrangement
of the Si atoms in crystalline silicon at the (100)
Si-SiO$_2$ interface resulting
in correlation length about
20Å and asperity 2-3Å.

4. Conclusions

High resolution electron microscopy studies of cross-sectional specimens
of native oxides on silicon have demonstrated that:

a. The native oxide on silicon is 20 ± 3Å thick, and this thickness
is independent of Si orientation.

b. Ellipsometry estimates of the native oxide thickness, 21 ± 4Å are
in good agreement with the HREM measurements.

c. The Si-SiO$_2$ interface for native oxides and high temperature oxides
grown on (100) silicon can be characterized by hill-and-valley structure

with asperity of 2-3Å and correlation length of about 20Å.

d. The Si-SiO$_2$ interface for 3° off (111) consists of atomically smooth (111) terraces connected by ledges of 3.14Å in height.

e. A morphological model of the Si-SiO$_2$ interfaces offers an alternative explanation for the nonstoichiometry region suggested by other techniques.

5. Acknowledgements

The work was supported by the Director, Office of Energy Research, Office of Basic Energy Sciences, Materials Sciences Division of the U.S. Department of Energy under Contract No. DE-AC03-76SF00098.

References

Ahn C (1979) private communication.
D'Anterroches C (1983) these proceedings.
Archer R J (1957) J. Electrochem. Soc. 104, 619.
Aspnes D E and Theeten B (1980), J. Electrochem. Soc. 127, 1980.
Chang Y C and Sullivan E A (1973), Surface Sci. 34, 717.
Flitsch R and Raider S I (1975), J. Vac. Sci. Technol. 12, 305.
Godfrey R B and Green M A (1979), Appl. Phys. Lett. 34, 790.
Ishizaka A and Iwata S (1980), Appl. Phys. Lett. 36, 71.
Krivanek O L and Mazur J H (1980), Appl. Phys. Lett. 37, 392.
Krivanek O L, Tsui D C, Sheng T T and Kagmar A (1978) in The Physics of SiO$_2$ and its Interfaces ed. S.T. Pantelides (Pergamon, New York) pp356-61.
Lukaš F (1972), Surface Sci. 30, 91.
Mazur J H (1983), Proc. 41th Annual EMSA Meeting (Phoenix) (to be published).
Plummer J D, ed., Computer-Aided Design of Integrated Circuit Fabrication Process for VLSI Devices, Stanford Electronics Laboratories, Research Report TR DXG501-82.
Raider S I and Flitsch R (1976) J.Vac. Sci. Tech. 13, 58.
Raider S I, Flitsch R and Palmer M J (1975) J. Electrochem. Soc. 122, 413.
Taft E and Cordes L (1979) J. Electrochem. Soc. 126, 131.
Twu B I (1979) J. Electrochem. Soc. 126, 1589.

Inst. Phys. Conf. Ser. No. 67: Section 2
Paper presented at Microsc. Semicond. Mater. Conf., Oxford, 21–23 March 1983

High resolution electron microscopy of Si-implanted and electron-beam annealed silicon-on-sapphire

David J Smith[1], L A Freeman[1], R A McMahon[2], H Ahmed[2], M G Pitt[3], and T B Peters[3].

[1]High Resolution Electron Microscope, University of Cambridge, Free School Lane, Cambridge, CB2 3RQ.
[2]University Engineering Department, Trumpington St., Cambridge.
[3]GEC Research Labs., Hirst Research Centre, East Lane, Wembley.

Abstract   The changes in morphology of silicon-on-sapphire films that were Si-implanted and e-beam annealed have been investigated by high-resolution electron microscopy at 500kV.   Ion implantation was used to amorphise the bulk of the silicon film, thereby breaking up the defects in the film, but leaving a relatively undamaged surface layer. E-beam annealing recrystallised the film from the surface seed back to the interface to give material with few defects compared to the as-grown film.   An amorphous layer of 10-20Å thickness was formed at the silicon-sapphire interface.   The observed changes in microstructure correlated well with RBS measurements of the film crystallinity.

1. Introduction

Silicon-on-sapphire (SOS) is widely used as a substrate material for CMOS integrated circuit manufacture.   However, the large lattice mismatch ($\sim$6%) between the epitaxially-grown (100) silicon film and the (01$\bar{1}$2) sapphire substrate results in a high density of defects in the silicon, particularly microtwins and stacking faults (Hutchison et al 1981; Ponce 1981).   The defects increase carrier scattering, lowering mobilities and hence reducing circuit speeds.   As lateral device geometries are reduced, it is also desirable to reduce vertical dimensions and so thinner silicon films than the usual 0.6µm layers are needed.   However, the active regions of MOS transistors are even closer to the silicon-sapphire interface where the density of defects is highest.   It then becomes essential to improve the quality of the film to avoid unacceptably low mobilities.   The improvement of film quality by a combination of ion-implantation to break up defect structures in the layer, followed by epitaxial regrowth by furnace annealing (Lau et al 1979, Amano, 1982) or laser irradiation (Golecki et al 1980) has been reported.   In the present study we have used instead the method of rapid isothermal electron-beam annealing (McMahon and Ahmed, 1979) to re-crystallise the silicon layer amorphised by implantation. High-resolution electron microscopy (HREM) and Rutherford back-scattering (RBS) measurements have been used to establish the success of this technique in producing high quality, crystalline films.

2. Film improvement by implantation and annealing.

The technique for improving the silicon film involves high energy implantation to break up the defects in the silicon layer, which are primarily

microtwins extending from the interface, as shown schematically in Fig.1a.
The nature and distribution of the resulting implant damage depends on the
energy and dose of the implant, as well as the implantation temperature.

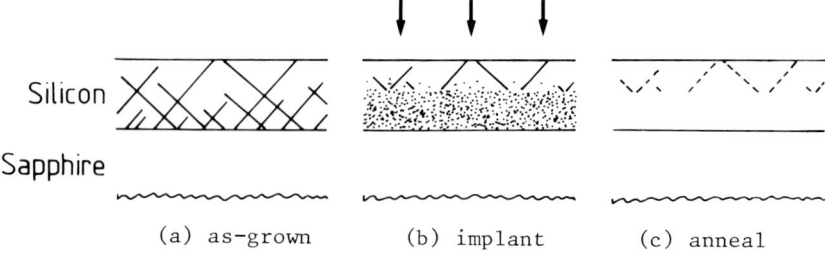

(a) as-grown   (b) implant   (c) anneal

Fig.1. Schematic showing implantation and annealing of 0.3μm SOS film.

The silicon layer should preferably be amorphous, except for a single
crystal surface layer needed as a seed for subsequent regrowth (see Fig.1b).
For the 0.3μm silicon films used in this work, such conditions are obtained
by a $10^{15}/cm^2$ Si implant at 190keV.   Characterisation by RBS confirmed
that higher doses did not leave an adequate surface seeding layer and that
lower doses did not amorphise the bulk of the layer.   Moreover, the
implant energy used locates the damage peak close to the interface.

The silicon layer is regrown from the surface seed to the interface, as
shown in Fig.1b.   Low temperature furnace annealing, at ∿550°C for about
1 hour , has been used to recrystallise films amorphised by Si implants at
liquid nitrogen temperatures (Lau et al, 1979; Golecki et al, 1980). (The
low implantation temperature being chosen to avoid generating thermally
stable defects).   In more recent work (Amano, 1982; Reedy and Sigmon,
1982) room temperature silicon implantation with high temperature
annealing has been used to achieve recrystallisation.   These techniques
improve the crystallinity of films compared to as-grown samples.

Rapid isothermal annealing by electron beam permits much shorter annealing
times (typically about 10s) than were previously attainable (McMahon and
Ahmed, 1979). This method should both recrystallise the layer and minimise
migration of aluminium or impurities from the sapphire into the silicon
layer (cf. Reedy and Sigmon, 1982), as well as limiting the growth of any
residual defects. The annealing conditions used in this work were a beam
power of $50W/cm^2$ with a heating time of 11.5s, which gave a peak temperature of ∿1250°C. The RBS results, described later, on materials annealed
at this and lower power densities (i.e. lower temperatures) established
that the best quality material was obtained under these annealing conditions.
Higher power densities caused blistering of the silicon films.

3. Results.

Specimens were prepared for electron microscopy in the standard manner
(Abrahams and Buiocchi, 1975) by mechanical thinning and ion-beam milling
of transverse cross-sections of the SOS interface.   Observations were made
at an accelerating voltage of 500kV with the Cambridge University HREM
(Smith et al, 1982); a double-tilting holder was used to orient the
specimens, normally with the [110] zone axis of silicon parallel to the
electron beam.

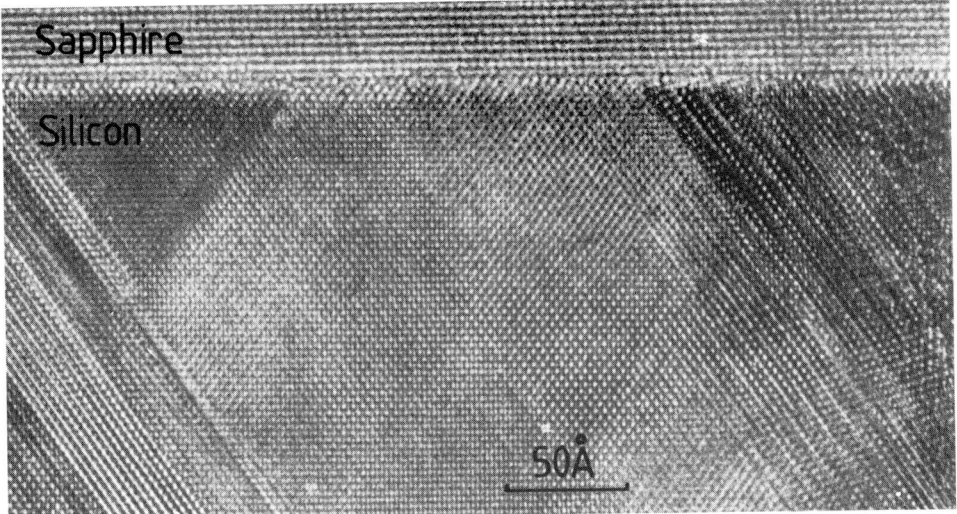

Fig.2. High-resolution lattice image of as-grown SOS film showing interface
between silicon and sapphire.Note stacking faults and micro-twins leading
away from interface.

Before implantation, the morphology of the samples closely resembled that
reported previously (e.g. Ponce and Aranovich, 1981; Abrahams et al, 1981),
as shown in Fig.2.   There was a high density of twins and stacking faults,
normally presumed to accommodate the interfacial mismatch, and the interface itself was observed to be planar and effectively without any amorphous
layer. The primary effect of implantation, as expected, was the amorphisation of the bulk of the silicon lattice. However, very little damage
was noticeable in the surface layer which was mainly single crystal to a
depth of about 500Å, although some twin boundaries were present. Fig.3.
shows a set of finely-spaced micro-twins emerging from the amorphous
region which almost reach the edge of the silicon. Fig.4. shows a
magnified image of the transition region between the crystalline surface
layer and the bulk amorphous layer. Small crystallites retaining the
silicon lattice spacing and orientation are clearly visible. Finally,it
was noted that the amorphous region did not extend fully to the SOS interface. Many small silicon crystals were retained in a layer of about 200Å
thickness near the interface which also seemed unaffected and there was no
obvious sign of ion damage in the sapphire.

After annealing, it was found that the silicon had re-crystallised but
most of the lattice defects observed in the original as-grown layer were
then absent.   Furthermore, as shown in Fig.5, the interface between the
silicon and sapphire now contained an amorphous layer which varied in
thickness between about 10 to 20Å - such a layer is consistent with recent
theoretical work on mismatch between two different crystalline materials
(Spaepen, 1978).   Finally, a novel type of small defect was also observed,
as shown in Fig.6.   These defects, which have not yet been fully
characterised, may be related to the (311) irradiation faults in silicon
recently observed at high resolution by Hirabayashi et al (1982).

Channelled Rutherford backscattering spectrometry was used to assess the
quality of the SOS films following the various heat treatments. 1.5MeV
helium ions were used for the measurements, with an energy-sensitive

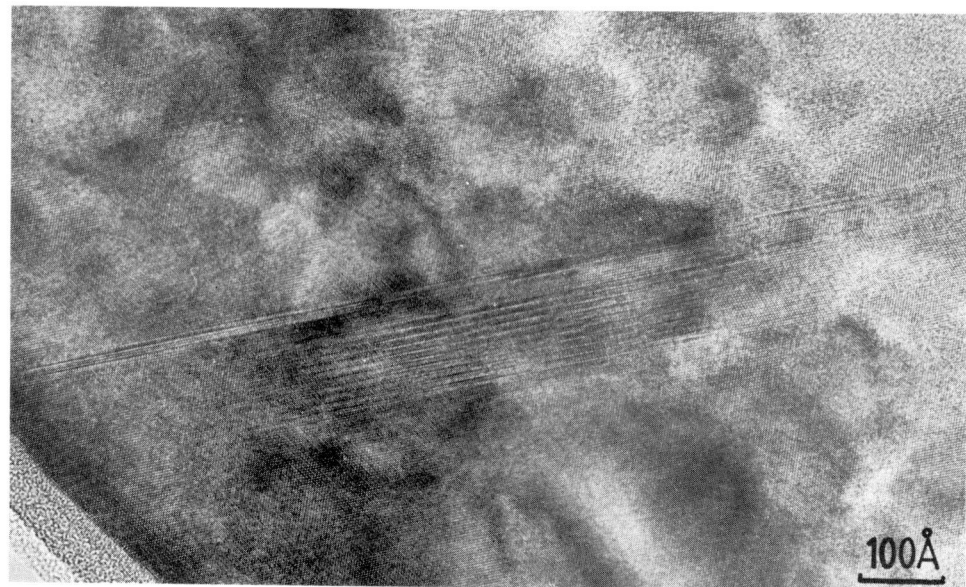

Fig.3. Region of silicon after implantation showing micro-twins emerging from amorphous region.

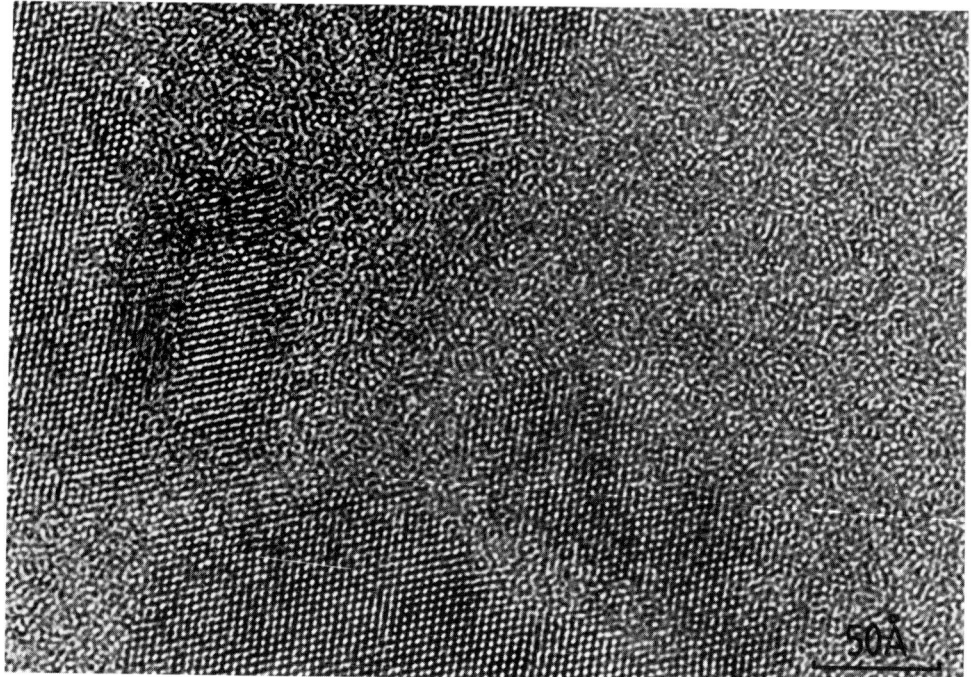

Fig.4. High resolution image of implanted sample showing transition between amorphous region and crystalline surface seed.

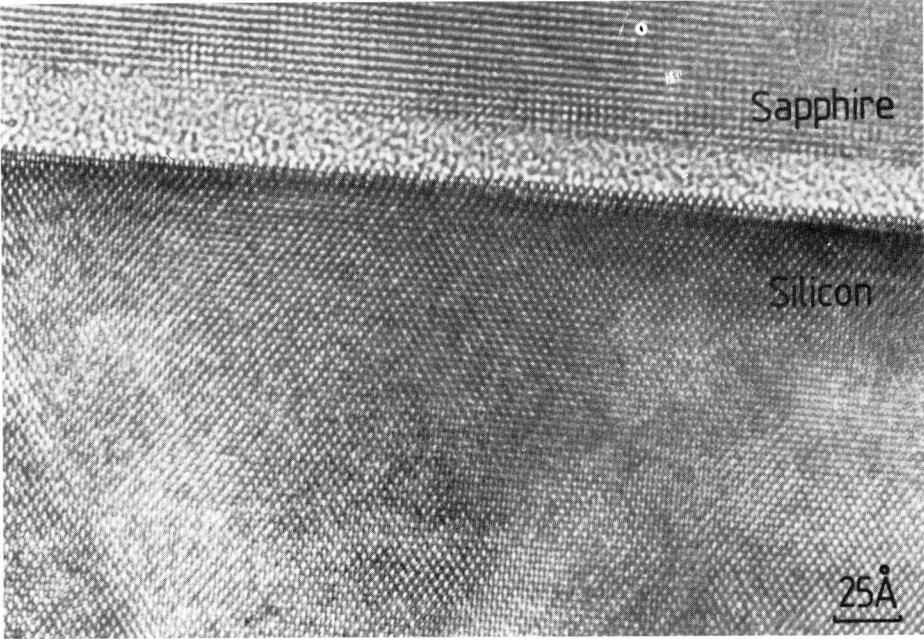

Fig.5. Interface region of annealed sample showing absence of defects in silicon and 10-20Å thick amorphous layer.

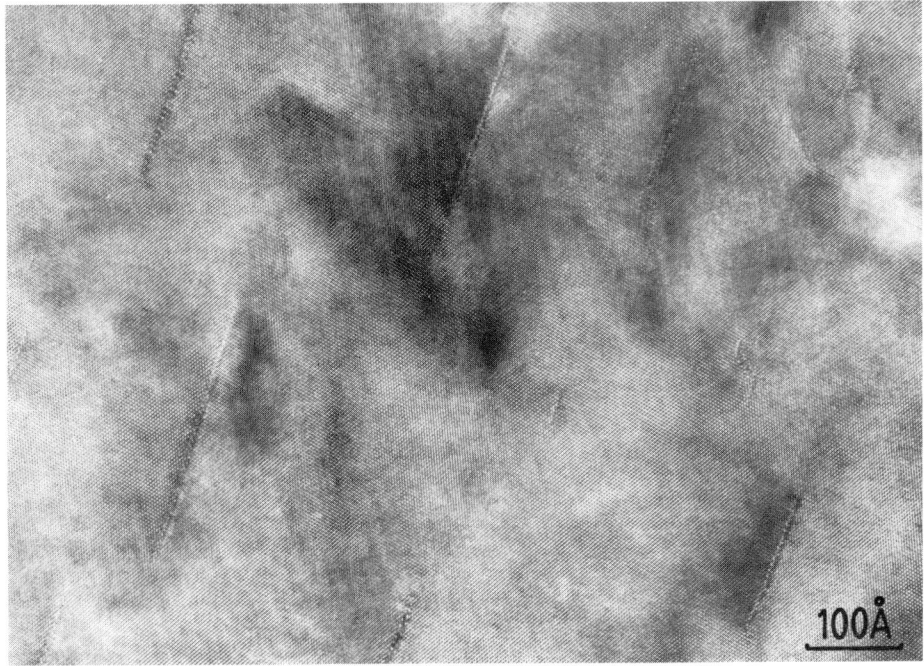

Fig.6. Small defects visible in silicon after annealing.

detector at an angle of 160° to the incident beam and a multichannel
analyser recorded the spectrum. Fig.7. shows channelled RBS spectra for
the as-grown SOS material and following implantation and annealing at
power densities of 7,20 and 50W/cm^2. Values of χ_o,χ_1,the channelling
yields, which were obtained at the silicon surface and the interface are
summarised in Table I. Substantial improvements in the crystal quality
are observed, especially at higher power densities.

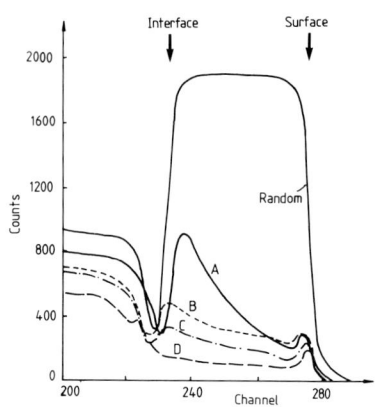

Table I

	χ_o	χ_1
Bulk silicon	.03	.05*
(A) 0.3µm SOS film	.10	.50

After implant and
e-beam anneal.

(B) 7W/cm^2; 37s	.12	.28
(C) 20W/cm^2; 22.5s	.07	.19
(D) 50W/cm^2; 11.5s	.05	.09

Channelling yields at silicon
surface, χ_o, and interface, χ_1,
for 0.3µm SOS samples.
(*measured at a depth of 0.3µm).

Fig.7 RBS spectra(see Table I)

4. Discussion and Conclusions

This study has shown that the combination of ion implantation of silicon
into silicon-on-sapphire films with subsequent rapid, high temperature
electron-beam annealing results in a significant improvement in the quality
of the silicon layer compared with the as-grown film. The defect density in
the regrown film is low, but there are some unidentified defects which will
be characterised in future work. The improved silicon films should permit
the fabrication of smaller devices with superior performance.

Part of this work has been supported by the S.E.R.C., U.K. and the
Procurement Executive, Ministry of Defence, sponsored by DCVD. We are
grateful to Dr. D. Fathy for his initial advice and assistance with ion-
beam thinning and to D.G. Hasko for electron-beam annealing.

References

Abrahams M S and Buiocchi C J 1975 Appl. Phys. Letts. 2 825
Abrahams M S, Hutchison J L and Booker G R 1981 Phys. Stat. Sol.(a)63 113
Amano J 1982 Rad. Effects 61 195
Golecki I, Kinoshita G, Gat A and Paine B M 1980 Appl.Phys.Letts.37 919
Hirabayashi M, Hiraga K and Shindo D 1982 Ultramicroscopy 9 197
Hutchison J L, Booker G R and Abrahams M S 1981 Microscopy of Semi-
    conducting Materials 1981 (A. Cullis, Ed) pp. 139
Lau S S, Matteson J W, Mayer J W, Revesz P, Cynlai J, Roth J, Sigmon T W
    and Cass T 1979 Appl. Phys. Letts. 34 76
McMahon R A and Ahmed H 1979 Electronics Letts. 15 45
Ponce F A 1981 In: Defects in Semiconductors (Narayan and Tan, Eds.)pp.285
Ponce F A and Aranovich J 1981 Appl.Phys.Letts. 38, 439
Reedy R E and Sigmon T W 1982 J. Cryst. Growth 58 53
Smith D J, Camps R A, Cosslett V E, Freeman L A, Saxton W O, Nixon W C,
    Ahmed H, Catto C J D, Cleaver J R A, Smith K C A and Timbs A E 1982
    Ultramicroscopy 9 203
Spaepen F 1978 Acta Met. 26 1167

Inst. Phys. Conf. Ser. No. 67: Section 2
Paper presented at Microsc. Semicond. Mater. Conf., Oxford, 21–23 March 1983

The atomic structure of the NiSi$_2$/(001) Si interface

C J D Hetherington, D Cherns and C J Humphreys
Dept of Metallurgy & Science of Materials, University of Oxford, Parks Rd,
Oxford, OX1 3PH, UK.

Abstract   The lattice imaging technique of transmission electron microscopy is used to examine (110) cross-sectional specimens of
NiSi$_2$/(001)Si. Comparison of theoretical and experimental images yields
an atomic structure consistent with a model of the NiSi$_2$/(001)Si interface where Si is everywhere tetrahedrally coordinated. It is shown that
interfacial steps in the (001) interface may be associated with misfit
dislocations of a/4<111> type inclined at 35° to the interface.
Experimental evidence for these dislocations is presented.

1. Introduction

The formation of intermediate silicide layers is of great technological
importance in the production of reliable and well-characterised silicon
devices. There is much current interest in the atomic structure of the
silicon-silicide interface which is believed to control electrical properties such as the Schottky barrier height (Schlüter (1982)). Moreover a
knowledge of this interface structure can throw light on the mechanism of
silicide growth (Cherns and Smith (1981)). In recent work atomic models
for the epitaxial NiSi$_2$/(111)Si interface were derived using the lattice
imaging technique of transmission electron microscopy applied to cross-
sectional specimens (Cherns et al (1982)). We have now extended these
studies to the NiSi$_2$/(001)Si interface. The NiSi$_2$/(001)Si interface contrasts with the relatively smooth NiSi$_2$/(111)Si interface in being highly
facetted. The (001)-interface is also of particular interest techno-
logically as device silicon is generally close to (001)-orientation. In
this paper we consider models for (001)-facets of the NiSi$_2$/(001)Si interface and the nature and role of interfacial steps.

2. Experimental

Ni/(001)Si was prepared by the deposition of nickel to a thickness ~100–
500Å on to cleaned high purity (001)-Si wafers at a substrate temperature
of less than 100°C in a bell-jar evaporator operating at 10^{-5} torr. The
Ni/(001)Si films were subsequently heated at 800°C for 1 hour in a separate
vacuum furnace to give NiSi$_2$/(001)Si. Cross-sectional specimens of
NiSi$_2$/(001)Si were prepared in (110) orientation as previously described
(Cherns et al (1982)). (001)-specimens were also produced either by chemical back-thinning or by mechanical polishing followed by ion-beam thinning
to perforation from the silicon side using 5keV ions at 25° incidence.
Cross-sectional specimens were examined in a JEOL 200CX electron microscope
operating at 200kV with a C$_s$ = 1.2mm. 001-specimens were examined in a
JEOL 100B electron microscope operating at 100kV.

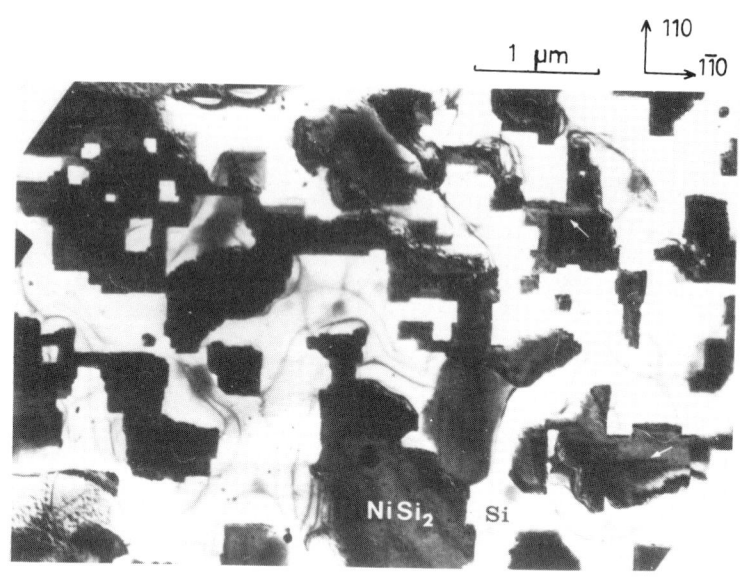

Fig. 1   Bright-field micrograph showing islands of NiSi$_2$ on (001)Si

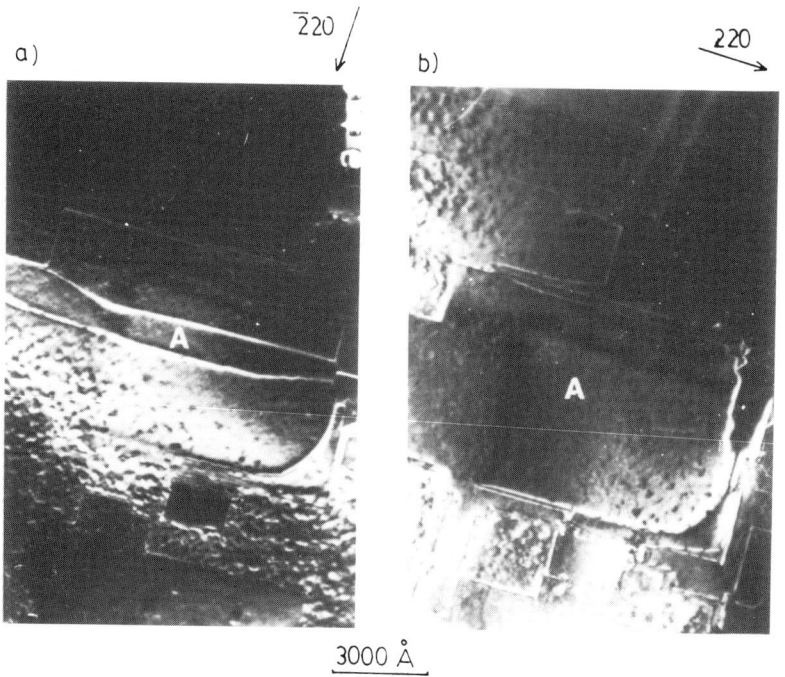

Fig. 2   Dark-field image showing interface dislocations in NiSi$_2$/(001)Si

The $NiSi_2$/(001)Si specimens showed discrete islands of deposit typically 1
-2μm across (Fig. 1). The deposit was epitaxial with the cubic CaF_2 unit
cell of $NiSi_2$ (a = 5.408Å) aligned parallel to the cubic unit cell of silicon (a = 5.428Å). Dislocations accommodating the 0.4% natural mismatch
between the $NiSi_2$ and Si crystals were sometimes but not always observed.
An example of misfit dislocations is shown in Fig. 2. The two dislocations
near A are visible in g = $\bar{2}$20 (Fig. 2a) and mostly out-of-contrast in g =
220 (Fig. 2b). This is consistent with a Burgers vector having a component
along the $\bar{2}$20 direction in the film plane such that the two dislocations
are predominantly of edge type and thus they may accommodate misfit
strains. The slight residual contrast of the off-edge segments in g = 220
suggests $\underline{g}.(\underline{b} \times \underline{u})$ contrast associated with an inclined Burgers vector.
Indeed a more detailed analysis, to be published elsewhere (Cherns et al
(1983)), suggested a Burgers vector lying along a <111> direction inclined
at 35° to the film plane. Further lines of contrast in Fig. 2 lying accurately parallel to $\bar{2}$20 and 220 directions are probably associated with
junctions between facets (see later).

Examination of cross-sectional specimens showed that the $NiSi_2$/(001)Si
interface had pronounced (001)- and (111)-facetting (Fig. 3), in agreement
with earlier studies (Foell et al (1982)). (110) lattice images taken near
the optimum defocus and under axial illumination with the objective aperture including the 4{111} and 2{002}-type reflections showed that the silicon-silicide interface was atomically abrupt and that facets were mostly
atomically smooth. Fig. 4 shows a magnified image of a junction between
(001)- and (111)- facets. The atomic structure of 001-facets was analysed
using lattice images near the optimum defocus and where the thickness of
both the silicon and silicide layers was found to be less than that corresponding to the first thickness reversal of contrast (Cherns et al (1982)).
However in order to understand the results we need first to consider
possible models for the $NiSi_2$/(001)Si interface.

3. Interpretation of Results

Silicon atoms in both $NiSi_2$ and Si lattices have tetrahedral coordination
with closely similar bond lengths ($a\sqrt{3}/4$ = 2.35Å). If we assume, following
Cherns et al (1982) that the atomic structure of the silicon/silicide interface should preserve the tetrahedral coordination and bond lengths of
the silicon atoms the atomic model in Fig. 5 may be constructed. Fig. 5a
shows a (110) projection with nickel atoms adjacent to the interface being
6-coordinated compared with 8-coordinated in bulk $NiSi_2$. Fig. 5b shows the
(1$\bar{1}$0) projection of the same model. The (110) and (1$\bar{1}$0) projections are
different because the tetrahedral coordination of silicon atoms in the
terminating silicon layer projects differently in the two orientations i.e
<1$\bar{1}$0> directions show only two fold symmetry. Of course, the (110) and
(1$\bar{1}$0) projections of bulk $NiSi_2$ and Si appear similar but are in fact related by a rigid body translation. This point is made clearer when the
nature of an atomic step in the (001)-interface is considered. Fig. 5c
shows that a step of height d_{004} in the interface is equivalent to a misfit
dislocation with a Burgers vector $a/4$ <111> with Burgers vector inclined
at 35° to the interface and effects a relative translation of the two lattices. Furthermore the interface structure on opposite sides of the step
corresponds to the (110) and (1$\bar{1}$0) projections in Fig. 5a and 5b.
Further models of the $NiSi_2$/(001)Si interface may be constructed having
tetrahedrally coordinated Si. These models are related to that in Fig. 5
by the same rigid body translation but differing interface composition.

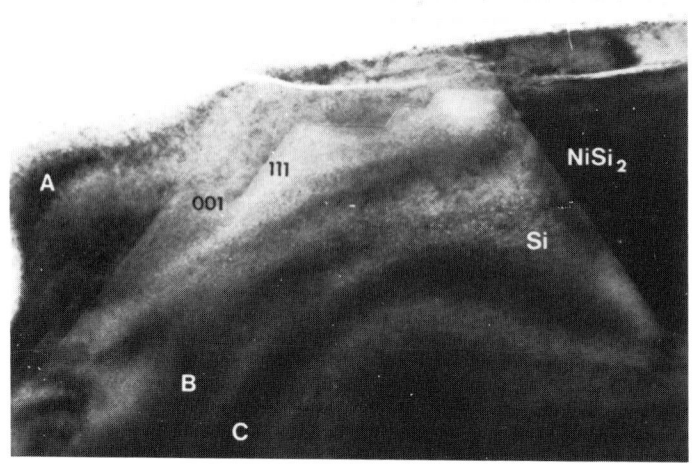

1000 Å

Fig. 3   Low magnification bright-field micrograph of a cross-sectional
NiSi$_2$/(001)Si sample showing facetting.

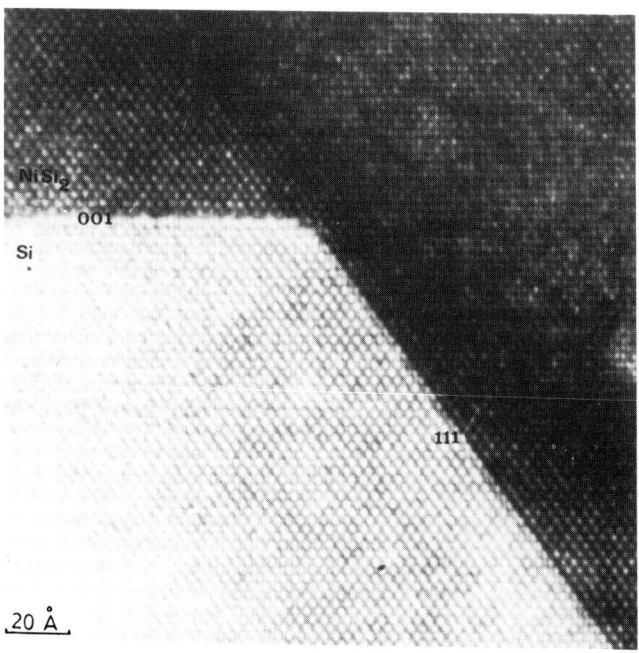

20 Å

Fig. 4   Lattice image showing a junction between (001)- and (111)-
facets

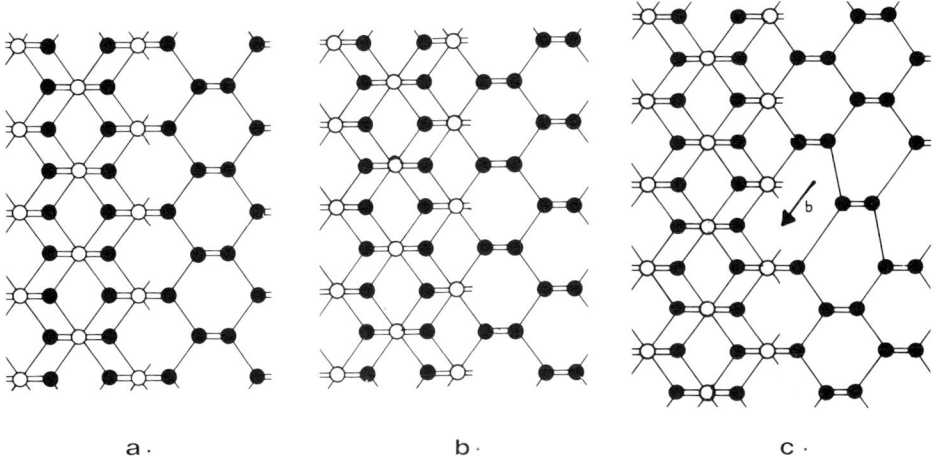

a .   b .   c .

Fig. 5   Atomic model for the $NiSi_2/(001)Si$ interface preserving
tetrahedrally-coordinated Si. a) ($1\bar{1}0$ projection b) ($1\bar{1}0$) projection,
c) schematic illustration of an interfacial step. o = Ni, ● = Si.

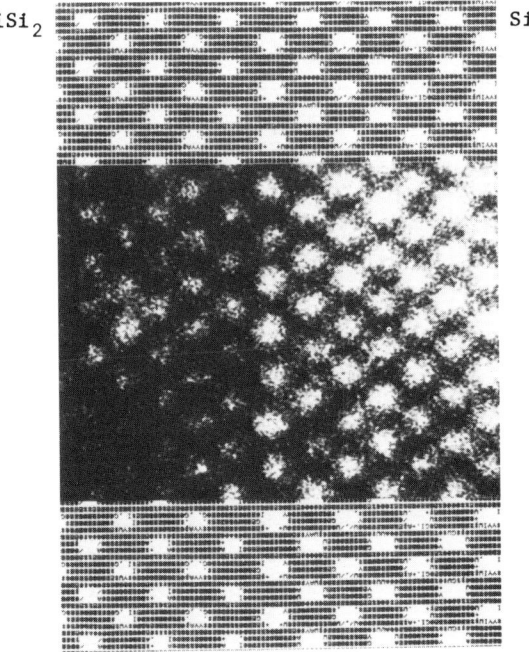

Fig. 6   Comparison of an experimental image from a thin crystal with a
simulated image for the interface in Fig. 5a. Assumed film thickness = 50Å,
defocus = -630Å, information cut-off radius = $0.39Å^{-1}$, beam divergence
semi-angle = 2 x 10^{-4} rad.

Fig. 6 compares an experimental image from a thin region (see earlier) with a simulated image for the model in Fig. 5a. Image computations were carried out using a multislice method and an array processor (Skarnulis et al (1981)). The computations shown assume atomic scattering factors for both Ni and Si atoms in both the Si and $NiSi_2$. Some additional computations were carried out substituting ionic scattering factors for Ni atoms (Ni^{2+}) to account for the ionic nature of the bonding in $NiSi_2$. These computations showed differences in image detail in the interface region while maintaining the same fringe displacement for regions well-separated from the interface. The details of these computations will be reported elsehwere. If we consider just the average displacement of fringes on crossing the interface the model in Fig. 5a should give an outward displacement of ½ d_{002} for (002) fringes or a displacement of $\{111\}$ fringes by 1/8 d_{111}. The corresponding displacements for the projection in Fig. 5b have the same magnitude but opposite sign. In Fig. 6 the shift of the $\{111\}$ type fringes in the theoretical images can be seen quite clearly and is in good agreement with the experimental result.

4. Conclusions

The results in Fig. 6 suggest the atomic structure of the $NiSi_2/(001)Si$ interface is consistent with a model where Si is everywhere tetrahedrally coordinated in agreement with the results for $NiSi_2/(111)Si$ (Cherns et al (1982)). It has also been noted that steps of height d_{004} in the (001)-interface should be associated with misfit dislocations of type $a/4$ 111 . The component of Burgers vector in the film plane is thus $a/4$ 110 compared with the value of $a/2$ 110 reported by Foell et al (1982); the latter are lattice dislocations and do not require an interface step. We believe from an extended analysis that dislocations in Fig. 2 are probably $a/4$ 111 type. For example this is consistent with the residual contrast in Fig. 2b (see earlier). In addition the single image in Fig. 2a (deviation parameter s~0) is consistent with a $|g \cdot b| = 1$ dislocation ($|b| = a/4[1\bar{1}0]$ rather than a $g \cdot b = 2$ dislocation ($|b| = a/2[1\bar{1}0]$). A fuller analysis of these dislocations will be given elsewhere (Cherns et al (1983)). A knowledge of the structure of $NiSi_2/(001)Si$ and $NiSi_2/(111)Si$ interfaces also enables the nature of junctions between facets to be analysed. A simple analysis shows that junctions may or may not be associated with misfit dislocations (Cherns et al (1983)). The images in Fig. 2 indicated that dislocation contrast may indeed be associated with such junctions.

It is worth noting that model proposed for the $NiSi_2/(001)Si$ interface has more dangling bonds per unit area than the $NiSi_2/(1\bar{1}1)Si$ interface in the ratio 2.5:1. One might therefore expect a greater interfacial energy per unit area for the (001)- than the (111)-interface. This may explain the greater facetting for $NiSi_2/(001)Si$ compared with $NiSi_2/(111)Si$.

References

Cherns, D, Hetherington, C J D and Humphreys, C J, 1983, to be published.
Cherns D and Smith D A, 1981, Proc. of the Materials Research Soc. Symposium, Boston, Nov. 1980, ed. J. Narayan and T Y Tan (New York. North Holland). Vol. 2, p.291.
Cherns, D, Spence, J C H, Anstis, G R and Hutchison J L, 1982, Phil. Mag. A46, 849.
Foell, H, Ho, P S & Tu, K-N, 1982, Phil. Mag. 45, 31.
Schluter, M, 1982, Thin Solid Films, 86.
Skarnulis, A J, Wild, D L, Anstis, G R, Humphreys, C J and Spence, J C H, 1981, Inst. Phys. Conf. Ser. No. 61, 347.

Inst. Phys. Conf. Ser. No. 67: Section 2
Paper presented at Microsc. Semicond. Mater. Conf., Oxford, 21–23 March 1983

High resolution TEM study of Al–Si 1%/Si interface

C D'Anterroches

CNET, Centre National d'Etudes des Télécommunications, BP : 42,
Chemin du Vieux Chêne, 38240 MEYLAN.

Abstract   Transmission Electron Microscopy has been applied to the study of sputtered Al–Si 1 % films on <001> silicon wafers. Analysis of diffraction patterns and lattice imaging of cross-sections have been carried out for a number of specimens. Epitaxial growth of Al and of Si is proved, in spite of the existence of an amorphous layer between the Al film and the Si substrate. Finally, the behaviour of the film during annealing is analysed.

1. Introduction

Sputtered Al–Si 1% films on silicon have several applications in the semiconductor industry. This work describes both the texture of very thin Al–Si 1 % films (600 A), and the structure of the deposited film/substrate interface. It is well-known that there is considerable diffusion of silicon from the substrate into the Al film (R. ROSENBERG et al., 1978) and there is a large variation with temperature of the solubility limit of Si in Al (0.48 % at 450° C in the post-metallization heat-treatment and 0.01 % at 221° C). Hence, during annealing, Si diffuses from the substrate into the Al film, and during cooling, Si precipitates from solution in Al forming agglomerates at the Al grain boundaries. Such alloy penetration may short-circuit the junction and destroy the device. To prevent this, the Al film is presaturated with Si by deposition of a binary Al–Si 1 % alloy. However, it has been shown that, using this technique, the distribution of Si through the film is very non-uniform (W. H. CLASS, 1979).

2. Experimental Method

The process is set on <100> Si wafers. To minimize the native oxide, wafers were etched with a H. F. solution just before deposition. A 600 A thick Al–Si 1 % film was then sputtered using a F. C. S. 3200 ERCO under the following conditions : deposition time 4 sec., pressure 6.10^{-3} Torr, applied power 10 kV, temperature 350° C. The cathode is an Al–Si 1 % binary alloy.

For T. E. M. study two kinds of specimen have to be thinned :

i) To study the texture of Al film, samples were thinned down to 30 μm by mechanical polishing starting from their underside followed by ion milling to achieve electron transparency. Transmission Electron

Microscopy (T. E. M.) was then carried out using a Philips EM 400 S. T. E. M. In the thick areas of the specimen there were both Al film and Si substrate and the specimen was tilted to align the <001> Si direction with the electron beam. The specimen was translated to areas without substrate then diffraction patterns and microdiffraction on individual grains were obtained.

ii) To gather information about the interface structure, cross-sections were cut normal to <110>, then polished and thinned by Ar^+ ion milling. High Resolution Transmission Electron Microscopy (H. R. E. M.) was carried out using a JEOL 200 CX having an ultra high resolution pole piece (C_s = 1.05 mm). Specimens were oriented so that <110> Si was parallel to the electron beam. High resolution images were obtained with 6 beams for Si (four {111} and two {002}). The case of Al is less straightforward, as the Al grains were generally disoriented. Nevertheless, two cases occured : sometimes [110] Al was aligned with [110] Si and two orientations of {111} Al planes were imaged, but more often only one family of {111} Al planes could be observed. As interplanar distances are different (2.34 A for {111} Al and 3.14 A for {111} Si) we had to work at the defocussing distance z = - 700 A to image both families of planes.

Size and distribution of Si precipitates were characterized by selective etching and Scanning Electron Microscopy (S. E. M.).

3. Experimental Results

Fig. 1 illustrates the structure of an Al film observed along the [001] direction of the Si substrate. The Al film is composed of grains of dimensions ranging from 0,1 μm to 0,3 μm. The inset of Fig. 1 shows a diffraction pattern of the same area. This pattern seems to show a texture : the <220> Al direction is roughly aligned with the <400> Si direction, within $\pm$ 10° degrees. Moreover, a heteroepitaxy of Al grains is observed as shown in Fig. 2 by the cross-sectional T. E. M. diffraction example. [002] Si and [T11] Al are aligned as well as [110] Al with [110] Si. One can see in Fig. 2 the corresponding image.

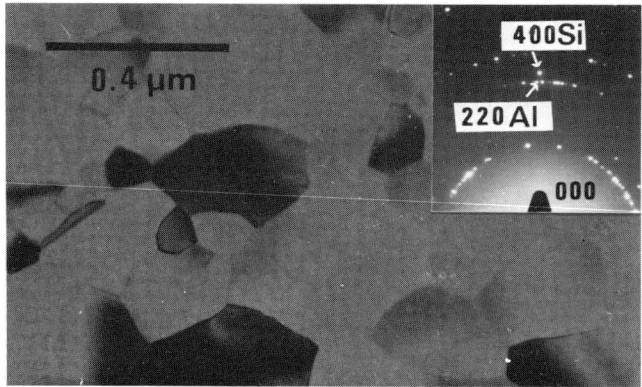

Fig. 1 : Image showing the structure of Al-Si 1% film obtained on samples thinned from their underside. The silicon substrate had disappeared. Inset shows a diffraction pattern of the same area.

Such orientation is very unusual. Generally {111} Al and (002) Si planes are disoriented within ± 10 degrees.

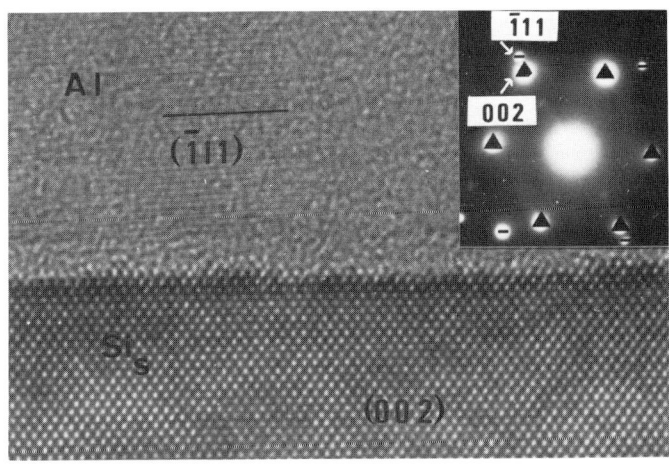

Fig. 2 : Cross-section [110] Si images illustrating alignment of (002) Si plane of interface with one family of {111} Al planes. Note amorphous layer 6 Å thick between crystallized Al and Si, probably due to initial native silicon oxide. The diffraction pattern shows [T11] Al//(002) Si ▲: Si beams −: Al beams.

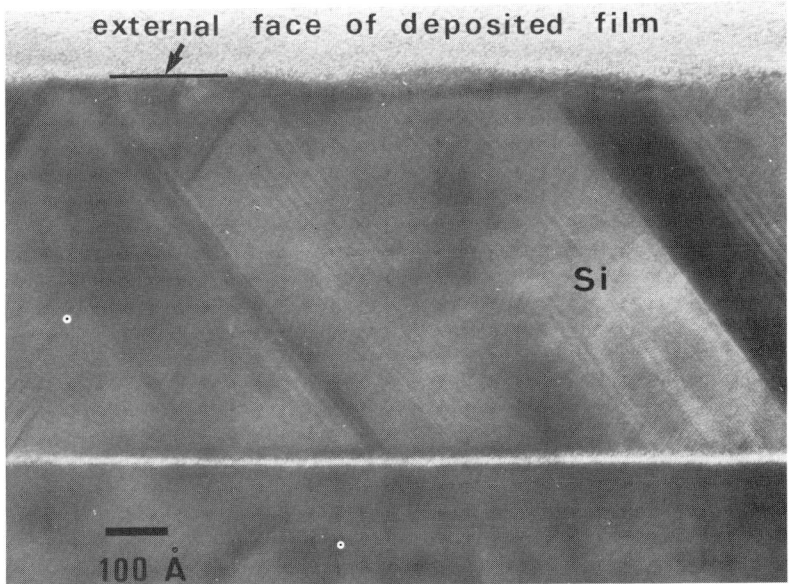

Fig. 3 : Image of Si grain on Si substrate. Orientation is the same as for the substrate at 2°.

Moreover, in this figure, one can see an intermediate amorphous 6 Å thick layer between Al and Si : this was usually observed, but sometimes a prolongation of Al planes reaching the Si atomic columns can be seen. Inside the Al layer we observed large Si grains, as shown in Fig. 3. In those grains there are many $\{111\}$ twin boundaries (Σ = 3) and stacking faults. One of these grains is shown in more detail in Fig. 4. It illustrates the epitaxy between the substrate and grain. The maximum disorientation observed is about 2° between $(1\bar{1}1)_S$ and $(1\bar{1}1)_I$. Atomic column contrast is the same when passing through the interface, so there is less than 0.1° misorientation between $[110]_S$ and $[110]_I$. The Si precipitates are in the 0.1 μm to 0.4 μm range, and always occupy the entire thickness of the Al film. For this reason these films have been studied in more detail.

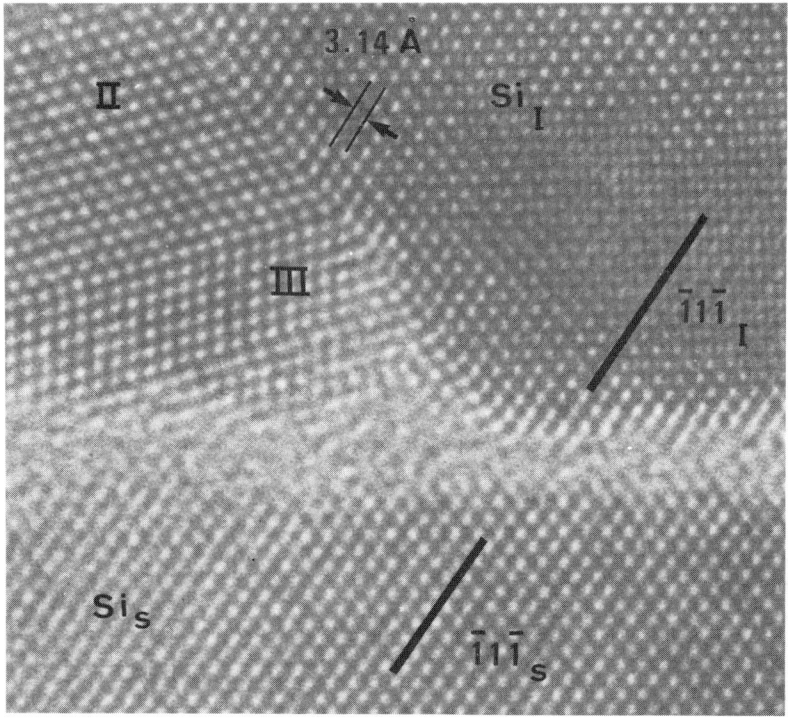

Fig. 4 : A view of one grain showing an epitaxial relationship between the substrate and grain I and the Σ = 3 twin-related grains II and I, and II and III leading to Σ = 9 twin-related grains I and III.

Fig. 5 illustrates non-uniform thickness of the amorphous film. Indeed, in area A for example, one can see a prolongation of the Si planes of the substrate in the precipitate. The prolongation is not very straight, but this may be an artefact due to the thickness variations of the specimen since the contrast of atomic columns is closely linked to that parameter.

In order to determine the origin of the Si precipitates, we had to know their density. As Si grains occupy the entire thickness of the Al film, selective etching seemed a suitable method for removing all grains. The samples were then observed using S. E. M., as shown in Fig. 6. The distance between grains was from 30 μm to 40 μm.

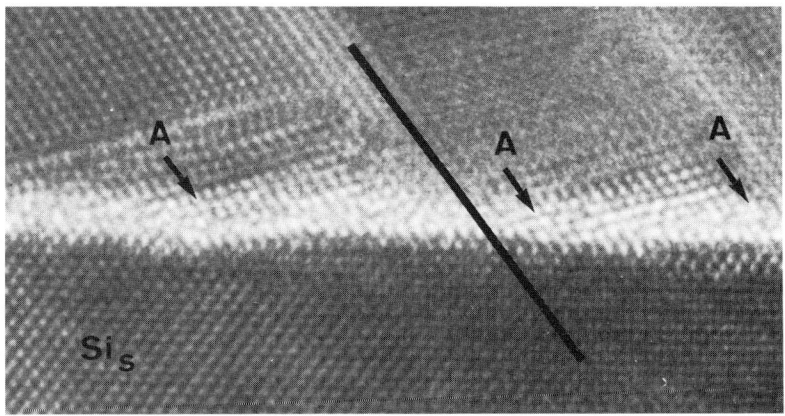

Fig. 5 : Area of interface between substrate and Si grain showing holes in amorphous layer which permit epitaxial growth of grain.

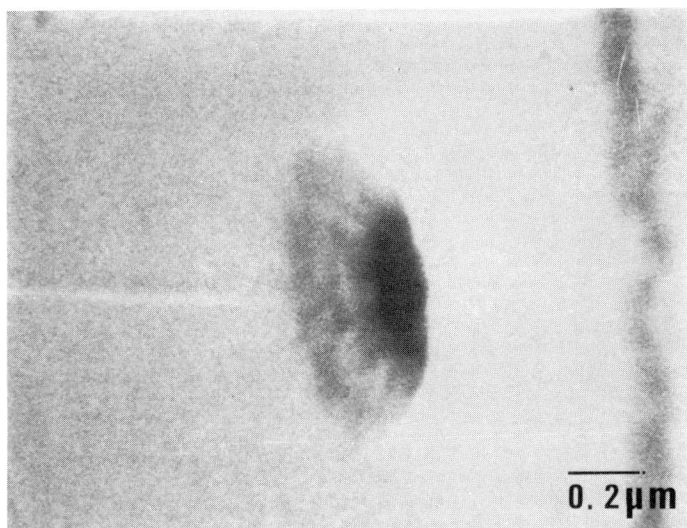

Fig. 6 : Image of a hole indicating the size of a Si precipitate in the Al film.

The last analysis of the Al/Si interface is done after the post metallization heat-treatment of the samples under following conditions : time : 1 hr, temperature : 450° C, one atmosphere. In Fig. 7 the new interface is shown. Its structure shows that there was diffusion of Si into the film. The interface is not flat as in the case without annealing (Fig. 2) and there are holes in the Si substrate (area A). In area B, one can see a moiré pattern with fringes separated by d = 16 Å. It might be interpreted as a double diffraction between one family of {220} Al planes and (002) Si planes parallel to the interface, the disorientation being about 1.5°. This suggests there could be an interdiffusion of Al and Si.

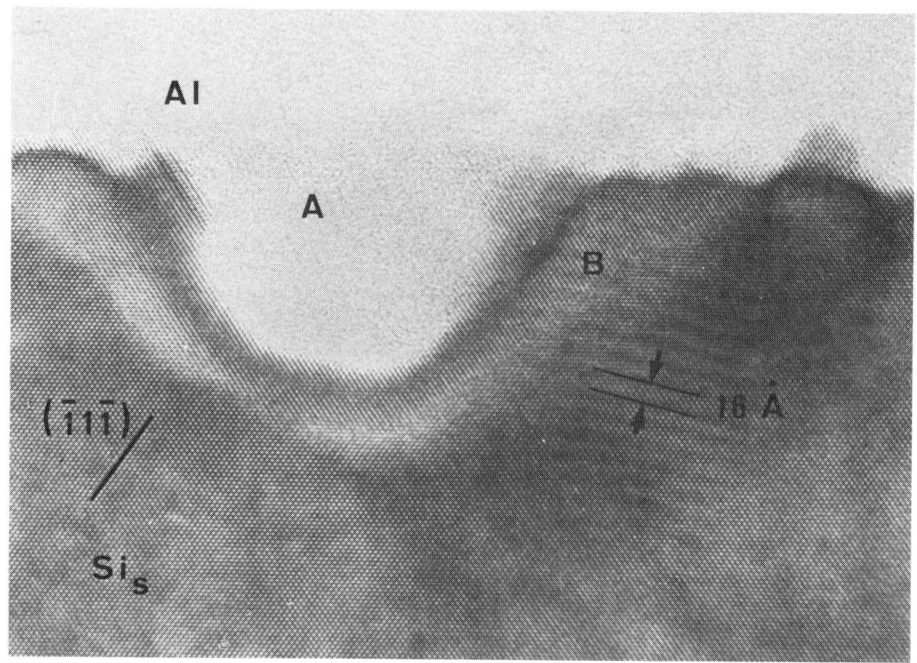

Fig. 7 : Transformation of the Al-Si interface after 1 hr at 450° C annealing. Holes show diffusion of Si of substrate in film; moiré pattern reveals presence of Al grain in substrate.

4. Discussion

We shall discuss the analysis of the texture of the Al-Si film, with special emphasis on the occurrence of the Si precipitates and their epitaxial growth.

Let us first consider the Al grains. Two preferential orientations have been observed : i) [111] Al parallel to (002) Si, ii) <220> Al//<400> Si.

i) The pile-up of {111} Al on (002) Si (Fig. 2) may be explained by the fact that the (111) is a dense plane of the f.c.c. structure. This would seem to be confirmed by their in-plane orientation which is randomly distributed. In some cases alignment of two directions has been observed : [110] Al// [110] Si and [1T1] Al//[002] Si (Fig. 2). The mismatch in the <110> direction is quite large : 25.4 %. However, an heteroepitaxial growth of $[110]$ Al on $[110]$ Si has been shown by M. ICHIKAWA et al. (1982) for ultra high vacuum (U. H. V.) deposition. The mismatch in the other direction is less marked : 16.12 %. So we may assume it would often occur if cleanest conditions for deposition are used.

ii) In addition there is <220> Al which is parallel to $[400]$ Si (Fig. 2) (mismatch less than 5.2 %). We assume that it corresponds to a possible heteroepitaxy direction. Although our Al deposition was not carried out in U. H. V., this alignment was observed for all our specimens. Therefore, the structure of Al film is linked to substrate in spite of the amorphous layer. This layer might result from the native oxide (H. R. PHILIPP, 1982).

R. N. THOMAS et al. (1969) have analysed the epitaxial growth of Si and established that homoepitaxial growth occurs at 350° C. Below this temperature amorphous structure was obtained. In our case there is an epitaxial growth of Si precipitates (Fig. 4) and the deposition temperature (Td)was about 350° C ; this result agrees with the previous one. This result can not be compared with Solid Phase Epitaxy (S. P. E.) experiments, as at the moment only high temperature annealings have been studied (typically 540° C, H. SANKUR et al., 1973).

The size of the precipitates could be explained by the diffusion of the Si coming either from the Al film, or from the substrate. Two facts seem to indicate that Si comes from Al film : their size which depends on Si concentration in Al, and their distribution.

First, let us consider the diffusion of Si in Al. The activation energy and the preexponential factor of this diffusion have been calculated by means of internal friction (M. OKABE et al., 1981) their values are 0.92 eV and 0.39 cm^2/sec. respectively. The diffusion coefficient at 650° K is 2.89 x 10^{-8} sec.$^{-1}$. As the deposition time is 4 sec., the diffusion length is 2 $\sqrt{Dt}$ = 6,8 μm. This implies that, during deposition, the density of 1 % Si in Al film may induce the creation of Si precipitates of about 0.2 μm which corresponds to the size of the observed precipitates. Their distribution is very dispersive : every 30 μm or 40 μm. Hence, the size and the distribution of the precipitates is explained by the diffusion of silicon coming from the film. So the film is composed by Si grains and Al grains having 0.35 % Si (solubility limit at 350° C).

We have demonstrated the presence of these areas poor in Si by observing the behaviour of the film during annealing. The heat treatment was 450° C for 1 hr. At this temperature the solubility limit of Si in Al is 0.48 %, so, if the film was uniformly doped with 1 % Si, silicon could not have diffused from the substrate into the film during annealing. The diffusion observed (Fig. 7) proved that the concentration of Si in numerous areas is less than 0.48 %.

Such a distribution of Si could be imputed to the structure of the amorphous layer between Al and Si. Indeed, if the thickness of this film was uniform, the Si precipitates would nucleate everywhere, as it has been proved by C. W. CLASS (1979) who had shown that the density of Si at room temperature into 2 μm thick sputtered films is more important near the substrate that near the top surface.

We have seen (Fig. 5) that there is   elongation of the Si planes of the substrate into the precipitates, proving that the amorphous layer is porous.   There are areas with a large density of pores which act as nuclei, and there are areas with an opaque amorphous film where the Si grains can not nucleate.   Such variations in the thickness of this film may be due to inhomogeneous etching or to sputtering effects (the energy of the sputtered particules would thin the wafer surface).

5. Conclusion

In this work it has been shown that the structure of the Al–Si film depends on the quality of the interface : substrate/deposited layer, and on the temperature of deposition.

If the thickness of the amorphous layer at the interface is uniform, the Si precipitates would be small and uniformly distributed inside the Al–Si film.   When the amorphous layer is very porous,   heteroepitaxy of Al and epitaxy of Si (for Td > 350° C) do occur. A preferential alignment of <220> Al and [400] Si has been shown. That has to be checked by using a U. H. V. technique for deposition. In addition, H. R. E. M. imaging allowed us to show a S. P. E. of Si during the cooling process due to the over–saturation of the Al film.   This has important implications for electronic device technology, since such precipitations can occur very quickly : only 4 sec. at 350° C suffice to generate large precipitates (0.2 µm). As demonstrated, such precipitation induces large areas poor in Si (30 µm^2). So, during post–metallization annealing there is always diffusion of Si from the substrate into the Al film, although we wanted to avoid this by depositing Al films oversaturated in Si. Therefore, we can conclude that there is a temperature range to be avoided during the deposition process : at the lower limit of this range, diffusion of Si is very low during the deposition time, and the upper limit is the post–metallization annealing temperature. In the later case, a fast quenching is needed after deposition.

Acknowledgements

The author would like to thank many of her colleagues at the CNET for stimulating discussions. In particular, thanks are due to D. BOIS, M. BRILLOUET and P. NORMANDON.

References

K. L. CHOPRA,   Thin   Film   Phenomena   : R. E. KRIEGER, Publishing Co. (1979) 25.
W. H. CLASS, Solid St. Techno. (1979) June 61.
M. ICHIKAWA and K. HAYKAWA, J. J. of Appl. Phys. 21 (1982) 154–163.
M. OKABE, T. MORI and T. MURA, Phil. Mag. A (1981) 44 1–12.
H. R. PHILIPP and E. A. TAFT, J. Appl. Phys. 53 (1982) 5224.
R. ROSENBERG,   M. J. SULLIVAN   and   J. K. HOWARD,   Thin   Films   Inter-diffusion and reactions, Ed. J. M. POATE, K. N. TU, J. W. MAYER (1978) 15–25.
H. SANKUR,   J. O. MC CALDIN   and   J. DEVANEY, Appl. Phys. Lett. 23 64 (1973).
R. N. THOMAS and M. H. FRANCOMBE, Solid St. Elec. (1969) 12 799.

Inst. Phys. Conf. Ser. No. 67: Section 2
Paper presented at Microsc. Semicond. Mater. Conf., Oxford, 21–23 March 1983

High resolution electron microscopy of II−VI compound semiconductors

R Sinclair[1], F A Ponce[2], T Yamashita[1] and David J Smith[3]

[1]Department of Materials Science and Engineering, Stanford University, Stanford, California, 94305, U.S.A.

[2]Hewlett Packard Laboratories, Palo Alto, California, 94304, U.S.A.

[3]High Resolution Electron Microscope, University of Cambridge, Free School Lane, Cambridge, CB2 3RQ, U.K.

Abstract   The compound semiconductors CdTe, ZnTe and ZnSe, as well as CdS grown on CdTe, have been studied by high-resolution electron microscopy at 100, 200 and 500kV.   A variety of extrinsic and intrinsic stacking faults were observed and dislocation Burgers vectors were determined directly from experimental images.

1. Introduction

Tetrahedrally-coordinated II−VI compounds are large band-gap semi-conductors which are useful for certain applications (e.g. photodiodes) and have the potential for primary technological importance. However, these materials exhibit high densities of defects, both electrical and structural.   In this paper we describe our recent studies of these lattice imperfections, using high resolution transmission electron microscopy (HREM) as the investigative tool.

The majority of II−VI semiconductors have the cubic zincblende (sphalerite) structure (e.g. CdTe, ZnTe,ZnSe) although some take the hexagonal equivalent, wurtzite (e.g. CdS,ZnO).   Their lattice constants are sufficiently large ($\sim$6Å) that satisfactory lattice images can be obtained at comparatively low levels of resolution (e.g. $\sim$3.5Å).   More structural information is available at higher resolutions and the character of an imperfection can then often be deduced from a single multibeam lattice image without recourse to a series of conventional diffraction contrast experiments. Ultimately, the atomic arrangements in the vicinity of the defect should be deduced by comparing experimental and computer-simulated images.

2. Experimental

Specimens of CdTe, ZnTe and ZnSe have been made from bulk single crystals by cutting slices parallel to a {110} plane, followed by mechanical thinning and finally by chemical polishing or ion-beam milling. The last procedure, which used Argon ions at 4-5kV incident at about 15° to the rotating sample, has generally provided the highest quality specimens.The CdS material used in this study has been in the form of a thin-film vapour-deposited on CdTe single crystal substrates.

A number of electron microscopes were used: a JEOL 200CX, operated at 200kV
with about 2.5Å resolution, a Phillips EM400 ST at 120kV with approximately
3.5Å resolution and the Cambridge University 600kV HREM (Smith et al, 1982)
with 2.0Å or better resolution. The last two instruments were equipped
with image pick-up and video-recording systems which were used to document
dynamic events which occurred during high resolution observation (Sinclair
et al, 1982).   Images were recorded in the <110> projection with at least
seven beams contributing to image formation.

3. Results and Discussion

3.1. Cadmium Telluride

The microstructure of CdTe is dominated by a variety of different stacking
faults. Whilst some may result from the dissociation of perfect dislocations, we believe that many arise during the crystal growth process,
which was by the Bridgemann technique.

High resolution images in the [110] projection reveal the stacking sequence
of the two sets of {111}-type planes, mutually inclined at 70.5° and
parallel to the electron beam. The image spots at medium resolution reveal
directly the arrangement of lattice points and, depending on the objective
lens focus setting, coincide either with the atomic column positions or
with the open channels in the structure. These situations can be distinguished by reference to the lattice image of a stacking fault (Ponce
et al, 1982).   More complex images are produced in through-focal series
and from thicker regions of the specimen (e.g. half-spacing fringes and
"dumbells") but it is extremely difficult to relate these features to the
separate atomic species (Yamashita et al, 1982).

The projected {111} stacking sequence allows the nature of the stacking
faults to be immediately identified. Similarly, a Burgers circuit drawn
around a dislocation determines the projected value of the Burgers vector,
from which the absolute value can usually be deduced. Examples of extrinsic stacking faults terminating at Shockley partial dislocations are
shown in Fig.1. A variety of stacking faults along {111} planes have been
observed, including single, double and multiple intrinsic stacking faults,
extrinsic stacking faults, and twins, as well as intersections of these
planar faults. The intrinsic stacking faults tend to have extensive
lengths (e.g. 0.1 - 10µm) whereas extrinsic faults are often quite short
(e.g. 100-500Å).   Extrinsic stacking faults terminating in a Frank
partial dislocation can be distinguished from those terminating in two
Shockley partials on adjacent {111} planes: the former are produced by an
additional {111} plane and are generally short whereas the latter are
associated only with a local lattice displacement and may be described as
a double intrinsic stacking fault.   An intrinsic stacking fault terminating in a Frank dislocation, equivalent to a vacancy loop, has only once
been seen.   Perfect dislocations are sometimes observed, which are generally dissociated into the traditional pair of Shockley partials separated
by an intrinsic stacking fault. The degree of dissociation can be used to
determine the stacking fault energy: a wider range of values has been
obtained compared to the data of Hall and Vander Sande (1978) from weak
beam microscopy but the average is similar to their value of 10mJm^{-2}.

During HREM observations, the defects can be induced to move by the local
heating effect of the electron beam (Sinclair et al, 1982; Yamashita and
Sinclair, 1983a).   The slip motion of individual Shockley partial dis-

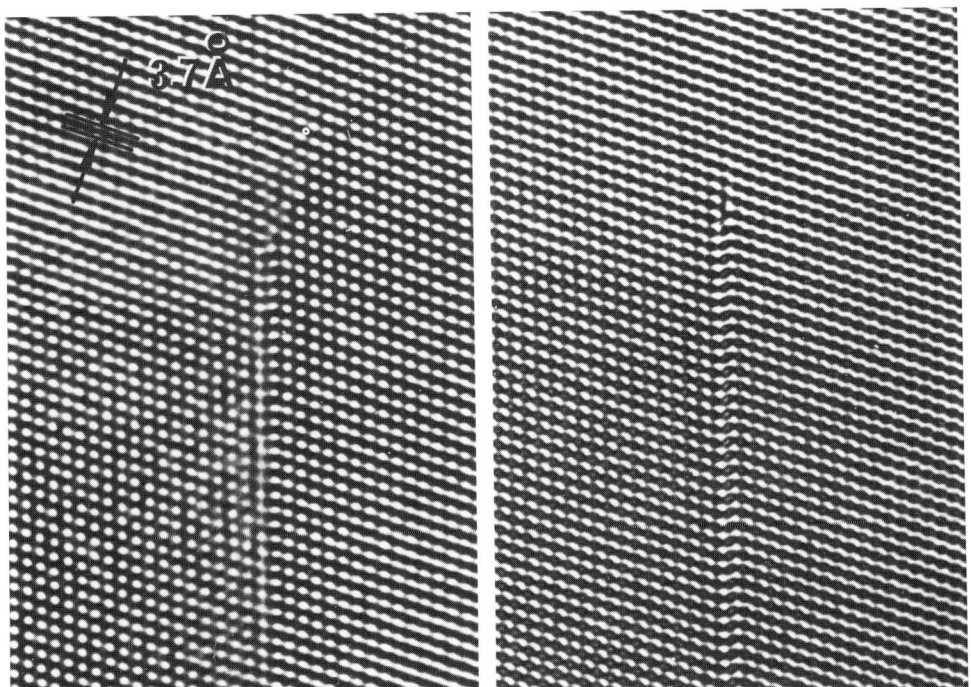

Fig.1 High-resolution lattice images of CdTe in (110) projection recorded at 500kV showing intrinsic stacking faults terminated by Shockley partial dislocations.

locations, the diffusional climb motion of Frank partial dislocations and a variety of dislocation reactions have been documented (Sinclair et al, 1982; Yamashita and Sinclair, 1983b). This capability permits the study of defect dynamics directly at lattice resolution thus providing fundamental information concerning annealing processes in CdTe, with obvious implications for the behaviour of tetrahedrally-coordinated solids in general.

3.2 Cadmium Sulphide

Cadmium sulphide-cadmium telluride heterojunctions have been fabricated under various conditions with the goal of producing efficient solar cells (Werthen, 1982) and we are currently correlating the electrical properties of the material with the structure of the interface. CdS itself contains a number of interesting defects and some of these have been noted elsewhere (Echigoya et al, 1982). Depending on the deposition conditions, CdS can be formed with the cubic (sphalerite) or hexagonal (wurtzite) structure, although the latter is the stable phase under equilibrium conditions. Cubic CdS can be produced in epitaxial orientation when deposited on a clean (cleaved) {110} CdTe surface. Numerous intrinsic stacking faults, each equivalent to a local wurtzite structure, occur on the {111} planes inclined to the interface; they are believed to relieve the strain associated with the lattice mismatch at the interface. When hexagonal CdS is formed on a {111} CdTe surface, stacking faults and partial dislocations are also present, but to a somewhat lesser degree than in the cubic phase.

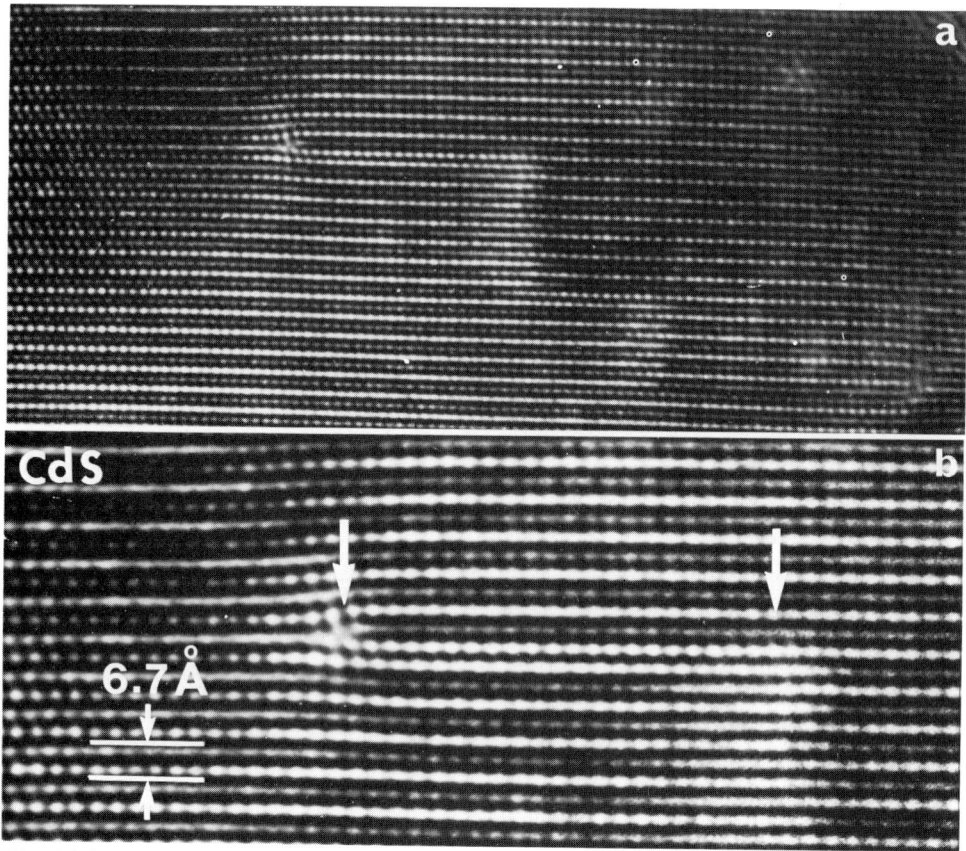

Fig.2(a) Region of hexagonal CdS in the [11$\bar{2}$0] projection
(b) Enlargement of (a). The terminating dislocation (arrowed) of the inserted layer has a Burgers vector of 1/6 [$\bar{2}$203]; the second arrow indicates the position of the 1/3 [1$\bar{1}$00] partial dislocation.

A typical example of a CdS defect is shown in Fig.2.   The usual hexagonal: cubic orientation relationship (i.e. hexagonal basal plane parallel to the cubic {111} plane) is achieved with a slight misorientation of about 2° which is necessary to accommodate the lattice misfit. Further details of this work will appear in due course.

3.3 Zinc telluride and zinc selenide.

Studies of zinc-based II-VI compounds have only just been initiated and, as with CdTe, numerous faults have been found on {111} planes. In ZnTe, intrinsic stacking faults terminating at Shockley partials have been identified as well as perfect dislocations of interstitial type (see, for example, Fig.3). It is interesting to note that no such defects have been observed by us in CdTe. However, the majority of faults remain as yet unidentified. There appears to be a preference for a particular {111}-type habit plane possibly associated with the growth process and fringe effects frequently occur which may indicate inclined faults or surface phenomena. The defects are usually quite short (e.g. ∿100Å), often lying on intersecting {111} planes. They may represent a clustering or pre-

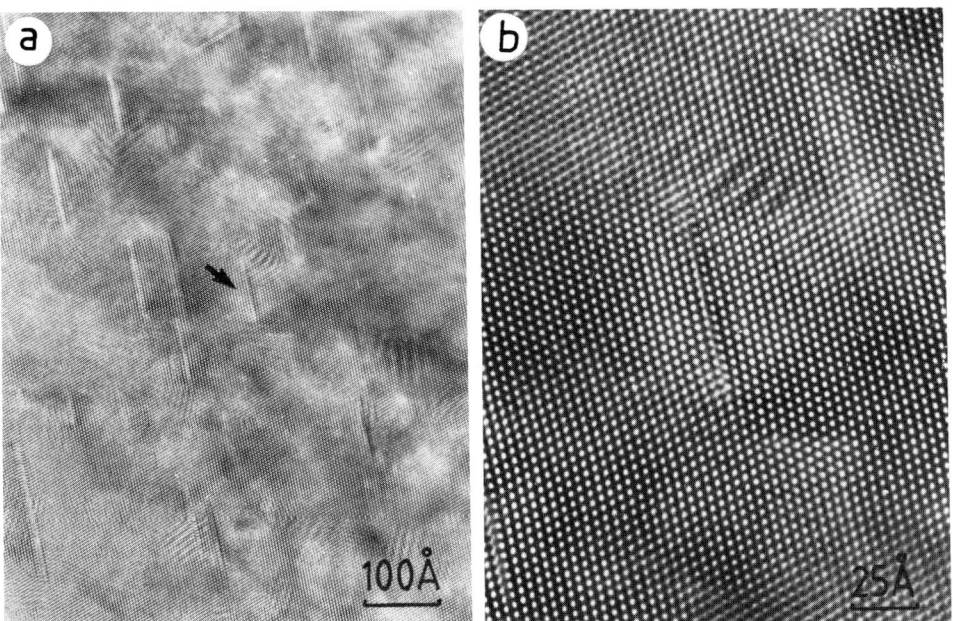

Fig.3 (a) Image of ZnTe recorded at 500kV showing typical short stacking defects. Note the central perfect dislocation loop of interstitial type shown enlarged in (b).

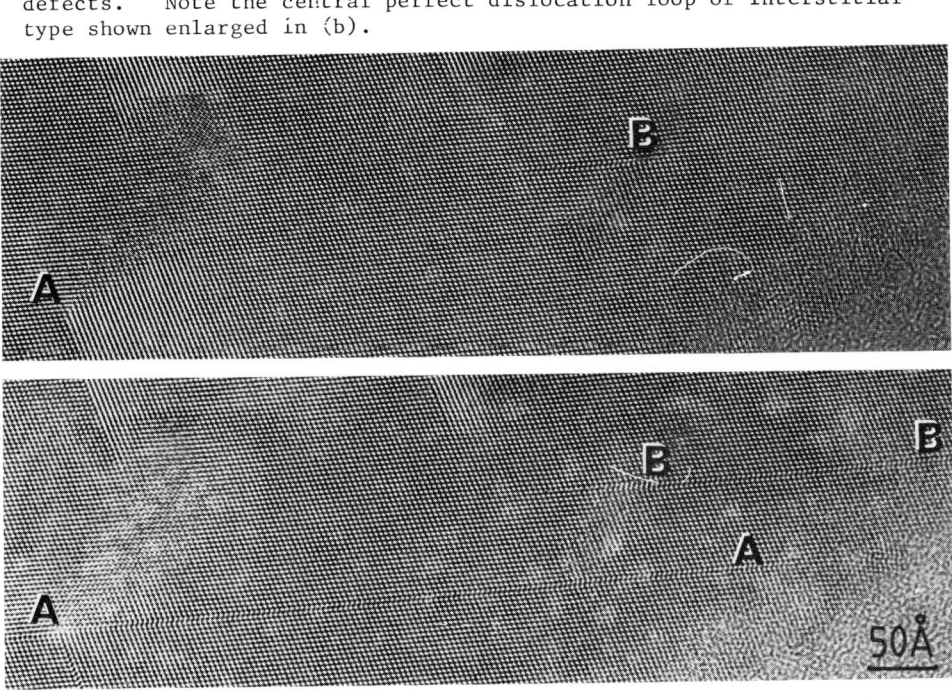

Fig.4. Appearance (A) and extension (B) of intrinsic stacking faults in ZnTe observed at 500kV. Time difference about 3 minutes.

cipitation process rather than being a lattice imperfection inherent to the stoichiometric, pure compound. More extensive analysis and further observations are required to characterise these microstructures fully. Finally, note that defect modification under the influence of the electron beam is also observed. In the particular case shown in Figs.4(a) and (b) the intrinsic stacking faults have been <u>extended</u> by movement of the terminating partials to the surface of the crystal.

4. Conclusions

High resolution electron microscopy has been applied to the study of defects in II-VI compound semiconductors. Various faults have been identified, predominantly stacking faults lying on {111} planes (cubic materials) or {0001} planes (hexagonal materials). Dislocation Burgers vectors have also been established directly from the experimental images and are of the type $\frac{1}{2}$<110> (normally dissociated), $\frac{1}{6}$<211> (Shockley partial) and $\frac{1}{3}$<111> (Frank partial) and their equivalents in the hexagonal phase. Various dislocation reactions have both been characterised and observed to occur during dynamic video-tape recording.

5. Acknowledgements

Financial support is acknowledged from the Basic Energy Sciences Division of the U.S. Department of Energy (RS,TY) and the U.K. S.E.R.C. (DJS).

References

Echigoya J, Pirouz P and Edington J W 1982 Phil. Mag. A45 455.
Hall E L and Vander Sande J B 1978 Phil. Mag. 37 137
Ponce F A, Yamashita T and Sinclair R 1981. Defects in Semiconductors
    eds. J Narayan and T Tan (New York: Elsevier) pp. 503-06
Sinclair R, Ponce F A, Yamashita T, Smith D J, Camps R A, Freeman L A,
    Erasmus S J, Nixon W C, Smith K C A and Catto C J D 1982 Nature 298 127
Smith D J, Camps R A, Cosslett V E, Freeman L A, Saxton W O, Nixon W C,
    Ahmed H, Catto C J D, Cleaver J R A, Smith K C A and Timbs A E, 1982
    Ultramicroscopy 9 203
Werthen J E 1982 Ph.D. thesis Stanford University
Yamashita T, Ponce F A, Pirouz P and Sinclair R, 1982 Phil. Mag. A45 693
Yamashita T and Sinclair R 1983a Proc. Mats. Res. Conf., in press.
Yamashita T and Sinclair R 1983b. Proc. 41st. Ann. Meet. EMSA, in press.

Inst. Phys. Conf. Ser. No. 67: Section 2
Paper presented at Microsc. Semicond. Mater. Conf., Oxford, 21–23 March 1983

Field ion microscopy and atom probe microanalysis of semiconductor materials

C R M Grovenor*, A Cerezo and G D W Smith

Department of Metallurgy and Science of Materials, University of Oxford, Parks Road, Oxford, OX1 3PH, England.

* Present address: IBM Inc, Thomas J Watson Research Centre, Yorktown Heights, NY 10598, USA.

Abstract   Improved methods are described for the preparation of field ion microscope specimens from single-crystal silicon. High-resolution mass spectra of silicon have been obtained for the first time using the pulsed laser atom probe technique. The growth of oxide films has been examined, and controlled field evaporation has been used to observe the interface between the oxide and the silicon substrate. Preliminary experiments on the deposition of metals (Pd and Ni) on silicon have been carried out. Atom probe spectra have also been obtained for the first time from CdTe.

1. Introduction

The techniques of field ion microscopy (FIM) and atom probe microanalysis (AP) have, until recently, been applied mainly to the study of metals and alloys. The combination of atomic-resolution imaging and single-atom sensitivity chemical analysis facilities has permitted many advances to be made, for example in the understanding of crystal defects, alloy chemistry, segregation processes, ion implantation and thin film formation (Muller and Tsong, 1969; Bowkett and Smith, 1970; Wagner 1982). Studies of semiconductor materials have been much slower to develop. The first FIM images of silicon were obtained by Muller (Muller, 1957). Shortly afterwards, more systematic studies of silicon and germanium were carried out by Allen (1961) and Arthur (1964). Overall image resolution was poor, with only major crystallographic poles being developed in the specimen end-forms. Further progress was made following the introduction of microchannel plate image intensification techniques, and somewhat improved micrographs have now been obtained from silicon (Melmed and Stein, 1975), germanium (Ernst, 1972; Ernst and Block, 1975), and GaAs and GaP (Ohno et al, 1977). Resolution of high index net planes is still not possible, however, and this difficulty now appears to be a fundamental one, associated with the properties of semiconductor surfaces. Extensive atomic rearrangement takes place as soon as high index orientations are exposed, due to the presence of high densities of unsaturated chemical bonds. It is notable that image quality is improved by the addition of a small quantity of hydrogen to the image gas; this is probably due to the adsorption of hydrogen at sites where unsaturated bonds are present, thus inhibiting the clean surface reconstruction process.

Atom probe microanalysis of semiconductors also proved difficult to carry out, initially. The response of the specimens to the nanosecond rise-time, high voltage pulses used to produce field evaporation was poor, especially in the case of materials of high resistivity. Only limited data were obtained from studies of silicon (Melmed et al, 1981a, b), GaAs (Ohno et al, 1978; Tsong et al, 1978) and GaP (Ohno et al, 1978; Yamamoto et al, 1982). However, a major advance has occurred with the introduction of the pulsed laser atom probe (PLAP) by Kellogg and Tsong (1980). With this instrument, the need for the application of high voltage pulses to the specimens is eliminated, and field evaporation is produced by a nanosecond laser pulse superimposed on the DC imaging voltage. This should permit the routine of a wide range of materials of limited electrical conductivity, including all the commonly used semiconductor materials (Kellogg, 1982a, b, 1983).

In this paper, we report the results of a number of exploratory studies of semiconductor materials, using both conventional voltage-pulsed AP and PLAP techniques. We demonstrate the improvement in mass spectrum resolution which is obtained using the PLAP, and show the applicability of the AP to the study of oxide film formation, and the deposition of metals onto silicon.

2. Experimental

Two main instruments were used in the course of this work. The first was an analogue output, low mass resolution, imaging AP, of the kind described by Panitz (1978). This had a large capture angle, and was ideally suited to the analysis of thin films. The second instrument was a computer-controlled, high resolution AP, with facilities for either voltage-pulsed or laser-pulsed operation. This instrument is shown schematically in Fig. 1. Further details are given by Miller et al (1979) and Smith et al (1982).

The main specimen material studied was silicon. Both n type phosphorous doped (6 x 10^{15} m^{-3}) and p type boron doped (2 x 10^{18} m^{-3}) samples were prepared from single crystals supplied by Wacker GmbH. Specimens were made by sawing blanks of dimensions 15 by 0.8 by 0.8mm from the bulk single crystals using an abrasive wire saw. These were first electropolished, and then chemically polished, in a 5:1 HNO$_3$/HF solution to give sharp FIM tips. The chemical polish tended to leave an amorphous oxide film on the surface of the specimens. This was removed if necessary by ion milling in a commercial Ion Beam Thinning unit. Ion milling was also successfully used to re-sharpen specimens after FIM imaging.

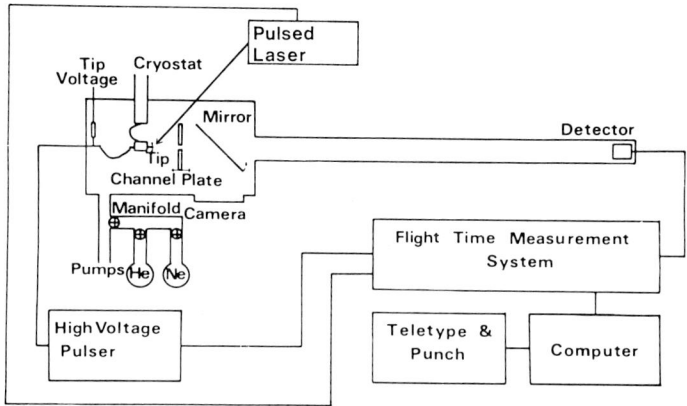

Fig. 1. Atom probe equipped for laser pulse evaporation of specimens.

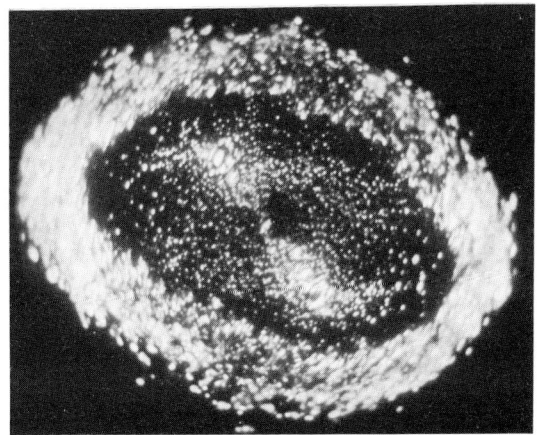

Fig. 2. FIM image of silicon at 25K, recorded using Ne-10% H_2 (20kV).

Fig. 3. Crystal projection corresponding to Fig. 2.

3. Results

A FIM images of a boron-doped silicon specimen is shown in Fig. 2, with a corresponding projection in Fig. 3. It can be seen that only the major crystallographic poles are resolved in the image. The edges of the specimen are covered by an amorphous oxide film, left after the chemical ·polishing of the FIM tip. There is a marked difference between the imaging behaviour of the oxide film and the silicon substrate, and the film-substrate interface can be clearly seen. The interface appears to be relatively smooth, at least to the ⁓1nm level observable. Video recordings were made, showing the progressive removal of oxide films. A sequence of still pictures from one such recording is shown in Fig. 4. In this case, the oxide film was very thin (⁓0.5nm), having been formed by exposing a previously field evaporated tip to a residual gas pressure of 10^{-5} torr for approximately one hour at room temperature. The removal of oxide begins at the centre of the imaging region, and progresses out towards the edges as the voltage is raised.

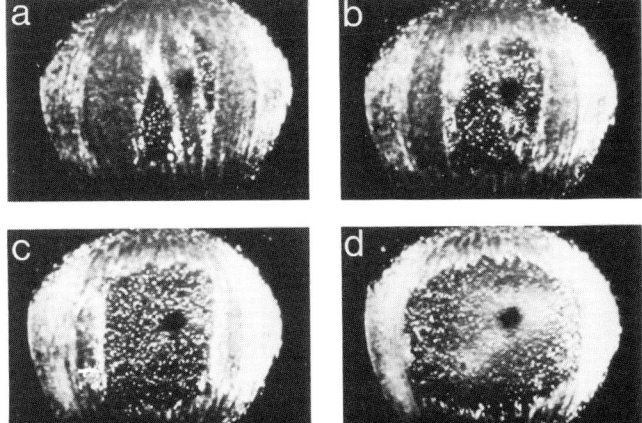

Fig. 4a-d. Field evaporation sequence showing the removal of a thin amorphous oxide layer from a silicon FIM specimen. (Recorded from videotape).

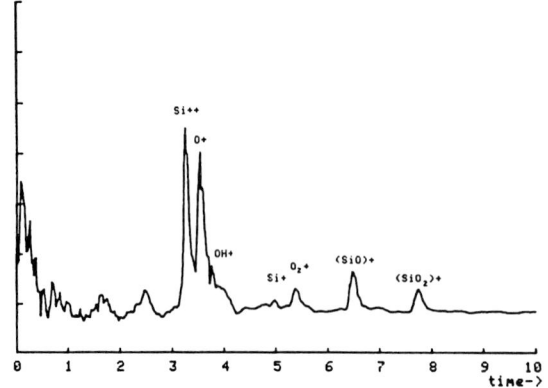

Fig. 5. Atom probe mass
spectrum obtained from
a specimen similar to
that of Fig. 4.

Atom probe mass spectra were taken from oxidised silicon specimens. The
imaging atom probe was employed, in order to maximise the number of ions collected from the surface layers. By using heavily boron-doped specimens,
enough conductivity was obtained to permit voltage-pulsed field evaporation
to be used. A typical mass spectrum from a very thin surface film is shown
in fig. 5. The peaks due to Si^{++}, O^+, Si^+, O_2^+, SiO^+, and SiO_2^+ are well
resolved. It may be noted that the total intensities of the Si and O peaks
are in the approximate ratio 1:1. Further work is required to establish
whether this corresponds to a definite sub-oxide composition.

A direct comparison between voltage-pulsed and laser-pulsed atom probe
spectra of phosphorous-doped silicon is shown in Fig. 6. Both spectra were
obtained in the high resolution atom probe of Fig. 1, and both were recorded
under similar experimental conditions. It is clear that the laser spectrum
shows greatly improved mass resolution; the minor isotopes of silicon are
resolved for the first time (Grovenor and Smith, 1982). There is also a
reduction in the level of hydride formation. It seems that the laser pulse
may induce thermal desorption of hydrogen from the shank of the specimen,
preventing it from reaching the apex region.

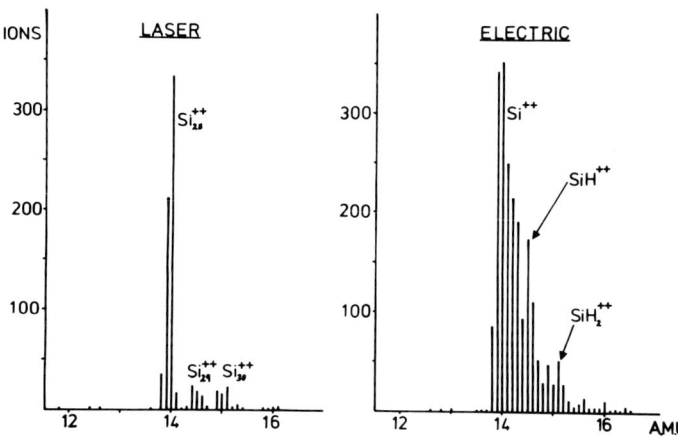

Fig. 6. Direct
comparison
between laser-
and voltage-
pulsed atom
probe mass
spectra of
silicon.

(Courtesy
Surface Science)

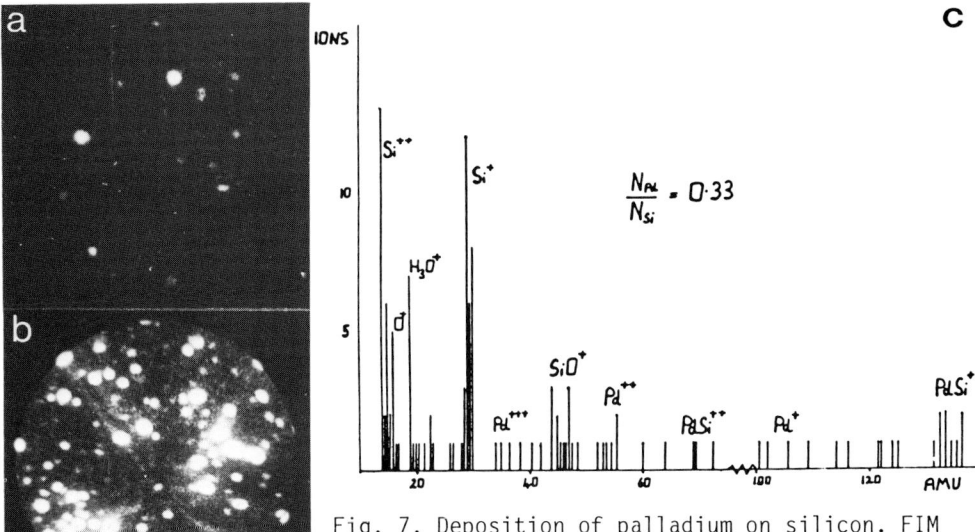

Fig. 7. Deposition of palladium on silicon. FIM images, (a) 6.6kV and (b) 9.7kV, show brightly imaging Pd atoms. A PLAP Mass spectrum (c) shows Pd, Si, PdSi and SiO peaks.

Preliminary experiments have also been carried out on the deposition of metals onto silicon FIM specimens. Both Pd and Ni have been deposited, under relatively poor vacuum conditions (10^{-5} torr). Under these conditions, adhesion of the metal layers to the substrate has proved to be poor, and in many cases the layers have desorbed catastrophically as the tip voltage was raised. A successful example of Pd deposition on a silicon specimen held at room temperature is illustrated in Fig. 7. The Pd atoms image brightly, against much fainter background of the silicon substrate, Fig. 7a, b. Mass spectra from such layers show Pd, Si and PdSi peaks, as well as oxide and water vapour peaks from the contaminated substrate surface. Nickel layers have also been studied, and show evidence of Ni_2Si formation after annealing at 350°C for 15 hours (Grovenor and Smith, 1983).

Other semiconductor materials have also been studied successfully in the atom probe. Fig. 8 shows a mass spectrum obtained from a CdTe specimen, using voltage-pulsed field evaporation, with the tip held at room temperature. Both Cd and Te have several isotopes, and hence the spectrum has a rather complex appearance, but all the major peaks can be readily identified from the spectrum. Further studies of 3:5 and 2:6 compound semiconductors are currently being planned.

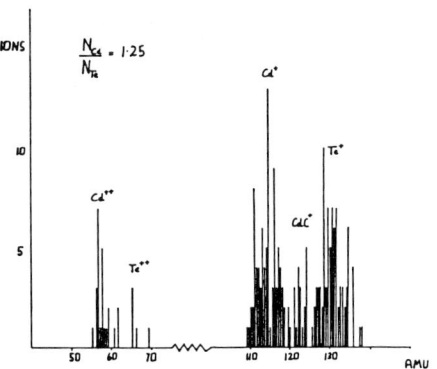

Fig. 8. Atom probe spectrum of CdTe.

4. Discussion

The results presented in this paper give some indication of the scope for
FIM and atom probe studies of semiconductor materials. FIM image quality is
limited, but features such as thin surface films can be studied with relative ease. Imaging of crystal defects may prove difficult, but dislocations
should be identifiable if they intersect the surface in low-index regions
such as (111) or (110). The prospects for atom probe microanalysis of semiconductors have been greatly improved by the new laser pulse desorption
technique. It is now possible to envisage atomic resolution studies of com-
position-depth profiles across metal-semiconductor and oxide-semiconductor
interfaces. These should be of value for the understanding of the properties
of ohmic contacts and Schottky barriers. Ion implantation and laser annealing processes should also be amenable to study. An essential requirement
for these studies is an atom probe with an ultra-high vacuum work chamber
attached, so that specimens can be prepared with a minimum surface contamination. We are currently developing such a system, which should be in
operation by the end of 1983. The potential applications for this instrument seem very exciting indeed.

References

Allen F G 1961 J. Phys. Chem. Solids 19 87
Arthur J R 1964 J. Phys. Chem. Solids 25 583
Bowkett K M and Smith D A 1970 Field Ion Microscopy (Amsterdam: N Holland)
Ernst L 1972 Surface Sci. 32 387
Ernst L and Block J H 1975 Surface Sci. 49 293
Grovenor C R M and Smith G D W 1982 Surface Sci. 123 L686
Grovenor C R M and Smith G D W 1983 in Defects in Semiconductors, eds.
  Mahajan S and Corbett J W (Materials Research Soc. USA) in press
Kellogg G L and Tsong T T 1980 J. Appl. Phys. 51 1184
Kellogg G L 1982a Appl. Surface Sci. 11/12 186
Kellogg G L 1982b J. Appl. Phys. 53 6383
Kellogg G L 1983 Surface Sci. in press
Melmed A J and Stein R J 1975 Surface Sci. 49 645
Melmed A J, Sakurai T, Kuk Y and Givargizov E I 1981a Surface Sci. 103 L139
Melmed A J, Martinka M, Girvin S M, Sakurai T and Kuk Y 1981b Appl. Phys.
  Lett. 39 416
Miller M K, Beaven P A and Smith G D W 1979 Surf. Interfac. Anal. 1 149
Muller E W 1957 Field Emission Symposium Pennsylvania State University,
  USA (unpublished discussion)
Muller E W and Tsong T T 1969 Field Ion Microscopy (New York: Elsevier)
Ohno Y, Nakamura S, Adachi T and Kuroda T 1977 Surface Sci. 69 521
Ohno Y, Kuroda T and Nakamura S 1978 Surface Sci. 75 689
Panitz J A 1978 Progress Surface Sci. 8 219
Smith G D W, Grovenor C R M, Delargy K M, Godfrey T J and McCabe A R 1982
  Proc. 29th Intl. Field Emission Symposium, eds. Andren H-O and Norden H
  (Stockholm: Almqvist and Wiksell) p.283
Tsong T T Ng Y S and Melmed A J 1978 Surface Sci. 77 L187
Wagner R 1982 Field Ion Microscopy (Berlin: Springer-Verlag)
Yamamoto M, Seidman D N and Nakamura S 1982 Surface Sci, 118 555

Acknowledgements

Financial support for this work was provided by the Paul Instrument Fund of
the Royal Society, SERC, and VG Scientific Ltd. CRMG wishes to thank IBM
(UK) Ltd for the award of a research fellowship. Mr M G Cutler assisted
with the studies of metal deposition.

Inst. Phys. Conf. Ser. No. 67: Section 3
Paper presented at Microsc. Semicond. Mater. Conf., Oxford, 21–23 March 1983

115

Electron microscopy and defect/impurity spectroscopy of CW beam-annealed semiconductors

J L Merz, M Mizuta[*] and N H Sheng[†]

Department of Electrical and Computer Engineering, University of California, Santa Barbara, CA, U.S.A. 93106

Abstract   CW laser and electron-beam annealing of implanted Si is reported. Results are described for three different but mutually complementary minority-carrier and point-defect characterization techniques: Electron-Beam-Induced Current, low temperature Photoluminescence, and Deep Level Transient Spectroscopy. It is found that laser-induced defects are easily produced at moderate and high laser power, and that the dominant deep trap observable by DLTS is interstitial Fe. Fewer defects and better minority-carrier behavior are obtained with electron-beam annealing than with laser annealing.

1. Introduction

CW beam annealing of the damage produced by ion implantation into semi-conductors has been extensively investigated in a number of laboratories because of the great technological advantages for device fabrication offered by this technique. Measurements of electrical properties and crystal quality have shown that annealing of implanted Si with a scanned CW Ar$^+$-ion laser produces good-quality, single-crystal material with high electrical activation (for example, cf. Gat et al. 1978, and Hess et al. 1979). Characterization techniques which are sensitive to minority-carrier effects and/or point defects, on the other hand, indicate that electrically and optically active defects are present in the laser-annealed material which can dominate the minority-carrier lifetime, and hence seriously degrade the performance of certain types of devices.

In this paper, results are described for three different but mutually complementary minority-carrier and point-defect characterization techniques: Electron-Beam-Induced Current (EBIC), low temperature Photoluminescence (PL), and Deep Level Transient Spectroscopy (DLTS). The results of extensive investigations of beam-annealed Si carried out in the author's laboratory using all three techniques are reviewed (Mizuta et al. 1980, 1981a,b; Sheng et al. 1981, 1982a,b, 1983). Electron microscopy is pivotal to these studies; however, a coherent picture of the nature of the defects induced by the laser emerges only when all three techniques are utilized.

Initial EBIC measurements show that the "window" for successful laser annealing (LA) is quite small; low charge-collection efficiency usually

*   Present address: Tokyo Institute of Technology, Tokyo, Japan.
†   Present address: Rockwell Intl. Science Center, Thousand Oaks, CA.

results, with a high-contrast pattern of light and dark lines running parallel to the laser scan. At low laser power, incomplete recrystallization of the amorphous implanted region occurs. At high laser power, laser-induced damage extending into the bulk of the wafer is observed. These measurements are complemented by PL investigations of luminescence from both the implanted As and the as-grown dopant B. The trapping energy of the dominant laser-induced defects is determined by DLTS. By comparison with detailed investigations in thermally-quenched, unimplanted material, these defects can be identified as interstitial Fe and Fe-B pairs.

Another role played by electron microscopy in this work is that of scanned electron-beam annealing (SEBA). It was found that SEBA was usually superior to LA with respect to lateral and vertical uniformity. Quite different carrier traps are produced by SEBA, and their concentration is considerably lower than those observed after LA. Because of the greater penetration of the 30 kV electrons used for SEBA than for the visible light used for LA, thermal gradients should be reduced in the SEBA case. These results suggest that other, more-penetrating sources may produce better results; an example of this is the use of incoherent lamps and strip heaters, whose use for annealing damage in compound semiconductors is currently receiving wide attention.

The experimental procedures used here are described in detail in the papers by Mizuta and Sheng and collaborators mentioned above. In brief, p-type Si (B doped) was implanted at 100 keV with doses of 5×10^{14}-2×10^{16} As^+ ions/cm^2. A raster-scanned Ar^+-ion laser with multiline output was used for LA. For SEBA, an ETEC SEM was modified for high-current by removing the final stage aperture from the beam column. Annealed samples were then characterized by EBIC, low temperature PL, and DLTS. In addition, each sample was used as a simple photo-detector, and the resulting short-circuit photocurrent, I(photo), was used as a measure of the minority-carrier collection efficiency.

2.   EBIC Evaluation of Laser-Annealed Si

A typical EBIC result of an As^+-implanted, laser-annealed sample is shown in Fig. 1. In the upper and lower portions of this figure, the EBIC mode has been switched off so that a standard secondary-electron micrograph results. The laser-annealed sample appears to be smooth and defect-free in this mode, as anticipated. However, with EBIC on (center of Fig. 1), it is clear that laser annealing has produced quite   non-uniform   minority-carrier effects: high contrast dark lines appear which are parallel to the direction of laser scan, with spacing of the order of the laser scan step. Most laser-annealed samples show such dark-line contrast. Optimal results were obtained for a large-

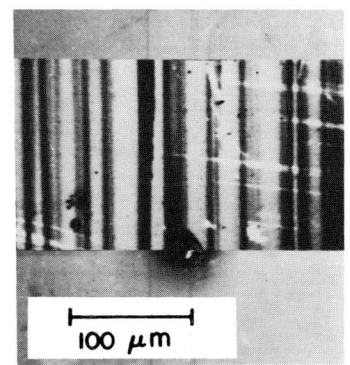

100 μm

Fig. 1(a).   EBIC micrograph (center) and secondary electron emission image (top and bottom) excited by 5-keV beam. The sample was implanted with 6×10^{14} As^+ ions/cm^2 and annealed with a 40-μm spot and 30-μm step. In the EBIC mode, dark regions correspond to low charge collection. Substrate temperature T_S = 250°C.

diameter laser beam (100μm), small x-direction scan step (6μm) and slow
scan speed (6 cm/sec), but some dark-line contrast was always observable.

To make these results more quantitative, the dependence of photocurrent
and sheet resistivity on laser power are shown in Fig. 2. The resistivity
decreases rapidly with laser power, reaching the value $\rho = 45$ $\Omega/\square$, and
then remains constant; this corresponds to 100% activation of the
implanted dose. However, even though complete activation is achieved, the
photocurrent falls off for laser power exceeding ~0.75 P_M, where P_M is the
laser power required to melt the Si. As the laser power approaches $\cong 0.9$
P_M, a dramatic decrease in I(photo) is seen, and a sharply cross-hatched
EBIC image is observed. This indicates the formation of slip dislocations, presumably due to the thermal gradient-induced stress at this high
laser power.

Thus, the true "window" for optimum laser annealing, as determined by
minority-carrier charge collection, is quite narrow, corresponding to
~70-75% of P_M. At lower P, recrystallization is incomplete. At higher P,
the charge-collection efficiency drops off rapidly. A very interesting
range of laser power occurs in the range 0.75 $P_M \leqq P \leqq 0.85$ P_M for which
stripe-like damage is induced but not slip dislocations. This is direct
evidence of laser-induced damage, and this damage is clearly observed in
the EBIC images at high accelerating voltage, suggesting that the damage
is deep.

The EBIC collection efficiency as a function of annealing laser power has
been measured using the method of Kimerling et al. (1977) for different
accelerating beam voltages, as shown in Fig. 3. For 20 and 30 kV accelerating voltages, the EBIC collection efficiency decreases at high laser

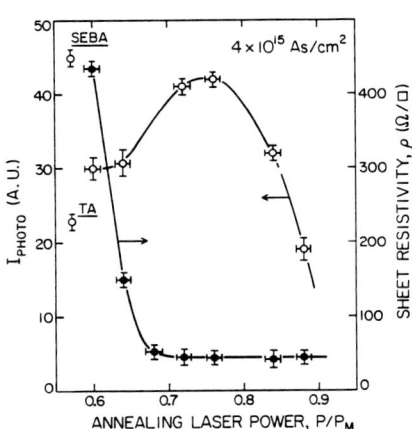

Fig. 2. Dependence of short-
circuit photocurrent and sheet
resistivity on annealing laser
power P for 4×10^{15} As ions/cm^2.
Laser power is normalized by the
melting power P_M. $T_S = 250°C$.
Also shown are electron-beam-
annealed (SEBA) and thermally-
annealed (TA) samples.

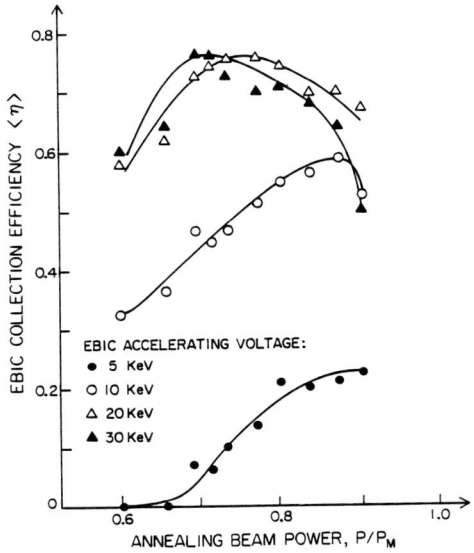

Fig. 3. Dependence of EBIC collection
efficiency $\langle\eta\rangle$ on laser-annealing
power for different SEM accelerating
voltages for 4×10^{15} cm^{-2} As ions/cm^2.

power, and the corresponding contrast of dark lines in the EBIC display becomes more obvious. On the other hand, for low accelerating voltage (5, 10 kV) the EBIC collection efficiency increases with laser power. The recyrstallization is more complete near the surface at high laser power due to the faster solid-phase epitaxial growth rate. Thus, the material near the surface appears to improve relative to the material deeper into the substrate as the annealing laser power is increased. Although it is difficult to extract accurate quantitative defect depth information from such experiments, an analysis of the EBIC efficiency vs. accelerating voltage strongly suggests that the laser-induced damage may extend considerably deeper than the implanted amorphous layer.

Figure 4 is a quantitative EBIC comparison of laser annealing done at two different values of laser power, optimal and high. (The SEBA results in this figure will be discussed in Section 5.) The crossing of the two laser-anneal curves suggests that there is a compromise between good surface recrystallization and laser-induced damage beneath the surface for optimal laser annealing.

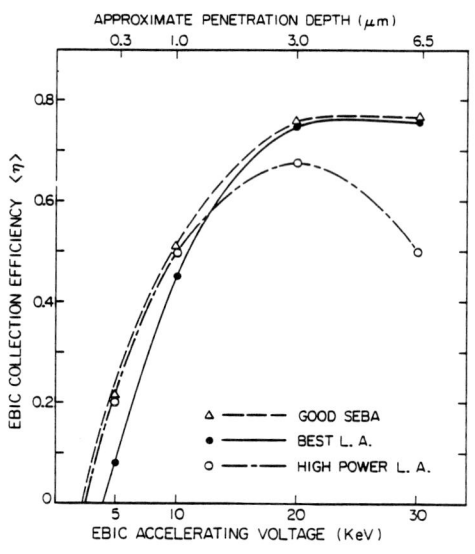

Fig. 4. Comparison of EBIC collection efficiency $\langle\eta\rangle$ as a function of accelerating voltage for SEBA, optimal LA ($P/P_M = 0.75$) and high power LA ($P/P_M = 0.9$).

3. Luminescence Evaluation of Laser-Annealed Si

Low-temperature PL has also been used to investigate minority-carrier recombination processes in low-dose implanted and laser-annealed Si. A typical luminescence spectrum obtained for a P-implanted and laser-annealed sample is shown in Fig. 5. The features of this spectrum are well known (Dean et al. 1967a,b; Kosai and Gershenzon 1974); one sees the recombination of excitons bound to neutral B and P, with no-phonon lines (designated B_{NP} and P_{NP}), bound multi-exciton complexes ($B_{NP}(b_i)$), and momentum-conserving phonon replicas (B_{LO}, B_{TO}, etc.).

Figure 6 shows the luminescence intensity of the P_{NP} line as a function of the annealing laser power. No luminescence signal is observed below 0.6 P_M. However, a sharp increase is observed between 0.6 and 0.7 P_M, which clearly shows activation of the implanted species. After reaching maximum intensity, the luminescence decreases with increasing laser power despite the fact that there is no change in the degree of electrical activation. This decrease in intensity must therefore result from a deterioration of the minority carrier lifetime and/or creation of non-radiative defects. A similar dependence on laser power was observed for the case of B luminescence, which implies that the laser-induced defects extend deep into the substrate, since most of the optically active B is in the substrate below the implanted layer. The behavior of the luminescence is nearly identical with the photocurrent (cf. Fig. 2). An important difference, however, is

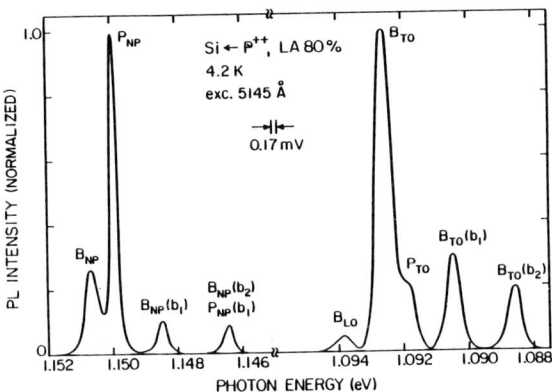

Fig. 5   Low temperature photoluminescence spectrum for P^{++} implants into B-doped Si, followed by laser anneal at $P = 0.8\,P_M$.

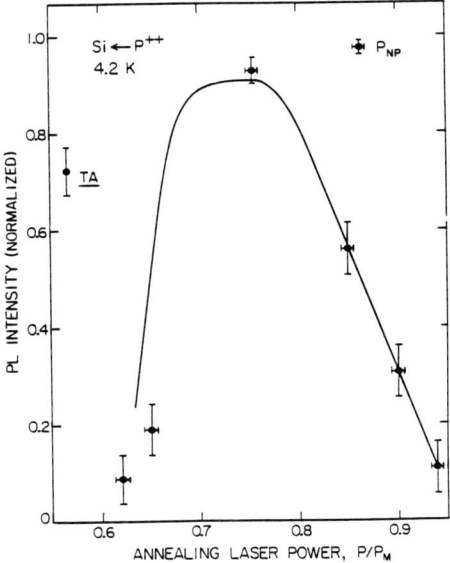

Fig. 6   Low temperature photoluminescence intensity as a function of annealing laser power for Phosphorus no-phonon line (P_{NP}). The intensity of these luminescence lines for thermally-annealed material is indicated by TA.

in the dose of the implanted species. The resistivity and photoconductivity measurements of Fig. 2 were obtained for implant doses of 4×10^{15} As ions/cm^2 (corresponding to a concentration exceeding 10^{19} cm^{-3}); the luminescence results, on the other hand, were obtained for low-dose P^{++} implants (concentration ~10^{16} cm^{-3}), below the dose necessary to make the Si amorphous. The similarity of these results suggests that the implant dose does not have a strong influence on the laser power required for the activation of implanted impurities.

4. Deep Levels in CW Beam-Annealed Si

We have already seen that the EBIC efficiency (or equivalently, the photocurrent) of a CW laser-annealed sample reaches a maximum at P~0.75 P$_M$, whereas the sample resistivity is reduced to a minimum corresponding to 100% electrical activation, and then stays constant (cf. Fig. 2). This suggests that some form of defect or deep level is induced by the laser at higher values of the annealing laser beam. Some sort of deep-level measurement is clearly called for. DLTS as developed by Lang (1974) is perhaps the most popular of the capacitance methods used for the study of electrically-active defects in semiconductors. Capacitance spectroscopy is complementary to luminescence in that it is particularly good for the study of deep levels, whether they are radiative or not; shallow levels, on the other hand, (less than 0.1 eV from the band edges) are easily detected by photoluminescence but difficult to see with DLTS. Transient capacitance spectroscopy can distinguish between majority- and minority-carrier traps, and can monitor the kinetics of defect reactions.

Figure 7 shows a typical DLTS spectrum taken immediately after the samples were beam-annealed and processed. For CW laser annealing, a dominant hole trap, H(0.45), with energy level close to mid-bandgap, about 0.45 eV above the valence band, has been observed in slip-free samples. The concentration of this center increases rapidly with increasing laser power (compare Fig. 7a and b). When the laser power is sufficiently high to produce slip dislocations, the dominant defect shifts to lower energy, 0.43eV (Fig. 7c).

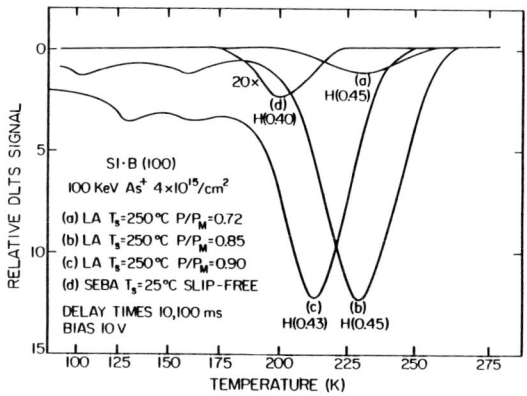

Fig. 7   Capacitance transient spectrum of p-type Si implanted with As$^+$ and annealed by either a CW Ar$^+$ laser (LA) using three different values of laser power, or annealed with a scanning electron beam (SEBA). T$_s$ is the substrate temperature during beam annealing.

In Fig. 8, the defect concentration is plotted as a function of beam power. For laser-annealed samples, the defect concentration increases more than one order of magnitude with increasing laser power until, for high-power laser-annealed samples, the defect concentration is comparable to the substrate dopant concentration. If this is compared with the EBIC results, shown by the broken curve in Fig. 8, it can be seen that the EBIC charge collection efficiency decreases as this defect concentration increases. Profile measurements show that the defect depth also increases with laser power.

The thermal behavior of the laser-induced defect, H(0.45), is extremely interesting, because it is not stable at room temperature. For slip-free samples the concentration of this defect decays dramatically with time; within ten days after annealing, it is not detectable. Moreover, this level reappears after heating to ~100°C, but completely disappears after annealing to 300°C for 30 min.

Another important defect is observable when DLTS measurements are performed at lower temperatures, using liquid He; this is a shallow hole trap only 100 meV above the valence band, H(0.10). Thermal annealing experiments demonstrate that H(0.10) and H(0.45) are closely related: creation of one of these defects is accompanied by the annihilation of the other. Immediately after laser annealing, H(0.45) is dominant; however, it gradually decays at room temperature, while H(0.10) simultaneously increases. The recovery of H(0.45) can be stimulated by low-temperature annealing (100°C-200°C), with an accompanying decrease in H(0.10). The sum of the concentrations of both traps always remains constant.

This unusual behavior of related hole traps is already well-documented in the literature. In 1977, a quenched-in defect was chemically identified by Lee et al., using electron spin resonance (ESR), as Fe on the tetrahedral interstitial site, although Fe was not expected to be present. Gerson et al. (1977) correlated the ESR result on quenched Si with DLTS studies, and found that the H(0.45) level arose from the presence of interstitial Fe, which was the dominant defect in thermally quenched Si.

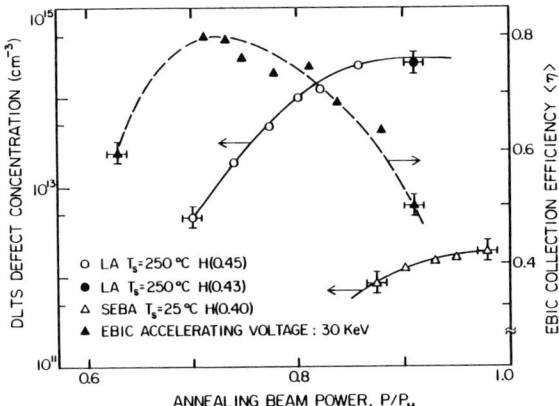

Fig. 8   Defect concentration measured by DLTS, as a function of annealing beam power. The broken curve displays the EBIC charge-collection efficiency as a function of laser power.

Both the ESR and DLTS signals due to interstitial Fe decay with time, if the samples were stored at room temperature. This was found to result from Fe diffusion in the lattice, and reaction with shallow acceptors such as B, Al, and Ga at room temperature to form pairs which are oriented in a <111>-direction. Graff and Pieper (1981a,b) confirmed the fact that the H(0.10) level results from Fe-B pairs in Fe-diffused silicon. Thus, at room temperature the H(0.45) level decreases and H(0.10) increases as interstitial Fe forms pairs with B acceptors present in the p-type material. The dissociation of Fe-B pairs to interstitial Fe, with consequent decrease in H(0.10) and recovery of H(0.45), can be stimulated by low-temperature annealing and light illumination (Graff and Pieper 1981a). Annealing above 300°C causes both interstitial Fe and Fe-B pairs to disappear due to precipitation of Fe.

Fe has therefore been identified as a most important defect in thermally-quenched but nominally pure Si. The presence of Fe in Si has been the subject of growing interest; the presence of Fe as the dominant deep level in laser-annealed samples of nominally Fe-free Si is probably the most significant result of the work reported here. It is important to determine the source of this Fe.

Because of its high solubility and large diffusion constant, Fe is easily introduced into Si during any form of heat treatment or chemical processing. For laser-annealed samples, it is unlikely that Fe diffuses into the sample from the environment because of the short dwell time involved in CW annealing (~1 msec). Therefore, lower Fe contamination in laser-annealed samples can be expected than in furnace-annealed samples. However, thus far we have been unable to completely eliminate these Fe-related laser-induced defects, despite care in processing. In an attempt to find the source of Fe in our CW laser-annealed samples, experiments on float-zone (FZ) Si having low Fe contamination ($\sim 10^{11}$ cm^{-3}) have recently been reported (Sheng and Merz 1982). As a result of these studies, we propose that the presence of Fe in laser-annealed samples may result from the following sources. Initially, the sample surface is contaminated during chemical processing - from the solvents, beakers and tweezers used for standard processing. During ion-implantation, this surface Fe is driven into the wafers by the high-energy incident ions. It has also been reported by Hass et al. (1978) that, for high-dose implantation, considerable sputtering occurs, introducing Fe and other heavy metals from various apertures of the implant accelerator. In both cases, the contamination of Fe increases with implant accelerating voltage, ion mass and dose. Since the surface of the implanted samples is usually amorphous, the Fe in the substrate is gettered to the damaged surface during annealing. All of these sources make it difficult to keep Si wafers free from Fe. The difficulty in eliminating Fe from Si devices produced with standard processing procedures, and the instability of these Fe-related defects, has important implications for Si device technology.

5.   Scanned-Electron-Beam-Annealing (SEBA)

In general, the use of the scanned electron beam from the electron microscope gives results that are somewhat similar but generally superior to those obtained with the scanned laser beam. EBIC images of electron-beam annealed samples are usually almost featureless, and the optimum value of the photocurrent is higher for electron-beam than for laser annealing. This implies that the electron-beam-annealed sample has better lateral uniformity of the minority-carrier collection efficiency.

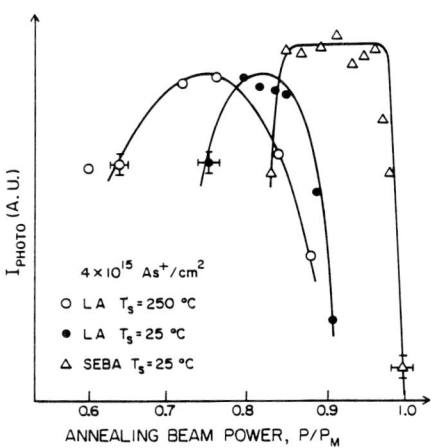

I_PHOTO (A. U.)

4 × 10¹⁵ As⁺/cm²

O LA T_s = 250 °C
● LA T_s = 25 °C
△ SEBA T_s = 25 °C

ANNEALING BEAM POWER, P/P_M

Fig. 9. Dependence of short-circuit
photocurrent on annealing laser
power for samples laser-annealed
(LA) at two different substrate
temperatures, and for samples electron-beam annealed (SEBA) at room
temperature.

The dependence of the photocurrent
on annealing beam power is compared
for LA and SEBA in Fig. 9. It is
found that the window for SEBA is
wider and flatter than that observed
for LA. The plot of I(photo) versus
laser power for the laser annealed
(LA) sample at 250°C is identical to
that shown in Fig. 2. However, for
comparison with the SEBA done at
room temperature, the LA curve
obtained with a substrate temperature of 25°C is more appropriate.
Two major differences between the LA
and SEBA cases are immediately apparent. First, electron-beam
annealing produces a wide, flat
window, compared to the laser-anneal
curve. Secondly, when the electron-
beam curve falls off, it does so
precipitously; that is, the sample
goes from good quality material to a
state of high slip-plane density for
a very small increase in beam power.
It is difficult to observe an intermediate region, where damage is
induced by the laser without the
formation of dislocations.

Figure 4 shows another quantitative EBIC comparison between LA and SEBA.
The EBIC collection efficiency after SEBA is higher than that obtained
after optimal LA sample, for low EBIC accelerating voltage. For high
accelerating voltage, the EBIC collection efficiency is comparable in the
two cases. This suggests that the sample quality near the surface is
better after SEBA than after optimal LA. Note that the SEBA results
represent the "envelope" of the best results obtained by laser annealing.

Finally, the deep levels observed in Si after SEBA are far less significant than after LA, as seen in Fig. 7. Only a weak signal is observed at
E_v + 0.40 eV, which does not depend strongly on electron-beam power (Fig.8).
The nature of this defect is presently unknown.

It is clear from both the EBIC and DLTS studies, that more satisfactory
results are obtained using SEBA than LA; however, the reason for this is
not presently understood. One rather obvious explanation is the greater
penetration of the electron beam compared to the laser beam; hence, thermal gradients are significantly reduced, so point defect and dislocation
generation is correspondingly reduced. This explanation would argue for
the use of incoherent lamps and strip heaters rather than LA or SEBA;
useful results are currently being obtained for the III-V semiconductors
using these techniques. On the other hand, similar dark stripes have been
observed by Baumgart et al. (1982) in CW CO_2 laser-annealed samples, which
is even more penetrating than SEBA. A fundamental understanding of energy
dissipation and electron-solid interactions is required to resolve this
problem.

6. Ackowledgements

The authors would like to thank A. Lietoila, R.B. Gold, and J.F. Gibbons for assistance during the early phase of these experiments, D. Zak and Z.L. Wu for considerable technical assistance, and L.C. Kimerling for supplying us with low-Fe FZ wafers, as well as technical discussions. This work was perfomed under Air Force Contracts No. F19628-79-C-0128 and F19628-82-K-0006, ROME/RADC, Hanscom AFB, Mass.

References

Baumgart H, Phillip F and Leamy H J 1982 to be published in Laser and Electron-Beam Interactions with Solids, (North Holland: New York)
Dean P J, Haynes J R and Flood W 1967a Phys. Rev. 161 711
Dean P J, Flood W F and Kaminsky G 1967b Phys. Rev. 163 721
Gat A, Gibbons J F, Devine V R, Williams P and Evans C A 1978 Appl. Phys. Lett. 32 276
Gerson J D, Cheng L J and Corbett J W 1977 J. Appl. Phys. 48 4821
Graff K and Pieper H 1981a J. Electrochem. Soc. 128 669
Graff K and Pieper H 1981b Semiconductor Silicon - 1981 (Electrochem. Soc.) p. 331.
Hass W, Glawischnig H, Lichti G and Bleier A 1978 J. Electron. Mat. 7 525
Hess L D, Forber R A, Kokorowski S A and Olson G L 1979 Proc. Soc. Photo-Opt. Instr. Engineers 198, 31
Kimerling L C, Leamy H J, Benton J L, Ferris S D, Freeland P E and Rubin J J 1977 Semiconductor Silicon/1977, (Electrochemical Society, Princeton, 1977) p. 468
Kosai K and Gershenzon M 1974 Phys. Rev. B9 723
Lang D V 1974 J. Appl. Phys. 45 3023
Lee Y H, Kleinhenz R L and Corbett J W 1977 Appl. Phys. Lett. 31 142
Mizuta M, Sheng N H, Merz J L, Lietoila A, Gold R B and Gibbons J F 1980 Appl. Phys. Lett. 37 154
Mizuta M, Sheng N H and Merz J L 1981a Appl. Phys. Lett. 38 453
Mizuta M, Sheng N H and Merz J L 1981b J. Appl. Phys. 52 6437
Sheng N H, Mizuta M and Merz J L 1981 Laser and Electron-Beam Solid Interaction and Materials Processing (Elsevier-North Holland: New York) 155
Sheng N H, Mizuta M and Merz J L 1982a Appl. Phys. Lett. 40 68
Sheng N H and Merz J L 1982b Laser and Electron-Beam Interactions with Solids (Elsevier-North Holland: New York) p. 313
Sheng N H and Merz J L 1983 Proc. Intl. Conf. on Defects in Semiconductors, Amsterdam, Sept. 1982 (North Holland).

Inst. Phys. Conf. Ser. No. 67: Section 3
Paper presented at Microsc. Semicond. Mater. Conf., Oxford, 21–23 March 1983

125

Electron microscopy of Se-implanted and electron-beam annealed GaAs

N J Shah[1], H Ahmed[1], L A Freeman[2] and David J Smith[2]

[1]Engineering Department, University of Cambridge, Trumpington Street, Cambridge, CB2 1PZ, U.K.

[2]High Resolution Electron Microscope, University of Cambridge, Free School Lane, Cambridge CB2 3RQ, U.K.

Abstract   Rapid isothermal annealing by the multiple scan electron beam method has been used in a study of the annealing behaviour of GaAs implanted with selenium ions in the dose range 10^{13} to 10^{15}cm^{-2}. TEM showed a markedly different defect morphology, dependent on both dose and annealing conditions, for specimens with equivalent electrical activity.   The transition from amorphous to polycrystalline GaAs has been studied by TEM, HREM and RBS for anneals lasting for only a few seconds at 1000°C.

1. Introduction

Ion implantation followed by the annealing of implantation damage are important steps in the fabrication of GaAs integrated circuits.   The ion damage induced in GaAs and the effects of conventional furnace annealing have been reported previously, for example by Mazey and Nelson (1969), Sealy (1975), Sadana and Booker (1979) and Hutchinson et al (1982). Recently, rapid isothermal annealing with the multiply-scanned electron beam method (MSEBA) (McMahon and Ahmed, 1979) has been used to study the annealing of co-implanted Se and Ga in GaAs (Shahid et al., 1983).   In the present work, we have used the MSEBA technique to anneal selenium implants in GaAs, in particular to investigate any significant differences compared with the standard furnace annealing methods.   Various implanted and annealed specimens have been analysed by sheet electrical measurements, Rutherford backscattering spectroscopy (RBS), conventional transmission electron microscopy (TEM) and high resolution electron microscopy (HREM).

2. Experimental

Semi-insulating (100)GaAs wafers, tilted off-axis to avoid channelling, were implanted with 400keV selenium ions.   The wafers were heated to 200°C for doses of 10^{13}, 10^{14} and 4×10^{14}cm^{-2} and held at room temperature for doses of 10^{15}cm^{-2}.   The hot-implant samples were then encapsulated by rapid chemical vapour deposition of Si$_3$N$_4$ at $\sim$700°C to minimise decomposition during subsequent annealing; the cold-implanted samples remained uncapped.   Scribed 5mm square chips were annealed in two ways: one used different beam power densities with constant exposure time giving different peak temperatures, whereas the other varied the time for a constant power density so that the sample remained at the peak temper-

ature for different periods. After annealing, the cap was removed in HF
and the samples were then evaluated electrically by measuring sheet
conductivity and mobility.   Specimens suitable for TEM and HREM were
then chemically and/or ion-beam thinned. Observations were made at 100kV
using a Siemens 102 and at 500kV using the Cambridge University HREM
(Smith et al., 1982).   RBS was mainly restricted to the 10^{15}cm^{-2} sample
since the implantation damage caused by the lower ion doses was insufficient to affect the spectra noticeably.

3. Results

None of the samples showed measurable electrical activity after implantation or after surface encapsulation. However, rapid isothermal annealing
of Se-implanted GaAs produced good electrical activity and high mobility
for a range of doses, even when the period of annealing was only a few
seconds(Shah et al, 1981). This is clearly shown by the values summarised
in Table I: these are very similar to those obtained from conventional
furnace-annealed samples.

TABLE I

Sheet electrical measurements in Se-implanted GaAs

Dose (cm^{-2})	Anneal	N_S (cm^{-2})	μ_S (cm^2/Vs)
10^{13}	24Wcm^{-2}, 5s	9.7 x 10^{12}	2700
	13Wcm^{-2}, 60s	8.2 x 10^{12}	2800
10^{14}	24Wcm^{-2}, 5s	3.4 x 10^{13}	1600
	13Wcm^{-2}, 60s	3.0 x 10^{13}	1600
4 x 10^{14}	24Wcm^{-2}, 5s	3.4 x 10^{13}	1000
	13Wcm^{-2}, 60s	2.3 x 10^{13}	1000

It is significant that equivalent electrical results were achieved for the
two extreme treatments, since annealing at 13Wcm^{-2} for 60s corresponded to
reaching a calculated peak temperature of $\sim$950°C for $\sim$45 seconds whereas
a 5s exposure at 24Wcm^{-2} was equivalent to reaching a transient temperature of $\sim$1100°C.

TEM of plan-view samples indicated a marked difference in the distribution
of defects depending on the annealing treatment. For example, Fig.1(a)
shows the sample implanted with 4 x 10^{14} Se ions/cm^2, before annealing.
Bright-field/dark-field contrast analysis indicated that the small defects
present were interstitial dislocation loops with Burgers vector a/2 <110>
in agreement with previous work, both with Se-implants (e.g. Sadana and
Booker, 1979; Hutchinson et al, 1982) and also, for example, with Ne
(Mazey and Nelson, 1969).   Moreover, electron diffraction patterns and
RBS spectra indicated that the material remained primarily crystalline,
despite being electrically inactive.   Figs.1(b) and (c) show the morphology after annealing for 5s and 60s respectively, and it appears that
the coarsening and coalescence of larger loops has removed many of the
smaller loops. Despite the identical electrical properties, it is also
interesting to note that the density of dislocation loops differs by about
an order of magnitude.   A similar TEM sequence for the 10^{14}cm^{-2} Se dose

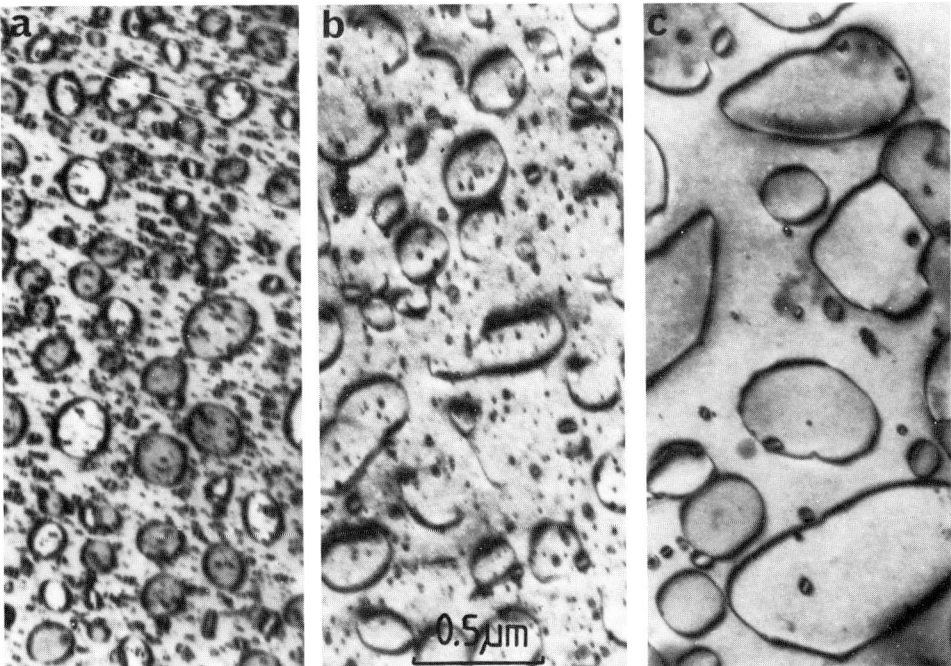

Fig.1. Low magnification images of $4 \times 10^{14} \text{cm}^{-2}$ Se-implanted GaAs; (a) unannealed; (b) annealed at 24Wcm^{-2},5s; (c) annealed at 13Wcm^{-2},60s.

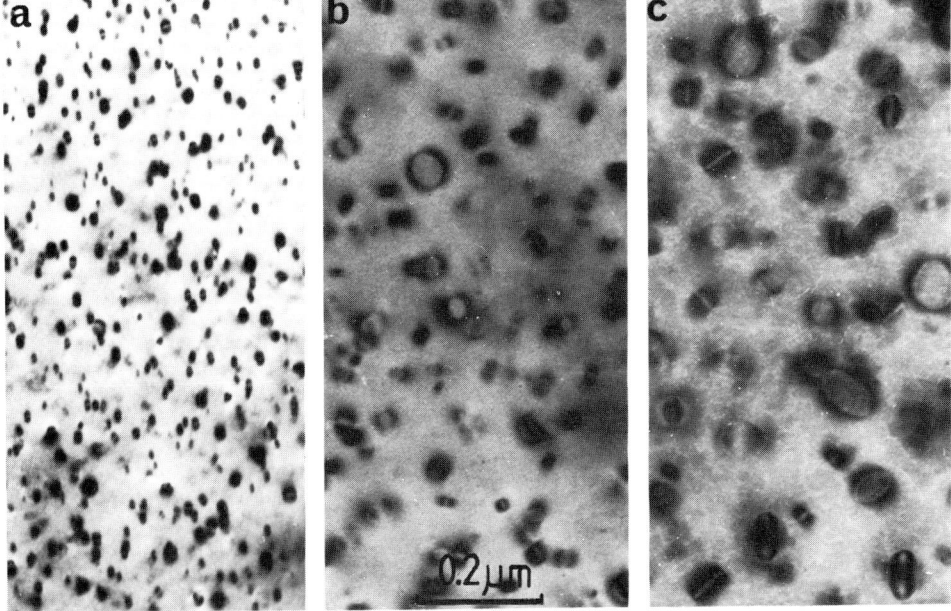

Fig.2. Low magnification images of 10^{14}cm^{-2} Se-implanted GaAs; (a) unannealed; (b) annealed at 24Wcm^{-2},5s; (c) annealed at 13Wcm^{-2},60s.

is shown in Fig.2.   The dislocation loops are smaller and the loop
density lower than for the high-implant dose, but the coarsening of loop
sizes shows the same trends as a result of heat treatment.   A high resolution lattice image showing one of these small interstitial loops is shown
in Fig.3.

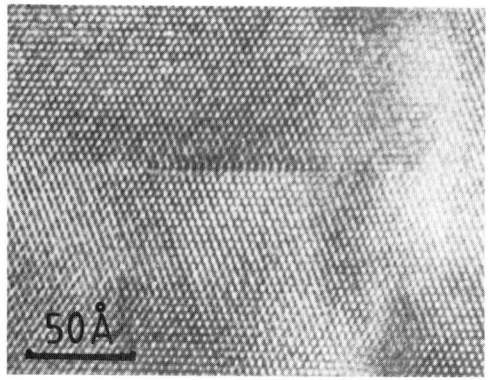

Fig.3. High resolution image of
small interstitial defect imaged in
<110> projection in $4 \times 10^{14} cm^{-2}$ Se-
implanted GaAs annealed at $13Wcm^{-2}$,60s.

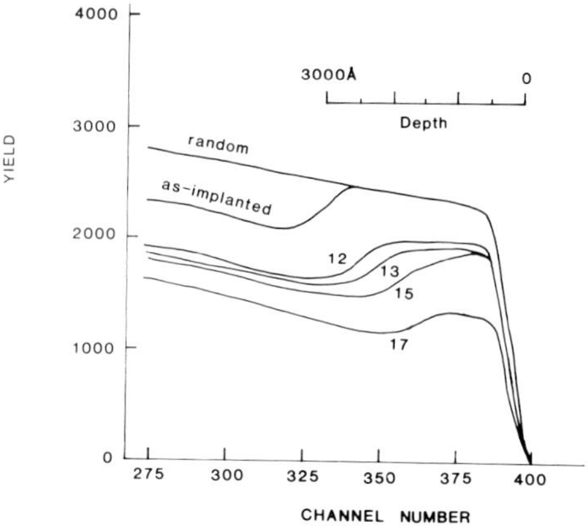

Fig.4. RBS of $10^{15} cm^{-2}$ Se-implanted
GaAs, and annealed at 12,13,15 and
$17Wcm^{-2}$,5s.

Implantation with a dose of 10^{15} Se ions/cm^2 resulted in an amorphous
GaAs layer to a depth of ∿3000Å as determined by RBS (see Fig.4).   This
is not equal to $R_p + \Delta R_p$ since R_p, the projected range of 400keV Se in GaAs,
is 1371Å and ΔR_p, the standard deviation,is 577Å although Sadana and
Booker (1979) also observed an experimental range well in excess of that
expected.   Using the MSEBA method, the implanted layer was regrown at peak
temperatures in the range 800°C to 1000°C for beam exposure times of 5s.
Beam powers greater than about $17Wcm^{-2}$ could not be used since, without
capping, GaAs decomposition degraded the surface.   The channelled RBS
spectra of Fig.4 indicated a substantial reduction in disorder with
increasing power density although a considerable amount of residual damage
remained even for the highest power density.   TEM of the sample receiving
$15Wcm^{-2}$ (Fig.5) revealed the development of a finely-grained polycrystalline structure which was confirmed by the predominant ring nature of the
corresponding electron diffraction patterns (see inset).   High resolution
images (e.g. Fig.6) revealed large numbers of moiré fringes arising from
the randomly-oriented, overlapping grains as well as extensive twinning.
It thus appears that some solid phase epitaxy develops from the substrate
but nucleation of polycrystallites within the damage region also occurs.

4. Discussion and Conclusions.

Because of constraints on thermal cycles, loop growth with furnace
annealing has previously been inseparable from implant activation.   The
MSEBA method has provided sufficient flexibility with heat treatments that
rate-phenomena can be separated.   In particular, we have established that
simple growth of loops is not the mechanism for obtaining activation.
Further work is required to identify the precise nature of annealing and
electrical activation in GaAs.

Cross-sectional specimen examination by TEM and HREM would be a worthwhile
extension of this study, both for the low doses and for the high amorphis-

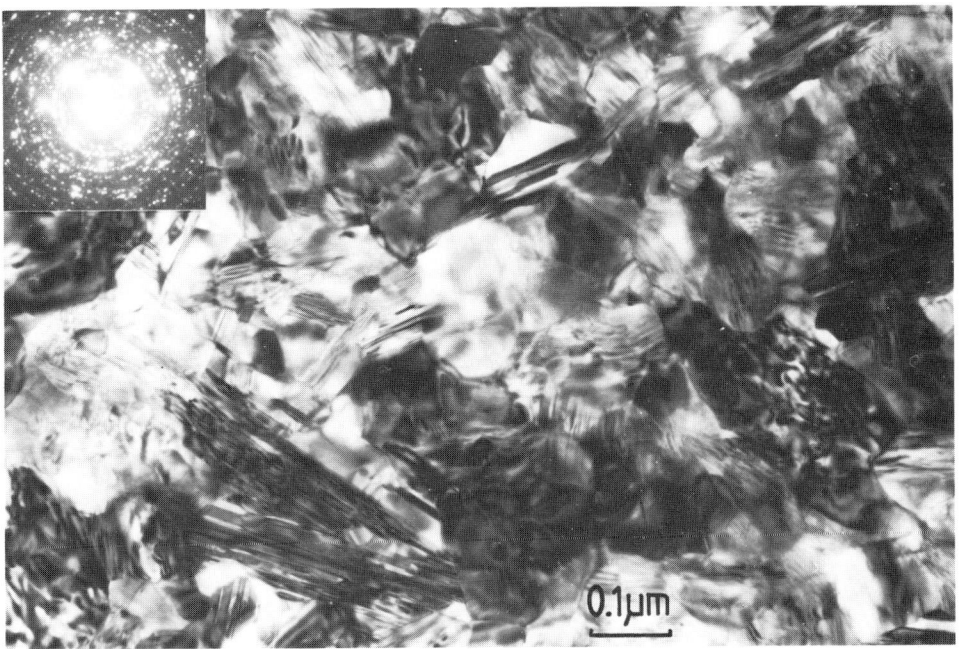

Fig.5. Typical low magnification image of $10^{15} cm^{-2}$ Se-implanted GaAs, annealed at $15Wcm^{-2}$,5s. Inset: electron diffraction pattern along <001> axis.

Fig.6. High magnification image of $10^{15} cm^{-2}$ Se-implanted GaAs, annealed at $12Wcm^{-2}$,5s.

ing dose.   This would allow comparison with theoretical range data, carrier concentration profiles and RBS.   It might then be possible to gain better insight into ion-damaged GaAs and the solid phase epitaxy of amorphous GaAs.

5. Acknowledgements

Financial support from the S.E.R.C., U.K. is gratefully acknowledged. We thank Dr. P.A. Leigh (British Telecom) for providing the implanted samples and Dr. A. Howie for access to the Siemens 102 at the Cavendish Laboratory.

References

Hutchinson P W, Ball R K, Dobson P S and Leigh P 1982 J. Mat. Sci. Letts. 1 457

McMahon R A and Ahmed H 1979 Electronics Letts. 15 45

Mazey D J and Nelson R S 1969 Rad. Effects. 1 229

Sadana D K and Booker G R 1979 Rad. Effects, 42 35

Sealy B J 1975 J. Mat. Sci. 10 683

Shah N J, Ahmed H and Leigh P A 1981 Appl. Phys. Lett. 39 322

Shahid M A, Moffatt S, Barrett N J, Sealy B J and Puttick K E 1983 Rad. Effects in press

Smith D J, Camps R A, Cosslett V E, Freeman L A, Saxton W O, Nixon W C, Ahmed H, Catto C J D, Cleaver J R A, Smith K C A, Timbs A E 1982 Ultramicroscopy 9 203

Inst. Phys. Conf. Ser. No. 67: Section 3
Paper presented at Microsc. Semicond. Mater. Conf., Oxford, 21–23 March 1983

Investigation of ion-implanted GaAs following electron-beam annealing

M A Shahid*, S Moffatt**, N J Barrett** and B J Sealy **

* Department of Physics, ** Department of Electronic & Electrical
Engineering, University of Surrey, Guildford, Surrey, UK

Abstract   (Ga+Se) implants in GaAs have been studied following
irradiation with a multiply scanned electron beam. A correlation
between sheet electrical properties, transmission electron microscopy
and Rutherford backscattering results with irradiation time was
observed.

1.   Introduction

Recently there has been increasing interest in the use of thermal pulses
to anneal ion implanted GaAs. The methods used are a graphite strip
heater (Surridge et al 1977, Sealy et al 1979, 1982), a multiply scanned
electron beam (Shah et al 1980, 1981, Bujatti et al 1982, Shahid et al
1983) and an incoherent light source (Arai et al 1981, Davies et al 1982).
All three have produced useful results but work is in a preliminary stage
for the latter two techniques. In this paper we will compare the results
of transmission electron microscopy, electrical and Rutherford backscattering measurements for samples implanted with (Ga+Se) ions followed
by annealing with a multiply scanned electron beam.

2.   Experimental

Gallium and selenium ions were implanted in a non-channelling direction
into (100) Cr-doped semi-insulating GaAs at room temperature. The ion
dose was 9×10^{13} cm^{-2} and the energy was 400 keV. 4mm x 4mm square
samples were coated with about 0.1 μm of pyrolytically deposited Si_3N_4 at
650°C. Samples were then exposed to a multiply scanned electron beam in
a Lintech Instruments Seza. The electron beam diameter was about 0.4mm,
its energy 30 keV and the power density was 25 W cm^{-2} which produced a
rise in sample temperature to 880°C in about 6 to 8 seconds (Shahid et al
1983). Specimens were examined in a JEOL STEM 200 CX operated at 160 kV.
Sheet electrical properties were measured on clover leaf shaped samples
using the van der Pauw technique having made electrical contact by
alloying small tin dots at about 300°C. Rutherford backscattering (RBS)
was performed using a 1.5 MeV beam of helium ions.

3.   Results

The density of dislocation loops decreases with increasing irradiation
time in the range 3s to 100s (Fig.1). In this figure there is included
also bright field micrographs of the same region of a specimen taken
under different reflection vectors for an irradiation time of 15s. This
procedure enabled us to identify the nature of the dislocation lines and
loops (Shahid et al 1983). The dislocation loops which were found to

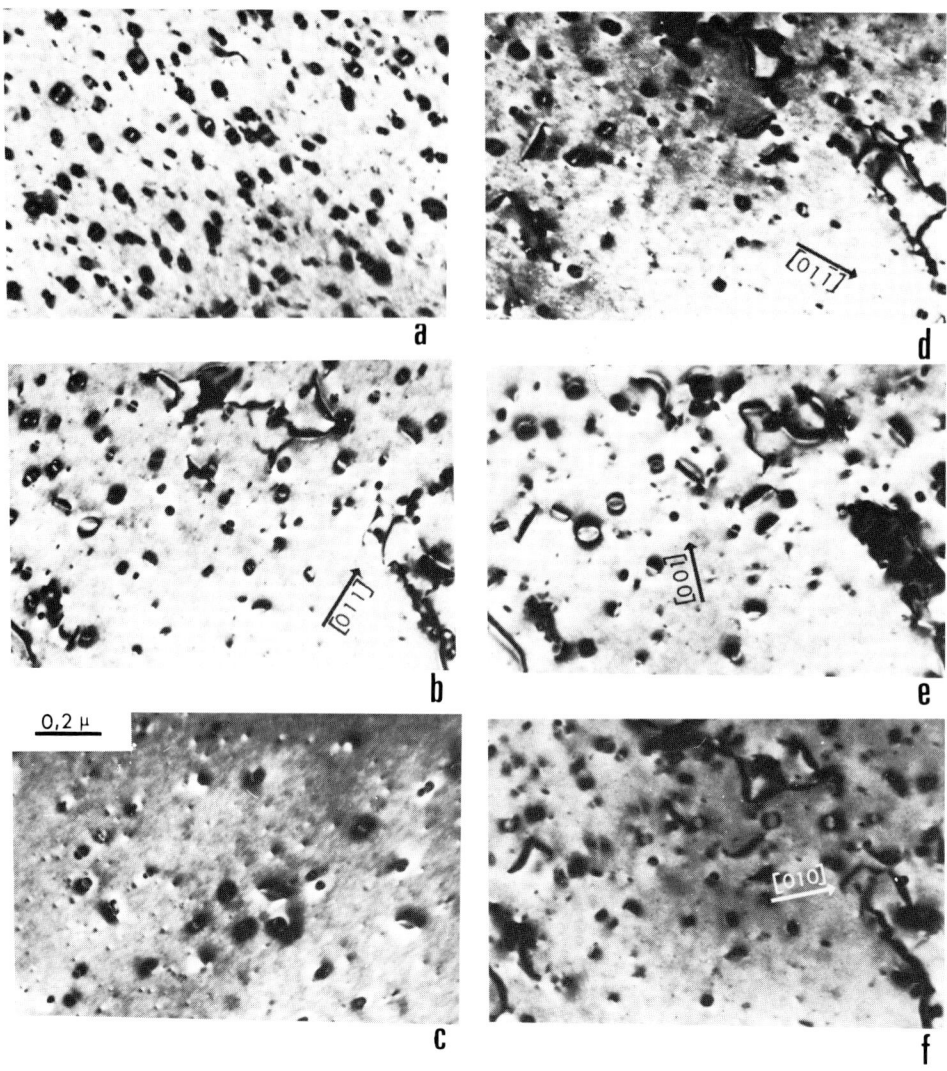

Fig. 1. Transmission electron micrographs showing samples annealed for
(a) 3s, (b) 15s and (c) 100s. The micrographs (d), (e) and (f) are from
the same area of the specimen as in (b) but have been produced using
different g- vectors. The following g- vectors have been used:-
(b) 022, (d) 02$\bar{2}$ (e) 004 and (f) 040.

lie on {110} planes had an interstitial character with Burgers vectors of
a/2 < 110>.The average size of the dislocation loops increases with
increasing irradiation time and dislocation lines possessing a/2 <110>.
Burgers vectors appear due to the interaction of closely lying expanding
loops with each other and the free surface. This deduction was made by
examining stereo micrographs which showed these effects more clearly
than conventional micrographs do. Stereo microscopy was also used to
demonstrate that the dislocation loops and lines shown in figure 1 are
confined to a narrow band close to the surface. Figure 2 illustrates the
depth distribution of damage for a specimen irradiated for 15s. The loop
density is approximately constant at a depth corresponding to the third
bright thickness fringe and for thicker regions of the specimen. This
implies that most of the dislocations lie within 0.35 µm of the surface.
Also it is possible to conclude from figure 2 that the dislocations are
largest near the surface and smallest deeper in the material.

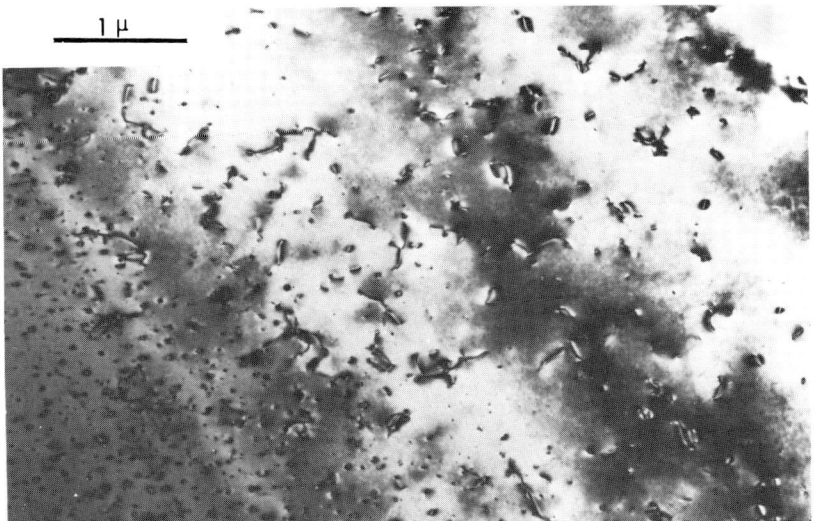

Fig. 2. Transmission electron micrograph of a sample annealed for 15s
showing thickness fringes. Variation of size and density of defects
with depth is clearly visible.

The RBS spectra show that the as implanted samples has an amorphous
surface layer of 0.25± 0.02 µm thick (Fig. 3.). The Si_3N_4 deposition
process caused a significant amount of recrystallisation. For example,
the damage parameter χ_{min} reduces from 71% for the as implanted sample
to about 60% for a sample that had been coated with Si_3N_4 and had no
other thermal treatment. Following e-beam irradiation for 3s, 15s and
200s, χ_{min} reduces to 12.0%, 10.5% and 8.9% ± 0.5% respectively. This
shows that even after the relatively long irradition time of 200s, the
GaAs is still somewhat disordered, since χ_{min} for good crystalline GaAs
should be about 5%. Also it is not possible to resolve clearly the
gallium and arsenic peaks which confirms that the surface is imperfect.

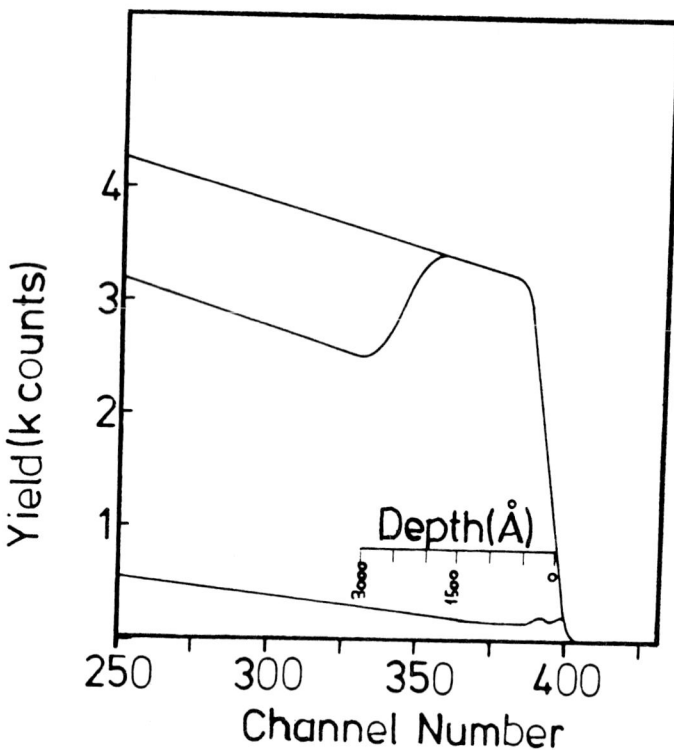

Fig. 3. R.B.S. spectra, channelled and random, for an as implanted
sample and a sample annealed for 200s.

The electrical sheet measurements for mobility, resistivity and carrier
concentration are shown as a function of irradiation time in figure 4.
Specimens exposed for 3 seconds were highly resistive   indicating that
although the temperature reached about $750^{\circ}C$, the implanted ions were not
electrically active. As the irradiation time was increased above 10s,
specimens became increasingly conducting but for times in excess of 300s,
the electrical properties did not vary significantly. For these longer
times, the specimens reached an equilibrium temperature of $880^{\circ}C$.
The electrical activity produced was about 35% with a mobility approaching 800 cm^{-2} V^{-1}s^{-1}.

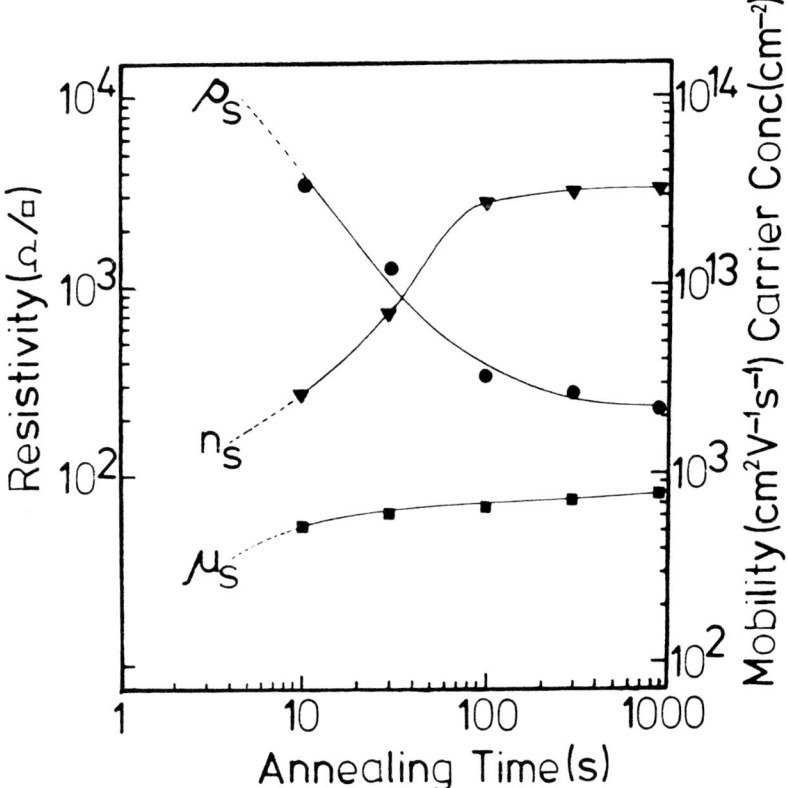

Fig. 4. Electrical sheet values for samples with varying annealing
times; resistivity (ρ_s), carrier concentration (n_s) and mobility (μ_s).

4. Discussion

The loop density, the damage parameter, χ_{min} and the sheet resistivity
decrease with increasing irradiation time. Thus as the lattice recrystal-
lises and the defects are annealed out the resistivity decreases.

The temperature of the sample saturates at 880°C for irradiation times
over 10 seconds. There is also a saturation of electrical properties
for times in excess of about 100 seconds. Thus, even though the maximum
electrical activity is only about 35%, this value cannot be improved by
annealing for longer times. This is in accord with our previous results
of thermal pulse anneals obtained using a graphite strip heater (Sealy
et al 1979, 1982). The only way to increase the electrical activity and
mobility is to raise the sample temperature (Bensalem et al 1983).

The fact that the minimum χ_{min} value was about 9% indicates that the
implanted amorphous regions did not recrystallise perfectly. The presence
of this residual damage is consistent with the rather low mobility of
about 800 cm^{-2} V^{-1}s^{-1}.

The Rutherford backscattering spectra indicate that the as-implanted layer is essentially amorphous , since the yield near the surface for the aligned spectrum reaches the yield obtained for the 'random' spectrum. The thickness of this grossly damaged region is estimated to be about 0.25 μm (figure 3). The TEM results (figure 2) and stereo micrographs (not shown here) indicate that following irradiation with the e-beam, recrystallisation takes place but a band containing a large density of dislocations remains near the surface. This band is approximately 0.35 μm thick. Thus the dislocations occur at depths up to at least three times the projected range (Rp = 0.1 μm). This is similar to published data (Sadana and Booker, 1978) for hot implants of selenium ions into GaAs which showed that dislocations occur at depths of up to four times the projected range. It is interesting to note that the residual damage in terms of dislocations extends to depths greater than the amorphous layer thickness. This suggests that there is a tail to the implant extending beyond the depth corresponding to the end of the amorphous layer.

In conclusion, there is a distinct correlation between the three techniques used as a function of e-beam irradiation time. The residual defects occur in a region close to the surface, but penetrating to a depth of at least three times the projected range which is beyond the thickness of the grossly damaged (amorphous) layer produced by the implant. Despite the presence of dislocations and residual damage (RBS), the electrical activity is relatively high but the mobility is low. Better results should be obtained by raising the annealing temperature.

5.   Acknowledgements

The authors thank the SERC and GEC for financial support and members of the Ion Implantation Group and Micro-Structural Studies Unit for experimental assistance.

References

Arai M, Nishiyama K and Watanabe N 1981 Jap. J. Appl. Phys. 20 L124
Bensalem R, Barrett N J and Sealy B J 1983 Elec. Lett. 19 112
Bujatti M, Cetronia A, Nipoti R and Olzi E 1982 Appl.Phys.Lett. 40 334
Davies D E, McNally P J, Lorenzo J P and Julian M 1982 IEEE Elec. Dev.
    Lett. EDL-3 102
Sadana D K and Booker G R 1978 Proc. IBMM (Budapest) p 1131
Sealy B J, Surridge R K, Kular S S and Stephens K G 1979 Inst. Phys. Conf.
    Ser. No.46, p476
Sealy B J 1982 Microelec. J. 13 21
Shah N J, Ahmed H, Sanders I R and Singleton J F 1980 Elec. Lett. 16 433
Shah N J, Ahmed H and Leigh P A 1981 Appl. Phys. Lett. 39 322
Shahid M A, Moffatt S, Barrett N J, Sealy B J and Puttick K E 1983
    Rad. Eff. 70 291
Surridge R K, Sealy B J, D'Cruz A D E and Stephens K G 1977 Inst. Phys.
    Conf. Ser. No.33a, p161

Inst. Phys. Conf. Ser. No. 67: Section 3
Paper presented at Microsc. Semicond. Mater. Conf., Oxford, 21–23 March 1983

137

Electron-beam annealing of Co and Cr implanted polycrystalline silicon

M N Kozicki and J M Robertson

Edinburgh Microfabrication Facility, Department of Electrical Engineering, University of Edinburgh, Kings Buildings, Edinburgh. EH9 3JL

Abstract   Co and Cr disilicides were created by first implanting the metal ions into polycrystalline silicon films at 5×10^{17} ions/cm^2 and 350 keV and then annealing using furnace or electron-beam techniques. The results are compared in terms of sheet resistance and surface morphology of the resulting silicide films. It was found that Co implanted material which had been e-beam annealed gave the lowest sheet resistance of 0.9 Ω/square. Furnace annealing produced higher sheet resistance values due to diffusion effects in the poly-Si.

1. Introduction

Polycrystalline silicon (poly-Si) films are currently used as gate electrodes and for short interconnections within MOS integrated circuits (Faggin and Klein 1970). However, even when heavily doped with boron or phosphorus (Seto 1977), the resistivity of these films is relatively high and precludes more general use of poly-Si as a connection medium. This high resistivity (around 10^{-3} Ω-cm or 20 Ω/square for a 0.5 μm film) is also a contributory factor in limiting the speed of MOS circuits and devices.

A low resistivity film may be formed by combining the poly-Si with a metal to create a silicide (Murarka 1980). Silicides are particularly attractive in integrated circuit technology because of this low resistivity and also because of their high temperature stability. The choice of metal largely depends on the properties of the resulting thin film silicide as regards physical and chemical compatability with the device fabrication process.

At present, silicides for use in integrated circuits are generally formed by cosputtering the metal with silicon onto poly-Si and then heating in a furnace (Murarka and Fraser 1980) to sinter or anneal the resulting film. This method is time consuming and requires expensive high purity target materials. The sputtered film may be sensitive to contamination introduced during sputtering either from the targets, the equipment (poor vacuum) or from the surface of the poly-Si (e.g. oxygen). The long furnace anneal may also result in high resistivity intermetallic compounds and uneven silicide films due to non-uniform diffusion in the poly-Si.

To overcome the problems of sputtered silicides, the metal ions may be implanted into the poly-Si and then electron-beam (e-beam) annealed. This new method allows direct control of stoichiometry and therefore

control of the resulting silicide phases. In the following sections details of cobalt (Co) and chromium (Cr) silicide formation by implantation
are described. These metals were chosen as they cannot reduce SiO_2 , a
factor which would be important if the material is to be used as a gate
electrode. The films were annealed using a 'low-energy' electron beam
and by a conventional furnace process and the results compared.

2. Preparation of samples

Silicon wafers, 75 mm diameter and 350 µm thick, were cleaned and then
oxidised at $950^{\circ}C$ using a 'burnt hydrogen' system to produce a 0.5µm high
quality SiO_2 film. A 0.5 µm layer of poly-Si was deposited on this at
low pressure by the pyrolytic decomposition of silane (SiH_4) at 600ºC.
The resulting structure is shown in Fig 1. SEM examination of the surface of the poly-Si showed a fine 'grainy' texture with typical feature
sizes of around 0.2µm (Fig 2). The wafers were then diced into 2 cm x 2
cm samples for ease of loading in the implantation equipment.

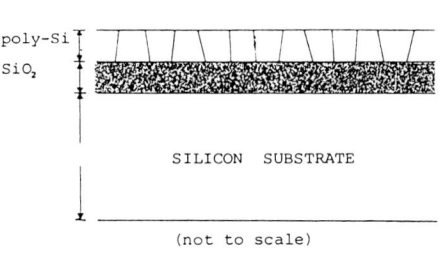

Fig 1: Structure of samples   Fig 2: As-deposited poly-Si

Implantation was performed at AERE Harwell using a 'sputter source' with
metallic Co or Cr. Note that source purity is unimportant as the implanter has a mass selection capability and thus can reject contaminants.
Both Co and Cr implants were performed at 350 keV and 5×10^{17} ions/cm^2
dose. The projected range and standard deviation of the resulting gaussian distribution are as shown in Table 1.

At 350 keV	Co	Cr
PROJECTED RANGE (µm)	0.22	0.25
PROJECTED S.D. (µm)	0.08	0.09

Table 1: Implant distribution at 350 keV.

The implant current was kept at 3 µA and the beam scanned at 30 Hz (200
Hz line) over 1 cm^2 to reduce beam heating effects. A dose of 5×10^{17}
ions/cm^2 at 350 keV will give an average stoichiometrically correct distribution for Co or Cr disilicide ($CoSi_2$, $CrSi_2$) over a region of around
0.2 µm centred approximately halfway into the poly-Si film, (Fig. 3);
i.e. the ratio of metal to silicon atoms in this region is 1:2. SEM examination of the surface showed no discernible features. Four point

probe measurements were performed on the as-implanted samples and the results are detailed in section 4.

3. Annealing

Annealing was carried out using conventional furnace treatments in a dry N_2 ambient and by an experimental low-energy electron-beam system. The e-beam system was developed by the authors in the Edinburgh Microfabrication Facility specifically for annealing poly-Si layers on SiO_2. Fig.4 shows a schematic of the system.

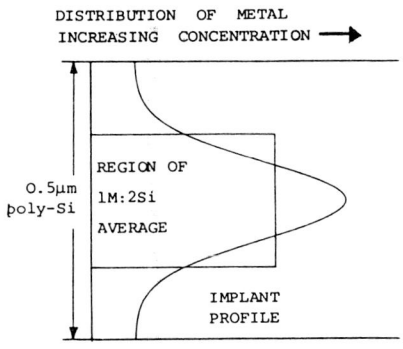

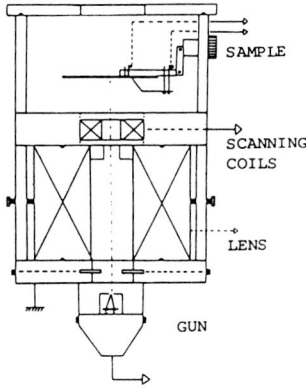

Fig 3: Position of implant

Fig 4: E-beam system

The e-beam treatments all used a 500 μm spot scanned for 10 seconds along 5 mm strips. Three values of beam power were used; (i) 7.5W (5kV), (ii) 10W (5kV) and (iii) 12.5W (7.5 kV). Parallel single scans were repeated to provide an annealed area large enough for 4 point probe measurements. Repeated scans were also performed on some samples to assess stability of the annealed material.

Furnace anneals were done at 900°C, 950°C and 1025°C for times varying between 5 mins and 60 mins (including 30 sec push and pull times) with the samples free standing in the N_2 ambient. Great care was taken to achieve uniformity of push and pull times in and out of the furnace especially for short overall treatments. At various points during the anneal, the samples were removed and 4 point probe sheet resistance measurements made.

4. Results

The 4 point probe measurements on the as-implanted unannealed samples revealed an uneven sheet resistance over the implanted area. This was generally small for the Co samples (around 5%) but could be as high as 50% for Cr. Typical as-implanted values ranged from 6 - 16 Ω/square between different Co samples, which even before annealing is very low, and around 50 - 150 Ω/square for different Cr samples. Because of the variations across the areas of the individual samples, regions of near constant sheet resistance were marked out for annealing to maintain consistency of results.

In general, the furnace anneals proved less effective than the e-beam anneals as regards both sheet resistance and surface morphology. Table 2 shows the sheet resistance for 5 min anneals at 900°C, 950°C, and 1025°C and Table 3 shows the sheet resistance for extended anneals at 950°C.

	5 mins	
	Co	Cr
As-imp	16	94
900°C	2.8	754
As-imp	6	105
950°C	1.7	880
As-imp	13	45
1025°C	2.7	367

	30 mins	
	Co	Cr
As-imp	16	150
950°C	3.6	1060

	60 mins	
	Co	Cr
As-imp	16	150
950°C	3.5	960

Sheet resistance in Ω/square.

Table 2: Results of 5 min furnace anneal

Table 3: Results of 950°C extended furnace anneal

For the prolonged treatments, the sheet resistance increased up to 30 minutes and then remained stable for longer times. The non-uniformity of un-annealed values creates difficulty in comparing results. However, certain points of interest arise when the surface is examined. SEM analysis of the samples shows a 'mottled' effect most likely caused by diffusion of the metals along the grain boundaries to the surface as shown in Fig. 5 (Co) and Fig. 6 (Cr). This occurs even for the 5 min anneals and becomes more apparent for longer times.

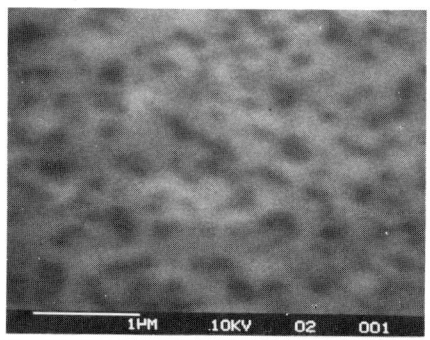
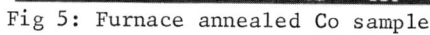

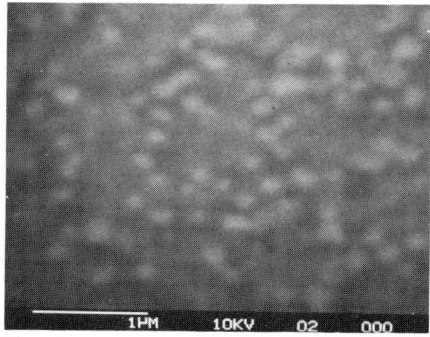

Fig 5: Furnace annealed Co sample   Fig 6: Furnace annealed Cr sample

The e-beam anneals on the Co samples proved highly successful with none of the problems of the furnace treatments. Table 4 shows the results for the 10 second e-beam treatments for the various beam power values on the Co samples.

As-imp	7.5W	As-imp	10W	As-imp	12.5W
7.5	1.6	6.9	0.9	9.6	0.9

Sheet resistance in Ω/square.

Table 4: E-beam anneals of Co samples.

These values remained stable for repeated anneals up to 60 seconds total
anneal time. However, the Cr samples proved more unstable under repeated
anneals for all beam values. Table 5 shows the variation for the 7.5W
beam over 3 repeated anneals.

As-imp	1 scan	2 scans	3 scans
56	120	220	230

Sheet resistance in Ω/square.

Table 5: Repeated scans at 7.5W on Cr samples.

SEM examination showed a smooth surface after annealing for all samples
with no evidence of 'precipitation'. On the regions along the axis of
the scan, large but 'flat' features of around 2 μm were just visible,
suggesting considerable regrowth of the material.

5. Discussion

It is interesting to note that the as-implanted material already has a
relatively low sheet resistance especially in the case of Co. An implant
of such a magnitude should convert the poly-Si film to a near amorphous
state. However, some short range order must exist after implantation for
the sheet resistance to be so low. The relative disorder plus small
variations in the concentration of implanted material will result in the
apparent large variations in sheet resistance between samples.

The problems created by furnace anneals are likely to be mainly due to
the apparent rapid diffusion of the implanted metals in the poly-Si especially in the grain boundaries. This will lead to the formation of
'unwanted' silicide phases or intermetallics and a 'loss' of metal due to
precipitation on the surface. The increase in sheet resistance value is
evident in the case of Cr because of the higher resistivities of the
various intermetallics which appear to form. The sheet resistance falls
from the as-implanted value for Co and rises for Cr after annealing.
This indicates a lack of control over the formation of the higher resistivity phases for Cr which leads to a high overall sheet resistance.
This does not occur with Co as it does not tend to form high resistivity
phases. For the implant parameters described, the sheet resistance
values calculated from the resistivity values given by Murarka (1980) for
the disilicides should be 0.9 Ω/square for Co (18 $\mu\Omega$-cm) and 25 Ω/square
for Cr (600 $\mu\Omega$-cm). Furnace annealing does not give these values. Temperature dependence of the resulting film resistivities is difficult to
assess due to the diffusion effects in furnace anneals. For both Co and
Cr, the sheet resistance stabilises at around 30 mins which suggests an
equilibrium condition is reached at this time.

The e-beam anneals produced good results with Co but the 'optimum' Cr
results were still not obtained, illustrating that the high resistivity
phases are formed within the first 10 seconds and increase with further
scans. The optimum Co value of 0.9 Ω/square was observed for the 10W and
12.5W beams but not for the 7.5W beam suggesting full annealing of the
samples occurs at around 10W or above. This remained stable for repeated
anneals up to 5 scans and then began to rise slightly for more scans as

diffusion effects became significant. For both metals after e-beam annealing the surface remained free of obvious precipitation due to reduced diffusion effects.

6. Conclusions

The low sheet resistance and good surface morphology of the e-beam annealed Co film makes the process very suitable for integrated circuit applications. Since the high conductivity silicide lies within the poly-Si film, it would be suitable for use as a gate electrode as well as an interconnect. The composite film 'appears' as poly-Si on its surfaces and thus will not significantly alter the threshold voltage of a MOS device.

Electron-beam annealing produces the best results for the implanted profiles although furnace annealing could possibly be used with higher implant doses to compensate for 'loss' of metal due to diffusion effects. E-beam annealed implanted Cr silicides are much less suited to use as interconnections but could be used for 'implanted resistors' within integrated circuits, the exact value being 'trimmed' using the electron-beam.

Although it is recognised that implantation equipment capable of such high implant doses is not widely available at this time, new equipment is being developed within the semiconductor industry which will make the process more viable for the near future.

Acknowledgements

The authors would like to thank Dr G. Dearnaley and his colleagues at AERE Harwell for providing the necessary implants. This work forms part of a CASE project supported by SERC and Hughes Microelectronics Limited, Glenrothes.

References

Faggin F and Klein T 1970 Solid State Electron. 13 1125
Murarka S P 1980 J. Vac. Sci. Technol. 17 775
Murarka S P and Fraser D B 1980 J. Appl. Phys. 51 350
Seto J Y W 1977 Proc. Electrochem. Soc. 77 241

Inst. Phys. Conf. Ser. No. 67: Section 3
Paper presented at Microsc. Semicond. Mater. Conf., Oxford, 21–23 March 1983

143

Heatpulse annealing of ion-implanted silicon: structural characterization by transmission electron microscopy

D K Sadana, S C Shatas* and A Gat*

Lawrence Berkeley Laboratory, University of California, Berkeley, CA
*AG Associates, 1052 Elwell Court, Palo Alto, CA

Abstract   A detailed structural investigation of Heatpulse rapid thermal annealed ion-implanted Si was carried out using TEM. Defect-free material was obtained after an 1100°C/10 sec. anneal for 80 KeV 2×10^{15} ^{75}As implanted Si, while dislocation loops were still present after lower temperature anneals. When compared with arsenic, 35 KeV ^{11}B implanted Si in the dose range 5×10^{14} exhibited markedly different behavior under the same annealing conditions, in that residual defects in the form of line dislocations, loops, and rods were observed. Comparison with 1000°C/30 min. furnace annealed samples showed differences in defect structures and impurity redistribution behavior as revealed by SIMS measurements. A model explaining the annealing behavior of implanted layers in silicon under rapid annealing conditions is proposed.

1.   Introduction

Rapid annealing techniques employing incoherent light sources as an alternative to conventional furnace annealing of ion-implanted semi-conductors have received a great deal of interest. Using arc lamps [1], graphite heaters [2], or tungsten-halogen lamps [3] as light sources and 1-100 second heating cycles, this technique offers similar advantages of scanned CW laser annealing: namely, high activation efficiency of the implanted species with minimal redistribution of the initial implanted profile [4]. Furthermore, the simplicity and energy efficiency of Heat-pulse rapid thermal annealing compared with scanning laser or electron beam sources makes this process applicable to high volume manufacturing of integrated circuits. Extensive electrical measurements of rapid thermal annealed ion-implanted silicon have been reported [3,4], but studies of residual defects after annealing have so far been limited [6]. In this paper results of a detailed investigation of defect structures after Heatpulse annealing of boron and arsenic implanted silicon are presented.

2.   Experimental

Czochralski grown (100) Si wafers, doped to a background concentration of $\sim5\times10^{14}$ cm^{-3} were implanted at room temperature in a non-channeling orientation. N-type phosphorous doped wafers were implanted with 35 KeV ^{11}B$^+$ in doses of 5×10^{14} and 5×10^{15} cm^{-2}, and p-type boron doped wafers were implanted with 80 KeV ^{75}As$^+$ to a dose of 2×10^{15} cm^{-2}.

Implanted samples were annealed in a Heatpulse apparatus, consisting of a water cooled reflective chamber containing upper and lower banks of tungsten halogen lamps. The stationary wafer, positioned equidistant between the lamp arrays, was rapidly heated at 50–100°C/sec. with up to 20 kW of radiant energy trapped between the reflectors. The lamp intensity, controlled by a microcomputer, was correlated with wafer temperature by using a type K thermocouple attached to the wafer. In a typical 1100°C, 10-second anneal cycle the lamp intensity was increased at a constant rate of 1.5 KW/sec., followed by a constant 76 percent intensity anneal of 10 seconds, and then decreasing the intensity at the same rate. The annealing ambient was air at atmospheric pressure, and a very small (<20Å) amount of oxide was grown during the anneal cycle. Samples furnace-annealed in a nitrogen ambient at 1000°C for 30 minutes served as a standard for comparison.

The implanted and annealed samples were then characterized by secondary ion mass spectrometry (SIMS) and TEM analysis. SIMS measurements were carried out with a Cameca IMS 3f ion microprobe analyzer using O_2^- primary ion bombardment with positive ion detection for boron analysis. Sensitivity factors were obtained by integrating the as-implanted SIMS profile and equating this value to the implanted dose, and the depth scale for analysis was obtained by a Dektak measurement of the sputtered crater depth for each sample.

TEM analysis was performed on both plan view and cross-section specimens using bright-field diffraction conditions and micrographs were recorded for a (220) type diffraction vector.

Results

Figure 1a-c shows the TEM plan view micrographs from furnace and Heatpulse annealed Si sampes that were implanted with boron to a dose of 5×10^{14} cm^{-2} at 35 KeV. All of the samples showed dislocation loops of a/2 <110> and a/3 <111> types. In addition, elongated rod and mix-

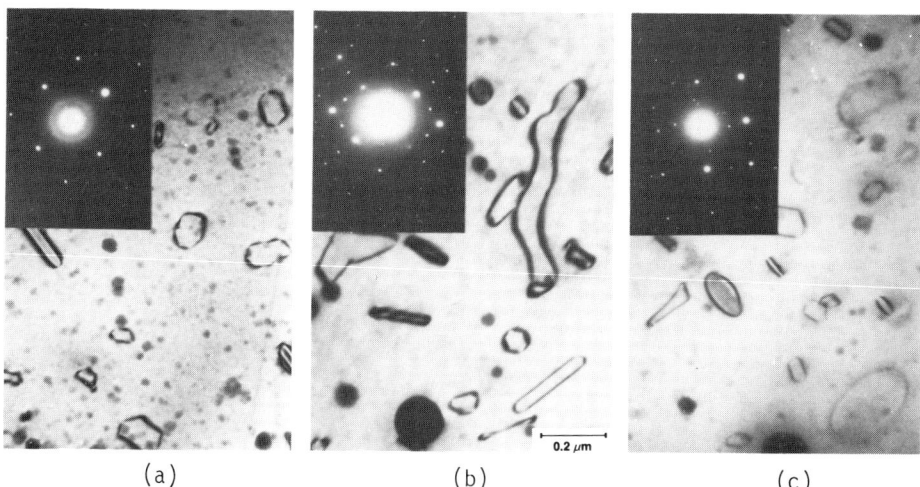

(a)	(b)	(c)

Fig. 1   TEM plan view micrographs of 35 KeV ^{11}B implanted (100) Si, 5 x 10^{14} cm^{-2}; (a) furnace 1000°C/30 min. (b) Heatpulse 1100°/10 sec. (c) Heatpulse 1100°C/30 sec.

shaped defects were present in the Heatpulse annealed samples. A large
increase in the dimensions of the dislocation loops was observed when the
Heatpulse annealing time was increased from 10 seconds to 30 seconds
at 1100°C.

Figure 2a-d shows the annealing sequence for Si implanted with 5×10^{15} cm^{-2}
B at 35 KeV. The as-implanted sample (Fig. 2a) exhibits a single-
crystalline diffraction pattern, indicating that no amorphous layer is
produced by the implanted boron even at this high dose. The furnace
annealed sample (Fig. 2b) showed a high density of irregular shaped loops
in addition to a dislocation network. In contrast, the sample Heatpulse
annealed for 10 sec. (Fig 2c) showed the presence of a dislocation network
still at its embryonic stage of formation. The extra spots in the
diffraction pattern of this sample disappeared when thicker specimen
foil areas were examined, giving no indication of the presence of polycrystalline Si in the implanted region. After a 30 sec. 1100°C Heatpulse
anneal a well defined cross-grid dislocation network was observed to form
in the implanted region (Fig. 2d).

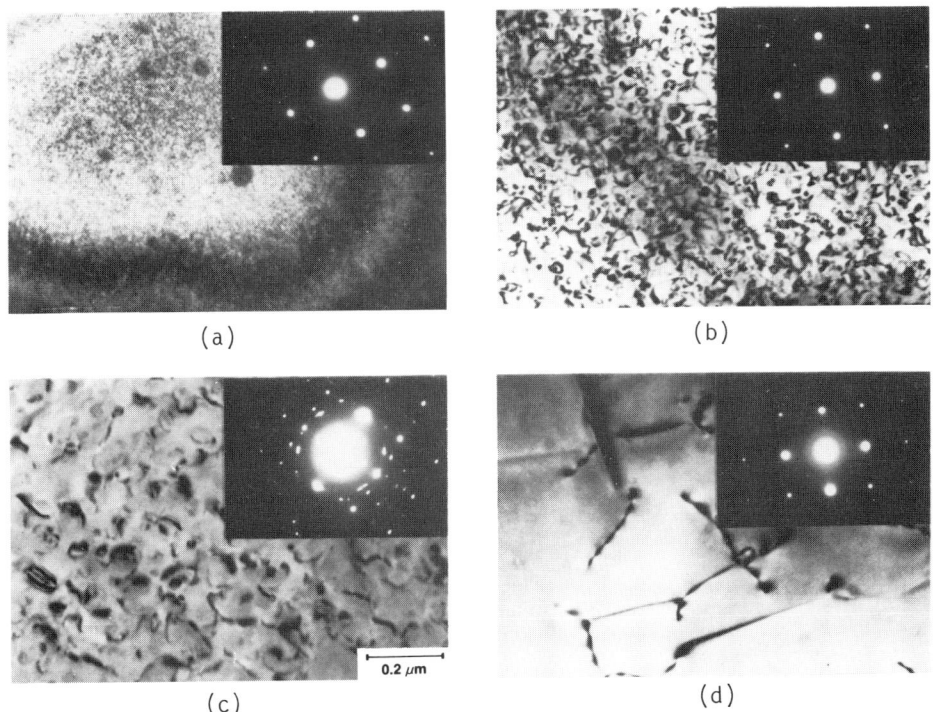

(a)   (b)

(c)   (d)

0.2 μm

Fig. 2.   TEM plan view micrographs of 35 KeV ^{11}B implanted (100) Si,
5×10^{15} cm^{-2}; (a) as-implanted   (b) furnace 1000°C/30 min.   (c) Heatpulse
1100°C/10 sec.   (d) Heatpulse 1100°C/30 sec.

The cross-section micrographs corresponding to Fig. 2a-c are shown in
Fig. 3a-c. The as-implanted sample contained a 0.12 μm wide band of
damage clusters at a mean depth of 0.09 μm from the surface. This depth

corresponds to 0.8 times the projected range (Rp) of 0.11 μm for boron at 35 KeV. Upon furnace annealing, the region containing damage clusters in Fig. 3b, converted into a dense band of dislocation loops (Fig. 3c). In addition, large loops and rods were found to be present below the initial damage layer that extended down to a depth of 0.38 μm from the surface. In contrast, the secondary defects in the Heatpulse annealed sample (Fig. 3a) were confined to a narrow band at a mean depth of 0.15 μm and did not show any deeply extending defects.

The SIMS profile of B from the as-implanted sample showed a characteristic skewed distribution (Fig. 3d). On subsequent furnace annealing broadening of the initial profile ocurred with much bigger effect in the tail of the profile (Fig. 3d). Furthermore, pinning of B in the vicinity of Rp was observed. These results are in agreement with the earlier published results of Hofker [5]. The Heatpulse annealing for 10 seconds at 1100°C produced a B profile similar to that of above, however, the broadening in the tail region reduced markedly. When the Heatpulse annealing time was increased to 30 seconds, the pinned fraction of B near Rp disappeared and significant diffusion in the tail region comparable to that of the furnace annealed sample was observed.

In contrast to the annealing behavior of boron implanted silicon, Heatpulse annealing of arsenic implanted Si at 1100°C resulted in defect-free material (Fig. 4c). Dislocation loops with mean diameters of 0.04 μm remained after 1000°C 1-10 sec. anneals (Fig. 4a,b). When both these samples were re-annealed at 1100°C for 10 sec., the dislocation loops disappeared and defect-free material was once again obtained. Similar results have recently been reported for flame annealing of ion-implanted Si [6].

(a)

(b)

(c)

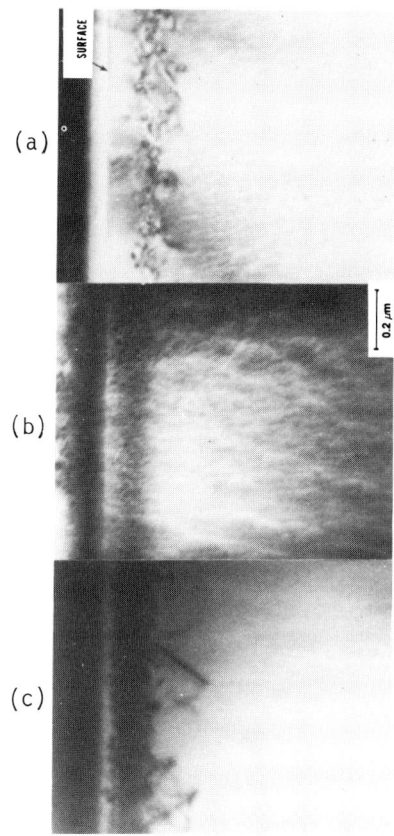

(d)

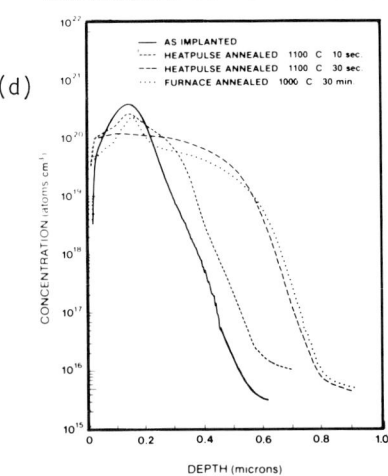

Fig. 3. Comparison of cross-section TEM micrographs corresponding to Fig. 2 with SIMS profiles of ^{11}B; (a) Heatpulse 1100°C/10 sec. (b) as-implanted (c) furnace 1000°C/30 min. (d) ^{11}B SIMS profiles.

Discussion

It appears from the results of Fig. 4 that perfection in recrystallization of the implanted region by Heatpulse annealing depends on the
degree of amorphicity of the initial damage region and the annealing
temperature. When the initial implant damage consists of a continuous
amorphous layer that extend to the surface, as in the case of a high dose
arsenic implant, the recrystallization on subsequent annealing can
produce defect-free material. However, in the case of B where the
initial damage structure consists of a high density of small damage
clusters in the crystalline matrix, the recrystallization produces a high
density of secondary defects. Impurities are known to play an important
role in nucleation and stabilization of secondary defects [8]. They can
segregate relatively easily to dislocations and be bound up with the
dislocation core with some characteristic binding energy. At lower
temperatures the impurities trapped at dislocations are not expected to
escape and they retard the annealing out of the dislocations. Taking
into consideration these factors, the results of Figs. 1-4 can be
explained as follows. In the case of arsenic, when the Heatpulse
annealing was carried out at 1000°C for 10 seconds, the amorphous layer
is expected to recrystallize perfectly by solid phase epitaxial growth.
The dislocation loops observed in the micrograph (Fig. 4a,b) probably
correspond to the region just below the original amorphous/crystalline
interface. If the As atoms in the vicinity of the loops segregate
during the annealing process, the dislocations become pinned and their
annealing will be retarded. At 1100°C, the binding energy of As to
a dislocation is such that it is released from the core, thereby
allowing complete annealing out of dislocation loops as observed in
the 1100°C re-anneal of the samples initially annealed at 1000°C.
Thus, defect-free material is obtained only after annealing at temperatures of 1100°C or higher.

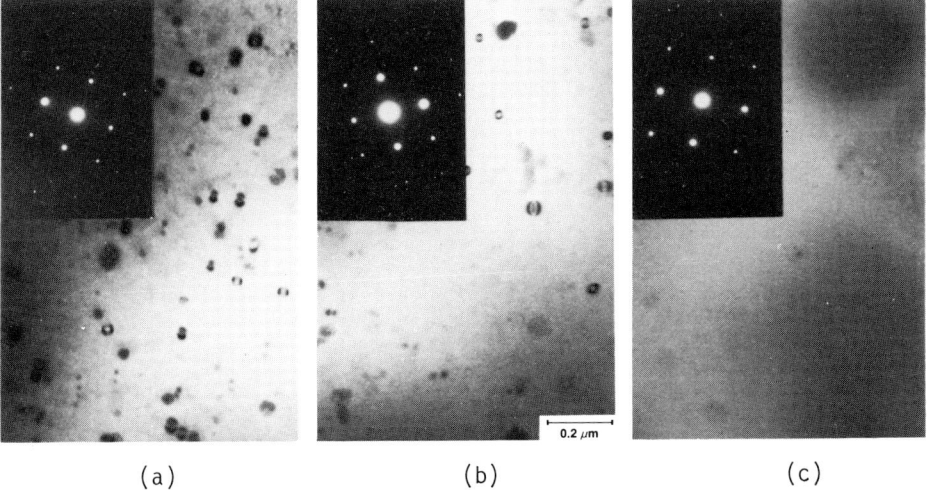

      (a)   (b)   (c)

Fig. 4. TEM plan view micrographs of 80 KeV [75]As implanted (100) Si,
2×10^{15} cm[-2]; (a) Heatpulse 1000°C/1 sec. (b) Heatpulse 1000°C/10 sec.
and (c) Heatpulse 1100°C/10 sec.

In the case of boron implanted Si where the implant damage does not consist of an amorphous layer, the annealing process does not occur by simple solid phase epitaxial growth. However, the annealing behavior can still be explained qualitatively in a similar manner as that of arsenic implanted Si. Comparison of the cross-section TEM micrograph and B profile (Fig. 3) from the 1100°C/10 second sample shows that B segregates to the damage, but this segregation is less pronounced than that observed in the furnace annealed (1000°C/30 minutes) sample. When the Heatpulse annealing time was increased to 30 seconds, the segregated B atoms diffused away from the dislocations, allowing the loops to coarsen into a cross-grid dislocation network which does not anneal out. When the Heatpulse annealing was carried out in an argon ambient, defect-free material was observed after an 1100°C/10 second anneal for the 5x10^14 cm^-2 boron implanted sample, in sharp contrast from the results shown in Fig. 1b. This difference in defect structure due to the annealing ambient may be explained by considering that the Si interstitials injected from the Si/SiO$_2$ interface during oxidation [7] can interact with defects already present and cause them to grow larger. Much longer diffusion lengths are observed in the Heatpulse annealed samples than is expected from the normal diffusion of B in Si. The cause of this enhanced diffusion during rapid thermal annealing is currently under investigation.

Acknowledgement

The authors would like to thank Professor Jack Washburn of the University of California, Berkeley, for useful discussions and Craig Hopkins of Charles Evans and Associates for the SIMS measurements. This work was supported by the Director, Office of Energy Sciences, Materials Research Division of the US Department of Energy under Contract No. DE-AC03-76SF00098.

References

1. Gat A 1981   IEEE Device Lett EDL-2(4) 85
2. Wilson SR, Gregory RB, Paulson WM, Hamdi AM and McDaniel FD 1982
   App Phys Lett 41 978
3. Benton JL, Celler GK, Jacobson DC, Kimerling LC, Lischner DJ, Miller
   GL and Robinson McD 1982 Laser and Electron Beam Interaction with
   Solids ed BR Appleton and GK Celler (Elsevier, North Holland)
   pp 765-770
4. Benton JL, Celler GK, Jacobson DC, Kimerling LC, Lischner DJ, Miller
   GL and Robinson McD 1982 Laser and Electron Beam Interaction with
   Solids ed BR Appleton and GK Celler (Elseiver, North Holland)
   pp 771-776
5. Hofker WK 1975 Philips Res Rept Suppl 8
6. Narayan J and Young RT 1983 App Phys Lett 42 466
7. Lin AM, Antoniadis DA and Dutton RW 1981 J Electrochem Soc 128 1131
8. Sadana DK, Washburn J and Booker GR 1982 Phil Mag B 46 611

Inst. Phys. Conf. Ser. No. 67: Section 3
Paper presented at Microsc. Semicond. Mater. Conf., Oxford, 21–23 March 1983

Microstructure of photodeposited thin films

C J Chen, H H Gilgen and R M Osgood Jr

Department of Electrical Engineering and Applied Physics,
Columbia University, New York, N.Y. 10027

Abstract   We describe the deposition of thin films with UV laser light
and the physical processes which contribute to the deposit microstructure. TEM studies of the early stages of film growth show the importance
of the light interaction with the deposit nuclei.

1. Introduction

Amorphous thin films are a basic structural form in the fabrication of many
optical and electronic components. Traditionally these films have been
grown by vacuum evaporation, sputtering, or chemical vapour deposition.
Each deposition technique imparts its own set of distinctive film properties. Thus, sputtered films are generally denser than evaporated films
because of physical bombardment by the depositing species during film
growth. It is possible to alter these properties drastically by changing
some important experimental conditions of film growth; for example, the
use of a high background pressure of inert gas results in a loose or porous
film as grown by vacuum evaporation.

Recently, it has been demonstrated that thin films of either conductors
or semiconductors can also be grown by photochemical decomposition (Ehrlich
et al 1982; Osgood 1983). This method of deposition relies on dissociat-
ing a gas-phase molecular encapsulant with a one- or several-photon optical
absorption process, i.e.

$$ABC + nh\nu \rightarrow A + BC,$$

thus depositing a single atomic species. Since a direct electronic transition is utilized substrate heating is minimized. Ultraviolet laser
sources have been used in most of these experiments. Feature dimensions
in the film of $\sim 0.5 \mu m$ have been produced with a very low-power focused
laser beam. Large-area growth has been achieved with weakly or unfocused
radiation from a pulsed laser.

Although the conditions for deposition are very similar to those for low-
pressure CVD film growth, the film microstructure is very different. The
interaction of the laser radiation with the substrate and the growing film
permits control of both the overall film pattern as well as the film microstructure and texture. In this paper, we will briefly review the physics
of photochemical film growth. We will then describe recent experiments
to study the very early stages of photochemical growth in the vicinity of
metal particles. The experiments relied on both optical and electron
microscopy for diagnosis and for deposition.

2. Background

Photodeposited films can be grown with pulsed or cw laser sources. Most of the experimental work discussed in this paper was obtained with tightly focused cw lasers. In this case, photodeposited films have been generally grown using the geometry shown in Fig. 1. A molecular carrier gas is enclosed in a chamber in which a semiconducting, conducting or insulating substrate is mounted. The carrier gas pressure and laser intensity are adjusted such that photodeposition produces sufficient atom flux in the vicinity of the substrate that substrate nucleation can occur. Typical pressures are from 1-100 Torr. The carrier gas is usually an organometallic species, although halide- or hydride-based carriers have also been used. Buffer gas can be added; however, increased gas-phase nucleation may result. Such relatively high gas pressures are often conducive to the formation of thin adsorbed molecular layers on the substrate (Ehrlich and Osgood 1981; Chen and Osgood 1983). Depending on the condition of the substrate and the nature of the molecular gas, these layers can be physi- or chemisorbed to the surface.

Most molecular bonds can only be severed with short wavelength, ultraviolet light. However, since the photodissociation process is direct very low laser powers, $1kW/cm^2$, are sufficient to produce rapid deposition rates in a small spot size. Under these conditions substrate heating is negligible, and the deposition occurs at room temperature.

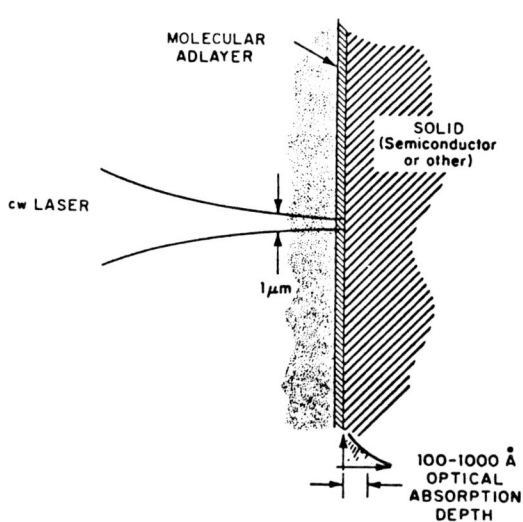

Fig. 1 Geometry of photodeposition of solid films.

It has been established that photodeposition can result from both gas- and surface-phase photodissociation. The relative contribution of the two processes depends on the thickness of the surface adlayer, the gas pressure, and the spectral match of the laser wavelength to the gas and adsorbed molecules. In the important case of Cd deposition from dimethylcadmium, which we will discuss extensively below, the cadmium film growth is mainly by gas-phase photodissociation; however, the process resolution derives largely from surface nuclei which are formed by photodissociation of the surface adlayer (Ehrlich et al 1980).

3. Structure of Photodeposited Films

Both the macro- and microstructure of photodeposited films depend critically on the interaction of the UV light with material on the surface of the supporting substrate. Figure 2 shows an optical micrograph of a Cd film which was formed by photodissociating dimethylcadmium with the output of a frequency doubled Ar^+ laser at 257nm. In this case the contouring in the film macrostructure is attributable to the interference pattern in the laser beam at the substrate surface.

The origin of this highly resolved patterning provides insight into the growth mechanism of the photodeposited films. Irradiation of the surface with UV light causes the formation of small metal nuclei. These nuclei form through the photodissociation of the relatively high-density surface film. Since the sticking coefficient of gas phase Cd atoms is much higher on the metal nuclei than on the base substrate, film growth occurs preferentially on the nucleation pattern on the surface. This pattern is determined exclusively by the spatial modulation of the optical beam on the surface. As the film thickness increases, the original macrostructure is retained until stray light scattered from the film structure degrades the image.

The microstructure of the metal film in its early stages reflects the properties of the metal nuclei. These nuclei are formed by photodissociation of surface adlayer which has approximately liquid density. As a result, the metal atoms produced in the molecular films by photodissociation have extremely small surface mobility, compared say, to atoms deposited under high-vacuum conditions. The nucleation pattern formed with low atom mobility will be characterized by small islands with very close spacing. TEM micrographs show (see Fig. 3) that the grain size for very thin films grown from these nuclei is $\sim 20 - 30\overset{\circ}{A}$, the grain spacing is of comparable dimensions.

As the film continues to grow in thickness, the film microstructure begins to be influenced by the interaction of the photodissociation light with molecules adsorbed on or in the close vicinity to the metallic grains (Chen and Osgood 1982). For polarized light interacting with small metal grains, the optical electric field in the vicinity of the particle is perturbed such that growth from atoms in the thin film will proceed preferentially along the axis of polarization. This effect will be described more completely in the experimental section which follows. In addition to the perturbation of the local surface electric field by individual grains, the thin film as a whole can cause collective changes in the surface optical field. In particular, for certain combinations of materials and laser light, surface polaritrons are launched on the surface of the deposited film. These surface waves can interfere with the incident light wave to produce an oriented microstructure for the film; an example of this structure is shown in Fig. 4 (taken from Osgood and Ehrlich 1982).

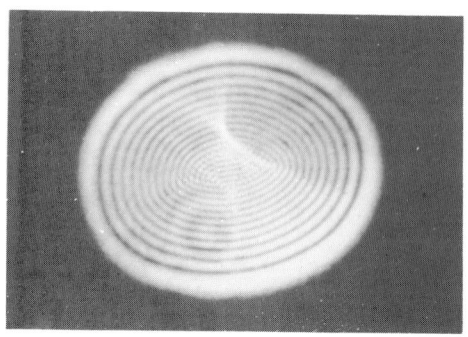

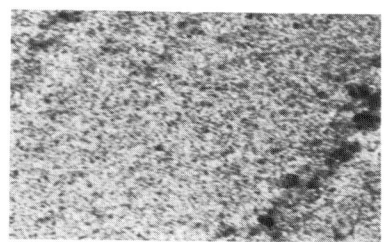

Fig. 2 Optical micrograph of Cd deposit, the long axis is 150mm.

Fig. 3 TEM micrograph of nuclei in very thin photodeposited film: the nuclei are 10 - 20Å in dia.

The oriented film growth shown in Fig. 4 is only seen under certain conditions. First, the UV laser light must be polarized. Second, the film must reach a certain critical thickness. Third, the film must be produced under sample conditions which enhance the growth of the film from the adsorbed molecules rather than gas-phase growth. When these conditions are present, a regular structure of ripples is seen with a period comparable to the wavelength of light. The explanation for this effect has recently been given by Brueck and Ehrlich (1981). They show that, as the film grows in thickness, random scattering launches surface waves on the deposited film. These waves interfere with incident light to produce a modulated optical field which forms periodic patterns of growth in the film. As the film grows, its optical dispersive properties cause surface waves within a certain range of wave vectors to dominate and the regular periodicity seen in Fig. 4 appears.

For thicker films, the ripple pattern evolves into a form of columnar growth characterized by high surface area. This pattern is similar to the growth of evaporated films at relatively high background pressure, since surface mobility is also reduced in that case.

Although we do not discuss it in this paper, we believe that high laser powers can substantially modify the film microstructure. For example, at higher laser powers, film melting or sintering can clearly occur. However, even at lower powers, rapid photodissociation of surface adlayers can enhance surface mobility, also leading to larger grain size.

4. Experimental Observations

As mentioned above, the perturbation of the optical electrical field in the vicinity of small particles can influence the growth of material on these spheres. We have begun an experimental investigation of this phenomenon which uses laser deposition through a high power optical microscope and a TEM study of the cadmium deposit (Chen and Osgood 1982), (Chen and Osgood 1983).

We have recently published a theoretical study of the optical field near the surface of microscopic, $\sim 200\text{Å}$, conducting spheres. The calculations showed that the overall electric field is strongly enhanced if the optical frequency is close to a resonant frequency of the sphere. In addition, the electric field is strongest at the poles of the sphere, where the orientation of the poles is determined by the polarization vector of the light; see Fig. 5. Both the overall and directional enhancements are dependent on the shape, size, and composition of the particle. The growth of the particle by photodeposition is dependent on the electric field at its surface. Thus, there is a coupling between the particle's shape and size and the photodissociating electric field. The theoretical study indicates that a spherical particle should grow most rapidly at its poles, transforming itself into an ellipsoid in the process. The final shape of the particle for a relatively short growth period will depend on how close the particle's resonant frequency matches that of the optical field during growth. The final size of the particle is governed not only by the match in frequencies, but also by the optical losses for both very large, $500 - 1000\text{Å}$, and very small, $0 - 50\text{Å}$, particles.

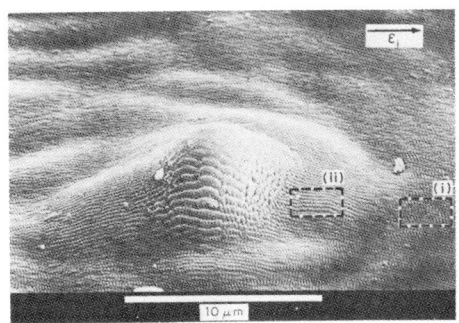

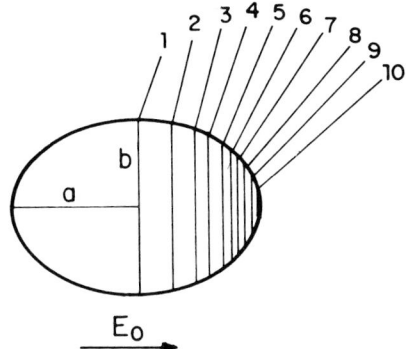

Fig. 4   SEM of photodeposited cadmium film (from Osgood & Ehrlich 1982).

Fig. 5 Electric-field strength near the surface of a conducting ellipsoid, illuminated by ultra-violet light.

In order to observe these effects, we used a collinear observing and illumination microscope, which allowed us to accurately position the sample and observe the deposited film thickness. With this system we observed deposition of Cd from $Cd(CH_3)_2$ molecules adsorbed on Cd spheroids, $10 - 500\text{Å}$ in diameter. These small metal spheres had a resonant frequency close to the optical frequency of the UV laser at 257nm.

The metal spheres were made by vacuum depositing the metal in an evaporator filled with a relatively high pressure of argon to encourage gas-phase nucleation. The spheroids were round for small diameters, but incipiently crystalline for larger diameter; see Fig. 6. Prior to irradiation with the UV laser light, the spheres, which were mounted on a thin carbon film, were positioned on a fine copper mesh for examination with a TEM. This sample was then placed in a cell chamber and exposed to sufficient $Cd(CH_3)_2$ gas to form an adsorbed film of these molecules on the spheres.

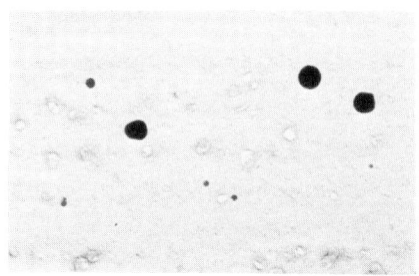

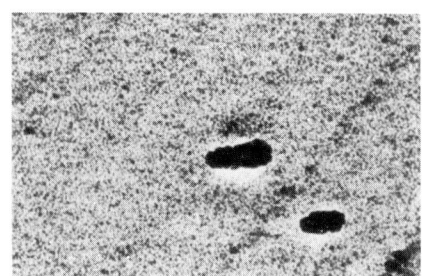

Fig. 6   Spheroids used for deposition (the larger spheroids have
a radius of 900Å).

Fig. 7   Spheres after irradiation,
showing ellipsoidal growth (same
scale as Fig. 6).   The electric
field is horizontal.

The results of the irradiation are shown in Fig. 7.   First, it is apparent
that, as predicted, the small Cd spheroids grow into ellipsoids.   This
growth pattern is in accord with the high electric field at the poles of
the particles.   Second, although not clearly shown in the figure, the
smaller particles do not exhibit rapid growth until after they grow by a
much slower process to a critical size of ∿50Å.   This latency period is
in accord with the optical losses of the small particles.   Similarly we
find that very large particles, which are lossey because of reradiated
light, do not grow rapidly.   Figure 8 shows a recent calculation of this
effect based on the theoretical model described above.

In this experiment, we have studied film growth from separately fabricated
tiny spheres.   However, it is clear that the results obtained are also
applicable to the evolution of small nuclei into macroscopic grains during
photodeposition.

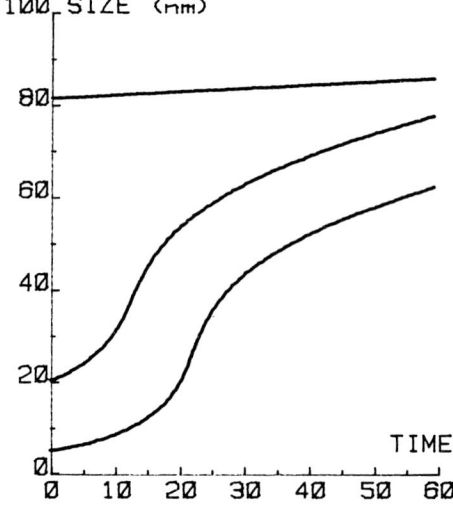

Fig. 8   Growth of metal spheres (major axis versus time).

Summary and Acknowledgement

Photodeposited films have a microstructure which, particularly in the early stages of growth, is controlled by the interaction of the light field with the deposited material and the photochemical medium. An interesting consequence of this phenomnenon is that the orientation of this micro-structure can be controlled by the polarization of the photodissociating laser beam.

We would like to acknowledge support for this project from the Department of the Army (ARO) and the Joint Services Electronics Program (U.S. Army, U.S. Navy and U.S. Air Force) under contract DAAG 28-82-K-0080. Also we thank Bob Kissinger for help with the TEM.

References

Brueck S R J and Ehrlich D J 1982 Phys. Rev. Lett. 48 1678
Chen C J and Osgood R M 1982 Paper 96 delivered at 184th Annual American Chemical Society Meeting Kansas City MO
Chen C J and Osgood R M 1983 Chem. Phys. Lett. to be published
Chen C J and Osgood R M 1983 Phys. Rev. Lett. to be published
Ehrlich D J, Osgood R M and Deutsch T F 1980 IEEE J. Quant. Electron. QE-16 1233
Ehrlich D J and Osgood R M 1981 Chem. Phys. Lett. 79 381
Ehrlich D J, Osgood R M and Deutsch T F 1982 J. Vac. Sci. Technol. 21 23
Osgood R M 1983 Ann. Rev. Phys. Chem. (in press)

Inst. Phys. Conf. Ser. No. 67: Section 3
Paper presented at Microsc. Semicond. Mater. Conf., Oxford, 21–23 March 1983

MeV ion backscattering spectrometry applied to the analysis of beam processed semiconductors

S S Lau
University of California, San Diego
La Jolla, California 92093

Abstract
In this report pulsed laser beam and ion beam induced interactions between metals and semiconductors are summarized. For laser induced reactions, emphasis is placed on the conditions under which amorphous alloys can be formed. Experimental results seem to suggest that the cooling rate necessary to quench a liquid binary mixture into a glass state is highest for compositions close to those of congruently melting compounds. For ion induced reactions, both equilibrium compound and amorphous alloy formation are discussed. The phase formation characteristics of these two techniques are compared.

1.Introduction

The techniques of Rutherford backscattering come directly from the field of low-energy nuclear physics. In principle, energetic ion beams in the range of a few hundred keV to several MeV are produced in the accelerator and analyzed, magnetically or electrostatically to give an energetically well defined beam of particles. The ions are then passed along a beam tube through collimating apertures to the target chamber (at typical pressures of ~10^{-6} to 10^{-7} torr) with a typical beam spot size of the order of 1 mm in diameter. The backscattered particles from the sample are energy analyzed, usually with a surface barrier semiconductor detector. The energy spectrum (so called backscattering spectrum) contains chemical information as a function of depth into the sample. The typical sampling depth by MeV $^4He^+$ particles is about 1 μm; therefore, backscattering spectrometry is well suited for near surface chemical analysis. When channeling and blocking techniques are used in conjunction, structure as well as lattice location can be investigated. For further information on these techniques, the reader is referred to two books (Chu et al. 1978, Feldman et al. 1982) devoted to the subject of backscattering spectrometry. The advantages of using backscattering spectrometry to beam processed semiconductors come from the fact that (i) the probed depth is commensurate with the thickness of the reacted layer induced by beam processing, and (ii) backscattering is a non-destructive, convenient and fast technique for depth profiling with a typical depth resolution of 300 Å for most solids. Although backscattering spectrometry is capable of structural analysis, for amorphous phase identification diffraction techniques (such as x-rays and electron diffraction) are necessary. In the following, we give examples of backscattering spectrometry applied to the analysis of beam processed semiconductors. The focus is placed in the area of metastable and equilibrium phase formation between metals and Si.

2.Glass Formation by Laser Pulse

The research activity of metastable phase formation was first initiated by Pol Duwez in early 1960 (Duwez, 1967). Duwez and his co-workers obtained solid solubility extension and new metastable crystalline or amorphous phases in certain binary alloy systems by rapid cooling from the liquid to the solid state. These rapid solidification techniques are commonly known as splat cooling techniques. The quenching rate of splat cooling generally ranges between 10^4 and 10^6 K/sec. With these quenching rates, Duwez and other have found that metastable alloy phases can be formed near the eutectic composition in binary systems where deep eutectic troughs are present. In Duwez's original thinking, metastable phases are formed as a result of "fooling" the atoms by freezing them into unconventional positions by rapid cooling from the liquid state. Subsequently, Turnbull and others (Cohen and Turnbull, 1961; Davies and Lewis, 1975; Sinha et al., 1976) have dealt with the theoretical and experimental aspects of splat cooling, and the field of rapid solidification has progressed in the interim to a very active, well documented area of research.

With the advent of pulsed laser processing of materials, quenching rates on the order of 10^{10} K/sec are now achievable. These much faster rates open up new dimensions for the formation of metastable phases. In a typical experiment, the sample is a planar, laterally uniform structure containing some distribution of two elements A and B (typically Si and a metal) as a function of depth. The sample surface is heated by a heat pulse with a duration in the range 10^{-9} to 10^{-6} sec. Heat flow in this time regime is essentially one-dimensional, since the thickness of the heated region (on the order of micrometers or less) is small compared to the lateral dimensions of both the sample and the beam. The absorbed fluence is on the order of 1 J/cm^2, enough to melt several 100 nm of the sample while creating thermal gradients in the 10^6 to 10^8 K/cm range. The maximum surface temperature may exceed the melting point by several 100 K, but it should not approach the boiling point.

Figure 1 shows a one-dimensional heat flow calculation for the position of the liquid-solid interface as a function of time. The film is deposited on a substrate of much higher melting point.

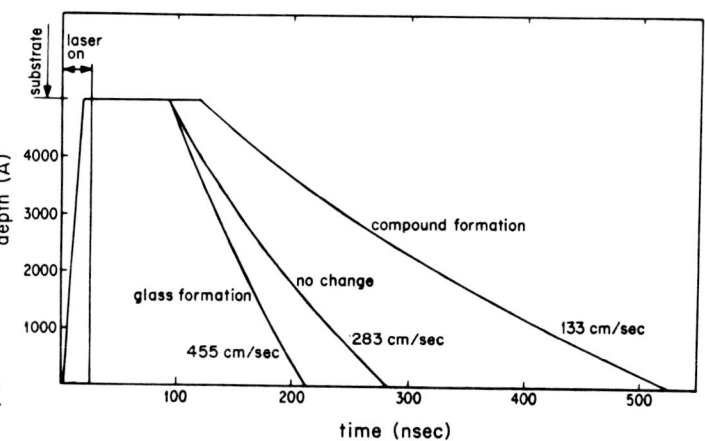

Figure 1. Calculated melt front position as a function of time in a laser-irradiated 5000-Å alloy film on top of an inert substrate. The three curves result from different assumptions on the nature of the solidifying phase, as explained in the text. (From von Allmen et al. 1981.)

For this particular calculation, film thickness and laser pulse duration were chosen to be 500 nm and 25 nsec, respectively; the values for conductivity, specific heat, and melting heat were selected to be about halfway between those of Si and those of Pt. Since the function of the calculation was to obtain some insight into the relevance of various parameters (rather than to obtain accurate temperature values), the melting temperature was arbitrarily set at 1000°C. For the substrate, a constant conductivity κ of 0.24 W/(cm K) was assumed. The three different solidification curves in Figure 1 result from three different assumptions about the solidification process:

(i)   For the curve labeled "no change", the same values for transition temperature and latent heat as for melting were used. This corresponds to the case of an elemental sample or, alternatively, to recrystallization of a mixture without a reaction between the components.

(ii)   For the curve labeled "compound formation", the latent heat for freezing was 1.5 times that for melting. This describes a case in which a compound is formed with a heat of formation of 50% of the average melting heat of the elements. (The corresponding value for formation of PtSi would be 46%; see Figure 5 for the Pt-Si phase diagram.) Further, the freezing temperature of the compound was set at 800°C. The consequence of this reduction in liquid-solid interfacial velocity due to compound formation is discussed in more detail later.

(iii) For the curve "glass formation", it was assumed that latent heat of two-thirds of the melting heat was liberated upon solidification. The freezing temperature was 1000°C in this case. (This procedure neglects the presence of the freezing interval.)

As is evident from Figure 2, rather different values for the interface velocity during solidification can result under otherwise identical conditions, depending on the thermodynamics of the solidification process. As a rule, the interface velocity is roughly proportional to the inverse of the total latent heat liberated at the interface if all other parameters are left unchanged.

Processing times in heat pulse annealing are usually small fractions of a second, thereby making the solid state diffusion negligible. However, the atoms in the melt have high mobilities, corresponding to diffusivities on the order of 10^{-4} cm^2/sec as compared to 10^{-9} to 10^{-12} cm^2/sec in a solid. This allows mass transport over tens of nanometers even for heat pulses in the nanosecond range, and composite samples will necessarily undergo redistribution of the elements.

The diffusivities D in liquids depend on composition as well as on temperature, but generally far less so than in solids. If D is taken to be a constant, concentration distributions due to diffusion in the melt can be readily calculated.

Figure 2 gives concentration distributions for three different initial sample configurations often found in practice:

(i)   The Gaussian (describing a profile obtained by ion implantation of a dopant B in a semiconductor A).
(ii) The rectangular distribution (occurring in the case of a deposited

layer B on top of a substrate A).
(iii) The square wave profile (describing a multilayer of alternating deposited films of A and B).
Cases (ii) and (iii) will also be referred to as "unlimited supply" and "limited supply" of the element A, respectively.

It can be seen that the initial concentration peaks are broadened by an amount of approximately $2\sqrt{Dt_1}$. For more accurate calculations, t_1 is allowed to vary within the molten layer as a function of depth, taking into account the finite velocity of the liquid-solid interface (Lau et al., 1979).

As the conditions governing heat flow during melting establish a boundary condition for atomic diffusion (via the duration of existence of the melt), the latter, in turn, results in an initial condition for solidification:

It defines the local thermo-physical properties (melting point, latent heat) of the molten layer at the instant of solidification and influences the structure of the solid, as well as the velocity of the l-s interface, as demonstrated in Figure 1.

Experiments on silicide and metastable phase formation are usually based on sample configurations shown in Figure 2b and 2c.

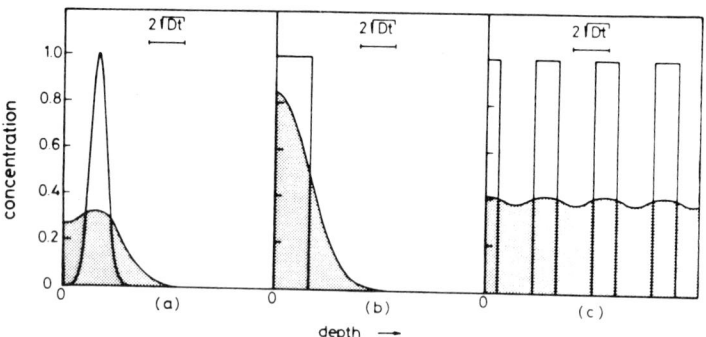

Figure 2. Calculated diffusion profiles in the melt for three different initial concentration profiles. (a) Gaussian (as produced by ion implantation of an impurity into a substrate). (b) Rectangular profile (vapor-deposited thin film). (c) Square wave profile (alternating multilayer).

In principle, an alloy melt can be cooled too fast for either nucleation or growth to take place. If the undercooling is great enough, the melt will, in this case, retain its structure indefinitely and be called a glass (Turnbull, 1969). The question then is, what cooling rate (or what interface velocity) is required for glass formation from a melt of given properties?

These two limiting situations (nucleation and growth limited) have been analyzed by numerous workers and are summarized recently for thin film cases by Von Allmen and Lau (1982). In the following we will discuss two cases: (i) metal semiconductor eutectic systems (Au-Si as an example) and (ii) silicide forming systems (Pt-Si and Pd-Si as examples).

2.1 Metal-Eutectic Systems

Since the first demonstration of an Au-Si metallic glass made by splat cooling in 1960, a multitude of metastable phases (amorphous or crystalline) have been obtained by rapid quenching of the melts. With the extremely high cooling rate made available by short laser pulses, it was a natural consequence to investigate the glass-forming characteristics of classical deep eutectic systems under laser irradiation. The configuration of a metal layer deposited on a semiconductor substrate is suitable for the study of glass formation at the metal-semiconductor interface under <u>unlimited supply conditions</u>. Figure 3 shows the schematics and backscattering spectra for an Au-Si sample before and after laser irradiation (Lau et al., 1981). The thin Si layer (~300 Å) on top of the Au layer (~2500 Å) was to facilitate the absorption of laser power (Nd-glass, 30 nsec pulse). After irradiation, the Au signal spread in width in both directions, indicating that the Au layer had reacted with crystalline as well as amorphous Si. From the decrease in the

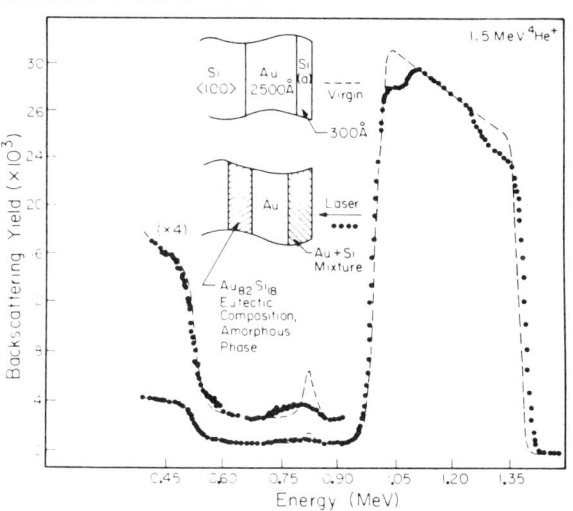

Figure 3. Backscattering spectra of an Au-Si sample (unlimited supply) before and after pulsed laser irradiation. (From Lau et al., 1981.)

height of the Au signal near the Au-Si substrate interface, the mixed layer has a composition of $Au_{82}Si_{18}$ with an amorphous structure. It is interesting to note that the composition of the mixed layer, $Au_{82}Si_{18}$, is the eutectic composition of the Au-Si system.

With the limited supply approach, the extent of glass-forming ability of Au-Si can be investigated at extremely fast quench rates which are induced by pulsed laser irradiation and are not accessible with splat cooling techniques. Experiments of this kind have been done using samples where multiple layers of Au and Si were vacuum-deposited on sapphire substrates. The thicknesses of the layers were adjusted such that the average film composition ranged between $AuSi_{10}$ and $Au_{10}Si$, i.e. from 9 to 91 at .% Au. The individual layer thickness was not more than a few hundred angstroms, with a total layer thickness of about 2000 Å. The surface layer was chosen to be Si and served the purpose of an antireflection coating.

The composition of the irradiated samples remained the same as the initial

average film composition, as monitored by backscattering. As an example,
Figure 4 shows backscattering spectra for a sample with the composition
AuSi₅ before and after irradiation with a 30-nsec pulse. In general, the
layers mixed by 30-nsec pulses showed some residual waviness in
composition versus depth, whereas those obtained with 300-μsec pulses were
very uniform.

X-ray diffraction
analysis of films
irradiated by 30-
nsec pulses
revealed an
amorphous
structure in all
but the most Au-
rich (91 at.%)
samples, usually
together with
traces of a
metastable Au-Si
compound. The
films irradiated
with 300-μsec
pulses were
polycrystalline
and consisted of
the same
metastable
compound,
generally together
with precipitates
of the predominant
component.

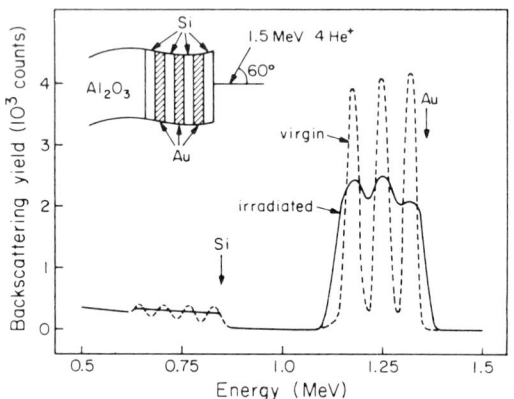

Figure 4. Backscattering spectra of an AuSi₅
sample before and after irradiation with a
Q-switched pulse of 30-nsec duration, Nd-glass
laser. (From von Allmen et al., 1980a.)

The compositional range of Au-Si glasses induced by laser processing
greatly exceeds previously established limits for glass formation by
liquid quenching. Glass-forming ability is believed to be related to
melting point depression (Donald and Davies, 1978), which is largest at
the eutectic point. The only Au-Si glasses reported so far have eutectic
compositions (Klement et al., 1960).

2.2 Silicide-Forming Systems

Laser annealing as a means of forming metal silicide contacts on Si
provides the advantage of transferring energy to only a very localized
region instead of heating up the whole wafer. Because of this practical
aspect, a wealth of experimental investigations have been reported in the
literature (for a summary, see von Allmen and Lau, 1982). For the
limited-supply case, a variety of compounds for a number of metal-Si
systems, usually not observed under thermal annealling conditions, have
been found. However, glass formation is usually not induced by pulsed
laser processing.

The application of the limited-supply techniques to the Pt-Si and Pd-Si
systems lead to interesting phase transformation under the influence of
pulsed laser processing.

Unlike the case of Au-Si, not all the compositions tested could be quenched into an amorphous state. Generally speaking, polycrystalline films resulted from samples with compositions at or close to those of the congruently melting compounds, Pt$_2$Si, PtSi and Pd$_2$Si, PdSi, respectively. Figure 5 shows the phase diagrams of the two systems along with arrows marking the compositions of the films studied. The letters beneath the arrows give the structure of the as-irradiated films.

Intimate mixing of the components by laser irradiation was found to be essential in obtaining amorphous films; incompletely mixed films containing local compositions too close to that of a congruently melting compound were not amorphous after irradiation but rather showed the presence of the congruently melting compound. The reason for this could lie in the liberation of the heat of formation of the compound, which reduces the cooling rate as discussed previously. The fact of a composition-dependent cooling rate notwithstanding, it seems clear that the cooling rate necessary to quench a liquid binary mixture into a glassy state is highest for compositions close to those of congruently melting compounds, as well as to those of the pure components.

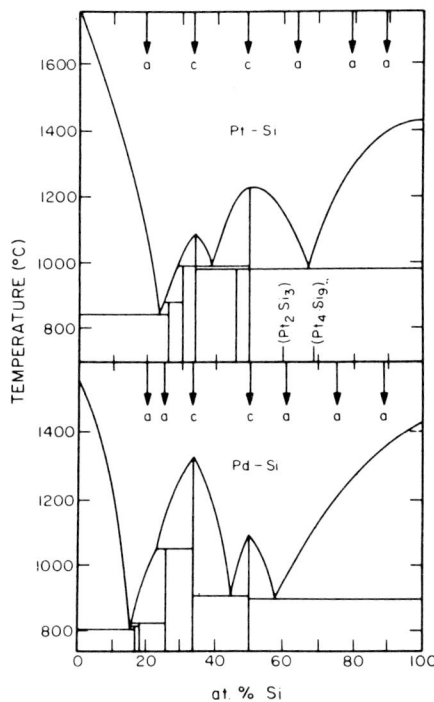

Figure 5. Phase diagrams of the Pt-Si and Pd-Si systems. The arrows mark the compositions of films used in this study. Small letters indicate the structure of the As-quenched films (a, amorphous; c, polycrystalline). (From von Allmen et al., 1980b.)

3. Ion Mixing Experiments

Ion-induced mixing not only can lead to stable compound formation, but also to metastable alloy formation. In some metal-metal systems, terminal solubilities can be greatly extended by ion mixing. In other cases, where the two constituents of the system have different crystal structures, extension of terminal solubility for both sides of the phase diagram eventually becomes structurally incompatible and a glassy (amorphous)

mixture can form. The composition range where this bifurcation is likely
to occur is in the two-phase regions of the phase diagram. These concepts
are potentially useful guides in selecting metal pairs that form metallic
glasses by ion mixing.

3.1 Equilibrium Compound Formation by Ion Mixing

The formation of equilibrium compounds by ion mixing at room temperature
has been investigated rather extensively in silicide forming systems using
bilayered structures. The ion-induced silicide formation characteristics
can be compared with those obtained by thermal steady-state annealing. It
can be clearly seen that the <u>first
silicide phase</u> induced by ion mixing
in unlimited supply samples is the
same as that obtained by thermal
annealing. Although some of the
layers that were ion mixed at room
temperature (RT) do not exhibit a
distinct crystal structure due to
weak x-ray diffraction patterns, the
composition of the mixed layer
deduced by backscattering is
invariably the same as that obtained
by thermal annealing. As the
temperature of irradiation is
increased to above RT, the crystal
structure of the phase can be
identified even in those cases as
being the same as that obtained by
thermal annealing. If we assume
that ion mixing at room temperature
is in the thermally activated regime
for all systems studied, we can
conclude that there is a direct one-
to-one correlation between thermal
annealing and ion mixing in the
thermally activated regime, at least
in silicide forming systems (Tsaur
1980). This is in marked contrast
to the case of pulsed laser
processing of <u>bilayered</u> samples
where the induced reactions are of a
rather complex nature (von Allmen
and Lau, 1982).

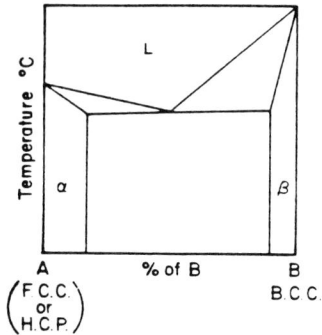

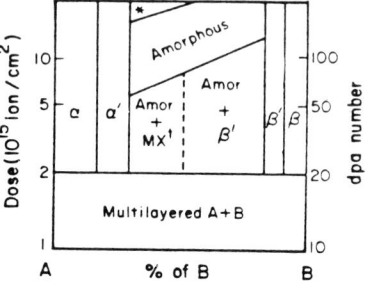

* sometime amorphous phase
dissociates upon relatively high
dose irradiation.

† MX is of h.c.p. structure but is
different from the H C P
metal A in size

3.2 Glass Formation by Ion Mixing

Previous experimental investigations
on ion-induced reactions using
multilayered samples (limited
supply) showed that ion mixing is
well-suited for metastable phase
formation in metal-metal systems
(Mayer et. al, 1981). In
particular, when the two components
of the binary system have the same
crystalline structure, but exhibit a

Figure 6. See text for
explanation.

miscibility gap in the phase diagram, supersaturated solid solutions are obtained upon ion mixing (for example, Ag-Cu, f.c.c. structure). On the other hand, when the crystal structures of the two components are different (say f.c.c. for component A and b.c.c. for component B), a tendency to form amorphous alloys by ion mixing can clearly be detected.

Figure 6 shows a schematical phase diagram with different crystal structures of the elements and the sequence of events as a function of dose (or dpa). Intermixing of the two phases is observed at relatively low doses ($<2 \times 10^{15}$ Xe/cm^2 or <20 dpa). For samples with compositions near either A or B, extended solid solutions α' and β' with structures identical to their parent phases are observed at increasing doses (α' has the structure of α, β' that of β). As the composition approaches from the A side (f.c.c. or h.c.p.) to the middle of the two-phase region, the mixed layer ($20 \lesssim \text{dpa} \lesssim 50 - 100$) consists of a mixture of an amorphous phase and a metastable crystalline phase of h.c.p. structure (but with a different c/a ratio compared to that of A if A is also h.c.p.). On the B side, a mixture of an amorphous phase and β' (b.c.c.) is observed in the mixed layer. Increasing the ion dose leads to complete amorphization of the mixed layer in a composition range near the eutectic composition. At even higher doses, the amorphous phase may dissociate into metastable or equilibrium phases, possibly due to ion-beam demixing effects.

It is found that in eight metal-metal binary systems with multilayered sample configuration, irradiation leads to uniform mixing of the interposed layers. An amorphous alloy is formed in every case, irrespective of the atomic size and the electronegativity properties of the constituents, as long as the constituents have different crystal structures (Liu et al. 1983).

4. Summary

In summary, glass formation in binary systems requires (i) uniform intermixing between different components in a given system and (ii) fast quench rates after the mixing is completed. MeV backscattering spectrometry provides a very convenient way to examine the mixing phenomenon after energetic beam processing. There are numerous other applications of backscattering spectrometry to material analysis which are well discussed in the two books and many articles reported in the literature on this subject.

Acknowledgement

The financial support of DARPA (S. Roosild) is greatly appreciated.

References

Chu W K, Mayer J W and Nicolet M A 1978 Backscattering Spectrometry (Academic Press)
Cohen M H and Turbull D 1961 Nature (London) 189, 132
Davies H A and Lewis B G 1975 Scr. Metall. 9, 1107
Donald I W and Davies H A 1978 J. Non-Cryt. Solid 30, 77
Duwez P 1967 ASM Trans. Q. 60, 83
Feldman L C, Mayer J W and Picraux S T 1982 Materials Analysis by Ion Channeling (Academic Press)
Klement W, Willen R H and Duwez P 1960 Nature (London) 187, 869

Lau S S, Tsaur B Y, von Allmen M, Mayer J W, Stritzker B, White C W and
    Appleton B 1981 Nucl. Instr. Meth. 182/183, 79
Liu B X, Johnson W L, Nicolet M A and Lau S S 1983 Appl. Phys. Lett.
    42(1), 45
Mayer J W, Tsaur B Y, Lau S S and Hung L S 1981 Nucl. Instr. and Meth.
    182/183, 1
Sinha A K, Giessen B C and Polk D E 1976 Treatise on Solid State
    Chemistry (N B Hannay Ed. Plenum, NY) Vol.1 pp.1
Tsaur B Y 1980 Proc. Symp. on Thin Film Interfaces and Interactions,
    (J. E. E. Baglin and J. M. Poate, Eds.) The Electrochemical Society,
    Princeton Vol.80-2, pp.205
Turbull D 1969 Contemp. Phys. 10, 473
von Allmen M, Lau S S, Maenpaa M and Tsaur B Y 1980a Appl. Phys. Lett.
    36, 205
von Allmen M, Lau S S, Maenpaa M and Tsaur B Y 1980b Appl. Phys. Lett.
    37, 84
von Allmen M, Affolter K and Wittmer M 1981 Laser and Electron Beam Solid
    Interactions and Materials Processing (J. F. Gibbons, L. D. Hess and
    T. W. Sigmon, Eds. North-Holland, NY) pp.559
von Allmen M, Lau S S 1982 Laser Annealing of Semiconductors (J. M. Poate
    and J. W. Mayer, Eds. Academic Press, NY) pp..439-478

Inst. Phys. Conf. Ser. No. 67: Section 3
Paper presented at Microsc. Semicond. Mater. Conf., Oxford, 21–23 March 1983

Ultra-rapid solidification of transiently molten laser-annealed silicon

A G Cullis, H C Webber and N G Chew

Royal Signals and Radar Establishment, St. Andrews Road, Malvern,
Worcs. WR14 3PS, England

Abstract   Very short ruby laser pulses are used in studies of the high
speed melting and solidification of ion implanted Si. Cross-sectional
transmission electron microscopy is used to study annealed samples and
demonstrates for the first time the orientation dependence of rapid
crystal growth from the melt. Crystal of (001) orientation exhibits
a growth velocity maximum which is much higher than that of (111) crystal.
Under the most extreme conditions, undercooling in the transient melt
can become so great that amorphous Si can form directly by solidification. The melting temperature of amorphous Si itself is estimated to
lie in the range 1185-1385K.

1. Introduction

Short pulses of radiation from Q-switched lasers can be used to transiently
melt the near surface regions of a solid. The thermal gradients in the
material are very steep so that the velocity of the solidification interface
can be extremely high. This feature of the process is exploited in the
present investigations to probe the very high speed crystal growth behaviour
of semiconductor Si. By use of 2.5ns, ruby laser pulses it has been
possible to melt ion implanted layers and to achieve final solidification
velocities of $\gtrsim 10m/s$. Predictions (Jackson 1983) concerning the orientation
dependence of crystal growth velocities in this regime are shown to be
generally consistent with observations. There has also been considerable
speculation about the thermodynamic properties of amorphous Si - in particular concerning the magnitude of its melting temperature (T_{al}) in relation
to the melting temperature (T_{cl}) of perfect Si crystal (Spaepen and Turnbull
(1979), Bagley and Chen (1979), Donovan et al (1983)). The present work
employs computer modelling of the melting sequence and demonstrates that
T_{al} is a few hundred Kelvins below T_{cl}, in accord with earlier electron-
beam measurements (Baeri et al 1980). It is shown that amorphous Si melts
at this reduced temperature to give a low viscosity liquid which exhibits
normal impurity diffusion behaviour.

2. Experimental

Radiation pulses of 2.5ns duration and 694nm wavelength were obtained from
a pulse-chopped ruby laser system. The pulses, which were rendered spatially
uniform using a light guide diffuser, were used to anneal a range of samples
of (001) and (111) ion implanted Si. The latter were prepared by implantation of one of the following ion species: 10^{16} In$^+$ ions/cm^2 at 170keV,
3×10^{15} Ar$^+$ ions/cm^2 at 70keV or 5×10^{15} As$^+$ ions/cm^2 at 80keV. The various
as-implanted and annealed samples were examined in cross-section in a

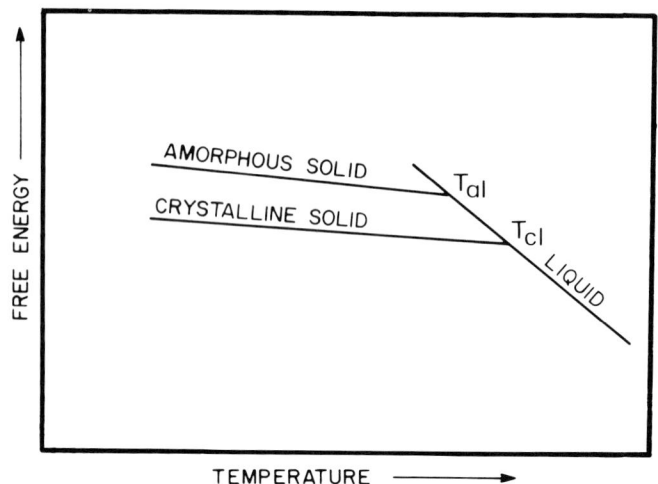

Fig. 1 Schematic diagram of free energy as a function of temperature for Si in crystal, amorphous and liquid states.

transmission electron microscope (TEM).   Specimens for structural study
were thinned to electron transparency by sequential mechanical polishing
and low voltage ion beam milling.

3. Results and Discussion

The high temperature phase transition behaviour likely to be exhibited by
amorphous Si is best deduced from free energy considerations. A schematic
diagram (Bagley and Chen 1979) of free energy vs temperature for the
different forms of Si (amorphous, crystal and liquid) is shown in Fig. 1.
The curve for amorphous Si lies above that for crystal and it is predicted
that amorphous material undergoes a melting phase transition at a temperature T_{al} which is below T_{cl}. The magnitude of T_{cl} - T_{al} has been in
dispute but it can be estimated from melting behaviour investigated in this
work, as described later.

When amorphous Si is irradiated with very short laser pulses, it is possible
to melt it (to give an undercooled liquid) for such a short period that
no recrystallization takes place and amorphous material reforms upon solidification. The fact that melting has actually taken place is most easily
revealed by the study of impurity segregation phenomena during the melting
sequence. Figure 2a shows an amorphous layer formed by implantation of
10^{16} In^+ ions/cm^2: the In is atomically dispersed and the amorphous layer
is very uniform. When this layer is melted (for less than $\sim$10ns) with a
laser pulse of $\sim$0.25J/cm^2, the melt front does not reach the crystal interface and the transient undercooled liquid resolidifies rapidly in an amorphous state. However, as shown in Fig. 2b, the In present segregates out
to give a precipitate band near the centre of the layer (Cullis et al 1982a).
The fine trails protruding from the band are the result of constitutional
supercooling which occurs in the liquid and yields a cellular precipitate
structure in the plane of the band. The general characteristics of the
impurity diffusion which occurs during the resolidification indicate that
amorphous Si is not formed from the liquid by e.g. a glass transition. The

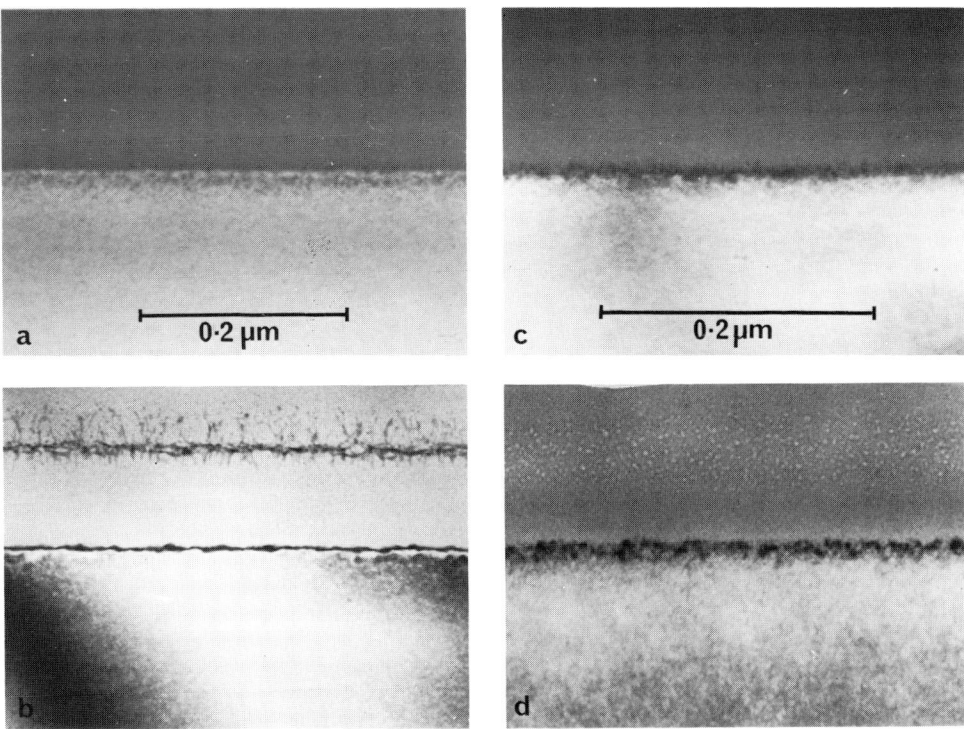

Fig. 2   Cross-sectional TEM images of ion implanted and laser
annealed Si layers: 10^{16} 170keV-In$^+$/cm^2 a) as-implanted and
b) 0.25J/cm^2; 3×10^{15} 70keV-Ar$^+$/cm^2 c) as-implanted and
d) 0.25J/cm^2.

behaviour of other impurities allows the same conclusion to be deduced.
For example, as shown in Figs. 2c and d implanted Ar can redistribute in
transient undercooled liquid Si to give an array of bubbles which become
trapped in the reforming amorphous material.   Once again, the bubble size
and distribution indicates that the undercooled liquid does not have a
viscosity much higher than the normal high temperature liquid.

When the laser pulse energy density is high enough to allow the transient
melt to penetrate to underlying crystalline Si, epitaxial recrystallization
can then take place upon resolidification.   However, the nature of the
crystal growth is markedly dependent upon the orientation of the substrate
lattice.   This is clearly demonstrated in Fig. 3 which compares (001) and
(111) pairs of As$^+$ ion implanted layers that were each annealed under
identical conditions.   When the energy density of the irradiating 2.5ns
pulse was ∿0.3J/cm^2 the melt penetrated to the crystal interface but,
whereas extensive, explosive-like recrystallization events had occurred
on an (001) substrate (Fig. 3a) only a very shallow (∿200Å) layer of new
crystal had formed on a (111) substrate.   In both cases, when crystal
formation in the resolidifying liquid ceased, solidification of remaining
liquid yielded an amorphous overlayer.   With increasing energy density of
irradiation (to 0.4 and 0.45J/cm^2), as Figs. 3b, c, e and f show, the (001)
substrate allowed complete recrystallization to occur while on the (111)

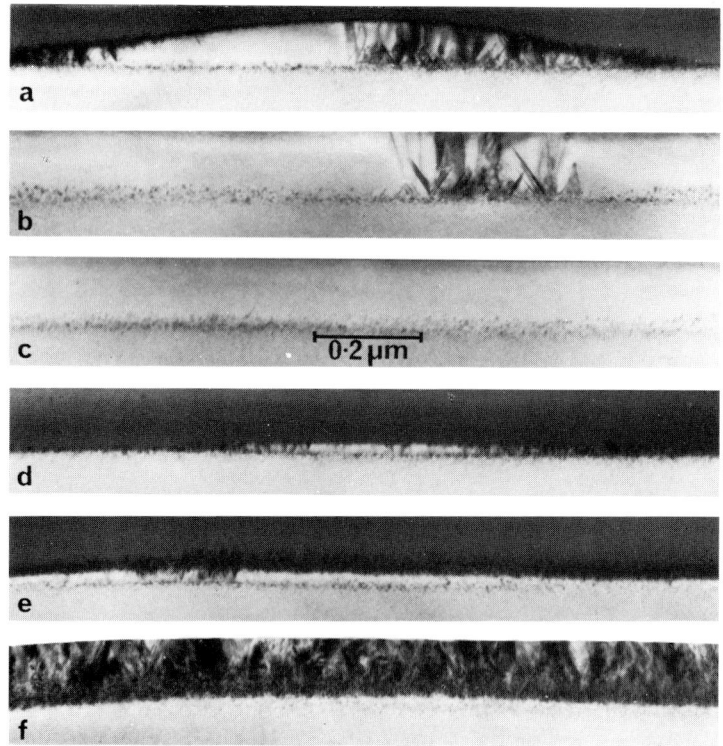

Fig. 3 Cross-sectional TEM images of As$^+$ ion implanted and laser annealed Si layers: (001) Si a) 0.3J/cm^2, b) 0.4J/cm^2 and c) 0.45J/cm^2; (111) Si d) 0.3J/cm^2, e) 0.4J/cm^2 and f) 0.45J/cm^2.

substrate still only a few hundred Ångströms of good crystal formed at the original interface.

From these results it can be deduced that the maximum crystal growth rate in the (001) direction is considerably greater than that in the (111) direction. Given this situation, if layers of the two orientations are transiently melted under otherwise equivalent conditions, the slower re-crystallization for the (111) substrate yields a slower rate of release of stored latent heat and, hence, a much faster decrease in the temperature of the remaining melt. When this temperature falls below T_{al} amorphous Si can form directly by solidification. Therefore, this condition occurs more readily during (111) crystal growth and can lead to the formation of relatively thick amorphous layers during the later stages of resolidifica-tion. It is also interesting to note that Fig. 3f shows that at 0.45J/cm^2 melting of the (111) implanted layer led to profuse twin formation after a narrow band of crystal had reformed upon resolidification. This is a consequence of an intermediate transition to defective crystal (Cullis et al 1982b) which is exhibited only by (111) material when the transient melt undercooling does not quite fall to T_{al}. A more detailed description of these orientation-dependent fast crystal growth phenomena will be given elsewhere (Cullis et al 1984).

The general differences in maximum growth rates for (001) and (111) crystals

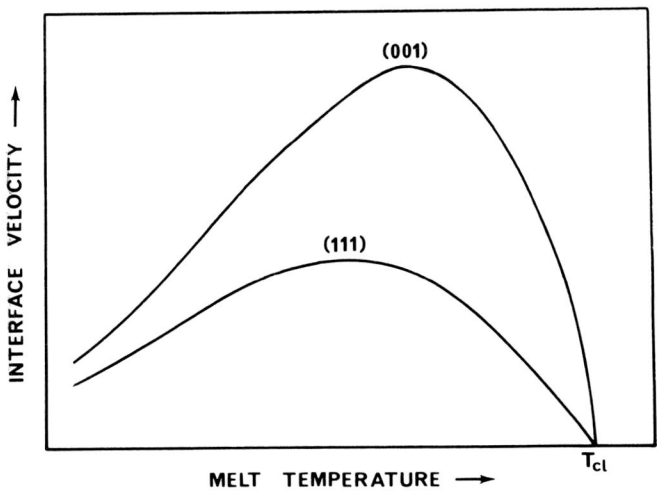

Fig. 4   Theoretically predicted growth velocities of (001) and
(111) Si as functions of temperature for a range of melt under-
coolings. (After Jackson (1983).)

have been previously predicted on a theoretical basis (Jackson 1983). The
overall form of the likely crystal growth velocity vs melt undercooling
relationships is shown in Fig. 4. Although growth velocity increases as
the temperature falls below T_{cl} , it reaches a maximum and ultimately
decreases due to effects of the finite melt viscosity. The velocity maximum
is predicted to be lower for (111) crystal (in accord with the experimental
observations outlined above) due to the relative difficulty of plane
nucleation on the faceted (111) surface.

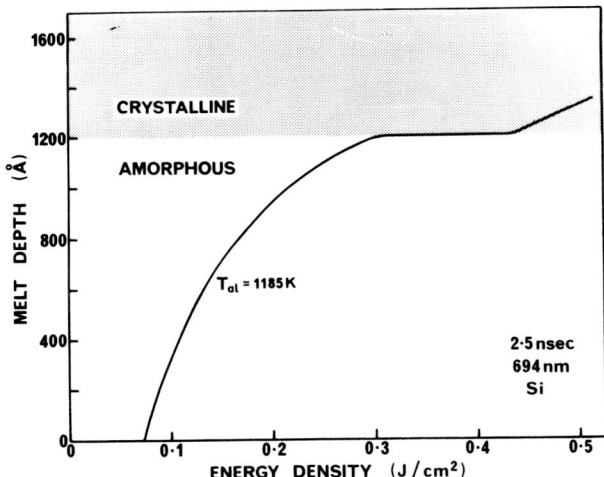

Fig. 5 Computed melt depth as a function of pulse energy density
for 2.5ns ruby laser pulses incident on a 1200Å amorphous Si
layer (T_{al} = 1185K).

It is also clearly of importance to determine the actual magnitude of T_{al}. This can be deduced from our work on computer modelling of the maximum melt depth as a function of laser pulse energy density - this work does not depend upon details of the final solidification sequence. Taking optical and thermal parameters as given elsewhere (Webber et al 1983), computations have been carried out for values of T_{al} between T_{cl} (1685K) and 1185K. Results of calculations for 2.5ns laser pulses incident on a crystal sample with a 1200Å thickness amorphous surface layer and with T_{al} = 1185K are shown in Fig. 5. The most important feature of the curve is the plateau region at the depth of the amorphous/crystal interface. The length of the plateau is critically dependent upon the exact value of T_{al} (for example, the plateau length is zero if T_{al} is artificially made equal to T_{cl}) and it is this parameter which is matched to the experimental work in order to obtain an estimate of T_{al}. Experimental results obtained for sample irradiations corresponding to those specified in the calculations show a good correlation with the theoretical curve for T_{al} = 1185K in Fig. 5. However, highly accurate determination of T_{al} is not at present possible due to uncertainties in a number of the parameters used in the calculations. Nevertheless, particularly when related to other recent studies (Donovan et al 1983), this modelling indicates that T_{al} is most likely to lie in the range 1185-1385K.

4. Conclusions

The present laser melting studies demonstrate for the first time that (001) Si crystal grows from undercooled liquid Si at velocities which are much higher than those attained by (111) crystal under corresponding conditions. This result is in accord with theoretical predictions. If the temperature of the undercooled melt falls below T_{al} amorphous Si can form directly by a phase transition which is most likely to be first order. The best estimate for T_{al} is 1185-1385K.

References

Baeri P, Foti G, Poate J M and Cullis A G 1980 Phys. Rev. Lett. 45 2036
Bagley B G and Chen H S 1979 Laser-Solid Interactions and Laser Processing - 1978 eds S D Ferris, H J Leamy and J M Poate (New York: Amer. Inst. Phys.) pp 97-101
Cullis A G, Webber H C and Chew N G 1982a Appl. Phys. Lett. 40 998
Cullis A G, Webber H C and Chew N G 1984 J. Crystal Growth (in the press)
Cullis A G, Webber H C, Chew N G, Poate J M and Baeri P 1982b Phys. Rev. Lett. 49 219
Donovan E P, Spaepen F, Turnbull D, Poate J M and Jacobson D C 1983 Appl. Phys. Lett. 42 698
Jackson K A 1983 Proc. NATO Institute on Surface Modification and Alloying eds J M Poate and G Foti (New York: Plenum)
Spaepen F and Turnbull D 1979 Laser-Solid Interactions and Laser Processing - 1978 eds S D Ferris, H J Leamy and J M Poate (New York: Amer. Inst. Phys.) pp 73-83
Webber H C, Cullis A G and Chew N G 1983 Appl. Phys. Lett. (in the press)

Inst. Phys. Conf. Ser. No. 67: Section 3
Paper presented at Microsc. Semicond. Mater. Conf., Oxford, 21–23 March 1983

173

SEM studies of structural defects induced by thermoelastic stresses during pulsed electron beam annealing in silicon

M Pitaval, M Tholomier, M Ambri, G Chemisky*, D Barbier* and A Laugier*

Département de Physique des Matériaux, Université Claude Bernard - LYON I
43 Boulevard du 11 Novembre 1918, 69622 VILLEURBANNE Cédex - FRANCE

*Laboratoire de Physique de la Matière, Institut National des Sciences Appliquées de Lyon, 20 Avenue A. Einstein 69621 VILLEURBANNE Cédex - FRANCE

Abstract   Crystal regrowth of As$^+$ implanted silicon by pulsed electron beam annealing is studied by means of scanning electron microscopy in the electron channelling imaging mode. Dislocations and subgrain boundaries are induced during the recrystallisation process by stresses associated with the high thermal gradient. The depth affected by dislocations is measured on bevelled samples. Results obtained for 1.0 and 1.2 J.cm^{-2} fluences are compared with computed profiles of temperature. It is concluded that subgrains extend over the layer reaching the melting temperature.

1. Introduction

Ion implantation in semiconductor devices is an attractive process to obtain shallow junctions. It must be followed by annealing to restore crystallographic order and to active dopant atoms in substitutionnal sites. Since few years, submicrosecond electron beam pulses irradiating a large area of the wafer are used for superficial annealing. For crystal regrowth and dopant activation, the molten layer needs to extend deeper than the layer containing implantation defects. Dopant redistribution occurs in the molten region, it is strongly dependant on the freezing rate.   Associated with high temperature gradient at the limit of the melting layer, induced stresses are relaxed in silicon by dislocations and subgrains network for a deposited energy density (fluence) in the 0.8 J.cm^{-2} to 1.2 J.cm^{-2} range. For higher fluences, thermal stresses are sufficiently high for surface cracking as predicted by Schoen (1980). These defects were previously observed by electron microscopy (Foti et al 1981, Tholomier et al 1983, Narajan et al 1980); we studied them with a SEM, taking advantages of the electron channelling imaging mode which allows to image crystallographic defects near the surface of solid specimens. We correlate their extension in the material with the computed temperature kinetics for 1.0 J.cm^{-2} and 1.2 J.cm^{-2} fluences.

2. Experimental procedure

P type silicon wafers (001) oriented were implanted with 10^{15} As$^+$ cm^{-2} at 140 keV. The range of penetration is 80 nm and silicon is amorphized in a layer 160 nm thick as deduced from Rutherford backscattering data. Specimens were annealed with a SPIRE 300 pulsed electron beam processor. The pulse had a duration of about 50 ns. The annealed area is limited by the cathode diameter (19 mm in these experiments). The poly-energetic beam electrons had a mean energy of 15 ± 2 keV, the fluence was varied from 0.8 J.cm^{-2} to 1.2 J.cm^{-2}, it was measured by means of a graphite calorimeter with a 6% accuracy. Annealed wafers were then cut in 2 mm × 2 mm squares. To observe the depth distribution of defects into the sample, the specimens were bevelled by polishing half of their surface with a NaOH solution and colloïdal silica mixture on an altuglass sheet with a lapping-polishing machine. The edge of the bevel lies roughly along the [110] axis. The 2°52' angle between original and polished surface leads to a 1 μm variation in depth when moving 25 μm along the bevelled surface.   Last, specimens were carefully rinsed to remove silica grains. These specimens allow SEM observations either on the original surface or over the whole annealed layer. The SEM was operated in the ECI mode to image extended crystallographic defects at the surface of solid specimens. Contrast occurs when the curvature of atomic planes near the core of a dislocation modifies the electron channelling (Booker 1970). The microscope is described in previous papers (Pitaval et al 1977 (a, b) ; Morin et al 1979) and some examples of applications showing the ability of this method is overviewed by Fontaine et al in this conference. In these observations, the specimen is carefully oriented in the Bragg position for a selected diffraction reciprocal vector $\vec{g}$. The orientation is achieved by means of the corresponding line on the electron channelling pattern obtained at low magnification with these samples (Coates 1967). Defects in contrast with the selected $\vec{g}$ are imaged at higher magnification.

3. Computed melting and freezing kinetics

The spatial energy distribution deposited in the wafer by the pulsed electron beam is computed with a Monte Carlo simulation taking into account the energy spectrum of the incident beam (Chemisky et al 1983). The temperature profiles are deduced from the density of energy deposited. At any time, they are obtained by solving the one dimensional heat flow equation taking into account the temperature dependance of the specific heat and the thermal conductivity. In this model, the latent fusion heat is 1800 J.g^{-1} for crystalline silicon, this value is lowered by 40 % for the amorphous silicon layer.

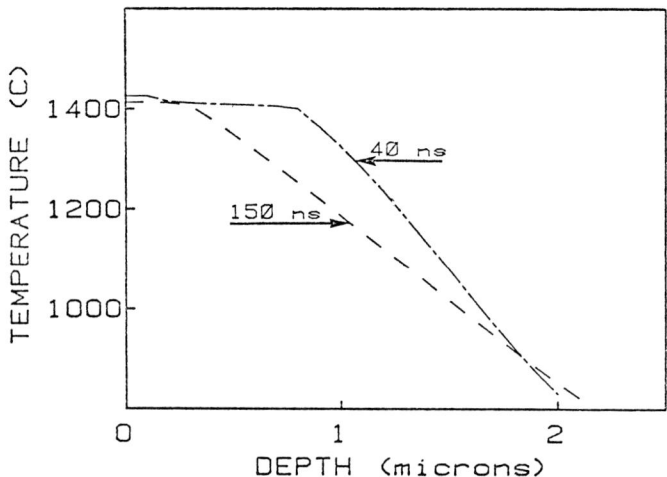

Fig. 1   Temperature profiles at the end of the electron pulse (t = 40 ns) and evolution with time (t = 150 ns) for 1.2 J.cm^{-2} fluence

Profiles are given on Fig. 1 for a fluence of 1.2 J.cm^{-2} at the end of the electron pulse (t = 40 ns) and later (t = 150 ns). The first curve shows three

regions from the surface : the first one receives an energy exceeding the energy
needed for melting. Due to the low latent heat of amorphous silicon, it is limited to the implanted zone. The second region reaches the melting temperature,
but the energy deposited is lower than the latent heat : it corresponds to a
layer where atoms are fast diffusing due to the broken bonds of silicon.
The third zone, the deepest one, does not reach the melting point, it is subject
to a temperature gradient of about 500 K μm^{-1} involving high thermoelastic
stresses. This temperature gradient is little dependent on the beam fluences.
Melting temperature is just reached at the limit between the second and third
zones. The depth position of this limit is given versus time on Fig. 2, its
maximum depth increases with fluence. It reaches 0.7 μm for 1.2 $J.cm^{-2}$, 0.4 μm
for 1.0 $J.cm^{-2}$. It is less than the amorphous layer for 0.8 $J.cm^{-2}$. It returns
to the surface with a speed decreasing with fluence (from 3.6 ms^{-1} for 1 $J.cm^{-2}$
and 2 ms^{-1} for 1.4 $J.cm^{-2}$). The whole recrystallisation process duration is
given on Fig. 2 by the intersection of curves with the time axis.

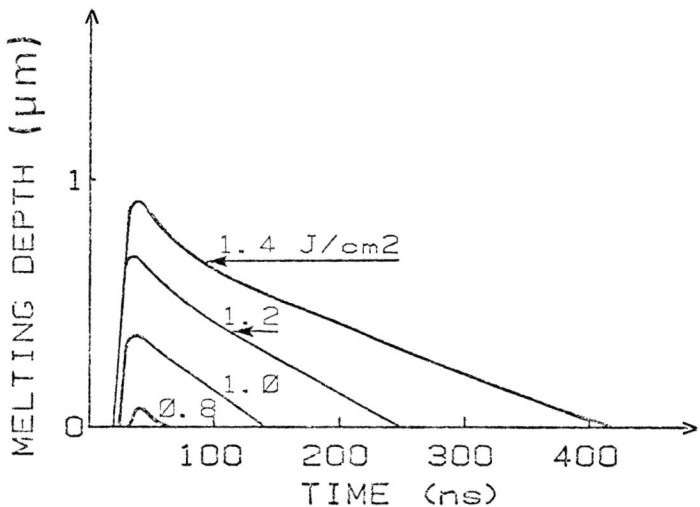

Fig. 2   Depth of the melting zone limit with time for different fluences

4. SEM observations

To have an efficient backscattered detection, specimens are 45° tilted with
respect to the SEM beam due to the orientation of the filter acceptance angle.
Consequently, on photographs, magnification is reduced in the Y direction by
a $\sqrt{2}$ factor. All photographs are recorded when specimen is oriented in Bragg
position for [220] reciprocal vector.
SEM observations in the ECI mode were made on bevelled samples (Fig. 3).

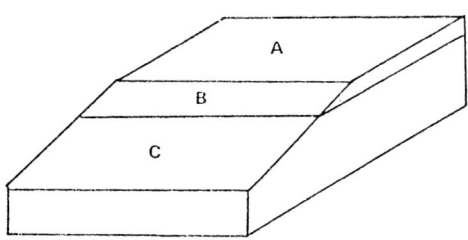

Fig. 3 Schematics of the specimen :
A is the original surface, B and
C are bevelled surface. Defects
extended in a layer intersecting
the bevelled surface in B. C is
the surface of perfect crystal.

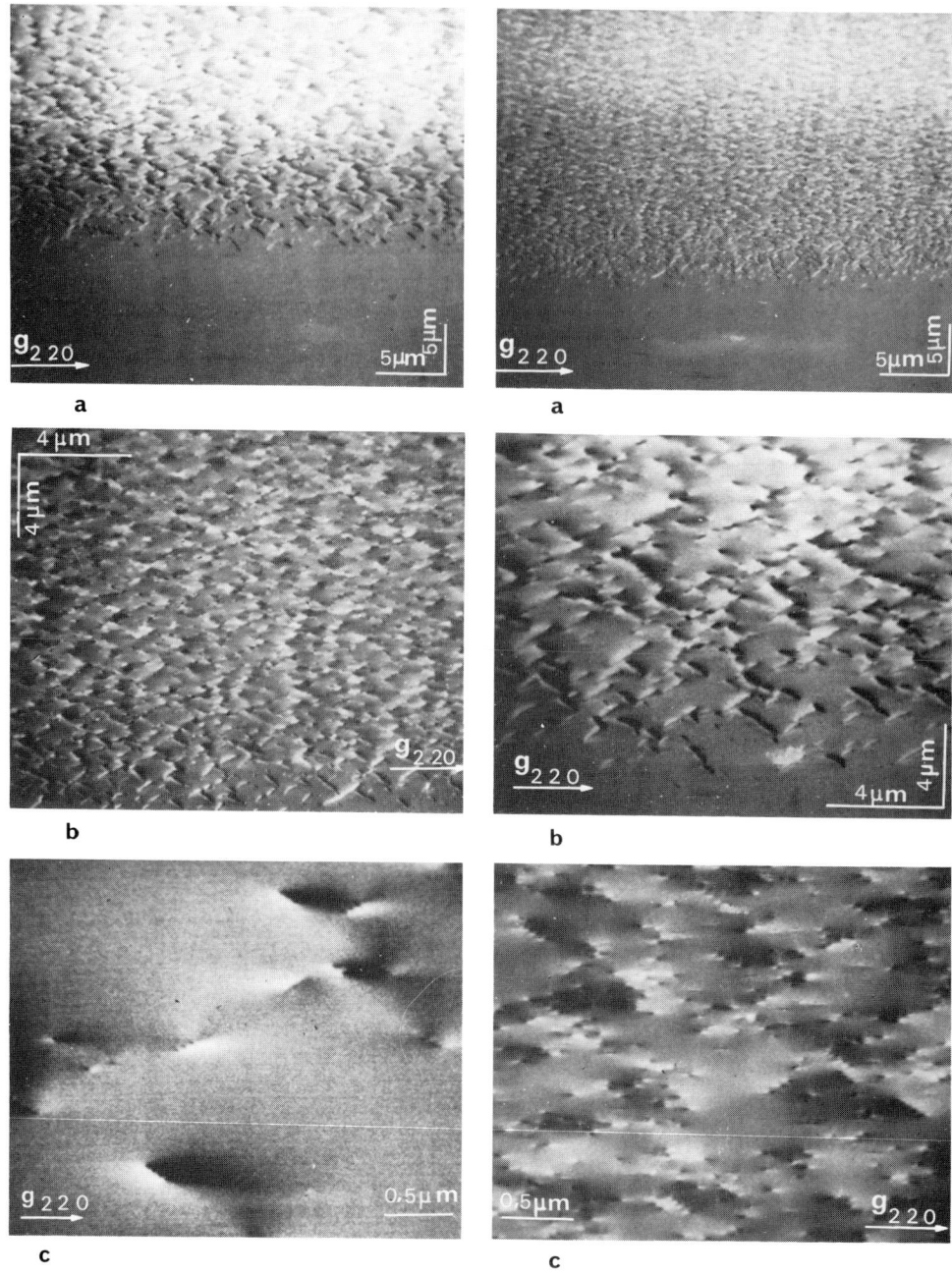

Fig. 4   1.0 J.cm⁻² annealed
         specimen

Fig. 5   like Fig. 4 with
1.2 J.cm⁻² annealed sample

(a) Depth extension of defects is measured on
    the bevelled surface
(b) subgrains exhibits crystallographic contrast
(c) Dislocations piled up in subgrain walls

These observations are presented at different magnifications on Fig. 4 for
1.0 J.cm^{-2} fluence and on Fig. 5 for 1.2 J.cm^{-2} fluence. For an easier comparison, photographs are given at the same magnification for each fluence. No
extended defects were observed on 0.8 J.cm^{-2} annealed samples.
Fig. 4a and Fig. 5a are low magnification photographs of the level edge region.
For each image, the upper part A is a view of the unlapped surface. Parts B
and C correspond to the bevelled zone as shown on Fig. 3. Surface A exhibits
a bright contrast with respect to B and C correlative to the variation of the
incidence angle.
Crystallographic defects in A and B regions are pointed out on Fig. 4a and
Fig. 5a. Their depth extension is measured from the distance between A-B and
B-C lines. It is found a 0.45 µm thickness for 1.0 J.cm^{-2} fluence and 0.7 µm
for 1.2 J.cm^{-2} fluence. By comparing these results with computed temperature
profiles, it is concluded that for these fluences the defects extend over the
depth of the so called second zone (§.3) which just reaches the melting temperature.
The whole B area is viewed at higher magnification on photographs Fig. 4b and
Fig. 5b. With 1.2 J.cm^{-2} fluence annealed specimen (Fig. 5b), subgrains slightly
desoriented exhibits a low crystallographic contrast, they make a regular array
with boundaries lying along the [010] and [100] directions. Grains have a homogeneous size of about 0.5 µm over the specimen surface. When the fluence is
lowered to 1.0 J.cm^{-2} (Fig. 4b) subgrains network is still present but with a
larger cells size (3 µm), the boundary direction remaining unchanged.
Individual dislocations are resolved Fig. 4c and Fig. 5c recorded on the original surface A at high magnification. They look as mall dots because they
are inclined with respect to the surface and are viewed end-on. Almost all
dislocations are arranged in cell walls for 1.0 J.cm^{-2} (Fig. 4c) while isolated dislocations can be observed in cells for 1.2 J.cm^{-2}. The difference in
defect densities is obvious from the two photographs, we estimate a mean value
of 2.5×10^9 dislocations.cm^{-2} for 1.2 J.cm^{-2} and 3.10^8 dislocations.cm^{-2} for
1.0 J.cm^{-2} fluence.
Subgrain limit with the remaining perfect crystal is shown at high magnification on Fig. 6.

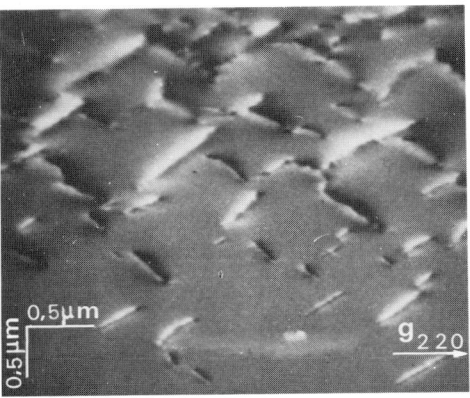

Fig. 6   1.2 J.cm^{-2} annealed sample. Dislocation lines at
the limit of perfect crystal are parallel to the surface

Dislocations lines are no longer imaged as small dots but appears as lines.
Since the ECI mode images dislocations up to 1000 Å from the surface, these
dislocations lies nearly parallel to the surface. This corresponds to the lower

part of extended dislocations with a "half loop" form as observed with TEM on cross section by Foti et al (1981).

For higher fluences, excessive thermal stresses lead to transgranular surface cracking as observed on micrograph Fig. 7 recorded on the original surface of a 1.4 J.cm^{-2} annealed specimen.

Fig. 7   Transgranular surface cracking on 1.4 J.cm^{-2} annealed specimen

5. Conclusion

In this work, pulsed electron beams with 15 keV of mean energy were used to anneal As implanted (140 keV 10^{15} cm^{-2}) (100) silicon. The associated defects were observed with SEM in the ECI mode. Dislocation lines and subgrains boundaries are present on all treated samples with fluences above the recrystallisation threshold (0.8 J.cm^{-2}). Dislocation density increases with fluence : It is 3.10^8 cm^{-2} for 1.0 J.cm^{-2} and 2.5×10^9 cm^{-2} for 1.2 J.cm^{-2} respectively. Moreover subgrains are present : they have an homogeneous size of about 3 μm for 1.0 J.cm^{-2} and 0.5 μm for 1.2 J.cm^{-2}. The depth defect distribution has been measured on bevelled samples. It well agrees with the computed extension of the zone where the melting temperature is just reached, as determined by solution of the time dependant heat flow equation.

Aknowledgements

The authors thank "Commissariat à l'Energie Solaire" (COMES) for financial support.

References

Barbier D, Laugier A and Cachard A 1982 Journ. Phys. 12 43 C5 411
Booker G R 1970 Modern Diffraction and Imaging Techniques in Material Sciences
Chemisky G, Barbier D and Laugier A 1983, to be published
Coates D G 1967 Phil. Mag. 16 1179
Foti G, Grimaldi M G, Cullis A G, Poate J M and Chew N G 1981 Microsc. Semicond. Mat. Conf. Inst. Phys. Conf. Ser. 60 2 79
Morin P, Pitaval M, Besnard D and Fontaine G 1979 Phil. Mag. A 40 4 511
Narajan J 1980 "Laser and electron beam process. of Mat." Acad. Press. 397
Pitaval M, Morin P, Baudry J and Fontaine G 1977a J. Mic. Spect. Elec. 2 185
Pitaval M, Morin P, Baudry J, Vicario E and Fontaine G 1977b SEM IITRI 1 439
Schoen N C 1980 J. Appl. Phys. 51 4747
Tholomier M, Barbier D, Pitaval M, Ambri M and Laugier A 1983 J. Appl. Phys. 1588.

Inst. Phys. Conf. Ser. No. 67: Section 3
Paper presented at Microsc. Semicond. Mater. Conf., Oxford, 21–23 March 1983

Transient annealing of implanted silicon: microscopic analysis and comparison with electrical characteristics

M C Boissy, P Ruterana* and G Nouet*

RTC Caen, BP 6025 - 14000 Caen, France
*Laboratoire de Cristallographie, Chimie et Physique des Solides, LA 251,
Equipe Matériaux-Microstructure et Equipe Physique du Solide, ISMRa,
Université de Caen, 14032 Caen Cedex, France

Abstract   Silicon layers, implanted with conventional doses of ions,
have been annealed using four rapid methods. The crystalline structure
of these layers was analysed by TEM. We have studied and compared the
electrical characteristics of diodes, for which the regrown layers
constituted the emitter regions. The main result of this study is:
There will be uniformity problems, if small beam pulsed lasers are to
be used in integrated circuit processing. In addition, thermal strain
at interfaces with oxide layers may be difficult to overcome. Heat
pulse methods (lamp annealing and rapid isothermal annealing) seem to
be most appropriate to those layers that have not been heavily damaged
by ion implantation.

1. Introduction

In recent years, a large amount of work has been carried out in order to
study the effects of rapid annealing on semiconductor layers damaged by
ion implantation (Kachurin et al 1970, Gat et al 1978, Young et al 1978,
Lau et al 1979, Nishiyama et al 1980 and Fulks et al 1981). We have
investigated four methods:

- pulsed laser annealing
- CW laser annealing
- heat pulse : arc lamp
                rapid isothermal

Our aim was to determine the rapid annealing method which may be most
appropriate for integrated circuit processing. The pulsed laser redistrib-
utes the implanted impurities. The junction depth depends on the pulse
duration and the beam energy density. In order to reduce the beam distortion, we used a small power laser and we focused the beam down to 120μm
diameter. In CW laser annealing, the layer regrows by solid phase epitaxy;
the laser energy is absorbed in a thin superficial layer. Therefore, the
substrate remains cool and is not affected by the laser generated heat.
Impurity profiles do not change, since diffusion in solid phase is too slow
at $\sim 10^{-12}$ cm^{2}s^{-1} (Grove 1967). During heat pulse annealing, the temperature
of the whole wafer is raised up to 1100°C for less than 10s. Published
results (Fulks et al 1981) show that the electrical characteristics of the
annealed layers are similar to those of conventionally annealed samples.
In this work, we found that very high doses of implants ($>10^{16}$ at/cm^2) are
difficult to regrow.

2. Experimental

High frequency transistors are made with very thin emitter and base regions, we have fabricated p/n junctions by ion implantation into 1 to 20Ωcm [111] and [100] silicon substrates.

2.1 Sample preparation and annealing

The bases were formed by implantation of a low dose (6.10^{13} at/cm^2, 60keV) of Boron ions. The fabrication of emitters was carried out in the following way:

Ion	Energy keV	Dose at/cm^2
As$^+$/p$^+$	100	7.10^{15}
		5.10^{16}
	10	7.10^{15}
B$^+$	10	7.10^{15}

Samples were implanted at 100keV, in order to yield a junction depth of about 0.2µm. They were afterwards annealed by CW laser, or heat pulse (10 to 20s). The lower energy implants (10keV) were diffused in liquid phase by a pulsed laser.
The annealing conditions were:

A) Pulsed Laser: We have used 1.5, 1.9, 2.25 and 2.55J/cm^2 energy density. The laser wavelength is 0.53µm and the pulse duration, 100ns. At 1.5, 1.9, 2.25J/cm^2, the distance between annealed spot centres was varied (120µm, 30µm, 8µm). In this way, we were able to study uniformity in the annealed layer.
The beam has a Gaussian shape and the spot overlap plays a crucial role in the formation of the junction.

B) CW Laser (spot diameter = 100µm): While the substrates were held at 300°C, the laser parameters were varied according to the density of defects in the layer. For the 5.10^{16}at/cm^{2}As$^+$ implanted layers, the best anneal was obtained for a laser power of 10W, and a scanning velocity at 1cm/s. The other samples were annealed at 10W, 14.5cm/s.

C) Heat Pulse: We have used only one condition for rapid isothermal annealing (1200°C, 10s). As for the lamp annealing, the [111] samples were annealed at 1150°C for 20s and the [100] samples at 1100°C for 10s. We used these different conditions because the growth rate in solid phase is higher in the [100] than in the [111] direction.

2.2 Experimental results

A) Pulsed laser: TEM analysis shows that, for the smallest overlap, amorphous and polycrystalline layers remain in the outer part of the laser spot (Fig. 1). The centre is entirely monocrystalline. The spectral response (Fig. 2) gives clear evidence that the electrical properties of the bulk are not affected. The I/V characteristics show that the junction quality depends strongly on the spot overlap. If we take: $I \sim \exp(qV/nkT)$ where $q = 1.602 \ 10^{-19}$C; $k = 1.38 \ 10^{23}$J.K^{-1}; T is the temperature; n = 4 for the smallest overlap (120µm) and it comes down to less than 2 when the distance between spot centres is only 8µm.

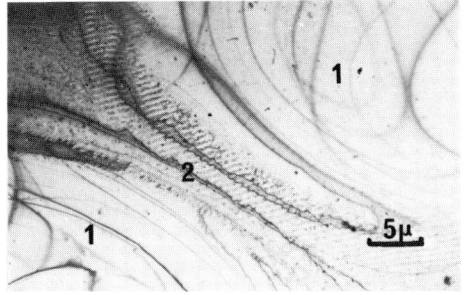

Fig. 1 Pulsed laser annealed area
2.25J/cm , 100ns, 0.53μm.

1. spot   centre,   monocrystalline

2. between spots, polycrystalline

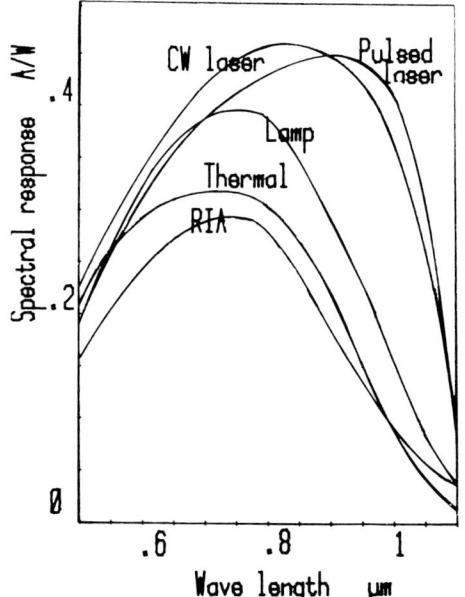

Fig. 2 Spectral response of the diodes
from all the annealing methods.

B) CW Laser: Lower concentrations
of dopants have been completely
activated, the diode quality factor,
n=1.5 (7.10^{15}at/cm^2). For high dose
implanted dopants (5.10^{16} at/cm^2)
the layer becomes polycrystalline
(Fig. 3).   For annealing durations
∿1ms, the impurity profiles do not
change.   The   spectral   response
(Fig. 2), shows that, in this case
too, the diffusion length of minority carriers in the bulk material
is not affected.

C) Heat Pulse:
C.1. Lamp: The [111] samples which
were annealed at 1150°C, 20s have
a larger junction depth (there is
a 2000Å difference). The lower concentration (7.10^{15}at/cm^2) implanted
layers are completely recrystallized.
The highest dose (5.10^{16} at/cm^2)
implanted   samples   contain   large
numbers of precipitates (Fig. 4).
The quality factor of the best
diodes is close to 1.5.

C.2. Rapid isothermal annealing: The annealed layers contain large numbers
of dislocation loops (Fig. 5);   the spectral response shows a drastic
reduction of the diffusion length (down to 25μm). This result indicates
that the conditions for the best anneal were not attained. However, other
work (Fulks et al 1981) in this area was done on lower dose implants
(<3.10^{15}at/cm^2). Therefore we presume that it may be difficult to anneal,
in 10s at 1200°C, very highly damaged layers (7.10^{15} - 5.10^{16}at/cm^2, 100keV,
As, P), by means of solid phase epitaxy.

3. Discussion and Conclusion

We have studied the various rapid annealing methods which can be applied
to silicon device processing.   A uniform junction is difficult to attain
in small beam pulsed laser annealing. In addition, as the overlap of laser

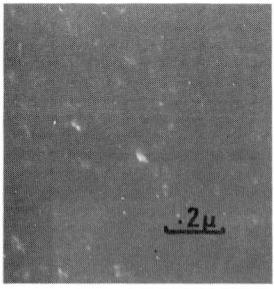 

Fig. 3   High dose arsenic implant after CW laser annealing, 10W, 1cm/s: substrate temperature 300°C.

Fig. 4   High dose arsenic implant after lamp annealing 1150°C, 20s.

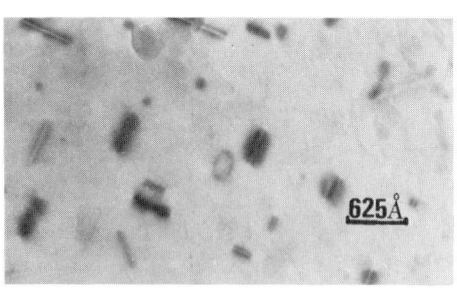

Fig. 5 7.10^{15} arsenic implant after rapid isothermal annealing (1200°C, 20s).

spots increases, the surface of the material is damaged. In this work we found that, even at our strongest overlap (8µm), the junction is not a plane. CW laser annealing in solid phase did not lead to concentrations above the solid solubility limits and very highly damaged layers became polycrystalline. CW and pulsed laser techniques seem to be appropriate to solar cell processing, where the diffusion length of the minority carriers in the bulk material plays a very important role.

Heat pulse annealing is a promising technique. Our results seem to point to its limitations for the recrystallization of highly damaged layers but, of course, a wide range of conditions would need to be analyzed in order to obtain a conclusive result.

References

Fulks R T, Russo C J, Hanley P R and Kammings T I 1981 Appl. Phys. Lett. 39 604
Gat A and Gibbons J F 1978 Appl. Phys. Lett. 32 142
Grove A S 1967 Physics and Technology of Semiconductor Devices (New York: John Wiley & Sons) p 38
Kachurin G A, Nidaev E V, Khodyachikh A V and Kovaleva L A 1970 Sov. Phys. Semiconduct. 10 1128
Lau S S, Allmen M Von, Golecki I G, Nicolet M A, Kennedy E F and Tseng W F 1979 Appl. Phys. Lett. 35 327
Nishiyama K, Arai M and Watanabe N 1980 Japan. J. Appl. Phys. 19 L583
Young R T, White C W, Clark G J, Narayan J, Christie M H, Murakami M, King P W and Kramer S D 1978 Appl. Phys. Lett. 32 139

Inst. Phys. Conf. Ser. No. 67: Section 3
Paper presented at Microsc. Semicond. Mater. Conf., Oxford, 21–23 March 1983

Laser alloying of nickel gold germanium contacts to InP

P J Topham, D K Skinner and B J Sealy*

Plessey Research (Caswell) Limited, Allen Clark Research Centre, Caswell, Towcester, Northants NN12 8EQ, U.K.

* Department of Electronic Engineering, University of Surrey, Guildford, U.K.

Abstract   A Q-switched ruby laser has been used to alloy a nickel gold germanium layer on InP employing energy densities in the range 0.035 to 1.54 Jcm^{-2}. The metal melts at an energy density of about 0.11 Jcm^{-2} and the InP melts at an energy density of around 0.3–0.4 Jcm^{-2}. This melting is correlated with the formation of an ohmic contact. At high energy densities some of the deposited metal is removed whilst the gold diffuses into the InP alloying with the indium.

1. Introduction

Laser annealing of deposited metal contacts have several advantages for making contacts to compound semiconductors. Only a thin surface layer is heated but the greater temperatures attained at the surface may increase dopant solubility and hence reduce contact resistivity. Also the morphology of the contacts is improved (Eckhart et al 1979). Laser alloyed contacts have already given good results on GaAs (Margalit et al 1978). This work extends a previous study on GaAs (Topham et al 1981) to cover InP and discover if laser alloying is a useful technique of making ohmic contacts to this semiconductor. The metallisation used is similar to that employed for making conventional furnace alloyed contacts, which have given low specific contact resistivities (Kuphal 1981) suitable for microwave devices.

2. Experiment

Two sets of samples were employed: firstly, tin doped InP (electron concentration ~ 1-2.10^{18} cm^{-3}), and secondly iron doped, semi-insulating InP. The material was all (100) orientation, polished by the manufacturer. The three metals were deposited by thermal evaporation during one pump down cycle of the vacuum system. Firstly a thin layer of nickel was deposited to act as a "wetting" agent followed by about 1,000 Å of gold and 400 Å of germanium. The germanium was placed on the top face of the contact to act as an absorbing layer and reduce the reflectivity of the contact. Prior to laser alloying the n-type samples had 200 μm square contacts defined by photo-engraving. Both these and the iron doped samples were diced into 5 mm squares.

The laser alloying was performed in air with a single pulse of 25 nS duration from a Q-switched ruby laser. A quartz waveguide homogeniser was used to produce a uniform energy density profile across the 8 mm diameter beam. Energy densities ranged from 0.035 to 1.5 Jcm^{-2}, varied by introducing

filters of differing optical density onto the beam prior to the homogeniser. After laser alloying some of the iron doped samples were etched in
I_2KI solution to remove the gold and HF/HNO_3 to remove the germanium and
nickel. This enabled the indiffused metals to be observed by Rutherford
back-scattering (RBS) using a 1 mm diameter beam of 1.5 MeV He^+ ions.

The contacts to the n-type samples were analysed in a JEOL 50B electron
microprobe fitted with an energy dispersive x-ray analysis system. The
electron beam was scanned over a 100 μm square with an energy of 15 keV.
The beam was de-focused to give a cylindrical analysis volume.

The chemical composition as a function of depth of various annealed contacts
was obtained using Auger depth profiling. This technique which utilises
surface analysis in combination with Ar^+ sputtering measures the surface
composition as the sample is slowly eroded. The profiles were carried out
using Ar^+ ions incident at 78° from the surface normal to reduce ion beam
induced artificats which can limit the quantitative accuracy of the profiles
(Skinner 1983).

Contact resistivity measurements were made using a vertical geometry similar
to that of Cox and Strack (1967), but employing an additional probe to eliminate back contact resistance. The back contact was made with In/Ga eutectic
alloy.

3. Results and Discussions

The results will be presented in three bands of laser energy density: less
than 0.3 Jcm^{-2}, between 0.3 and 0.6 Jcm^{-2}, and greater than 0.6 Jcm^{-2}.
These three groups will highlight three distinct behaviours of the contact:

(a) Laser Energy Densities less than 0.3 Jcm^{-2}

The Auger depth profile illustrates that no change occurs in the contact at
the lowest energy density of 0.035 Jcm^{-2}. A slight change in the appearance
of the contact is visible in the Nomarski micrograph for a irradiation of
0.098 Jcm^{-2} but a laser energy density of 0.20 Jcm^{-2} is necessary for clear
evidence of a reaction occurring in the contact (fig.1a). The etched
sample at the same energy density shows no change in the InP surface (fig.1b)
so melting of the semiconductor does not seem to have occurred. The strong
pull-back of the metal from the deposited edges suggests that the contact
has melted. The diffusion of the nickel into the InP can be seen in the
Auger results which also suggests a reaction (Erickson et al 1979).
If the InP melts it is to be expected that the gold would diffuse in rapidly
so the absence of a gold signal on the RBS spectrum for 0.27 Jcm^{-2} (fig.2a)
is further evidence for the lack of melting of the semiconductor for this
band of laser energy densities.

(b) Laser Energy Densities of 0.3 to 0.6 Jcm^{-2}

When the laser energy density is increased to 0.38 Jcm^{-2} the appearance of
the contact is entirely different to that at the lower energy densities just
considered. The metal is smooth (fig.3a) with no discernable pull back from
the original edges. The underlying InP, revealed by etching, (fig.3b) is
finely rippled so it can be taken that the semiconductor has melted. This
phase change causes the indium and phosphorus to out-diffuse (fig.4) and the
nickel signal forms a broad peak centred on the metal-semiconductor interface.

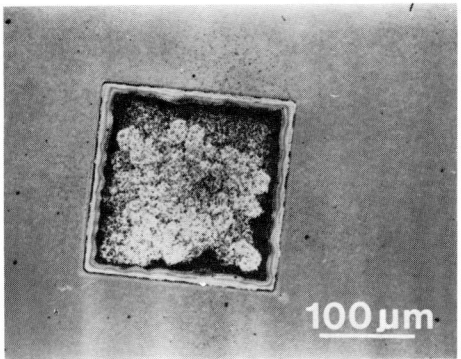

Fig.1a. Optical micrograph of contact to InP irradiated with 0.20 Jcm^{-2}

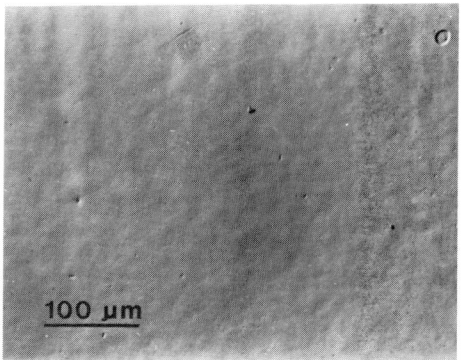

Fig.1b. Optical micrograph of etched, semi-insulating InP irradiated with 0.20 Jm^{-2}.

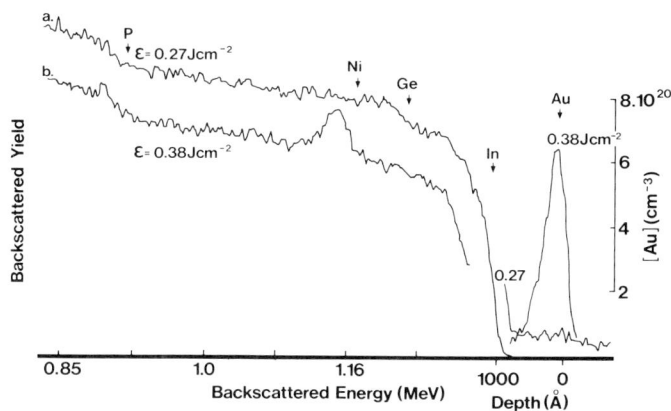

Fig.2. RBS spectra of etched, semi-insulating samples.

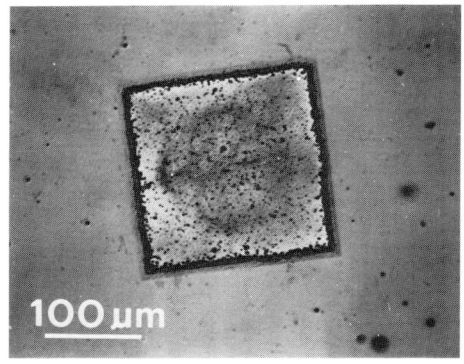

Fig.3a. Optical micrograph of contact to InP irradiated with 0.38 Jcm^{-2}.

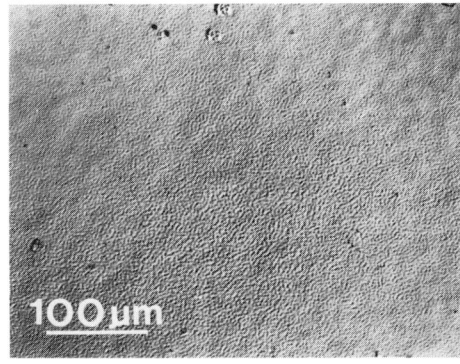

Fig.3b. Optical micrograph of etched, semi-insulating InP irradiated with 0.38 Jcm^{-2}.

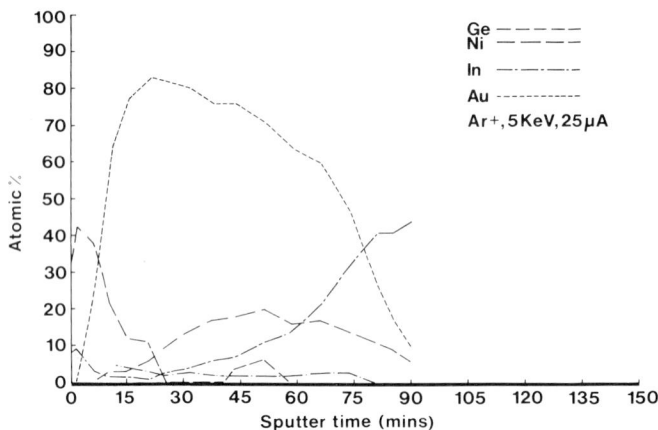

Fig.4.   Auger depth profile of contact
irradiated with 0.38 Jcm^{-2}.

After laser annealing at this energy density (0.38 Jcm^{-2}) the gold diffuses
into the InP, as seen by RBS for the etched samples (fig.2b).   The peak gold
concentration exceeds 6.10^{20} cm^{-3} but the gold is confined to a thick surface
layer less than 1,000 Å thick.   Another notable feature of this spectrum is
the large nickel peak (roughly 10^{22} cm^{-3}) right at the surface.   Possibly
associated with this is a peak in the phosphorus signal at the surface but
due to the low sensitivity for phosphorus and the high background due to
the indium signal no positive identification can be made.   Very similar
spectra were obtained for samples laser alloyed with up to 0.55 Jcm^{-2}.

Within this band of laser irradiance
(0.3–0.6 Jcm^{-2}) the specific contact
resistivity falls sharply (fig.5).
The contact resistance actually rises
after laser alloying with 0.058–0.096
Jcm^{-2} but then falls with increasing
laser energy density to a value of
about 10^{-5} Ωcm^2 corresponding to
irradiance with 0.55 Jcm^{-2}.   It seems
plausible that the reduced resistivity
is associated with the melting of the
underlying InP, particularly if it is
noted that the edge of the contact is
alloyed at lower energy densities
than the centre due to melting of the
InP around the contact.

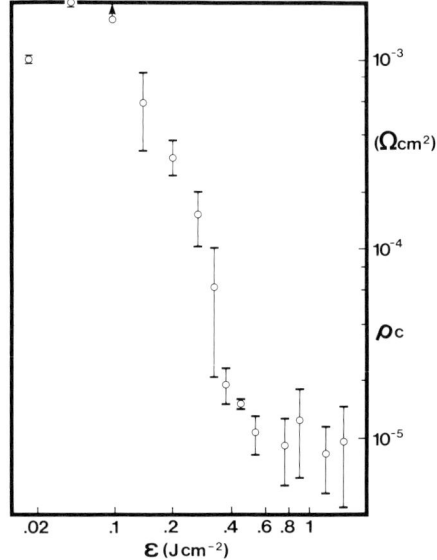

Fig.5.   Specific   contact   resistivity
as a function of laser energy
density.

(c) Laser Energy Densities of 0.6-1.54 Jcm^{-2}

After these higher energy pulses the surface of the contacts is more rippled and so Auger profiling is not practical. Furthermore the AES shows the contact metals to be heavily oxidised. Rutherford back-scattering is not so troubled by surface condition and the spectrum for the unetched, semi-insulating sample irradiated with 0.75 Jcm^{-2}, is given in figure 6a. The gold is present at the surface, having alloyed through the germanium, and the gold and germanium levels are lower, possibly due to the presence of phosphorous, indium and oxides in the metal layer. At a still higher laser energy density of 1.24 Jcm^{-2} the gold is present in large quantities at the surface (fig.6b) and a large peak is present at a mass corresponding to indium. The interface is diffused indicating the formation of indium-gold alloys. At the highest energy density used, 1.54 Jcm^{-2}, the RBS spectrum is of gold-indium alloy at the surface, with no germanium being detected.

The increased diffusion of the gold is also visible in the RBS spectra of the etched samples with significant diffusion occurring at 0.2 Jcm^{-2} and at the highest energy density the gold is dissolved to a concentration of about 6.10^{21} cm^{-3} strongly indicating the formation of an alloy.

The loss of gold is quite marked for samples irradiated with more than 0.5 Jcm^{-2}, as shown by the x-ray microprobe (fig.7). The rise in the gold signal for laser energy densities up to 0.5 Jcm^{-3} may be due to "balling-up" of the gold. The germanium is removed preferentially and considerable loss takes place for energy densities in excess of 0.14 Jcm^{-2}. No germanium is detected in the spectrum for the sample irradiated with the highest energy density, in agreement with the RBS observations.

The visible effects of laser alloying with 1.54 Jcm^{-2} are dramatic. The contacts are still clearly visible but frozen "waves" of InP are breaking over the edges whilst the underlying InP is furrowed by deep ripples. In view of the great physical changes in the metal-semiconductor interface for laser energy densities above 0.6 Jcm^{-2} some effect might be expected in the specific contact resistivity. However, the resistivity falls only slightly (fig.5) to values around 8.10^{-6} Ωcm^2 indicating that the high concentrations of gold in the surface layer have no effect either beneficial or detrimental upon the surface carrier concentration. Similarly the presence of

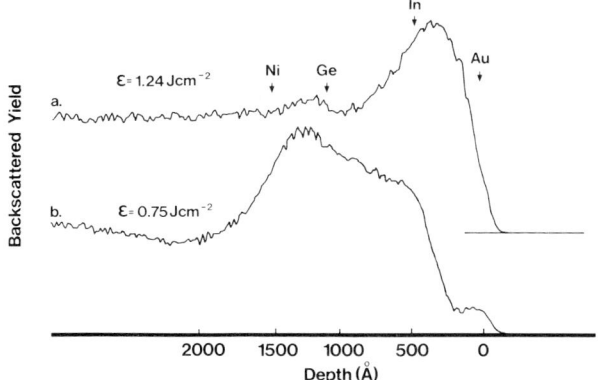

Fig.6. RBS spectra of metals on semi-insulating samples.

nickel at the interface does not seem
to influence the resistivity to any
significant extent.

4. Conclusions

Two distinct phase changes occur in
the NiAuGe-InP system. The first is
when the metal melts at an irradiance
around 0.1 Jcm^{-2}, and the second when
the semiconductor melts at around
0.3-0.4 Jcm^{-2}. In the first case
very little metal diffuses into the
semiconductor and alloy regrowth does
not appear to occur. Probably as a
result of this lack of diffusion the
contact resistivity remains high. In
the latter case of the semi- conductor
melting, diffusion does take place and
the contact resistivity falls sharply
to values less than 10^{-5} Ωcm^2.
However the gold is contained in a
layer about 1,000 Å thick and there is
some evidence of the formation of a
nickel phosphide layer which may act
as a diffusion barrier. At high laser
energy densities (greater than 0.75
Jcm^{-2}) this layer is absent and the
gold diffuses in to a greater extent,
alloying with the indium from the substrate. Another problem when using
high laser energy irradiance is the loss of the metallisation. Despite
these problems smooth, shallow, low resistance contacts can be obtained by
laser alloying with energy densities between 0.4 and 0.5 Jcm^{-2}.

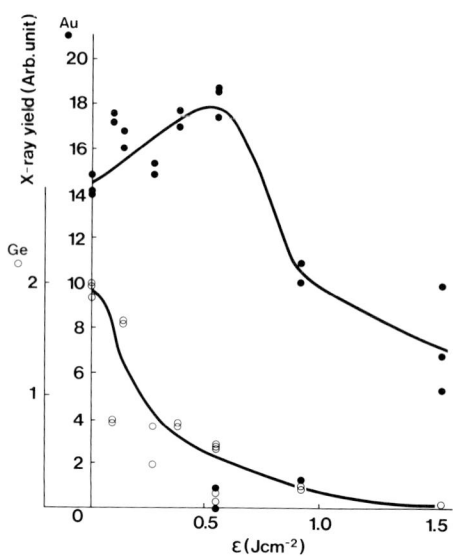

Fig.7. Microprobe yield
of gold and germanium vs
laser energy density.

Acknowledgements

The authors wish to thank R. Varma for depositing the metals. The SERC and
Plessey Research (Caswell) Limited are thanked for their financial support.

References

Cox R H and Strack H: 1967 Solid State Elec. No.10, 1213

Eckhart G, Anderson C L, Hess L D, Crum C F: 1978 AIP Conf. Proc. No.50,
641

Erickson L P, Waseem A, Robinson G Y: 1979 Thin Solid Films No.64, 421

Kuphal E: 1981 Solid State Elec. No.24, 69

Margalit S, Fekete D, Pepper D M, Lee C P, Yariv A: 1978 Appl. Ph. Lett.
No.33, 346

Skinner D K, Swanson J G, Haynes C V: 1983 Surf. Interface Anal. No.5. 1

Topham P J, Shahid M, Sealy B J: 1981 Inst. Phys. Conf. Ser. No.60, 133

Inst. Phys. Conf. Ser. No. 67: Section 3
Paper presented at Microsc. Semicond. Mater. Conf., Oxford, 21–23 March 1983

Phase conversion of electrophoretically deposited cubic CdS layers for use in CdS–Cu$_x$S solar cells

G J Russell, P C Pande, J Woods, I D McInally* and S P Gibbons*

Department of Applied Physics and Electronics, University of Durham,
Durham, DH1 3LE
*Thorn EMI plc, Central Research Laboratories, Trevor Road,
Hayes, Middlesex UB3 1HH

Abstract   Very thin layers (1-3 μm) of CdS exhibiting the sphalerite
structure and a resistivity > 10^7 Ω cm have been produced on SnO$_2$ coated
glass by electrophoresis. To improve their suitability for use as substrates for photovoltaic devices, they have been given thermal and laser
annealing treatments, the effects of which have been monitored using the
RHEED technique. This shows that annealing at 475°C and above produces
both a phase transformation from cubic to hexagonal and a reduction in
resistivity by two orders of magnitude. The characteristics of photovoltaic devices prepared on this improved material are briefly described.

1. Introduction

In the preparation of thin films of CdS for a range of applications, there
has been a number of studies concerned with the recrystallisation of the
deposited layers with the aim of producing material with properties nearer
to those of single crystals than to those of thin films. Several workers
(see for example Gilles and Van Cakenberghe 1958, Vecht and Apling 1963,
Addiss 1963 and Herinckx et al 1972) have employed Ag or Cu to promote
recyrstallisation as these particular group 1B elements lower the threshold
for this process from temperatures in excess of 700°C for undoped films
(Kahle and Berger 1970) by between 100 - 200°C, so that glass based substrates may be employed. In fact Addiss (1963) recorded a threshold temperature for crystallisation as low as 350°C for such doped films when the
heat treatment was performed in a closed chamber which ensured that there
was an appreciable overpressure of cadmium and sulphur during the process.
To minimise the deleterious electrical effects of Ag and Cu in these films,
Herinckx et al (1972) reduced the Ag doping level to 40 ppm and still
achieved recrystallisation of CdS at a threshold below that for undoped
films.

Fraas et al (1975) avoided the group 1B dopants and recrystallised
evaporated CdS films by heating in H$_2$S at 500°C. This technique was
extended to films containing Ag by Srivastava et al (1979) who observed
recrystallisation at temperatures as low as 300°C. While the use of H$_2$S
makes it possible without introducing impurities to recrystallise CdS
films on a wide range of substrate materials, including SnO$_2$ coated glass
of the type used in photoactivated displays and in solar cell applications,
it does tend to produce material with a high resistivity (Fraas et al
1975). This can be attributed to the reaction of H$_2$S with the excess Cd

in the vacuum deposited films, which reduces the donor density and makes the films unsuitable for use in photovoltaic devices.

In all the work described above the incidence of recrystallisation was established using one or more of the techniques of x-ray diffraction, polarising light microscopy or scanning electron microscopy. In the present work using the RHEED technique, it is shown that annealing electro-phoretically deposited CdS films in an argon ambient in the absence of any group 1B additives leads to the phase transformation from sphalerite to wurtzite starting at temperatures as low as 475°C. In a separate set of trials, this same phase transformation has been produced by laser anneal-ing. Thus, while grain growth resulting from these treatments may not be significant, considerable reordering obviously occurs within the crystal-lites causing the phase transformation to take place at a temperature 200°C below that necessary to produce recrystallisation of evaporated CdS films (Kahle and Berger 1970). Furthermore, in contrast with the effects of treatments involving the use of H_2S or group 1B additives, the resis-tivity of our films decreased by at least two orders of magnitude with the transformation to the hexagonal phase. This change in the electrical properties of films makes it possible to use them in CdS-Cu$_x$S photovoltaic devices. However, to make them more suitable for this purpose, it has been necessary to evaporate a 100Å layer of CdCl$_2$ onto the films before the heat treatment to increase the donor density even more. The characteristics of devices fabricated on this improved substrate material are briefly discussed.

2. Experimental

Thin films of CdS were deposited on SnO$_2$ coated glass using the electro-phoretic process described by Williams et al (1979). Most of the layers produced were made from CdS sols with no binder added, but recently by using 1% by weight of poly vinyl pyrrolidine (PVP), the thickness of the deposited layers has been increased without introducing the problems of cracking. The structural assessment of the films was made largely using a Cambridge Instrument S180 scanning electron microscope and RHEED with a JEM 120 transmission electron microscope operating at 100 KV. The relative intensities of the reflections in the recorded RHEED patterns were measured using a Hilger and Watts type L502-2 microdensitometer.

Thermal annealing in the range 300 to 530°C was carried out in open silica tubes with an argon flow rate of 20 ml/min, except for films containing P V P which were first heated in air at 350°C for 90 min to oxidise the polymer. This oxidation treatment had no adverse effects on the structural or electrical properties of the films. Laser annealing was performed using an argon ion laser operating at powers ranging from 0.3 to 2 W. Evaporated circular contacts 1 mm in diameter of indium and of gold provided ohmic and Schottky barrier electrodes to the films respectively. The dry barrier technique first discussed by Te Velde (1973) was used to form heterojunc-tions of CdS-Cu$_x$S with circular junction areas of 1, 2 and 4 mm diameter on other films which were first treated with a 100Å layer of CdCl$_2$ (Pande et al 1983). Current voltage characteristics of these devices were recorded both in the dark and under 100 mW/$_{cm}$2 (AM 1) incident illumination and their spectral responses were measured using a Barr and Stroud prism monochromator type VL2.

3: Results. <u>Structural Observations</u>

The structure of the electrophoretically deposited films was demonstrated most clearly by examining their fractured edges in the SEM. Micrographs of such edges of layers with and without PVP are shown in figures 1b and 1a. When attempts were made to increase the thickness of binderless layers beyond 1 μm, cracking usually occurred on drying. This problem was overcome by using PVP.

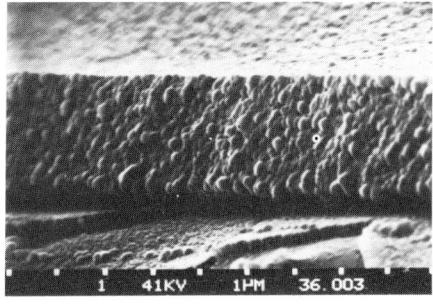

(a) Without PVP   (b) with PVP

Fig. 1 SEM micrographs of electrophoretically deposited CdS

The crystal structure of the as-deposited films was shown by RHEED examination to consist mainly of the sphalerite cubic phase (figure 2a). When the sol was prepared using $CdCl_2$ instead of $Cd.(CH_3CO_2)_2$, a fraction of the wurtzite hexagonal phase became evident, but it always remained the minor component in as-deposited layers. However on annealing at 475°C and above for 30 min in argon, the hexagonal content of all films increased, and annealing at 530°C produced a film that was predominantly hexagonal as shown in figure 2b.

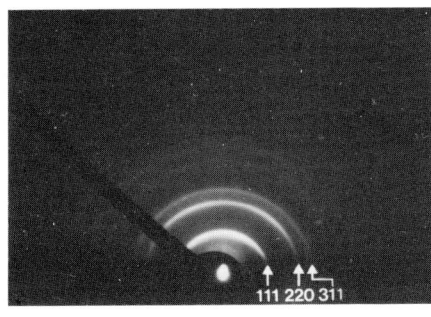
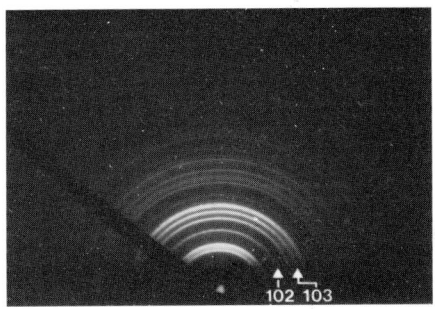

(a) as-deposited   (b) after thermal anneal-
                                     ing at 530°C for 10 min.

Fig. 2 RHEED Patterns from electrophoretically deposited CdS

The effect of the annealing temperature on this phase conversion is shown more clearly by the densitometer traces in figure 3a which were recorded from a number of 100 KV RHEED patterns. The traces clearly show that the (102) and (103) reflections of the hexagonal phase differ from most of the other hexagonal reflections in that they do not coincide with reflections from the cubic phase. Consequently they provide the best indication of the hexagonal content of the films and they increase in intensity relative to the other reflections when the annealing is carried out at 475 and 530°C.

In a similar way, the effects of annealing with different laser powers is shown by the densitometer traces of 100 KV RHEED patterns in figure 3b. With increasing laser power, the intensities of the hexagonal (102) and (103) reflections increase relative to those of other reflections from both the cubic and hexagonal phases. As with thermal annealing at the temperatures used here, no significant increase in grain size could be detected by SEM examination of these laser annealed samples. This demonstrates that the cubic to hexagonal phase transformation precedes grain growth with both types of anneal.

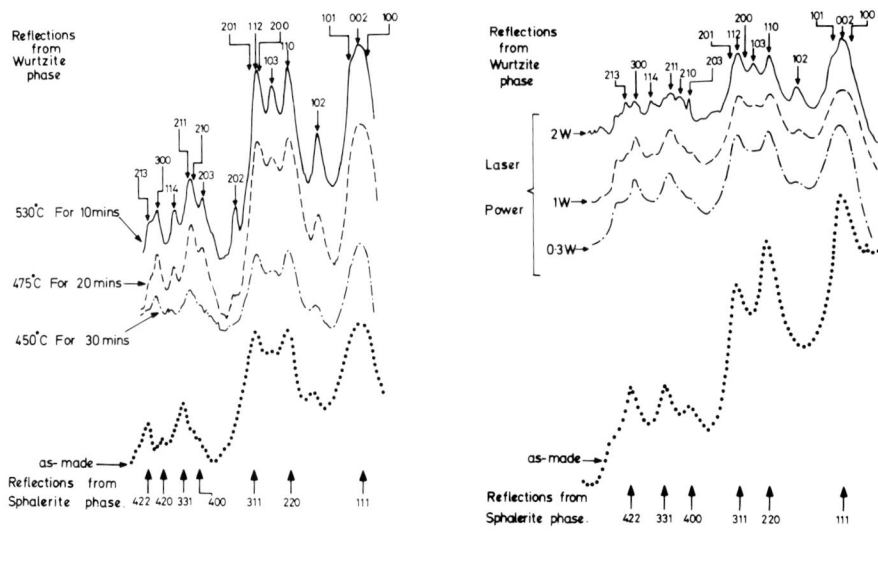

(a)   thermally annealed   (b)   laser annealed

Fig. 3   Densitometer traces taken from RHEED patterns of differently annealed films

Electrical Properties.   The resistivity of each film was measured using the ohmic contacts provided by the evaporated indium dots. It was found to be greater than 10^7 Ω cm for as-made films, but after annealing at 530°C for 10 min was reduced by two orders of magnitude. This reduction did not occur abruptly with annealing temperature but occurred gradually over the range from 400 to 530°C on annealing for the same period. When gold

Schottky contacts were applied to as-made films, it was not possible to
pass any significant current in either forward or reverse bias because of
the high resistivity of films in this form. However Schottky behaviour was
obtained from films annealed at 530°C for 10 min and the I-V characteristic
of a film after such treatment is shown in figure 4a. The forward bias
characteristic of this device provides confirmation of the improved resistivity of the annealed film. In an attempt to quantify this improvement in
resistivity with annealing temperature, capacitance-voltage measurements
were made on the Schottky barriers. From plots of capacitance^{-2} versus
voltage, the donor concentrations of the films annealed at 475 and 530°C
were determined to be 5.5 x 10^{14} and 8.8 x 10^{15}/cm^3 respectively.

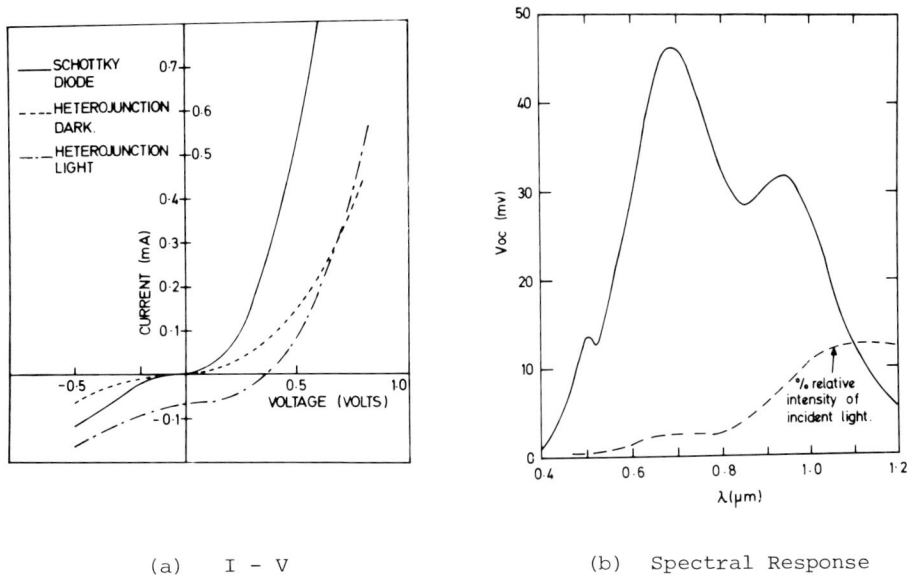

(a)   I - V   (b)   Spectral Response

Fig. 4   Characteristics of a CdS-Cu$_x$S device fabricated on improved
         substrate material.

Even after reducing the resistivity of the films by two orders of magnitude as described above, they were still not sufficiently conducting for
photovoltaic device fabrication. Consequently a series of experiments
was undertaken to dope the films with chlorine (Pande et al 1983). Under
optimised conditions, it was then possible to reduce the resistivity of
films to about 5 x 10^3 Ω cm, and the I - V characteristics measured in
the dark and under AM 1 illumination for a CdS-Cu$_x$S heterojunction formed
on such a film are also shown in figure 4a. The relatively poor short
circuit current and fill factor of the device can be attributed to the
series resistance of the CdS film, which is still too large. The spectral
response of the device in figure 4b shows peaks at about 0.7 and 0.9 μm.
From our earlier work using single crystal CdS substrates (Caswell et al
1975), these peaks can be associated with the djurleite (Cu$_{1.96}$S) and
chalcocite (Cu$_2$S) phases of Cu$_x$S respectively.

4. Discussion

The important finding of this investigation is that the cubic to hexagonal
phase transformation of CdS in the form of electrophoretically deposited
thin films can be produced at temperatures some 200°C below that reported
to be necessary to promote grain growth in vacuum deposited layers
(Kahle and Berger 1970), thereby making it possible to effect the transformation when employing SnO_2 coated glass substrates. Indeed, preliminary
studies of polycrystalline sphalerite cubic layers of CdS, prepared by a
quite separate process in which single crystal slices are mechanically
polished using alumina powder of 1 µm particle size (Russell et al 1982),
have shown that this phase transformation is not confined to electrophoretically deposited CdS. The improvement in the conductivity of these
films which is produced by the same thermal treatment that promotes the
phase transformation and which is increased by the addition of chlorine
donors,makes it feasible to fabricate CdS-Cu_xS photovoltaic devices on
this low cost thin film CdS. It is interesting to note that this chlorine
further reduced the temperature necessary to promote the phase transformation to 400°C (Pande et al 1983).

The other important consequence of this phase transformation in the context
of the CdS-Cu_xS photovoltaic device relates to the formation of the Cu_xS
layer on the CdS substrate. As this layer is formed by the chemical displacement of Cd in the CdS lattice by Cu, the result is that the phase of
Cu_xS produced is closely related to the crystal structure of the CdS substrate. Since the cubic phase of Cu_xS (diginite or $Cu_{1.8}S$) has undesirable properties as far as photovoltaic devices are concerned and as the
most desirable phase for these purposes (chalcocite, Cu_2S) is closely
related to the hexagonal phase of CdS (Cook et al, 1970), it is obviously
essential that the heterojunctions are formed on a hexagonal based substrate. This has indeed been confirmed by the presence of chalcocite in
the devices produced.

References

Addiss, R. Jr. 1963 in Tenth National Symposium on Vacuum Technology Trans-
    actions, Pergamon, New York, 354.
Caswell, B. G., Russell, G. J. and Woods, J. J. Phys. D: Appl. Phys 8 1889.
Cook, W. R., Shiozawa, L. R. and Augustine, F. 1970, J. Appl. Phys. 41
    3058.
Fraas, L. M. Bleha, W. P. and Braatz, P. 1975, J. Appl. Phys. 46 491.
Gilles,J. and Van Cakenberghe, J. 1958 Nature 182 862.
Herinckx, C., De Sutter, W., Fourdeux, A. and Terao, N. 1972, Phys. Stat. Sol.
    A 10 387.
Kahle, W. and Berger, H. 1970, Phys. Stat. Sol. A 2 717.
Pande, P. C., Russell, G. J. and Woods, J. 1983 to be published.
Russell, G. J., Fellows, A. T., Oktik, S, Ture, I. E. and Woods, J. 1982,
    J. Mat. Sci. Lett. 1 176.
Srivastava, R. S. and Prakash, K, 1979, J. Appl. Phys. 50 7245.
Te Velde, T. S. 1973, Solid State Electronics 16 1305.
Vecht, A. and Apling, A. 1963, Phys. Stat. Sol. 3 1238.
Williams, E. W., Jones, K, Griffiths, A. J. Roughley, D. J. Bell, J. M.,
    Steven, J. H., Huson, M. J., Rhodes, M. and Costich, T. 1979 Solar
    Cells 1 357.

Inst. Phys. Conf. Ser. No. 67: Section 3
Paper presented at Microsc. Semicond. Mater. Conf., Oxford, 21–23 March 1983

195

SEM CL studies of II–VI powder layers transformed into heterojunction solar cells using transient heating techniques

T J Cumberbatch, I D McInally and S P Gibbons

THORN EMI plc, Central Research Laboratories, Trevor Road, Hayes, Middlesex, England UB3 1HH

Abstract   Thin films of CdS or CdZnS produced by electrophoretic deposition consist of microcrystallites (20 - 100 nm in diameter) within a closely packed powder layer. Pulsed and CW laser radiation have been used to promote recrystallisation of these layers and to fabricate Cu_xS–CdS heterojunctions. SEM CL, in conjunction with other techniques, has been used to monitor the material properties and provide an insight into the mechanisms involved during laser processing.

1. Introduction

Electrophoresis offers many advantages over other techniques for the preparation of thin ($\leqslant 3\mu m$) semiconductor films: principally, deposition at ambient temperature and pressure, very efficient material usage and pin-hole free coatings over large areas in short times (Williams et al 1979/80). The primary disadvantage being that the layer so produced consists of randomly orientated microcrystallites with predominantly cubic structure. For stable efficient solar cells this must be transformed into a poly-crystalline thin film with hexagonal columnar crystals which extend through the thickness of the layer and whose c-axis is perpendicular to the substrate – preferably without the addition of any extrinsic impurities (Hewig et al 1982).

The thermal and chemical properties of the substrates (tin oxide coated glass or stainless steel) preclude conventional thermal sintering since, at the temperatures required for grain growth (600 - 700C), the glass softens and CdS appears to react with stainless steel; phase conversion, however, can be achieved at lower temperatures (Russell et al 1983). Initial data from transient heating experiments revealed that both grain growth and phase conversion were possible with a 300 ns laser pulse but not a scanned electron beam. Here we report on significant grain growth achieved with long duration (3µs) pulses from a flashlamp pumped dye laser (λ_{pk} 505 nm) and a scanned, focussed Ar^+ laser operated in the all lines mode.

Clearly, many of the advantages of this approach are lost if vacuum procedures are introduced into the solar cell fabrication process. Since these layers are too thin to use the well established Clevite technique (Shiozawa et al 1968) for the production of cuprous sulphide layers, lasers have been employed to grow topotaxial layers of Cu_xS on CdS films immersed in aqueous cupric salt solutions.

2. Laser Recrystallisation

The behaviour of a powder layer when exposed to a high power beam of
optical energy is complex. The maximum surface temperature attained and
the temperature distribution throughout the layer are principally controlled by the power level (i.e. rate of temperature rise) and rate of intergranular heat transfer: the net result being that it is possible to melt
the CdS for a fraction of a second without sublimation. Pulses of
duration <1µs melt a region within 100 - 200 nm of the exposed surface
but leave the remainder of the layer unchanged; durations >∿2µs produce
significant homogenisation and initiate grain growth at the top surface
(Fig. 1). Although this can be accompanied by problems of particle
adhesion, ablation and microcracking - they are eliminated by raising the
temperature of the layer to ~200C just prior to laser exposure.

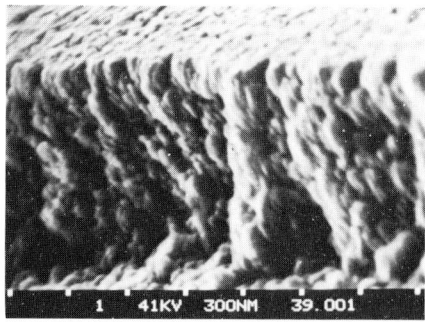
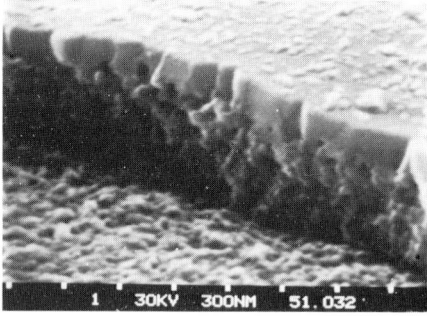

Fig. 1   2.3µs 220 mJ.cm^{-2} 20C (SEI)   Fig. 2   3x3µs 300 mJ.cm^{-2} 200C (SEI)

The result of such an approach after three pulses is shown in Fig. 2 -
further shots increase the thickness of the recrystallised film.

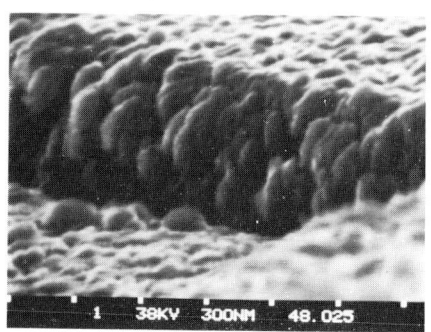

Fig. 3   Ar^{+} laser (3W), 100µs dwell
         (SEI)

A single pass with a scanned Ar^{+}
laser has also been used to promote
recrystallisation as illustrated in
Fig. 3. The type of behaviour exhibited by these layers under CW laser
exposure is related to that for pulsed
radiation although the power/dwell
time window for homogenisation/
recrystallisation without dissociation
is elusive and appears to be a function of CdS thickness.

Results to date suggest that recrystallisation occurs via a liquid phase
mechanism for pulsed radiation and
for a 'slowly' scanned beam
(>10^{5}µm.s^{-1}), the solid phase.

3. SEM CL Characterisation

It is well established that electrical data obtained from inhomogeneous
semiconductor layers require careful evaluation to distinguish between
intergranular and intragranular contributions to the measured parameters
(Ma and Bube 1977; Orton and Powell 1980). In contrast, the imaging and

spectral analysis modes of room temperature CL can provide detailed information (Davidson 1977), especially during the development of recrystallisation procedures.

As deposited films exhibit weak broadband luminescence that remains substantially unchanged for short duration pulses (Cumberbatch et al 1982) or
heat treatments in inert (N_2, Ar) ambients (Fig. 4). The origin of the
long wavelength tail is unknown but thermal annealing in a hydrogen ambient or exposure to an Ar^+ laser generate a substantial increase in the CL intensity (Figs. 5a and b) accompanied by a pronounced reduction in the width of the emission peak (Fig. 4). Although these laser treated regions revealed almost complete coalescence, significant grain growth was observed in neither layer suggest-ing that the narrowing and increased intensity are due to a reduction

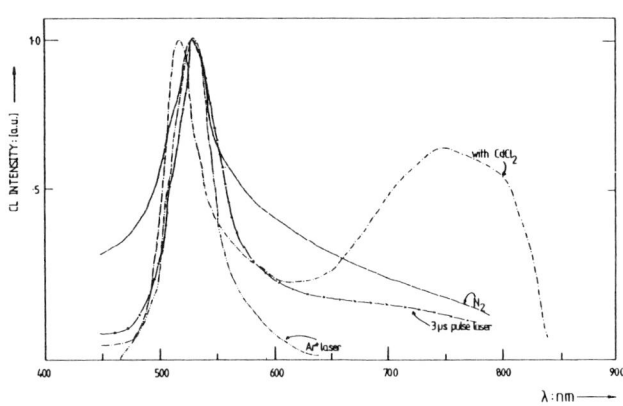

Fig. 4   CL spectra from CdS powder layers after
         different treatments

in the electrical activity of the surface - supported by the reversible
effects of hydrogen adsorption (Ambridge and Carter 1971).

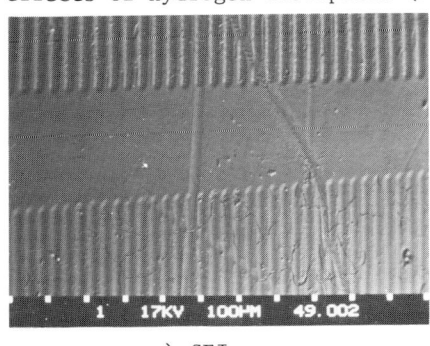
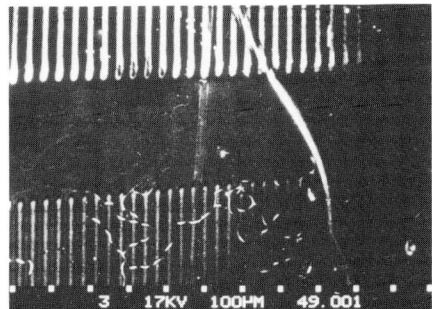

a) SEI   b) CL

Fig. 5 Ar^+ laser. 3W (top scan, .5ms dwell; bottom .2ms dwell (laser
         incident from rear)

For long duration pulses, the homogenisation illustrated in Fig. 1 is
associated with an increase in the CL intensity (Fig. 6), the conductivity
(Fig. 7) and reasonable narrowing of the peak width (Fig. 4); the reduced
improvement associated with the depth dependent effect of a laser pulse
and confirmed with PL spectra which sharpen considerably for short wavelength excitation.

Thus, the CL data indicate that significant progress towards a phase
change (confirmed by TEM and RHEED) and recrystallisation has been
achieved with laser processing and without the debatable side effects

introduced by the addition of fluxes such as $CdCl_2$ - responsible for the unwanted long wavelength emission (Hewig et al 1982) but, in this instance, a more complete phase change (Fig. 4).

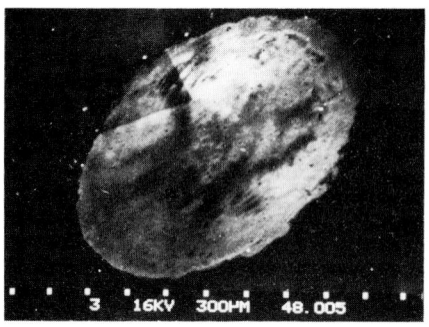

Fig. 6   3µs 300 mJ.cm^{-2} 20C (CL)

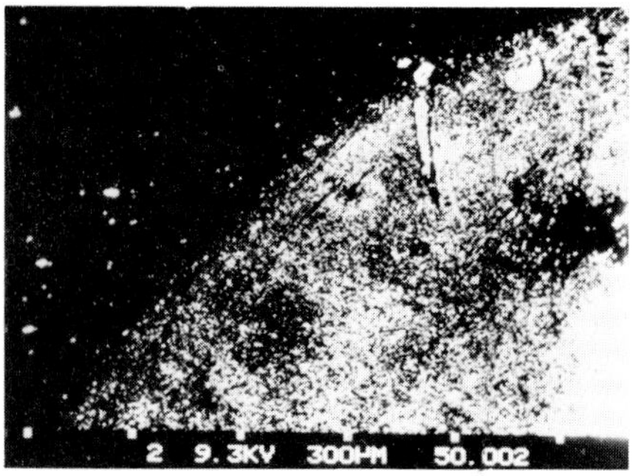

Fig. 7   3µs 350 mJ.cm^{-2} 20C
(specimen current)

4. Heterojunction Formation

Cuprous sulphide, although a very efficient photon absorber, is an undesirable semiconductor possessing many different phases, each of which has characteristic crystal structure and electrical properties. The desired phase, chalcocite ($Cu_{1.966}S$, E_C ~1.2eV), is normally grown as a topotaxial layer (.1 → .2µm) on the CdS through an ion exchange reaction; the high mobility of copper leading to accelerated growth down grain boundaries. In an attempt to fabricate abrupt junctions outside a vacuum, we have investigated the use of lasers to initiate the exchange reaction in aqueous cupric salt solutions at room temperature. RHEED patterns and CL spectra (Fig. 8) support the production of a djurelite layer (Cumberbatch et al 1982) with initial device characteristics shown in Fig. 9 - the shape of the spectral response attributed to a very thin Cu_xS layer.

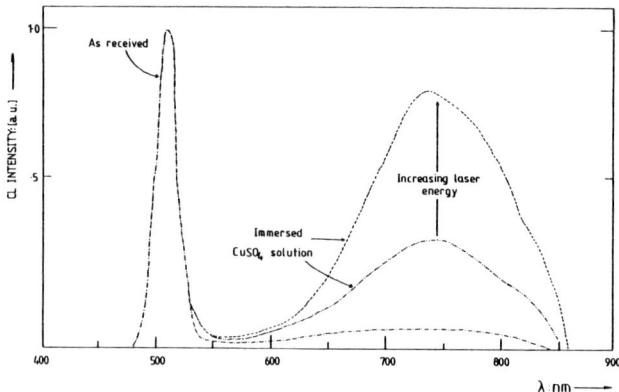

Fig. 8   CL spectra from evaporated CdS layer after formation of a Cu_xS layer with a laser pulse.

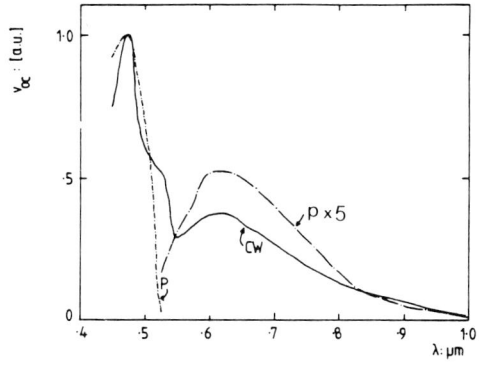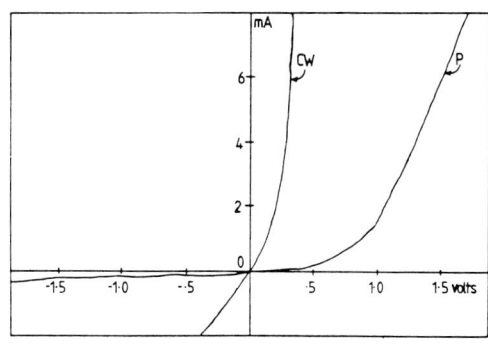

a) Relative spectral response

b) Dark current–voltage characteristic

Fig. 9   Laser Fabricated Cu_xS-CdS (evaporated) heterojunction: P-3 µs
         Laser Pulse; CW - Ar^+ laser.

The luminescence detected from this type of structure requires careful
evaluation since, even at the lowest useful accelerating voltages (~ 5 kV),
a majority of the electron-hole pairs are generated within the underlying
CdS. However, the high optical absorption coefficients of the various
Cu_xS phases (at ~ 510 nm) mean that a substantial proportion of the CL
emerging from the CdS will be absorbed in the topotaxial layer so providing a possible optical pump for photoluminescence. Thus, although the
excitation source for the 'CL' from the cuprous sulphide layer may be open
to debate, the NIR emission in Fig. 8 is believed to arise from a phase of
Cu_xS for the following reasons:

  i)   Laser exposure alone does not generate the defect responsible for
       the self-activated luminescence (see Fig.4)

  ii) The luminescence from copper doped CdS lies beyond 1 µm.

  iii) Complementary measurements on the same area reveal a hetero-
       junction.

At this stage of the work, evaporated layers of CdS from the University of
Stuttgart are being used as a test vehicle for the fabrication of heterojunctions with a laser - until we have completely recrystallised the
powder layers. Preliminary results from partially recrystallised films
demonstrate that this technique does work on very thin layers. Unfortunately, existing data for the different phases of Cu_xS are sparse; this,
in conjunction with the complexity of the phase diagram, precludes prediction of the resultant phase as a consequence of laser exposure.
However, the correspondence between the shape of the spectral responses
for pulsed and CW laser produced junctions suggests that the distribution
of phases in both instances is similar. The status of these devices after
further laser exposure has still to be investigated but the increased
optical absorption should lead to much higher surface temperatures which
may in turn promote the dominance of another phase, hopefully chalcocite!

5.   Discussion and Conclusions

The thickness ($\leqslant 3$ µm) and quality of the as deposited and partially recrystallised films require the use of high excitation conditions (5-15 kV, 100 nA). For this reason the CL spectra are acquired with the beam scanned over an area for which the optical collection efficiency is uniform. Variation of the beam energy produces predictable changes in the intensity whereas the non-linear increase with probe current would suggest saturation of non radiative recombination centres above a defined threshold; reducing the temperature to 100K produces little change in the intensity or shape of the spectra until substantial recrystallisation has taken place.

In conclusion, we have demonstrated that it is possible to fabricate cuprous sulphide-cadmium sulphide hererojunections from electrophoretically deposited CdS powder layers - using transient heating techniques to promote recrystallisation and the growth of topotaxial semiconductor layers. Although lasers are currently employed for transient heating, experiments to replace these with incoherent light sources or graphite strip filaments are in progress.

Acknowledgements

The authors would like to acknowledge financial assistance from the EEC and UK DoI; Dr. G. Russell (Durham University) RHEED measurements and Dr. N.M. Pearsall (Newcastle Polytechnic) optical measurements.

References

Ambridge T and Carter G 1971 J.Phys.D: Appl.Phys. 4 1630
Cumberbatch T J, McInally I D, Ke W K and Hamilton B 1982 Proc. 4th EC PV
    Solar Energy Conf. (London: D Reidel) p551 and refs. therein
Davidson S M 1977 J.Microscopy 110 117
Hewig G H, Pfisterer F, Schock H W, Arndt W and Bloss W H 1982 Proc. 16th
    IEEE PVSC (New York: IEEE) to be published
Ma Y Y and Bube R H 1977 J.Electrochem.Soc. 124 1430
Orton J W and Powell M J 1980 Rep. Prog. Phys. 43 1263
Russell G J, Pande P C, Woods J, McInally I D and Gibbons S P 1983
    this proceedings volume
Shiozawa L R, Sullivan G A and Augustine F 1968 Proc. 7th IEEE PVSC
    (New York: IEEE) p39
Williams E W, Jones K, Griffiths A J, Roughly D J, Bell J M, Stephen J H,
    Huson M J, Rhodes M and Costich T 1979/80 Solar Cells 1 357

Inst. Phys. Conf. Ser. No. 67: Section 3
Paper presented at Microsc. Semicond. Mater. Conf., Oxford, 21–23 March 1983

In situ TEM microscopy of α−Ge films in laser annealing conditions

J Marfaing*, P Pierrard+, W Marine*, B Mutaftschiev+ and F Salvan*

*Département de Physique, Faculté des Sciences de Luminy, Case 901 - 13288 Marseille Cédex 9, France
+CRMC², Campus de Luminy, Case 913 - 13288 Marseille Cédex 9, France

Abstract   In situ TEM observations of the crystallization of unsupported a-Ge films have been performed using nanosecond pulsed laser irradiations in the microscope. We analyze the formation of the crystallites versus laser energy density with single pulse irradiation: the formation of mono/polycrystallites. We determine the variations of the crystallization threshold and evaluate the temperature reached in the samples under laser irradiations: maximum temperature does not reach the melting point at the crystallization threshold and different crystallization mechanisms can be considered in the case of mono/polycrystallites.

1. Introduction

Under various experimental conditions, crystallization of amorphous materials can be described through different processes. One of them is the explosive crystallization which occurs by a rapid propagation of an exothermic crystallization wave through the amorphous material initiated by a localized energy impulse. Another process includes the possibility that melting occurs before crystallization and a liquid zone is then induced between the crystalline and the amorphous regions. But both experimental results (such as reflectivity, transmissivity, Raman scattering measurements...) and advanced theories have not yet concluded in a decisive way whether or not a thin supercooled liquid layer is always present during the crystallization process.

In this paper, we explore the explosive crystallization of a-Ge under nanosecond-pulsed laser irradiation: we irradiate in situ in an electron microscope unsupported a-Ge thin films in order to detect the amorphous-crystal transformation via transmission electron microscopy (STEM and TEM) observations. The aim of this experiment is to give an evaluation of the temperature reached in the film when crystallization takes place and to relate it to the structure of the observed crystallization patterns in order to distinguish different crystallization mechanisms.

2. Experimental Procedure

Amorphous Ge films of various thicknesses (150 to 1100Å) have been produced by thermal evaporation of Ge (99.999% pure) from graphite baskets, in an ultra high vacuum system (2×10^{-9} torr) on to freshly cleaved NaCl substrates. During the evaporation the pressure was lower than 2×10^{-7} torr) and the films were formed at the rate of about $0.5 Å s^{-1}$ controlled with a piezoelectric quartz balance. After deposition, the films were floated off in deionized water and picked up on 120μm square mesh grids. The amorphous nature of as-deposited films was controlled in situ.

The laser beam is introduced into the microscope via a lateral side window, reflected and focused by a lens on to the grid with a final spot 1mm diameter; the irradiations are made either with a ruby or a YAG laser (λ = 0.69μm, full width half maximum pulse duration τ = 40ns, λ = 0.53μm, τ = 14ns (FWHM) respectively). All the irradiations are made in situ through a 100μm aperture just in front of the grid specimen so that the mesh can be irradiated in its centre without touching the Cu edges of the mesh. Neutral filters with various transmissivities have been used to vary the irradiation energy density.

For each irradiation the temporal duration of the pulse is detected with a PIN diode and stored on a memory oscilloscope (Tektronix 7834). At the same time, the total energy of the pulse is measured with a Joulemeter (Scientech). The stability of the shots is about 5% and 8% respectively for the YAG and ruby laser. The spatial distribution of the excitation pulse a few cm behind the 100μm aperture has been measured with a 1024 photodiodes Reticon and exhibited a Gaussian profile.

3. TEM and STEM Observations

By in situ irradiation of the a-Ge films with single pulses of in-creasing incident energy density, we observe that no crystallization occurs below a minimum energy density defined as the crystalliz-ation threshold. Then regime 1 is observed: it is characterized by a typical crystallization pattern about 20 μm diameter in the irradiated mesh. From the centre radial dendrites grow as long crystallites in preferred directions. Figure 1 shows TEM and STEM images of such crys-tallization. A ring system of azimuthally tilted crystals surrounds the irradiated zone while the separation at the amorphous-crystalline interface is abrupt.

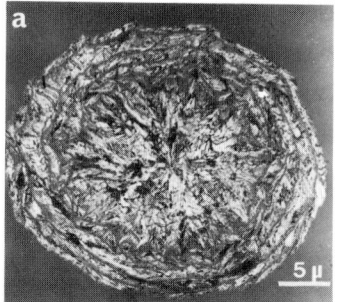

Fig. 1 a) TEM image obtained when irradiating a-Ge film at the crys-tallization threshold (regime 1). b) STEM image of the centre of the same crystallized area.

For irradiations with higher energy density, regime 2, there appears in the centre of the spot a concentric zone with numerous poly-crystallites of small dimensions corresponding to a high density of nuclei (Fig.2). The crystallized zone size increases with irradiation energy density. Then radial long crystals can develop growing in preferential directions while the concentration of obstructing nuclei decreases. The periphery of the ring is formed of tilted crystallites as in regime 1. The surface is wavy with monocrystalline composition on one side and polycrystallites on the other side.

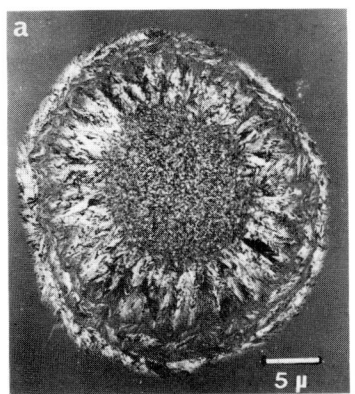

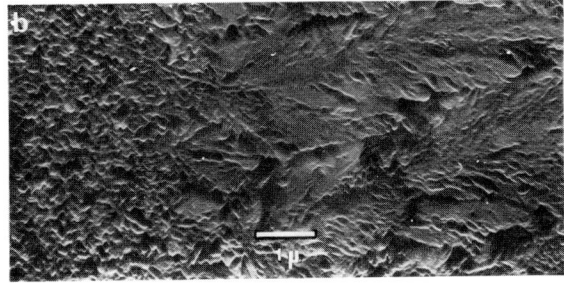

Fig. 2 a) TEM image when irradiating a-Ge film in regime 2. b) STEM image of the interface poly-monocrystallites in this regime.

For the highest irradiation energy densities used, the centre of the irradiated zone can be partially vaporized; regime 3. Independent crystallites sometimes with spherical shape are formed supported on a thin layer, as seen in Fig. 3. These particles are crystalline while the thin layer is amorphous in diffraction patterns. Experimentally, we determined the variation in the crystallization threshold versus thickness of the films for irradiation with the ruby and the YAG lasers, in each regime.

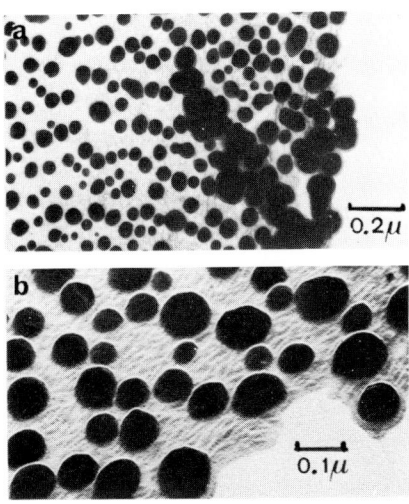

Fig. 3 a) TEM image of crystallization obtained in regime 3. b) STEM image in this regime. The film-support layer can be seen.

4. Discussion

Explosive crystallization has been characterized by Coffin and Johnston (1934) as an exothermic amorphous-to-crystalline phase transformation. They noted that once nucleated, the transformation propagates at high speed through the amorphous film. Most papers on explosive crystallization (Takamori et al (1972), Leamy et al (1981)) have proposed a model of the phenomenon which is based on the heat of crystallization released in the film; once initiated the crystallization front is self-propagating as long as the energy balance between the heat released during crystallization and the heat lost through thermal diffusion is maintained.

An analysis of the explosive crystallization of amorphous layers has been made by Gilmer et al (1980) who analyzed the possibility that melting occurs before crystallization and gave the thickness of the molten zone. More recent experiments (Wickersham et al (1978), Koba et al (1982)) have revealed that there are minimum film thicknesses and temperatures below which the explosive transformation cannot be initiated.

In our case we want to evaluate the temperature T^* reached in the film under laser irradiations at the crystallization threshold, i.e. in regime 1. As the studied samples are homogeneously irradiated in volume, the temperature T^* can be obtained from the following equation:

$$(1-R)(1-e^{-\alpha X})I_0 = X \int_{T=300}^{T^*} \rho \, c(T)dT$$

where I_0 is the energy density of irradiation at which crystallization occurs for a given thickness (X) film of density ρ, optical absorption α, reflectivity R and specific heat $c(T)$. This is the maximum temperature which induces crystallization.

In Figure 4, we plot the values of T^* versus thickness when the film is irradiated under regime 1.

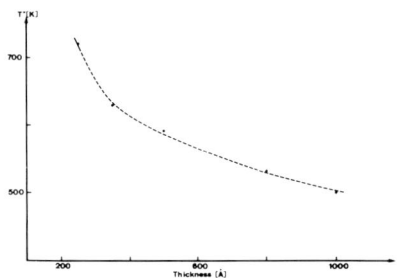

Fig. 4 Values of temperature reached at the crystallization threshold, (regime 1), versus thickness of the samples.

We note that the thick films crystallize at a lower temperature than the thin ones. This can be explained if we consider the role of internal stresses in the amorphous film. The free energy barrier for the amorphous to crystalline phase transition is ΔG. The internal stresses stored in the film have an energy γ and can reduce the height of the effective barrier of the a- to c-transformation. Paesler (1974) has determined that the bending force per unit width for amorphous germanium is proportional to X^2. Thus the minimum energy E for the a-c transformation can be reduced by the internal stresses and $E = \Delta G - \gamma$. As we observe for the thick films, this leads to a decrease of the minimum energy E necessary to obtain crystalliz-ation. We can also note that even for the thinnest films T* never reaches the transition temperature of amorphous-to-liquid germanium estimated at about 970K (Liu et al (1979), Zeiger et al (1980)). Taking into account the exothermic latent heat of crystallization ΔH, as measured by Fan et al (1981) after crystallization is induced, the maximum temperature T reached in the film is then given by $\Delta H \, \rho_a = \rho_c \, c(T^*)(T-T^*)$ where ρ_a and ρ_c are respectively amorphous and crystalline densities. In the regime 1 this temperature T is 1077K for films 250Å thick and 874K for films 1000Å thick, always lower than the melting temperature of crystalline Ge (1214K).

Now let us consider regime 2; the values of the energy density of irradia-tion necessary to obtain the polycrystalline zone in the centre of the spot are important. These values can lead to a further increase in the tempera-ture T of about 350° so that the melting temperature of amorphous Ge can be reached: a thin layer of liquid Ge can be formed under irradiation and induces crystallization via polycrystallites. Such a process is different from that analyzed in regime 1.

5. Conclusion

The crystallization mechanism in unsupported a-Ge films induced by laser pulses of a few nanoseconds duration has been studied. At the threshold of energy density irradiation (regime 1), large dendritic crystallites about 10μm long grow in radial directions from the centre of the impact. This central zone is surrounded by a system of rings with tilted crys-tallites. We observed variations in the crystallization threshold with thickness of the film. In this regime, the evaluated temperature of the films 250 to 1000Å thick, under threshold irradiation, is lower than the melting temperature in the centre of the crystallized zone. In this case, the induced crystallization under nanosecond pulses takes place without melting.

If the energy density of irradiation increases (regime 2) a thin layer of liquid Ge can be formed and the crystallization passes through a liquid phase. In this case, in the centre of the irradiated zones polycrystallites are obtained which are different from the dendrites formed at the threshold energy density for crystallization.

References

Coffin C C and Johnston S 1934 Proc. of the Royal Society A 146 564
Fan J C C and Anderson Jr C H 1981 J. Appl. Phys. 52 4003
Gilmer G H and Leamy H J 1980 Laser and Electron Beam Processing of
    Materials eds C W White and P S Peercy (New York: Academic Press) pp 227
Koba R and Wickersham C E 1982 Appl. Phys. Lett. 40 672
Leamy H J, Brown W L, Keller G K, Foti G, Gilmer G H and Fan J C 1981
    Appl. Phys. Lett. 38 137
Liu P L, Yen R, Bloembergen N and Hodgson R T 1979 Appl. Phys. Lett. 34 864
Paesler M A 1974 Amorphous and Liquid Semiconductors eds J Stuke and
    W. Brenig pp 229
Takamori T, Meissier R and Roy R 1972 Appl. Phys. Lett. 20 201
Wickersham C F, Bajor G and Greene J E 1978 Solid State Commun. 27 17
Zeiger H J, Fan J C C, Palm B J and Chapman R L 1980 Laser Electron Beam
    Processing of Materials eds C W White and P S Peercy (New York: Academic
    Press) pp 221

Inst. Phys. Conf. Ser. No. 67: Section 3
Paper presented at Microsc. Semicond. Mater. Conf., Oxford, 21–23 March 1983

207

Defect migration and temperature gradient effects in low-power laser annealing of α−Ge

G Vitali,U Zammit,M Marinelli

Istituto di Fisica-Facoltà di Ingegneria, P.le Aldo Moro,5,00185 ROMA.

Abstract   A particular RHEED technique allowed us to observe opposite characteristics in the crystalline-amorphous front motion of implanted and glow discharge deposited α-Ge samples, respectively, during low power multi superimposed pulsed ruby laser annealing.
Moreover a combination of etching of the implanted specimen,and double stage replicas, showed both that defect migration is a mechanism which occurs during low power laser annealing,and that temperature gradients enhance the crystallization process in such a regime.

Low power pulsed laser annealing of semiconductors has not been very widely investigated until now, and seems to provide evidence of the existence of processes which involve neither melting nor purely thermal recovery of the material as obtained in an ordinary furnace. We would like to present in this paper a review of some recent results we have obtained in this field (1),(2),(3).
Specimen structure analysis was carried out by means of the RHEED (Reflection High Energy Electron Diffraction) technique, where the electron beam incidence angle, ϑ , with respect to the specimen surface, was varied in order to detect structure changes at different depths (up to about 500 Å (1)) in the specimen. Moreover, the implanted specimen was first etched with superoxol (1 HF: $1H_2O_2$: 600 H_2O) which is selective for amorphous damage regions produced by the implanted ions with respect to the surrounding,less damaged matrix (4), and two stage Ni-Cr shadowed replicas of the etched surfaces were then performed. This enabled us to follow the behaviour of the damage regions during the laser annealing process (2). The thickness of the material removed by the etch action, as determined by interferometric measurements,was about 200 Å. Irradiation was performed with a Q-switched ruby laser with 30 ns pulse length operating in the 2-4 MW cm^{-2} power density range. Temperature values calculated in such a range were well below the material melting threshold (1),(3). The beam homogeneity was improved by means of a diffusing fused silica bent light pipe and by an azulene solution whose absorption coefficient increases with increasing light intensity (5). The examined specimens were glow-discharge films, about 500 Å thick, deposited onto <111> Ge single crystal substrates and <111> and <100>,<110> Ge specimens implanted with doses of $1\times10^{15}cm^{-2}$ and 5×10^{14} cm^{-2},46 KeV Te^+ ions, respectively. The implanted layers were also about 500 A thick. The influence of temperature gradients on the crystallization process was tested using   irradiation with a spatially modulated profile (interference fringes). In this case a defocusing optical system was used instead of the bent silica pipe in order to improve the homogeneity of each of the interfering beams, as the beams coming out of the pipe were too diverging.

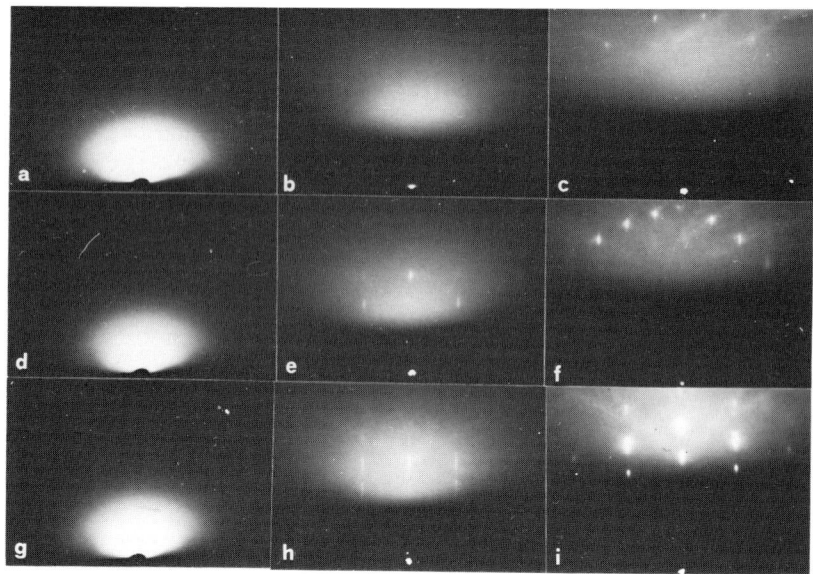

Fig. 1 RHEED patterns of a 500 Å thick glow discharge amorphous
Ge film deposited onto <111> Ge single crystal : (a),(d),(g),
$\vartheta = 0,8°$;(b),(e),(h), $\vartheta = 2°$; (c),(f),(i), $\vartheta = 4°$; (a),(b),(c)
after 10 2 MW cm^{-2}; (d),(e),(f), after 20 2 MW cm^{-2}; (g),(h),
(i), after 50 2 MW cm^{-2} superimposed laser pulses. Ref.(1).

Fig.1 refers to glow discharge deposited α-Ge specimen(1). Fig.1a,1b and
1c (corresponding to ϑ values of about 0.8°,2° and 4°, respectively) show
the RHEED patterns obtained after a specimen had been irradiated with 10
successive superimposed laser pulses of about 2 MW cm^{-2}: (irradiation with
just a single 2 MW/cm^2 pulse had not shown any perceptible change, for the
whole range of ϑ values, in the incoherent amorphous-like RHEED patterns).
Fig.1c shows the beginning of an amorphous-single crystal transition in the
specimen deep layers (300-400 Å), while intermediate (ranging between few
tens and few hundreds of Å) and surface (few tens of Å) layers still
remain amorphous (1b and 1c respectively). Fig.1d,1e,1f refer to the same
specimen irradiated with 10 additional 2 MW/cm^2 pulses. Fig.1f shows a
remarkable increase of the crystallized material in the specimen deep
layers while fig.1e shows that the intermediate layers have also turned to
single crystal. Surface layers still remain amorphous (1d). Finally fig.
1g,1h,1i obtained after having irradiated the specimen with 30 additional
2 MW/cm^2 pulses show a further increase of the crystallized material, the
surface layer still remaining amorphous. The fact that crystallization
takes place in the deep layers while the surface layer remains amorphous
is evidence of a non melting process, because such a process should start
from the surface. Fig.2 (6) refers to a <111> Ge sample implanted with a
dose of 10^{15} cm^{-2} of 46 KeV Te ions. Fig.2a,2b show that the specimen
surface and intermediate layer have turned to single crystal after the
sample had been irradiated with 31 2 MW cm^{-2} pulses, while the deep
layers have remained amorphous (Fig.2c). Fig.2d shows a RHEED double
pattern (7) of the specimen showing that the crystallized material and
the undamaged substrate have the same crystallographic orientation,
although they are separated by an amorphous layer (2c). The patterns

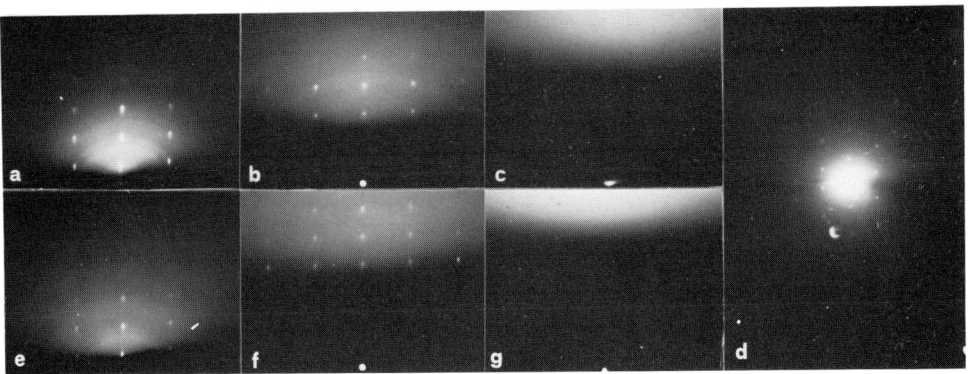

Fig. 2 RHEED patterns of <111> Ge implanted with 10^{15} cm^{-2} doses of 46 KeV Te$^+$: (a),(e), $\vartheta = 0,8°$; (b),(f), $\vartheta = 2°$; (c),(g), $\vartheta = 4°$; (d): double pattern;(a),(b),(c),(d), after 31 2 MW cm^{-2};(e),(f),(g), after 51 2 MW cm^{-2} superimposed laser pulses. Ref.(6).

obtained after having superimposed 20 additional 2 MW/cm^2 pulses (Fig.2e-2g) show that the amorphous-crystalline material front has moved deeper into the specimen, as can be appreciated by the fact that the single crystal pattern (2f) can be obtained using a probing beam with an inci-dence angle which is greater than in the corresponding previous case (Fig.2b). The presence of an amorphous layer beneath the crystallized surface should exclude the possibility of columnar crystalline regrowth from the substrate due to hot spots (8).
Fig.3a shows the replica of the etched surface of a 5x10^{14}cm^{-2} <100> implanted specimen before any laser treatment (2). The black spots, whose dimensions are about 70-80 Å, are the damage regions produced by the im-planted ions. They look evenly distributed in a less-damaged matrix. The RHEED insert shows that the specimen is amorphous. Fig.3b and 3c (2) show the same specimen after irradiation with ten pulses of power density of 1,2 and 2,8 MW/cm^2 respectively. The respective RHEED inserts show a gradual crystalline recovery. The defects are shown to have gradually gathered together in some places to form some 'clusters' and in fig.3c such clusters are at the boundaries of areas depleted from damage. The typical dimensions of the recovered zones in fig.3c are about 2000 Å, so

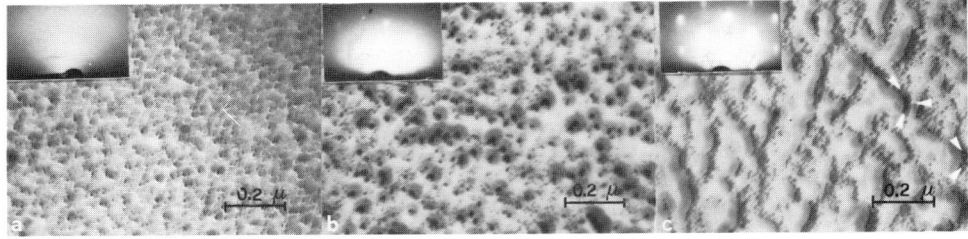

Fig. 3 Damage regions on <100> Ge implanted with 5x10^{14} cm^{-2} 46 KeV Te$^+$ together with respective RHEED patterns: (a) before irradiation; (b) after 20 1,2 MW cm^{-2}; (c) after 20 2,8 MW cm^{-2} superimposed laser pulses. Ref.(2).

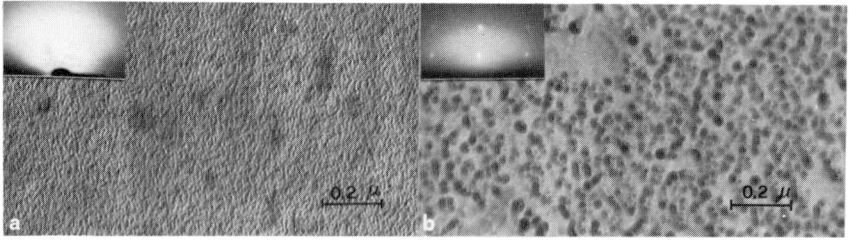

Fig. 4 Damage regions on <111> Ge implanted with 10^{15} cm^{-2} 46 KeV Te$^+$ (a) before irradiation; (b) after 100 2 MW cm^{-2} superimposed laser pulses, together with respective RHEED patterns. Ref.(2).

the possibility of attributing the recrystallization to melting caused by hot spots must be excluded. In fact intensity inhomogeneities such as those caused by multimode interference (9) should take place in areas of the order of the laser wavelength (0,69 μ) and effects caused by hot spots have been observed in areas of a few microns dimensions (8). The results shown in fig.4 and 5 indicate that defect precipitation is the process taking place at this early stage of crystallization (2). Fig.4a and 4b show a 10^{15} cm^{-2} <111> Te$^+$ implanted specimen with respective RHEED inserts, before and after 100 2 MW cm^{-2} pulses irradiation respectively. Some defect precipitation can still be observed. However, unlike the previous case (fig.3a,3b,3c) the overall area covered by dark zones after irradiation has decreased with respect to the original situation, as though the precipitate areas had "dissolved" in the sorrounding matrix (2). This can take place through the migration of simple or bound defects, similarly to what occurs in furnace annealing.

Fig.5a and 5b show the result obtained after irradiating a 10^{15} cm^{-2} <111> Te$^+$ implanted Ge specimen with 20 pulses of spatially modulated light (periodicity d = 13 μ and peak intensity I_p = 3 MW cm^{-2}) (3). The area indicated by A, corresponding to the fringe maximum (where thermal gradient is zero or very small) is practically unchanged with respect to the areas denoted by C where the intensity is very small. In areas denoted by B, however, where the temperature gradients are different from zero, crystallized areas start appearing. This can be explained assuming a temperature gradient contribution to the crystallization process, once the system has been provided enough energy (possibly to increase the defect mobility). In fact,a distance d/4 from the fringe peak, where temperature gradient is maximum, no crystallization is detectable, due

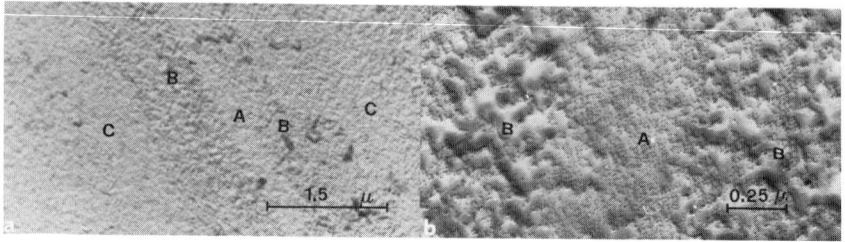

Fig. 5 (a) crystallized fringe from a 13 μ periodicity fringe pattern with peak intensity I_p = 3 MW cm^{-2};(b) enlargement of (a). Ref.(3).

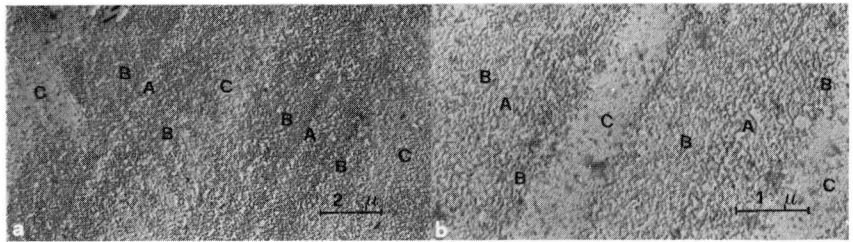

Fig. 6 crystallized fringes from fringe pattern of peak intensity I_p = 4 MW cm^{-2} and periodicity (a): d = 6μ; (b) d = 3μ. Ref.(3).

to the fact that too low an intensity is incident in that area. Fig.6a and 6b (3) show the result obtained after irradiating 5x10^{14}cm^{-2} <110> Te$^+$ implanted Ge samples with 20 spatially modulated pulses, I_p = 4 MW cm^{-2} and spatial periodicity of about 6μ and 3μ respectively. The central, less-annealed channel (A area) has progressively decreased in size, (it can just be detected in fig.6b), as the fringe periodicity and therefore the area with zero temperature gradients have also decreased in size. Moreover the B areas have become larger since the peak intensity was increased and the zones where the incident intensity was high enough to cause (together with the temperature gradient effect) crystallization has also become larger.

We can therefore conclude that low power laser annealing of semiconductors is a cumulative process and that ion implanted and glow discharge deposited amorphous materials have different topology (1) although they give the same incoherent amorphous-like RHEED pattern: in fact, they crystallize differently under the low power laser action. Moreover, defect migration is a mechanism which causes crystallization in low power pulsed laser annealing of semiconductors (10)(11). This can occur through a mobility increase of the defects, caused by thermal activation, and through plasma formation effects due to bond-electron photoexcitation (11),(12). Temperature gradients also enchance the crystallization process.

Acknowledgements

The authors would like to thank Prof.F.Scudieri with whose cooperation the fringe profile irradiation experiment was carried out.

References

1. Vitali G, Bertolotti M, Zammit U and Marinelli M 1982 Phys. Lett. 89A 199
2. Vitali G, Marinelli M, Zammit U and Scudieri F Phys. Lett. A in press
3. Vitali G, Marinelli M, Zammit U and Scudieri F to be published
4. Bertolotti M, Sette D, Stagni L and Vitali G 1971 Appl. Phys. Lett. 18 257
5. Aslanidi E B, Tikhonov E A and Shpak M T 1971 Opt. Spectroscopy 31 233
6. Vitali G, Zammit U, Marinelli M and Bertolotti M 1983 Appl. Phys. A30 in press
7. Vitali G 1976 Rev. Sci. Instrum. 43 276
8. Regolini S, Sigmon T W, Gibbons J F, Magee T J and Peng J 1979 Proceedings of "Laser and Solid Interactions and Laser Processing" MRS Conference Boston 1978 eds S D Ferris, H J Leamy and J M Poate AIP New York p 393

9. Von Almen M F July 13–25 Proc. of "Physical Processes in Laser Material Interaction" NATO-ASI, Villa Le Pianore, Versilia, Italy ed M Bertolotti Plenum Press

10. Troxell J R, Chatterjee A R, Watkins G D and Kimerling L C 1979 Phys. Rev. B19 5336

11. Van Vechten J A 1980 J. de Phys. C4 7

12. Van Vechten J A and Thurmond C D 1976 Phys. Rev. B14 3551

SEM developments in semiconductor characterization

G Fontaine, P Morin and M Pitaval

Département de Physique des Matériaux, Université Claude Bernard - LYON I
43 Boulevard du 11 Novembre 1918, 69622 VILLEURBANNE Cédex - FRANCE

Abstract   Electron-channelling imaging is a new mode of operation of the SEM
which has been developed in our laboratory and applied to the characterization of semiconductors. Images of extended crystalline defects in solid
samples are observed with low-loss backscattered electrons. The explored
depth beneath the surface is around 100 nm and the resolution about 10 nm.
Polishing the samples at a small angle from the original surface makes it
possible to examine different depths. The observations reported here concern
extended defects in various silicon samples and (Ga-In)As graded epitaxial
layers.

1. Introduction

The bending of lattice planes near crystalline defects is known to change slightly the electron absorption in the neighbourhood of Bragg reflections and consequently modify electron backscattering (Booker, 1970). The beam conditions
which make it possible to observe a well defined contrast are however rather
critical. For instance, the bending might be of the order of 10^{-3} radians at
a distance of 30 nm from a dislocation core. It is thus necessary to obtain
simultaneously a micrograph resolution of 30 nm and an electron-beam illumination aperture of 10^{-3} radians. The contrast is weak in the range of 1 %, so that
a relatively high probe current of the order of 10^{-8}A is necessary. This corresponds to a brightness B of 4.10^8 A cm^{-2} for a SEM operating at 50 kV, which
necessitates a field emission gun. In addition, on a thick sample, the broadening effect resulting from scattering of the electrons must be overcome. This
has been accomplished by energy filtering to select low-loss backscattered electrons (Wells, 1971). This filtering increases strongly the observed contrast
with a compromise between high contrast and low noise obtained with a filter
window of 500 eV at a primary beam energy of 50 keV.

2. Experimental conditions

The modifications of a SEM Cambridge Stereoscan MKIIa have been described by
Morin et al (1979) : field emission gun, ratio amplifier to eliminate residual fluctuations in the emission current, goniometer stage and high pass energy
filter of spherical grid type with a large collection angle (45°). Diffracting
conditions are determined by electron channelling pattern (ECP) recorded from
a selected area of 10 µm in diameter by the use of a rocking beam method. In order
to avoid contamination, the specimen is surrounded by copper walls cooled by
liquid nitrogen. The filtered backscattered electrons are detected by a scintillator. The photomultiplier output is connected either to the video preamplifier for specimen imaging, or to the discriminator amplifier of a photon

counting system to measure the absolute value of the detected current. The maximum count rate corresponds to 10^{-12} A.

3. Electron channelling contrast

The physical principles underlying the observation of crystalline defects in solid samples with a SEM are similar to those which are producing ECP (Coates, 1967). It is thus of interest to begin with some experimental and theoretical work which has been performed in our laboratory to check the influence of energy filtering on the contrast of ECP. Two observations are well-known at high incidence angles : inversion of the contrast in a band parallel to the greatest slope of the sample's surface (Fig. 1a) and dissymmetry of lines perpendicular to this direction (Fig. 2a). As shown Fig. 1b and 2b energy filtering with energy window ΔE reestablish the contrast observed at normal incidence.

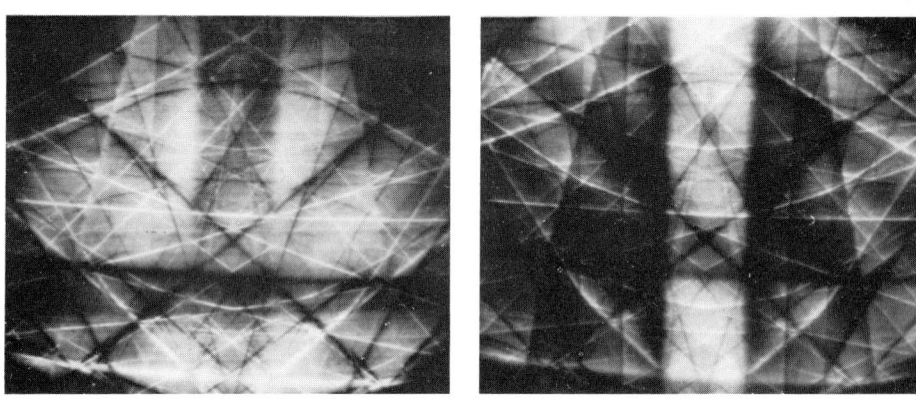

(a) no filtering   (b) ΔE = 0.5 keV

Fig. 1 Influence of energy filtering (ΔE) on the inversion of contrast of 220 band. Silicon at high incidence angle ($\simeq$ 70°).

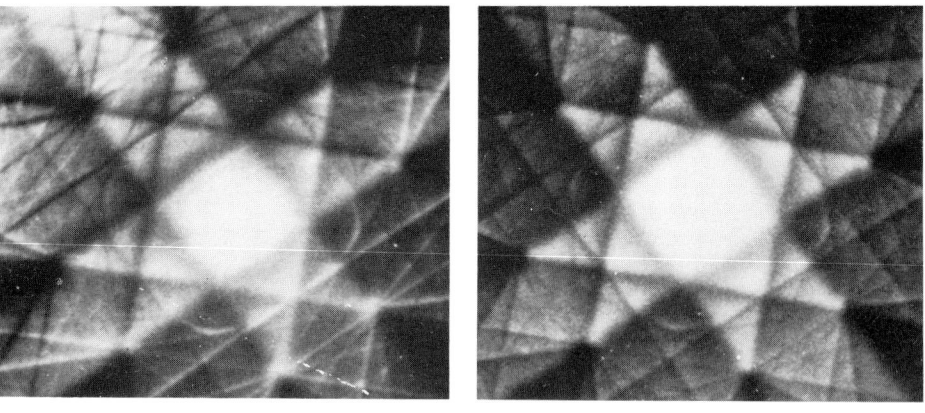

(a) no filtering   (b) ΔE = 0.5 keV

Fig. 2 Influence of energy filtering (ΔE) on the dissymetry of lines. Silicon at 55° incidence angle.

A silicon monocristal with a (1$\bar{1}$1) surface was oriented so as to have the ($\bar{1}$01) planes normal to the tilt axis.

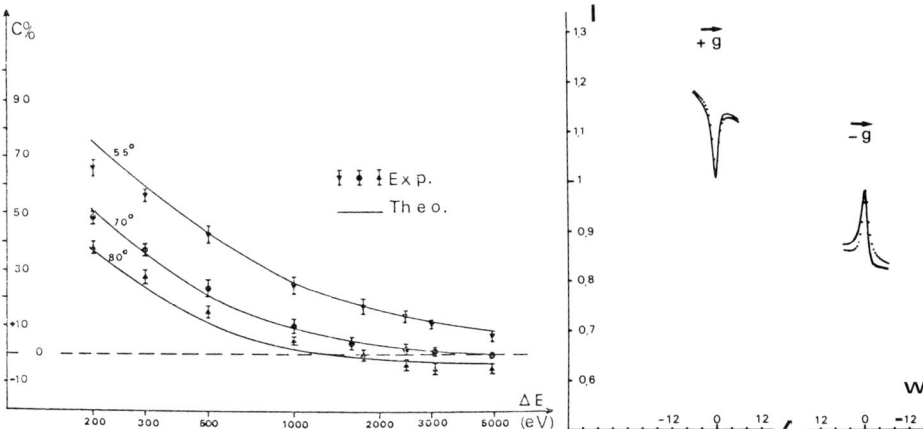

Fig. 3 Experimental and computed contrast for three incidence angles as a function of the energy window ΔE.

Intensity profiles of the $\overline{2}02$ band of the ECP were recorded with three tilt angles (55°, 70°, 80°) and for various widths ΔE of the filter window (Fig. 3). Contrast, defined as $\Delta I/I_{mean}$, was measured on the recorded profiles, where ΔI is the variation of the video signal and I_{mean} its mean value. The contrast is positive when the band has an intensity higher than I_{mean} and negative at high tilt angles without filtering. A theoretical three step model has been developed by Morin (1981) whose results are also shown Fig. 3. In this model primary electrons are described in a two-wave approximation. In the second step electrons are inelastically scattered by phonons, plasmons and individual excitations. The third step describes the escape of scattered electrons through the surface and their detection by the filter with a Monte Carlo calculation. Besides the fact that the inversion of the contrast is correctly predicted in this model it must be emphasized that

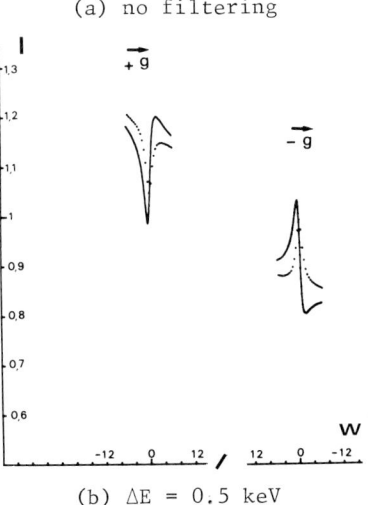

(a) no filtering

(b) ΔE = 0.5 keV

Fig. 4 Computed intensities of 440 and $\overline{4}40$ lines in silicon, with anomalous absorption (lines) and without (dots)

filtering strongly increases the contrast from a small percentage to approximately 60 % for ΔE = 0.5 keV and an incidence angle of 45°, which values are generally used to image crystalline defects. The inversion of contrast comes from the fact that at high incidence angles, as shown by Monte Carlo calculation, the probability of detecting an electron inelastically scattered at small angles at a depth z, presents a maximum as a function of z. Inelastic events occur nearer the crystal surface inside an ECP band than outside and thus, at high incidence angles, electrons scattered inside a band are less efficiently detected. The filtering of electrons scattered outside the band at a greater depth which have greater energy losses eliminates these electrons, leaving only those electrons scattered inside the band. If inversion of band contrast is thus due to anomalous absorption, the dissymmetry of lines Fig. 2 is on the contrary a purely geometrical effect. The line which corresponds to the greatest incidence angle is black because the diffracted beam is then directed towards the sample and the line which corresponds to the lowest incidence angle is white because the diffracted beam is then directed towards the surface. Inelastically scattered electrons from the transmitted beam have thus lesser or greater chances of being detected for each line respectively. Filtering eliminates those geometrical effects. The results of the computations are shown in Fig. 4 for the 440 lines in a silicon sample at an incidence angle of 55° with anomalous absorption (lines) and without (dots). It is clear on the Fig. 4a that anomalous absorption has practically no influence on the dissymmetry of ECP lines and that filtering reestablishes the black and white contrast of each line (Howie, 1978 ; Yamamoto et al, 1978).

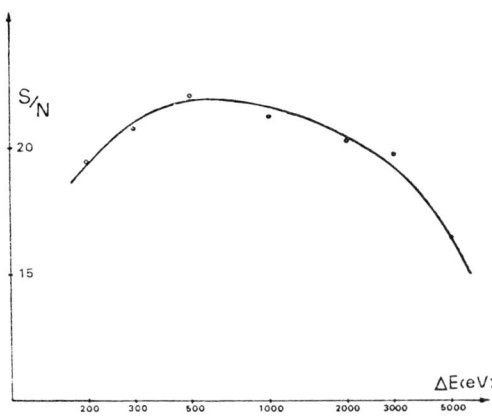

Fig. 5   Signal to noise ratio S/N for the 220 band in silicon as a function of the energy window ΔE.

The diminution of the width ΔE of the filter window strongly increases ECP's contrast but increases the noise simultaneously. The ratio R between mean detected intensity and incident beam current is linear with ΔE, being approximately 5.10^{-4} for $\Delta E = 0.5\,keV$. Signal to noise ratio S/N is equal to $C(Rn)^{1/2}$ where C is the contrast previously defined and n the number of incident electrons per image point. S/N is shown in Fig. 5 for an incident angle of 55°, a probe current of 10^{-8} A, and a frame scan time of 100 s. The optimal window is 500 eV, which value was generally used in the subsequent observations. This corresponds to approximately 100 nm explored depth beneath the surface of a solid Si specimen.

4. Microtwins in epitaxial silicon on sapphire substrate

One of the most evident applications of the method concerns the control of crystalline quality in epitaxial layers on substrates. We present here some results which have been obtained for (100) silicon deposited from the vapor phase at 980° by the pyrolysis of silane in H_2 on ($\bar{1}012$) CZ single crystal sapphire substrate. A growth rate of 2 μm min^{-1} was used and silicon film thicknesses in the range 0.1 - 5 μm were investigated. For film thicknesses lower than 0.4 μm, small dimensions and high density of defects approach the limit of resolution of the method (Zorilla et al 1981). At higher thicknesses individual micro-twins and dislocations can be resolved as shown in Fig. 6 for a 0.6 μm film thickness. Four systems of twinning occur. In Fig. 6a the specimen was oriented to display the center of the [111] pole in the ECP mode ; ($\bar{1}11$) twinning planes intersect the surface along the [$\bar{1}10$] direction. These microtwins have a [511] type axis parallel to the incidence direction and fewer electrons are backscattered than for [111] pole : the twins have a black contrast. In Fig. 6b the specimen is oriented in order that the beam impinges along the [212] axis of the matrix and [0$\bar{1}$0] axis of the microtwins which have their twinning axis along [1$\bar{1}$1]. The 202 band is excited for both the matrix and the microtwins which appear brighter than the matrix. By a tilt of a few degrees, the incidence is varied to correspond to the side band, the ($\bar{2}20$) planes are then in Bragg position and dislocations become visible, see edge on Fig. 6c. A long range modulation of contrast some μm in size, is also visible on Fig. 6 which is not a topographic effect and may probably be correlated to heterogeneous structures in the surface of the initial substrate.

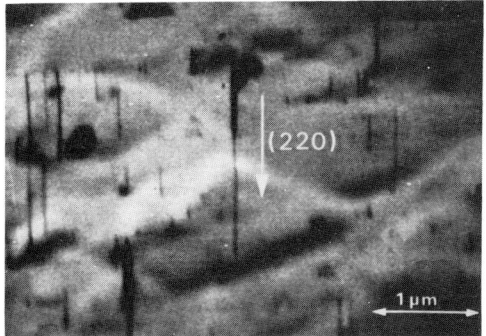

(a) beam along [111] matrix

(c) beam in Bragg position
[$\bar{2}$02] matrix

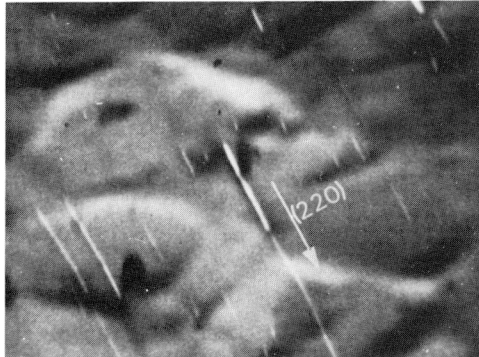

(b) beam along [212] matrix

Fig. 6 Microtwins in
epitaxial silicon on
sapphire substrates

5. Highly doped diodes

We have observed avalanche photodetectors made on silicon (111) wafers. These diodes have a p-n junction parallel to the surface at a depth of 0.5 μm. SiO_2 layer and Al rings were removed with concentrated HF acid, then the surface was cleaned with syton for a few seconds. In these diodes, due to high concentration of dopant and different thermal processes (oxidation, diffusion), it is possible to observe misfit dislocations, extrinsic stacking faults and precipitates. It is not the purpose of this paper to correlate these defects to the process but only to show different applications of the method.

Fig. 7 shows a dislocation network at the limit (Fig. 7a and b) of a highly doped region. The network is made of dislocations parallel to the junction and segments perpendicular to the surface originating at the junction. A better knowledge of the in-depth distribution of the defects is obtained by polishing the sample at a small angle from the original surface. This has been done in Fig. 8 for an angle of 2°52'. Below the horizontal mark 0, we have the original surface with emergence of dislocations viewed edge-on and coming from the network situated between B and C. From 0 to C, the sample has been thinned with increasing depth. At A the surface is 100 nm below its position at 0, at B 200 nm and 300 nm at C. It is thus apparent that approximately 100 nm are analysed beneth the surface with this method (between B and C).

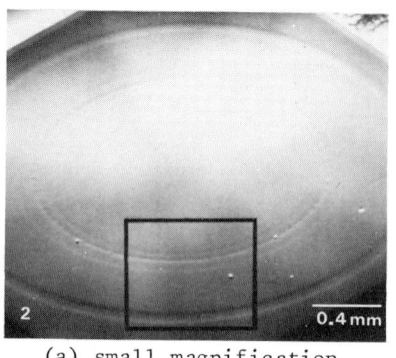

(a) small magnification

(b) limit of the diffused area

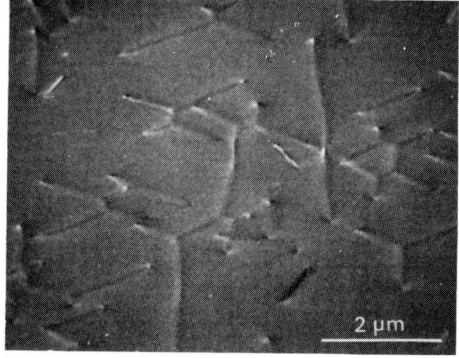

(c) dislocations near the junction

Fig. 7 Localization of dislocations at the limit of highly P doped silicon

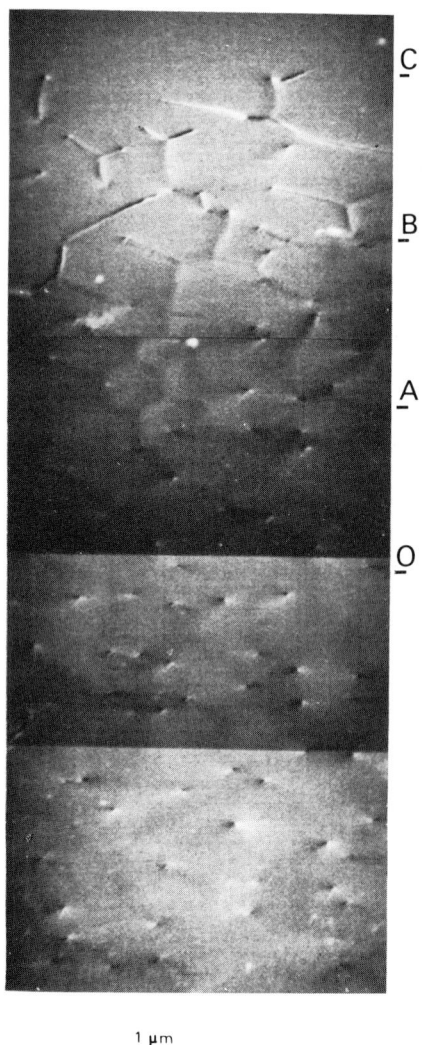

Fig. 8 Dislocations network near a p-n junction in silicon

The Burgers vector may be determined as in TEM by finding the reflection in which the dislocation becomes invisible. This is shown in Fig. 9a and b which shows the corresponding contrast for the edge dislocation in the middle of the figures. It should be noticed that there is no oscillating contrast in the images of dislocations inclined to the surface of the sample. This experimental result has been confirmed by the model previously described. Theoretical results for the contrast of an edge dislocation in silicon is shown in Fig. 10. As before this result takes into account the filter parameters but here the

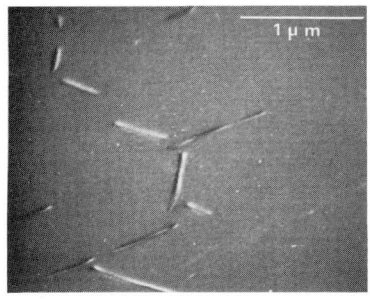

(a) $\vec{g}.\vec{b} = + 2$

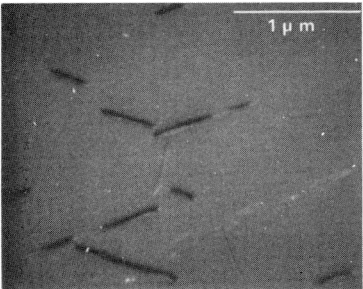

(b) $\vec{g}.\vec{b} = 0$

Fig. 9 Burgers vector's analysis

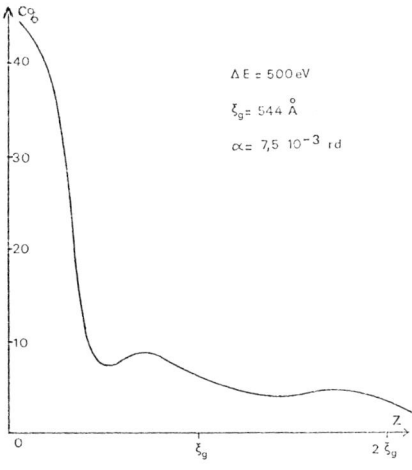

$\Delta E = 500 \, eV$

$\xi_g = 544 \, \overset{\circ}{A}$

$\alpha = 7,5 \, 10^{-3} \, rd$

Fig. 10 Computed value
of the contrast of an
edge dislocation as a
function of its depth Z

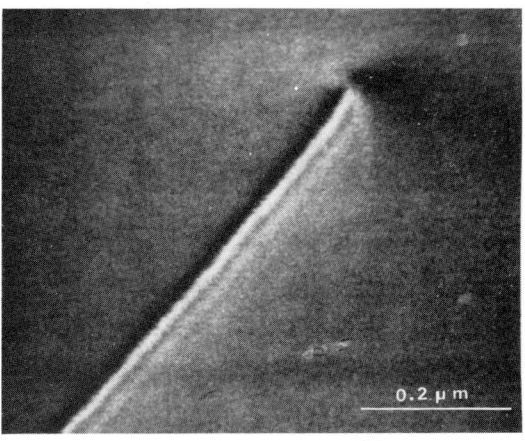

Fig. 11 Extrinsic stacking
fault in silicon

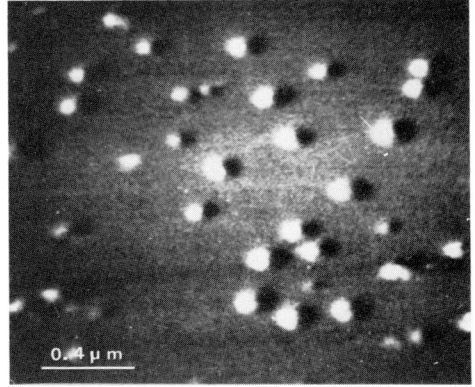

(a) $\vec{g} = [220]$

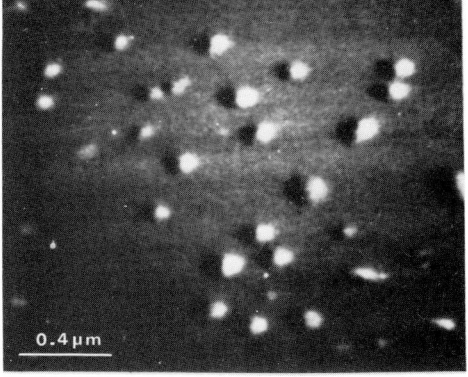

(b) $\vec{g} = [\bar{2}\bar{2}0]$

Fig. 12 Phosphorus precipitates in
silicon

half aperture angle α of the probe has also been taken into account. A strong decrease occurs in the first $\xi_g/2$ Å, after which the contrast decreases slowly with a mean value of approximately 5 % up to 100 nm depth. Oscillating contrast is however observed at a stacking fault as shown Fig. 11. Such a fault is observed in the preceding sample in some conditions, its extrinsic nature can be deduced from the contrast of the first fringe (complementary to that observed in TEM on thin films). Comparison with the computed profile and taking into account the geometry of the experience (incidence angle, fault plane) gives as the successive depth for the bright fringes : 41 nm, 95 nm, 150 nm and 204 nm. Once again one sees that it is possible to image defects in depth down to about 100 nm (3 fringes).

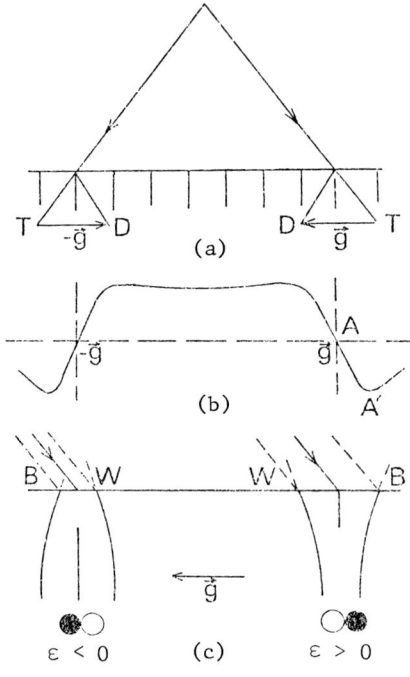

Fig. 13   Determination of the strain ε near a precipitate.

In some conditions of implantation with phosphorus ions, phosphorus precipitates may be observed in the samples as shown Fig. 12 a and b taken under $\pm\vec{g}$ diffracting conditions. As in TEM, the line of no contrast lies perpendicular to the direction of $\vec{g}$. Two strong black and white lobes appear in the direction of $\vec{g}$ whose contrasts are inversed with the sign of $\vec{g}$. The sign of the strain ε introduced by the inclusion is given by the same criteria as those used in TEM. Fig. 13 is a schematic illustration of the procedure used in this kind of orientation. The ECP is first recorded at low magnification. Fig. 13a shows the vector $\vec{g}$ between transmitted (T) and diffracted (D) beams used to image Fig. 12b. The ECP line A corresponding to $\vec{g}$ is centred in the middle of the cathode ray tube display by an appropriate tilt of the specimen. Going back to the image mode gives the contrast corresponding to $\vec{g}$ since the variation of the probe incidence angle can be ignored at high magnification. A center of compression ($\varepsilon < 0$)/dilatation ($\varepsilon > 0$) gives the black (B) and white (W) contrasts shown in Fig. 13c. Precipitates imaged Fig. 12b can thus be characterized as compression centers.

6. Annealing after implantation by pulse electron beam

Silicon samples with a (100) surface have been implanted with As ions of 140 keV energy at a dose of 10^{15} cm^{-2}. These samples have been thermal annealed by a pulsed electron beam. The results are presented by Pitaval et al elsewhere in this meeting. The purpose here is merely to illustrate another interest of the described method : the possibility of visualizing large areas at moderate magnification. Fig. 14 shows the specimen after pulsed (50 ns) electron beam (1,4 J cm^{-2}) annealing. Part I corresponds to the original surface of the sample, parts II and III correspond to a polishing of the surface at a small angle (2°52) starting at the limit between I and II. Residual defects made of small angle boundaries are localized in part II, of Fig. 14 where temperature has reached a value of the order of the fusion temperature down to a depth of 0.5 μm. Grain size is approximately 1 μm in diameter and individual dislocations in the boundaries can be resolved at higher magnification.

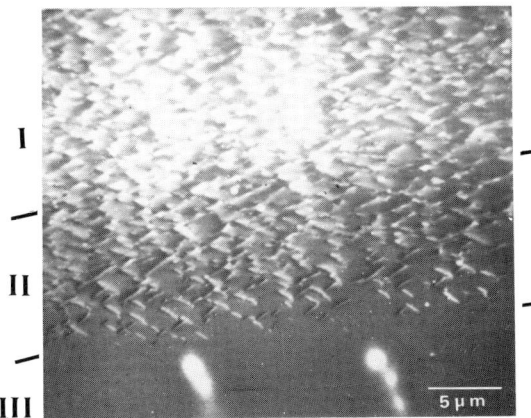

I

II

III

Fig. 14   Pulsed electron beam
annealed As implanted silicon

7. Dislocations in vapor-grown compositionally graded (Ga,In)As

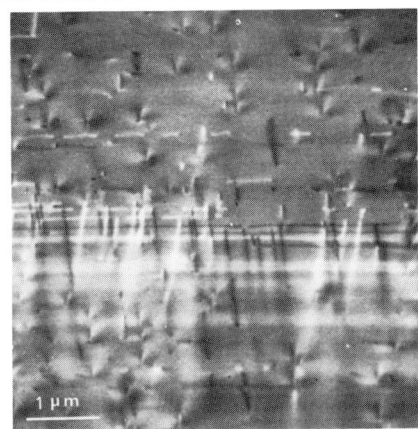
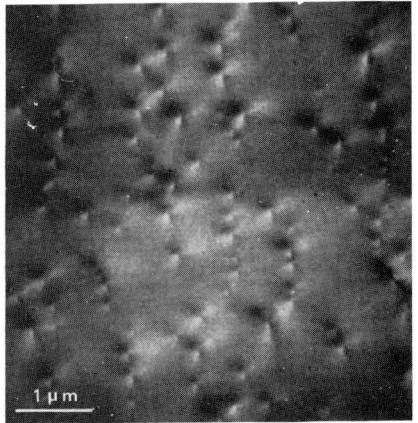

(a) at one interface   (b) between two interfaces

Fig. 15   Misfit dislocations in step graded (Ga,In)As

The chemical composition of epitaxial layers of (Ga,In)As was changed by step
grading from a substrate GaAs (110) to a final $Ga_{0.6}In_{0.4}As$ in seven steps,
with a variation in concentration of about 5 % in each step. Fig. 15a shows
misfit dislocation patterns at one of the interfaces and Fig. 15b dislocations
generated inside the succeeding step. There are three directions of dislocations in Fig. 15a which can be analysed by polishing at a small angle in different directions. For example, the horizontal dislocations (white contrast)
in Fig. 15a are edge dislocations lying along $[01\bar{1}]$. It is of interest to
note that these dislocations have not the usual black and white contrast and
that their image width is lower than the image width shown in Fig. 9. The
explanation is that Fig. 15a is obtained with an orientation of the specimen
slightly different from the exact Bragg angle, say the orientation A' in Fig.
13b, thus reducing dynamical effects in comparison to Bragg orientation A,

which gives Fig. 9. These kinematical images are promising for future observations, however mechanical vibrations are currently limiting the resolution to approximately 10 nm.

8. Conclusions

A new method of observing single defects near the surface of solid specimens has been developed in SEM which can be named electron-channelling imaging (ECI). Applications of ECI, to the characterization of semiconductors in different areas of interest has been presented in this paper. In the future we plan to correlate this technique with other characterization methods : Auger microanalysis and pulsed EBIC near the emergence of crystallographic defects.

Acknowledgments

The authors thank Mr. GOUBE (Centre de Recherche C.G.E. Marcoussis) for supplying photodetector samples and Mr. SCHILLER (L.E.P. Limeil-Brevannes) for supplying (Ga,In)As samples. We are grateful to D.G.R.S.T. (Aide N° 81.A.0755) and to G CIS (Contrat de programme N° 9028) for financial support.

References

Booker GR 1970 Modern diffraction and imaging techniques in material sciences
    (London : North Holland) pp 645 - 7
Coates D G 1967 Phil. Mag. 16 1179
Howie A 1978 Inst. Phys. Conf. Ser. 41 1
Morin P, Pitaval M, Besnard D, Fontaine G 1979 Phil. Mag. A 40 511
Morin P 1981 Thèse : Université de Lyon
Wells O C 1971 Appl. Phys. Lett. 19 232
Yamamoto T, Mori M, Yshida Y 1978 Phil. Mag. A 38 439
Zorrilla M L, Trilhe J, Pitaval M, Morin P 1981 J. Electrochem Soc. 128 385

Inst. Phys. Conf. Ser. No. 67: Section 4
Paper presented at Microsc. Semicond. Mater. Conf., Oxford, 21–23 March 1983

Defect structure of epitaxial films grown on porous silicon

H Baumgart[o], F Phillipp* and G K Celler

Bell Laboratories, Murray Hill, New Jersey 07974, USA
*Max-Planck Institute for Metal Research, D-7000 Stuttgart 80, W. Germany

Abstract   Porous silicon layers were formed by anodic dissolution
in hydrofluoric acid. The resulting low density porous material
keeps the single crystalline structure. We have investigated the
epitaxial growth of Si films on porous substrates by Nd:YAG laser
fusion and molecular beam epitaxy (MBE). Microscopical analysis
by TEM reveals single crystalline epitaxial films with grown-in
defects. The underlying porous substrate can subsequently be
oxidized to provide vertical dielectric isolation from the Si wafer.

1. Introduction

Because of the large application potential, the growth of device worthy
silicon thin films on insulating substrates is the object of much recent
efforts. Using dielectric material to isolate active devices in integrated circuits results in low parasitic capacitance and high voltage
isolation. The search for better device isolation techniques has included
such different materials systems as silicon-on-sapphire (SOS) and laser
recrystallized Si films on SiO_2. Another technique for dielectric
isolation (DI) can be realized by using porous Si as substrates for
epitaxial Si film growth. Most important among properties of porous Si is
its extraordinarily high chemical reactivity, which is a consequence of
its high surface-to-volume ratio. In particular porous Si may easily be
oxidized at high rates and low temperatures, thus providing the desired
dielectric isolation. During the epitaxial growth the individual pores of
the porous substrate have to be bridged to form a continuous Si film. Any
device performance will depend crucially on the quality of the epitaxial
Si layer and the number of defects contained in it. In this paper two
methods of crystal growth are discussed. We investigated Si molecular
beam epitaxy (MBE) on porous Si and the growth of single crystal silicon
films on porous substrates by laser fusion. The crystalline quality of
the grown epitaxial layers and the structure of the interface with the
underlying porous Si was evaluated by high voltage electron microscopy
(HVEM) and scanning electron microscopy (SEM) using cross sectional and
plan view specimens.

2. Experimental: (Formation of Porous Si)

Since Uhlir in 1956 and Turner (1958) reported the first experiments on
electropolishing of silicon, the conditions of the anodic dissolution
process of silicon in aqueous hydrofluoric acid (HF) have been extensively

---

[o] Present address: Philips Laboratories, Briarcliff Manor, N.Y. 10510.

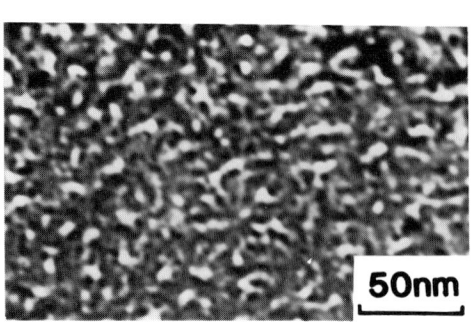

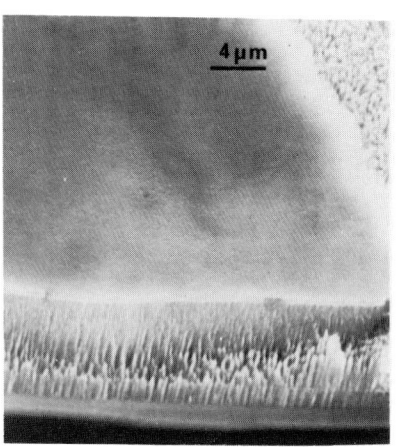

Fig. 1 Porous silicon structure at magnification of 300,000 (from TEM plan view specimen).

Fig. 2 SEM micrograph of laser fused surface on porous Si substrate.

studied (Arita et al. 1977 and Unagami et al. 1978). The silicon wafers used for this work were p-type 0.006 Ohm-cm wafers for both <100> and <111> surface orientations. The anodization process was carried out in an electrolytic cell, where the silicon wafer served as anode and a platinum sheet as cathode. Uniform porous silicon layers were formed by anodic dissolution in a 33% HF solution. A constant current density of 50 mA/cm^2 produced a growth rate of ~5 μm/min. During the anodic reaction a strong nitrogen gas flow through the HF solution was used to immediately remove the H_2 bubbles, that evolved on the porous silicon surface. The Si wafers were mounted in a Teflon fixture so that only the front surface was exposed to the electrolyte. Anodic treatment of single crystal Si wafers yields a low density layer that contains a large number of small pores. The final density of the porous film depends upon both the anodic curent density and the doping level of the silicon crystal. Anodic dissolution is governed by the supply of holes at the semiconductor surface (Turner, 1958). Since holes are necessary for the reaction to occur, the sample surface has to be illuminated in the case of n-type silicon to generate holes. The p-type Si wafers used for our experiments

were anodized without illumination. Immediately after anodization the wafer surface shows a black or dark blue color. The inner structure of porous silicon can only be revealed by high resolution transmission electron microscopy (TEM). Figure 1 represents a plan view (TEM) micro-

Fig. 3 TEM cross section of laser irradiated porous Si showing large grains embedded in partially sintered porous material.

graph showing the pores and the remaining silicon matrix at a magnification of 300,000 X. The pores are disorderly located and have various shapes. TEM cross sections demonstrate that the pores follow the current flow line perpendicular to the surface. Pore size is extremely small and is determined by TEM work not to be larger than 10-50 Å. Electron diffraction patterns indicate that the porous layer retains good single crystal quality despite the reduction in density. The porous layer also cleaves perpendicular to the surface. Since the low density porous material keeps the single crystalline structure it can be used as seeding material for subsequent

Fig. 4 Cross sectional TEM micrograph of laser fused surface layer which was liquid phase epitaxially regrown. Two different reflections are shown to image various grains.

epitaxy. In the case of standard vapor phase epitaxy the substrate is subjected to high temperatures over 900°C. However, Unagami & Seki (1978) observed that heat treatments over 1000°C changed the pore distribution and crystalline structure of the anodized Si layer, and recently Baumgart et al (1982) found a sintering effect after thermal treatment of 850°C in vacuum. The originally fine scaled porous structure was changed to a single crystal containing large voids. This effect decreases both the reactivity of porous Si to oxidation and the surface area. It is quite obvious that in this case alternative low temperature crystal growth techniques are required.

3. Results and Discussion

High temperature processing of the porous substrate has been completely eliminated by employing laser fusion for thin Si film growth. As pointed out it is essential to maintain the porous structure underneath the epitaxial layer for fast and complete oxidation (Watanabe et al 1975 and Unagami 1980) and spatially limited laser fusion with its inherent ultra rapid melt cycles appears to be a promising approach.

The experimental work was done in two parts. First a large number of experiments were performed with Nd:YAG laser irradiation at $\lambda = 532$ nm in order to grow Si films by liquid phase epitaxy and secondly we investigated Si molecular beam epitaxy on porous substrates. Figure 2 shows a

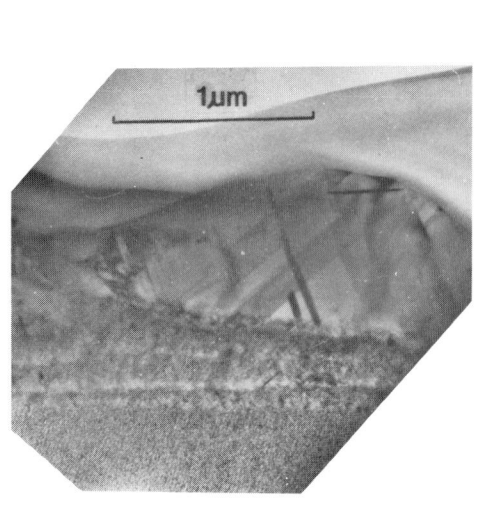

Fig. 5 TEM micrograph showing a cross section through Nd:YAG laser melted sample with nonuniform surface morphology at 50 mW laser output.

Fig. 6 TEM cross sectional view of recrystallized laser melted Si droplet obtained with a 95ns Nd:YAG laser pulse at 75 mW output power.

SEM micrograph of a cleavage edge after laser processing. The porous structure underneath the top epitaxial layer is clearly visible following brief etching in diluted NaOH. This etching process does coarsen the originally fine scale porous layer. We used a frequency doubled Q-switched Nd:YAG laser at 95 ns pulse duration. For the experiments we employed a scan speed of 30 mm/sec and a pulse repetition rate of 2.5 kHz to produce a spot overlap of about 80 % of the 50 μm spot diameter. Overlapping individual scan lines by a like amount produces large epi-taxially regrown areas. Prior to processing, the wafers were vacuum baked at 250°C for 1 h to remove HF left in the pores. The epitaxial film shown in Figure 2 is relatively thin because we used in this case a low laser output power of 15 mW to melt the surface. With increasing laser power, we get thicker and better liquid phase epitaxial layers. The limited resolution of the SEM does not allow a detailed investigation of the structure of the laser fused surface layer. Subsequent TEM cross section analysis revealed that the laser processed films, at these low laser power levels, are not yet single crystalline. The TEM cross sectional view in Figure 3 of a laser irradiated sample reveals large crystallites embedded in a partly sintered top porous layer. We used transmission electron microscopy to analyze the different stages of this development. At even lower laser powers, we observed just a very thin sintered surface layer of uneven thickness which extends locally along the pore direction. This creates a very nonuniform interface between the sintered surface layer and the underlying porous substrate. Higher laser intensities produce increasingly larger rod like grains, extending like dentrites in the partially sintered porous layer parallel to the pore direction. Finally at laser powers above 50 mW, we find large single crystalline areas which were liquid phase epitaxially reconstructed. Figure 4 shows an example of a 1 μm thick epitaxial layer of good crystal

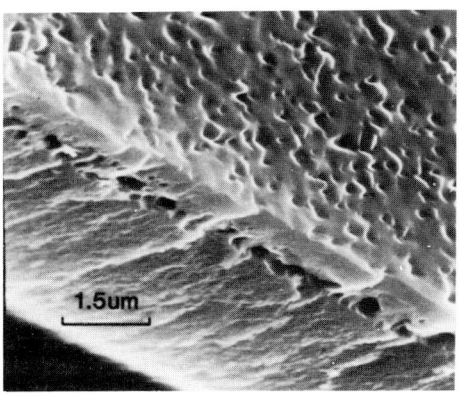

Fig. 7 SEM micrograph of cleaved sample showing the Si MBE layer grown on porous material.

quality containing only some stacking faults and grain boundaries. The presence of grain boundaries in Figure 4 is attributed to strain in the porous silicon substrate. Recently the existence of strain in porous Si films was confirmed by X-ray line broadening and X-ray topographic imaging (Unagami & Seki, 1978). We have a situation where already existing strain in the porous layer will induce some slight misorientations during recrystallization of the molten surface film which leads to the formation of the observed grain boundaries. However, at higher laser intensities another drawback is the formation of discontinuous surface layers with a tendency to bead up. Similar behavior is found in Si films on SiO_2, where insufficient substrate wetting causes the problem. Layers with uniform thickness, as shown in Fig. 4, cover only small areas of each sample while in the remaining part the molten Si recrystallized in little droplets. Bearing in mind that porous silicon has only half the density of single crystal Si, it is apparent that the laser irradiated surface layer has to shrink considerably upon melting. For laser powers equal or greater than 50 mW we always observe a nonuniform surface morphology. The surface layer shrinks during melting and beads up to form little single crystalline droplets. A typical example of such a recrystallized Si droplet is shown in cross sectional view in Fig. 5 and Fig. 6 corresponding to 50 mW and 75 mW laser power respectively. These epitaxially reconstructed areas are generally free of grain boundaries but do contain twins, dislocations, and stacking faults.

In the second part of our investigation we used a different approach and studied Si molecular beam epitaxy on porous films. Following brief Ar ion sputtering to clean the surface the Si MBE films were grown undoped on porous silicon in ultra high vacuum without high temperature preheating. A first analysis, however, showed that the Argon sputtering caused some slight damage to the porous Si surface which is readily recognized in Fig. 7. In this SEM micrograph of a cleaved sample the porous substrate and a 750

Fig. 8 TEM cross section of Si MBE surface layer with grown in defects at interface with porous substrate.

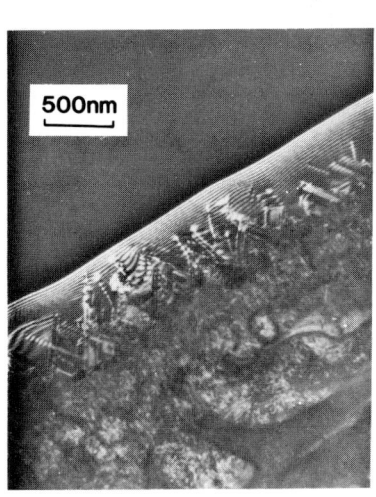

nm thick grown Si MBE layer on top of it can be clearly distinguished. It is interesting to note that the interface between both layers displays large voids which are attributed to the surface sputtering procedure and outgassing of the porous substrate. In spite of this interface problem the Si MBE growth proceeded remarkably well and bridged the voids. The initial surface roughness shown in the micrograph gradually disappears as the film grows. Thicker MBE layers display a very smooth surface morphology. The crystalline quality of the Si MBE films was then studied by TEM, and the cross section in Fig. 8 represents a typical result. The MBE surface layer is indeed single crystal Si. There are lattice defects in the layer but all the observed dislocations and stacking faults appear at the interface region, so that the top surface layer remains free of any defects. Directly underneath the MBE film is the substrate region showing sputtering damage. The initially present strain in the substrate is accommodated in the epitaxial layer by the formation of lattice defects in the first few hundred nanometers, so that the upper layers remain essentially strain free and without any defects. Single crystal Si films of this quality appear to be quite suitable for device fabrication and dielectric isolation applications.

Conclusion

In this study we have investigated new processing techniques for growing epitaxial films on porous silicon substrates. TEM analysis demonstrates that Nd:YAG laser processing can fuse the top surface of porous substrates The resulting crystal quality exhibits, however, a high density of lattice defects and an increasingly unstable surface morphology at higher laser intensities. Si MBE growth on porous substrates has been shown to produce excellent epitaxial layers with very smooth surfaces. Since the upper layers at the surface are defect free, this technique is capable of producing device worthy material. Both methods provide for low temperature processing of the substrate so that the original porous structure is not changed. Subsequent oxidation steps of the porous material offer interesting applications for dielectric isolation (Imai, 1981, Konaka et al, 1982).

Acknowledgments

The authors acknowledge H.J. Leamy for many helpful discussions and would like to thank R.C. Frye for providing the porous silicon samples and J.C. Bean for the Si MBE film growth.

References

Arita Y and Sunohara Y 1977 J. Electrochem. Soc, 124 285
Baumgart H, Frye R C, Trimble L E, Leamy H J and Celler G K 1982 Laser and
    Electron-Beam Interactions with Solids eds B R Appleton and G K Celler
    (New York: North Holland) p 609
Imai K 1981 Solid State Electronics 24 159
Konaka S, Tabe M and Sakai T 1982 Appl. Phys. Lett. 41 86
Turner D R 1958 J. Electrochem. Soc. 105 402
Uhlir A 1956 Bell System Tech. J. 35 333
Unagami T 1980 Jap. J. Appl. Phys. 19 231
Unagami T and Seki M 1978 J. Electrochem. Soc. 125 1339
Watanabe Y, Arita Y, Yokoyama T and Igarashi Y 1975 J. Electrochem. Soc.
    122 1351

Inst. Phys. Conf. Ser. No. 67: Section 4
Paper presented at Microsc. Semicond. Mater. Conf., Oxford, 21–23 March 1983

The nucleation and growth morphology of nickel impurity precipitates in silicon wafers

P D Augustus

Plessey Research (Caswell) Limited, Allen Clark Research Centre, Caswell, Towcester, Northants NN12 8EQ, U.K.

Abstract   Transmission Electron Microscopy has been used to identify the morphology of nickel impurity precipitates in inadvertently contaminated silicon wafers. The precipitates formed at the wafer surface during the cool down from high temperature processes. Ni is shown to precipitate as cubic $NiSi_2$ which is coherent with the Si lattice. The silicide occurs either as thin platelets lying on {111} planes, in which case they are twinned about the plane in which they lie, or as regular octahedra with {111} faces. Atomic models of the precipitate to matrix interface are proposed.

1. Introduction

The ability of Transmission Electron Microscopy (TEM) to detect and identify metallic impurity precipitates is a valuable asset in the investigation of the causes of low yield of semiconductor devices. This paper describes the use of a microscope equipped with an Energy Dispersive X-ray Analyser (EDX) to identify and characterise nickel bearing precipitates in epitaxial silicon layers. Semiconductor grade silicon is a highly refined material, however silicon is very reactive combining readily with any metal with which it comes into contact, and the utmost vigilance must be ensured at all stages of wafer processing to avoid contamination. An example of the effects of poor wafer handling was shown by Pearce and McMahon (1977). They demonstrated that metals resident on the susceptor of an epitaxial reactor could be responsible for crystallographic defects on the top surface of a wafer heated to 1120°C and allowed to cool. The defects were due to the diffusion of metal from the back to the front surface of the wafer. More recently Stacey et al (1982) have shown that mechanical contact with stainless steel or kovar followed by oxidation at 1100°C produced extrinsic stacking faults, decorated with Ni, on the front surface of a wafer. These results confirm the behaviour of Ni in silicon which was first investigated by plating Ni onto one surface of a wafer and then heat treating the wafer. Radiotracer techniques (Yoshida et al 1962) showed a high Ni concentration at the unplated surface which decays rapidly with depth in the specimen increasing again to reach a plateau in the interior of the sample. Using infra-red microscopy Iizuka et al (1963) showed that the plateau region contained hexagonal precipitates of up to 30 microns in size. Picker and Dobson (1972) investigated the surface region of wafers prepared in this way using TEM. They observed hexagonal shaped precipitates which they deduced to be $NiSi_2$. The present paper describes a study of the surface region of wafers which were inadvertently contaminated. It extends the observations of Picker and Dobson (1972) describing two classes of precipitate and proposes a reason for their distinction.

2. Experimental Details

The wafers used in this investigation were 40 mm diameter cut 2° off the (111) orientation towards a <211> direction. After chemical cleaning they were placed in an epitaxial reactor where they were given an HCl vapour etch at 1200 °C for 5 min. Epitaxial layers of 0.5 μm thickness were grown at 1025°C by deposition from silane. The perfection of epitaxial layers was routinely assessed by "Wright etching" (Jenkins 1977) and the etch features interpreted by TEM. 3 mm discs were cut from wafers using an ultrasonic drill and prepared for TEM by chemical jet thinning from the back face using 20%HF/80%HNO$_3$ acid. The resulting thin foils were examined at 100 kV in a JEOL 100CX TEMSCAN electron microscope.

3. Results

Contamination in epitaxial layers was usually detected in the first instance by the presence of "haze", the dense arrays of shallow pits (s-pits) which appeared after etching. The density of these pits on a contaminated wafer was around 10^8 cm^{-2} each pit being produced by a single impurity precipitate. TEM showed that as the level of contamination was reduced the precipitates became smaller the density remaining the same. With further reductions in contamination a break point was found at which the density of pits began to fall and eventually they became non-existent.

Wafers were examined at two stages of processing, either just after gas polishing or after gas polishing and epitaxial growth. In each case the precipitates found by TEM all intersected the top surface of the wafer. Wafer surfaces are known to provide nucleation sites for precipitation and nucleation is also expected to occur in the bulk of the wafer at dislocations and at point defect complexes. With most precipitation phenomena in silicon wafers there is a denuded region just below the surface. The sample preparation method used here did not select the region of bulk precipitation and precipitation will only be considered in the near surface region.

Figure 1 shows the surface region of an epitaxial layer which showed "haze" when etched. Small (50 nm) precipitates can be seen in the TEM image when strongly diffracting (s_g=0) two beam conditions are used and when the foil is less than 100 nm thick. Using EDX these precipitates were shown to contain Ni, on other occasions the contaminant was shown to be Fe. Although extra diffraction spots could be found originating from Fe precipitates of this size no extra spots could be found from Ni precipitates despite their single crystal appearance in exhibiting diffraction contrast.

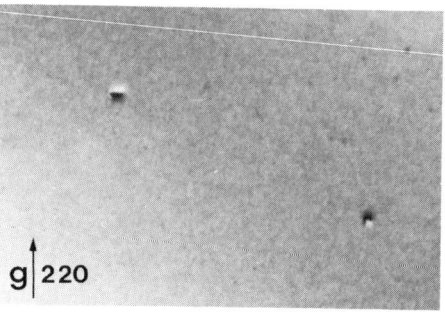

g 220

Figure 1

Transmission electron micrograph of small Ni bearing precipitates responsible for 's' pits or "haze"

1 μm

The opportunity to fully characterise Ni bearing precipitates arose in a series of wafers where there was gross contamination. The precipitates occurred in three distinct forms:-

1. octahedral precipitates bounded by {111} faces

2. planar precipitates lying on inclined {111} planes

3. hexagonal precipitates, flat platelets lying in the surface (111) planes.

Octahedral and planar precipitates are shown together in figure 2 labelled (1) and (2) respectively. In 2(a) the electron beam direction is normal to the surface of the wafer and in 2(b) the specimen has been tilted to bring the planar precipitate end on. Both precipitates can be seen to have produced dislocations in the silicon matrix, the mechanism for this will be discussed elsewhere (Augustus 1983). Figure 3 shows a thinner planar precipitate, in this case the lattice strain is accommodated and no dislocations have been produced.

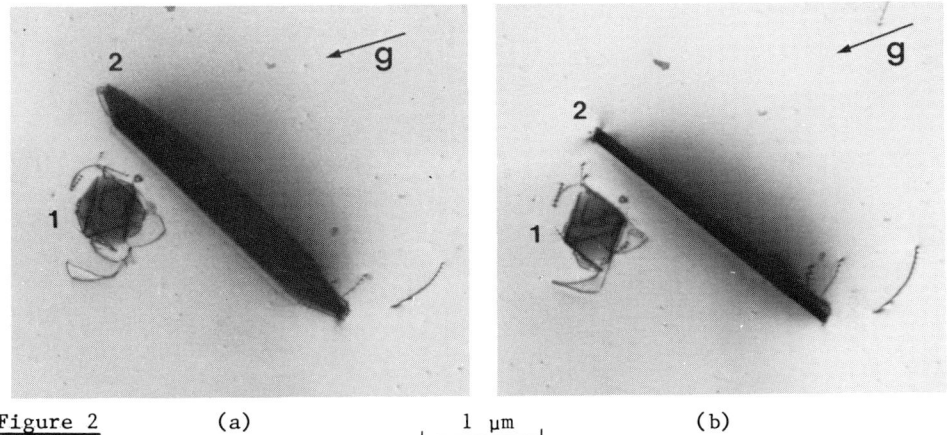

Figure 2   (a)   1 μm   (b)

TEM images of octahedral (1) and planar (2) precipitates. In (a) the beam direction is close to [111] and in (b) it is close to [110] in each case $g = 2\bar{2}0$.

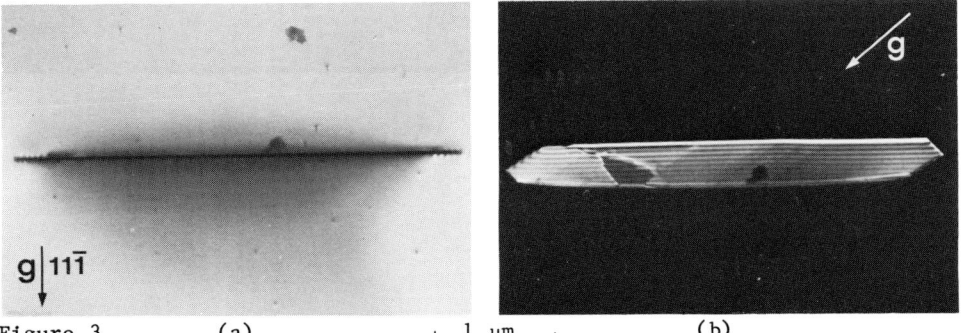

Figure 3   (a)   1 μm   (b)

TEM images of a 25 nm thick planar NiSi$_2$ precipitate imaged end on in (a) and in dark field in (b) $g = 1\bar{3}1$

Figure 4 shows a hexagonal platelet in bright and dark field. The dark field image was again formed from a twin spot. The platelet lies within the surface (111) planes. Diffraction patterns obtained with the beam direction parallel to the silicon matrix [111] direction showed only the Si(111) pattern. Additional spots from the precipitate could only be obtained by tilting about $<220>$ axes to find precipitate {323} and {211} poles in twin orientation to the silicon matrix. In the same way the image in figure 3(b) was formed from a 113 type twin spot with the beam direction between [111] and [121] in the silicon matrix. Precipitates lying in the surface (111) planes are twinned by rotating 180° within that plane in the same way that twinning occurs with double positioned silicon growth nuclei (Booker and Joyce, 1966). All the planar precipitates appeared to be twinned but despite repeated attempts no evidence could be found for twinning of the octahedral precipitates.

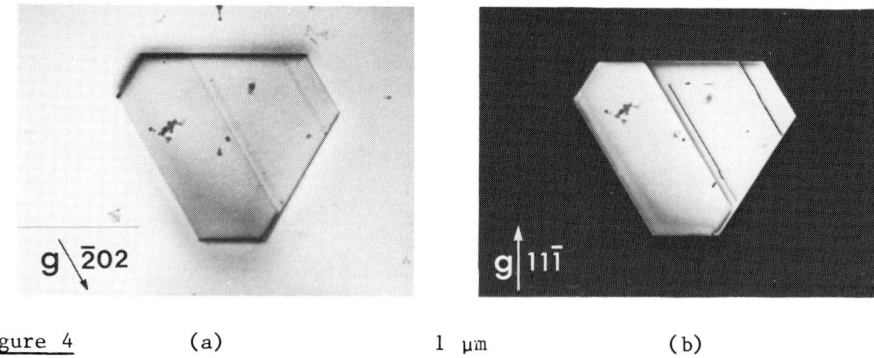

Figure 4   (a)   1 µm   (b)

TEM images of a hexagonal platelet lying in the surface (111) plane imaged in bright field 4(a) and dark field 4(b).

X-ray analysis of octahedral precipitates in thin regions of the matrix silicon gave a ratio of 2:1 Si to Ni. To check for the presence of low atomic number elements one specimen was examined in an SEM with a windowless X-ray detector, no elements other than Ni and Si were found, therefore the precipitates are most probably $NiSi_2$.

4. Discussion

Pearson (1958) reports that the solid solubility of nickel in silicon is very small and describes only one phase which would fit with the electron diffraction and X-ray results, $NiSi_2$. This has the cubic fluorite structure with a lattice parameter of 5.406 Å which is very close to the lattice parameter of silicon, a_o = 5.4282 Å. To understand the way in which $NiSi_2$ precipitates fit into the silicon matrix we need to consider the atomic positioning in the fluorite and diamond cubic structures. These are described in the atomic models of Wyckoff (1963). Silicon has a very open lattice structure, determined by the constraints of tetrahedral bonding, the unit cell containing eight atoms. Silicon is transformed into $NiSi_2$ by the addition of four nickel atoms at the corner and face positions of the unit cube, they displace silicon atoms from these sites to the vacant $\frac{1}{4}$ and $\frac{3}{4}$ positions. In the new arrangement the unit cell still contains eight silicon atoms but they all occupy the $\frac{1}{4}$ and $\frac{3}{4}$ positions the four corner and face positions being occupied by nickel atoms. The foregoing atomic arrangement conveniently allows the precipitation of nickel in silicon without an increase in the size of the unit cell. Figure 5 shows a model of such a coherent precipitate in the bulk of a silicon lattice. The precipitate has been drawn bounded by {111} faces which forms an octahedral structure. In the $NiSi_2$ structure nickel atoms will occupy one of the two interpenetrating fcc lattices which make up the diamond structure. The closest packed plane for nickel atoms will therefore be the {111} plane and in the arguments to follow nucleation will be considered to occur on this plane. The model shown in Fig.5 has been drawn in such a way that silicon atoms are conserved in the structure and the stoichiometric balance is maintained.

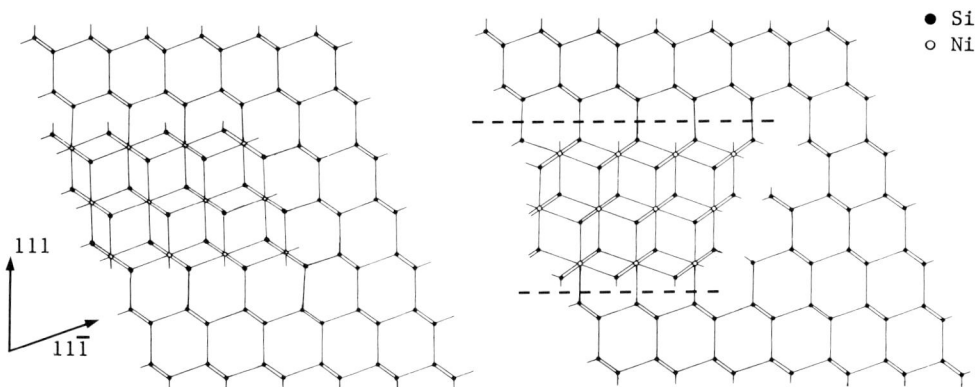

Figure 5. A coherent precipitate   Figure 6. A twinned precipitate

The alternative twinned structure which occurs in the case of the hexagonal platelets and inclined planar precipitates is shown in figure 6. The twin planes are shown as dotted lines. The good fit of nickel into the silicon lattice which has been outlined above would suggest that homogeneous precipitation without the need of nucleation centres might occur throughout the wafer. This does not in fact happen, even the bulk precipitates described by Iizuka et al (1963) occurred along lines as if they were nucleated on dislocations. Examination of nucleation on a single {111} plane shows that there are certain difficulties. These are illustrated in figure 7 where we can see the close packed {111} planes of the diamond structure stacked Aa, Bb, Cc, etc. A row of silicon atoms at "a" have been displaced to c' by a row of nickel atoms. There is now a close packed triple row c'aB in place of the double row aB. This triple row contains two silicon atoms for each nickel atom, hence $NiSi_2$. Note however that the displaced silicon atoms at c' are directly above the atoms at c whereas a rotation of the $NiSi_2$ by 180° in the plane xx would bring the atoms at c' above the vacant B position, producing a twinned precipitate. Twinning in the initial nucleating layer therefore allows a stoichiometric single plane of $NiSi_2$ to exist in a lower energy state. It now becomes clear that a nucleation centre is required to give a perturbation to the lattice which would allow either coherent or twinned precipitates to nucleate.

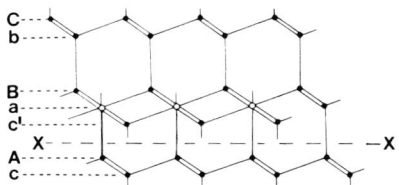

Figure 7

The nucleation of a single layer of Ni onto a (111) plane

● Si atom   o Ni atom

Further examination of figures 5 and 6 show that in the case of coherent precipitates a model can be made where the tetragonal bonding of the silicon lattice is completely undisturbed across the interface, however in the case of the twinned precipitate it is undisturbed across the (111) plane but highly dislocated across the other boundaries. This leads to the conclusion that on the gounds of minimising interfacial free energy a twinned precipitate lying on (111) would expand within the (111) plane becoming a flat platelet. On the other hand a coherent precipitate would expand equally in all directions, the constraint to its shape being the tendency to facet on the low energy {111} planes and hence take an octahedral shape.

Silicide to silicon interfaces have been the subject of several recent investigations using high resolution lattice imaging techniques (Föll et al 1981, Cherns et al 1982 and Gibson et al 1982). The interfaces studied here were created by standard contact formation techniques, they show two similarities with the precipitate to silicon interface, that is the twin relationship and the affinity to lie on low energy {111} planes. Similar reaction kinetics might be expected as d'Heurle et al (1982) have shown that nickel is the main diffusive species for the growth of $NiSi_2$.

Acknowledgements
The author would like to express his thanks to D. Jebb for supplying the specimens and for stimulating discussions. This work has been carried out with the support of Procurement Executive, Ministry of Defence, sponsored by DCVD.

References

P D Augustus. 1983. Proc. Electrochem. Soc. Symposium on Defects in Silicon, San Fransisco, May 1983 (to be published).

G R Booker and B A Joyce. 1966 Phil. Mag. 14, 301.

D Cherns, J C H Spence, G R Anstis and J L Hutchinson 1982, Phil. Mag. A46 849.

F d'Heurle, S Petersson, L Stolt, B Strizker. 1982 J. Appl. Phys. 53, 5678.

F Föll, P S Ho and K N Tu. 1981 J. Appl. Phys. 52, 250.

J M Gibson, J C Bean, J M Poate and R T Tung. 1982 Appl. Phys. Lett. 41, 818

T Iizuka, M Kikuchi and K Kanasaki. 1963 Japan J. Appl. Phys. 2, 309.

M W Jenkins. 1977 J. Electrochem. Soc. 124, 757.

C W Pearce and R G McMahon. 1977 J. Vac. Sci. Technol. 14, 40.

W B Pearson 1958. Handbook of Lattice Spacings and Structures of Metals, (London: Pergamon Press).

C Picker and P S Dobson. 1972 Crystal Lattice Defects, 3, 219.

W T Stacy, D F Allison and T C Wu. 1982 J. Electrochem. Soc. 129, 1128.

R W G Wyckoff 1963. Crystal Structures, (New York: Wiley).

M Yoshida, Y Yamaguchi and H Aoki 1962. J. Phys. Soc. Japan 17, 1676.

Inst. Phys. Conf. Ser. No. 67: Section 4
Paper presented at Microsc. Semicond. Mater. Conf., Oxford, 21–23 March 1983

235

TEM studies of $<111>$ and $<100>$ silicon slices implanted with B⁺ and BF₂⁺ ions after annealing and oxidation

M I J Beale, G R Booker

Dept. Metallurgy and Science of Materials, Parks Road
University of Oxford

Abstract   TEM studies were made of $<111>$ and $<100>$, n-type Si specimens that had been a) implanted with 1.10^{16} cm^{-2}, 40 keV B⁺ ions at room temperature, and either annealed or oxidised at 1100°C, and b) implanted with 1.10^{14} to 1.10^{16} cm^{-2}, 40 keV BF₂⁺ ions also at room temperature, and annealed at 900°C. In both cases the structural behaviour was strongly dependent on whether $<111>$ or $<100>$ specimens were used, and these behaviours correlated well with electrical and device data. The defects of main importance were dislocations for the B⁺ implants, and twins and dislocation loops for the BF₂⁺ implants.

B⁺ Implantation

n-type Si wafers of $<111>$ and $<100>$ orientation were implanted at room temperature with 1.10^{16} cm^{-2}, 40 keV B⁺ ions (peak projected range = 1300 Å) and either annealed in argon or wet oxidised, the standard treatments being at 1100°C for 30 min. Damage structures were determined by making TEM examinations of plan-view and cross-section specimens, the former being prepared by thinning from the unimplanted side.

TEM images of $<111>$ samples annealed at 1100°C (Figs 1 & 2) show a defect-free region at the Si surface, below which lies a tangled network of dislocations situated at a depth of 1300 Å. This depth corresponds closely to the peak projected range for 40 keV B⁺ ions in Si. Below the tangled network lies a less dense, geometric network, and this is more clearly revealed in plan-view if the surface 2000 Å of the Si is chemically removed prior to the TEM sample thinning (Fig 3). Analogous $<100>$ Si samples show similar features for the tangled network.

The depth of the tangled network suggests that it originates directly from the implant damage, and further evidence for this is obtained by comparing the damage structures after annealing at various temperatures. At 950°C fine-scale damage occurs (Fig 4), at 1050°C this develops into small interconnected loops (Fig 5), at 1100°C this develops into the more open tangled network (Fig 6), and at 1200°C the tangled network anneals out leaving only a geometric network (Fig 7). The geometric network is absent at 950°C, and at 1050°C small segments of it begin to form. At 1100°C the geometric network is well developed beneath the tangled network, and at 1200°C the geometric network is still well developed but the tangled network is now absent.

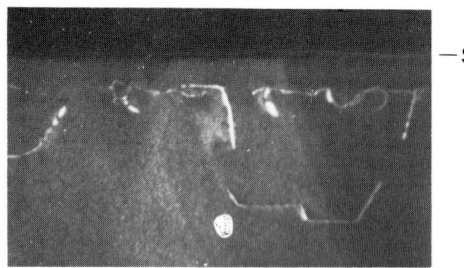 

Fig. 1   Fig. 2

B$^+$ implanted <111> silicon after 1100°C
anneal, dark field TEM.

Fig. 1 Cross-section.

Fig. 2 Plan view.

Fig. 3 Plan view after removal of
the surface 2000 Å.

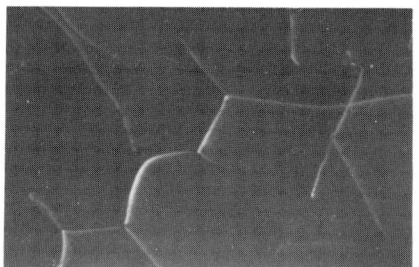

Fig. 3

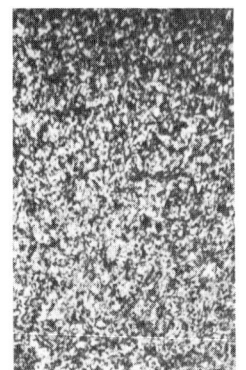 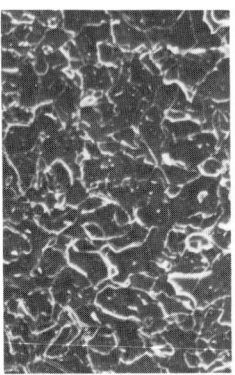 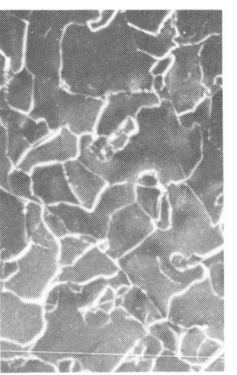 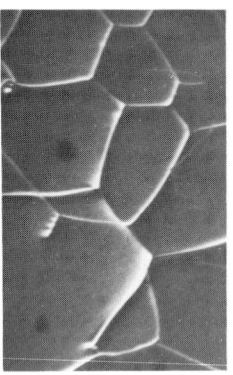

Fig. 4   Fig. 5   Fig. 6   Fig. 7

B$^+$ implanted <111> silicon after annealing at

Fig. 4 950°C

Fig. 5 1050°C

Fig. 6 1100°C

Fig. 7 1200°C

5000 Å

The dislocations present after oxidation at 1100°C (Fig 8, 9) are markedly different from those present after annealing. There is no dislocation-free surface layer and no tangled network. The dislocations form a geometric network extending from the surface into the bulk material. The oxidation process consumes 2250 Å of Si from the implanted surface, which therefore removes the implantation damage. The absence of the tangled network is therefore consistent with its having originated directly from the implantation damage.

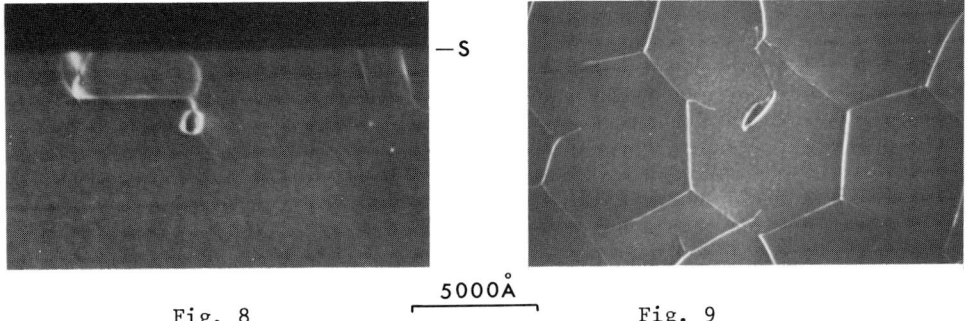

Fig. 8   5000Å   Fig. 9

B⁺ implanted <111> silicon after 1100°C wet oxidation, dark field TEM in cross-section (Fig. 8) and plan view (Fig. 9).

The dislocations in the geometric network, formed by either the annealing or oxidation treatments, are different in <111> and <100> samples. For both types of treatment, the dislocations in <111> samples exhibit a three-fold geometric symmetry. They are edge dislocations with a/2 <110> burgers vectors lying in the (111) plane parallel to the specimen surface. The dislocations in <100> samples exhibit a four-fold symmetry. They are either edge dislocations with a/2 <110> burgers vectors lying in the (100) plane parallel to the specimen surface, or 60° dislocations with a/2 <110> burgers vectors inclined to the (100) plane. For both types of treatment, the dislocations in the <100> samples extend to greater depths than in the <111> samples. The observed dislocation depths, together with p-n junction depths measured using a bevel and copper stain technique, are listed in Table 1. For <100> samples the dislocations often penetrate the p-n junction and depletion region, while for <111> samples this does not occur. It is suggested that it is this behaviour which is responsible for the device results reported by a number of investigators, namely that when diodes are fabricated in this way, the reverse-bias leakage currents are larger for <100> specimens than <111> specimens, MacIver et al (1977). Dislocations reaching the junction, particularly if decorated by metallic impurities, would have this effect.

The occurrence of the geometric network is considered to be due to the misfit-induced stresses arising from the high concentrations of boron present. Agreement between calculated and measured dislocation densities, and experimental determination of the sign of the dislocations, support this view. The different depths of the dislocations for the two specimen orientations is thought to be due to the different types of dislocations formed. Thus, in <111> specimens the 90° dislocations that are present can only move slowly down into the specimen by climb whereas in <100> specimens the 60° dislocations that are also present can move rapidly down into the specimen by glide. The boron present provides the necessary stress for this to occur due to its small tetrahedral radius compared with silicon.

Additional experiments showed that if the implantation damage was closer to the surface it was 'attracted' to the surface and was annealed out. Thus removal of the surface 1000 Å by chemical etching prior to 1100°C annealing resulted in the absence of the tangled network.

Silicon Orientation	Anneal 1100°C		Wet Oxidation 1100°C	
	Dislocation Depth (μm)	Junction Depth (μm)	Dislocation Depth (μm)	Junction Depth (μm)
<111>	0.9	1.65	0.3	1.25
<100>	1.5	1.65	0.8	1.40

Table 1

Maximum dislocation depths observed using cross-section TEM and p-n junction depth measured by bevel and stain.

BF$_2^+$ Implantation

n-type Si wafers of <111> and <100> orientation were implanted at room temperature with 1.10^{14} to 1.10^{16} cm^{-2}, 40 keV BF$_2^+$ ions (R$_p$ = 300 Å) and annealed in nitrogen at 900°C for 20 min. For doses below the amorphous threshold ($\cong 3 \times 10^{14}$ cm^{-2}) the post-anneal damage consists of dislocation loops and is similar for <111> and <100> samples. The sheet resistivities are also similar for <111> and <100> samples.

For doses above the amorphous threshold, as-implanted damage consists of an amorphous layer extending from the surface to a depth of $\cong$ 700 Å, together with a region of heavily damaged, single-crystal material extending for a further 100 Å below this (Fig. 10). Annealing of these samples results in the following behaviours (Figs. 11 to 14). For <111> samples, the amorphous layer gives rise to a highly defective layer, while for <100> samples, it gives a substantially damage-free layer. For both <111> and <100> samples, the heavily damaged region gives a buried layer of dislocation loops (not so readily seen in plan-view for the <111> samples because of the additional damage present).

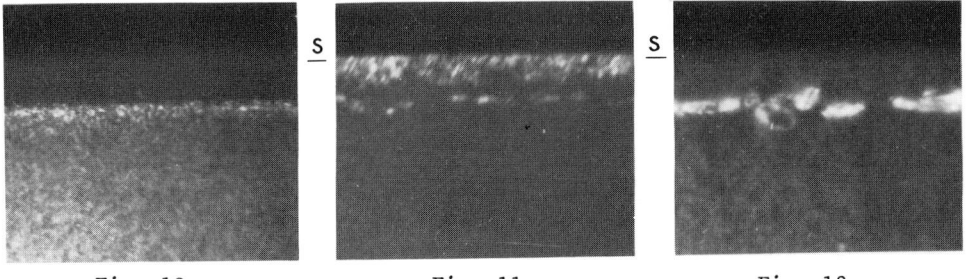

Fig. 10   Fig. 11   Fig. 12

Cross-section TEM dark field images of BF_2^+ implanted silicon.

Fig. 10 As implanted.

Fig. 11 <111> silicon after 900°C anneal.   1000 Å

Fig. 12 <100> silicon after 900°C anneal.

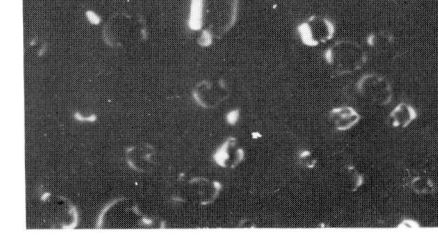

Fig. 13   Fig. 14

Plan view TEM dark field images of BF_2^+ implanted silicon after 900°C anneal,

Fig. 13 <111>   Fig. 14 <100>

Combined TEM/TED studies of the defective layer in <111> samples reveals twins of mean size 300 Å. The twinning occurs mainly on the (111) planes parallel to the specimen surface, and occasionally on the inclined (111) planes. Multiple twinning is also sometimes present. High-resolution micrographs (Fig. 15) and lattice-image micrographs show that the twins are often tetrahedral being bounded by (111) planes which are stacked in matrix-twin-matrix etc sequences. The (111) boundary planes appear to be atomically flat.

The different regrowth behaviours of the amorphous layer for <100> and <111> samples have been explained in terms of the different ways in which the Si atoms have to bond to the single-crystal Si surface for the interface to advance, (Drosd et al, 1982). For <100> specimens, this is 'easy', growth is rapid and relatively defect-free material arises. For <111> specimens, this is 'difficult', growth is slow and defective material arises. Experimental results from the present work show that for the highest dose used (1.10^{16} cm^{-2}), annealing of <100> samples produces a small amount of twinned material. Thus, even the 'easy' growth for <100> samples is adversely affected when large amounts of boron and fluorine are present.

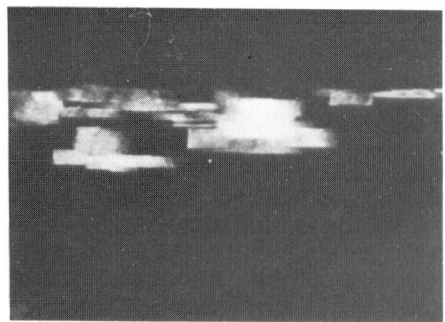 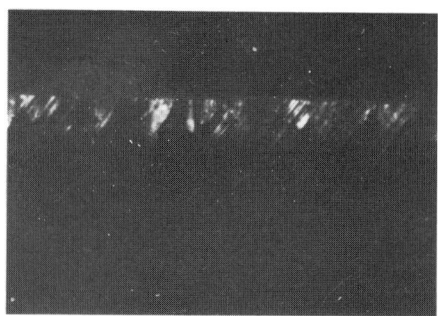

Fig. 15 1000Å Fig. 16

Cross-section dark field TEM images of BF_2^+ implanted silicon showing twins in <111> (Fig. 15) and <100> (Fig. 16) silicon.

For doses above the amorphous threshold, annealed samples show that the sheet resistivity is significantly higher for <111> samples, which contain the defective layer, than for <100> samples, which are relatively defect-free. This strong correlation suggests that for the <111> samples, some of the B is decorating the twins.

The SIMS technique was used to obtain atomic concentration depth profiles for the annealed samples. Both the boron and fluorine profiles for both the <111> and <100> samples, showed a peak at a depth corresponding to the implantation range, and a second peak at the depth of the buried dislocation loops. There was no significant loss of boron to the specimen surface. However, there was a small loss of fluorine to the surface for <111> samples, and a large loss for <100> samples. The latter suggests that for <111> samples, some of the F is decorating the twins and is not readily released. Additional experiments were performed involving implantations of B^+, BF_2^+, $Si^+ + B^+$, and $Si^+ + B^+ + F^+$ (the Si^+ being used to introduce damage only), and both furnace and laser annealing were used. These showed that the F, apart from its role as a constituent of BF_2^+ producing heavy damage during the implantation, played no directly active role thereafter during the annealing, ie had no significant effect on the resulting structures or electrical behaviours.

Acknowledgements

AERE Harwell for financial assistance for this work.
Chris Jackson AERE Harwell for SIMS analysis.

References

MacIver B A and Greenstein E 1977 J. Electrochem. Soc., 124, 273.
Drosd R and Washburn J 1982 J. Appl. Phys., 397 53(1).

This work is reported in greater detail in the D. Phil thesis of M I J Beale.

Inst. Phys. Conf. Ser. No. 67: Section 4
Paper presented at Microsc. Semicond. Mater. Conf., Oxford, 21–23 March 1983

241

Defects generated by oxidation of BF_2^+ implanted silicon

S Albin, R Lambert, S M Davidson and M I J Beale[*]

GEC Research Laboratories, Hirst Research Centre, Wembley UK
*Department of Metallurgy and Science of Materials, University of Oxford UK

Abstract Oxidation of silicon implanted with doses of 10^{14} BF_2^+ cm^{-2} and above generated a high density (about 10^8 cm^{-2}) of stacking faults and dislocations. There is a critical BF_2^+ dose between 5×10^{13} and 10^{14} cm^{-2} for stacking fault generation. The density of stacking faults generated during 950°C oxidation was found to be lower than that from 900°C oxidation. TEM studies revealed this to be due to fault annihilation by the interaction of orthogonal stacking faults. EBIC studies showed the implantation induced defects to be electrically active. Pre-oxidation annealing was found to be effective in reducing the density of implantation induced defects.

1   Introduction

There are several advantages in using ionised heavy molecules rather than the corresponding ionised atoms for dopant implantation. For example, molecules such as $^{28}Si^1H^+$ are used to implant Si to avoid mass equivalence with the undesired $^{28}N_2^+$ (Sansbury and Gibbons 1969). In the case of BF_2^+ implantation, boron atoms become the acceptor dopants while the fluorine produces amorphisation of the silicon surface. This amorphisation of the surface by BF_2^+ gives a higher electrical activation of dopants after low temperature (e.g. 600°C) annealing and lower reverse bias leakage current for p-n junctions, as compared to B^+ implantation (Tsai and Streetman 1979, MacIver and Greenstein 1977). Additionally, when BF_3 gas is used as the ion source, at least two orders of magnitude higher beam current can be obtained for BF_2^+ ions as compared to B^+ ions of equivalent dose and energy. Another advantage of BF_2^+ over B^+ ion implantation appears in a CMOS process where a local p-type doping self-aligned to the edge of the n-channel active area is required. A low implantation energy must be used if penetration of the silicon nitride mask by the boron ions is to be avoided. With the practical lower limit of 25 keV energy available from most commercial implanters, this requires a thick oxide/nitride layer for selective field oxidation, which is undesirable for a defect free field oxidation step. The use of molecular ion species such as BF_2^+ offers an attractive solution to this problem as, compared to boron ions of equivalent energy, their penetrating power is significantly lower. Thus BF_2^+ molecular ions were used in this work for a self-aligned p-type field doping. However, it has been found that oxidation of BF_2^+ implanted silicon generates stacking faults (SFs) and dislocations which are well known as major yield detractors in silicon devices. Consequently we have carried out experimental work to investigate the defects generated by the oxidation of BF_2^+ implanted silicon, the details of which appear in the following sections.

2   Experiment

Substrates used were 76 mm diameter, n-type 25 Ωcm (100) CZ Si wafers. BF_2^+ doses from 2×10^{13} to 1×10^{15} cm^{-2} were implanted at 40 keV, equivalent in penetrating power to 9 keV $^{11}B^+$. For comparison purposes, a 1×10^{14} cm^{-2} dose of 9 keV $^{11}B^+$ was also studied. A semi-recessed field oxidation step was carried out at 900 and 950°C for seven hours in a wet oxygen ambient to grow 0.5 and 0.8 μm thick SiO_2 films respectively. Additionally, annealing of the BF_2^+ implanted wafers prior to the field oxidation was carried out in a nitrogen ambient for various times at 950°C. The effect of the annealing step on the characteristics of p-n junctions was studied by measuring the reverse-bias breakdown voltage of diodes fabricated to these schedules. Oxidised wafers and diodes were examined using the TEM and SEM (EBIC) and by using Nomarski contrast interference microscope after preferential defect etching (Jenkins 1977).

3   Results and discussion

3.1 Ion implantation induced defects – BF_2^+ vs B^+

Optical micrographs of the wafers implanted with 10^{14} cm^{-2} B^+ and BF_2^+ are compared in Figure 1(a) and (b). The field oxidation was carried out at 950°C and the wafers were etched in Wright etch for one minute after the oxide strip. The unimplanted regions and the B^+ implanted regions are free of defects, whereas, as seen in Figure 1(b), a high density of SFs and dislocations is generated in the BF_2^+ implanted regions of the wafer. Both the wafers had received the same dose of dopants and had undergone the same processing. Thus the implantation damage due to BF_2^+ is responsible for the high density of defects.

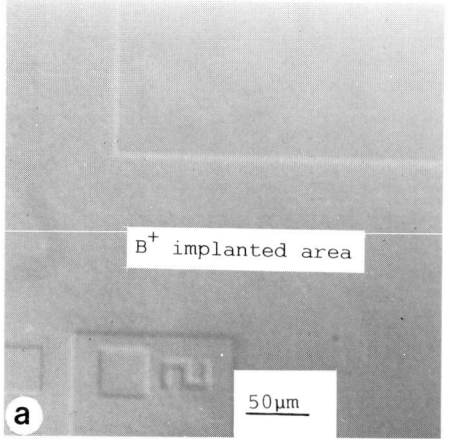
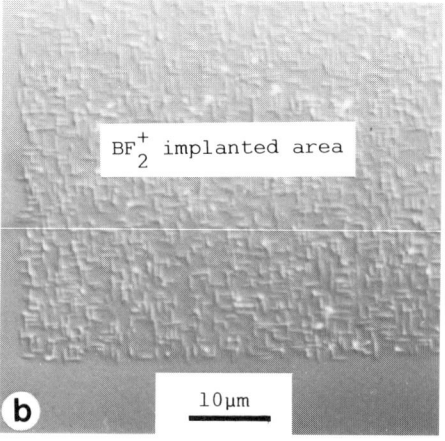

Fig 1   Optical micrographs after Wright etching of oxidised wafers implanted with (a) 10^{14} B^+ cm^{-2}, (b) 10^{14} BF_2^+ cm^{-2}

During the implantation, displacements of the target atoms result from the strong coulombic interaction between the projectile and the target atoms. When the energy of the projectile is low, hard sphere collisions take place. This assumption is valid until the energy E_A of the projectile equals the coulombic interaction potential of the two nuclei separated by the sum of the radii of their screening electron clouds. Thus,

$$E_A = 2 E_R Z_1 Z_2 (Z_1^{2/3} + Z_2^{2/3})^{\frac{1}{2}} (M_1 + M_2)/M_2 \qquad ....(1)$$

where E_R = 13.6 eV, Z_1 and M_1 are respectively the atomic number and mass of the projectile and Z_2 and M_2 are those of the target. From (1),

E_A = 8 keV for B^+ implantation into silicon. In the present experiments, the B^+ energy used is 9 keV. Thus the B^+ ions are expected to undergo hard sphere collisions with the silicon atoms.

When the projectile energy is less than E_A, the number of target atoms displaced, N_d, by a primary recoil is given by (Robinson and Torrens 1974),

$$N_d = [0.8 (E_o - Q)]/2E_d \qquad .....(2)$$

where E_o is the initial ion energy, Q is the amount of energy lost by the ions in scattering collisions with the target electrons and E_d is the displacement energy for the target atoms, which is 15 keV for silicon (Gibbons 1972). As (2) suggests, the 9 keV B^+ implantation produces very little damage to the Si lattice. Consequently there are not enough stable microdefects to interact with the excess silicon interstitials produced during the field oxidation and SFs are not generated, which is in accordance with the result shown in Figure 1(a).

A similar calculation is not possible in the case of BF_2^+ molecular ions. For F^+ implantation into silicon, E_A is approximately 18 keV. It is assumed that the BF_2^+ molecule splits into its components when it hits the target surface. Therefore, the simultaneous implantation of a boron and two fluorine atoms acts like a high dose rate implantation compared to the sequential implantation of individual components. Hence the damage produced by BF_2^+ is higher than that due to B^+ implantation with equivalent energy and dose. The high density of defects generated in the BF_2^+ implanted region of the wafer indicates that a high density of microdefects is available to interact with the interstitial flux generated during the field oxidation.

3.2 Effect of field oxidation temperature on implantation induced defects

The SFs generated during the 900°C oxidation are fairly uniform in length (0.8 μm) and 80 nm deep into the silicon substrate (Figure 2(a)). The SF density is about 5×10^8 cm^{-2}; in addition a dislocation density of the order of 10^7 cm^{-2} is observed. In comparison the 950°C oxidation generated SFs with lengths varying from 0.25 to 2 μm which were 150 nm deep into the substrate (Figure 2(b)). However the density of SFs is about 10^8 cm^{-2} which

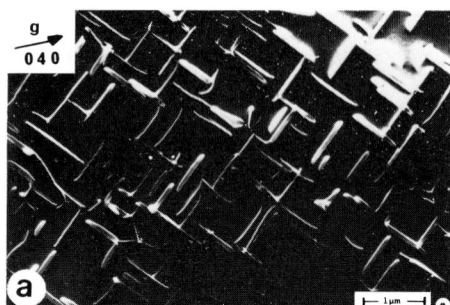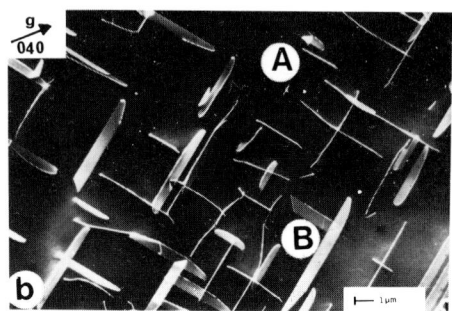

Fig 2   TEM of wafers implanted with 10^{14} BF_2^+ cm^{-2} after oxidation at
(a) 900°C, (b) 950°C

is only a fifth of the number produced by the 900°C oxidation. The
dislocation density is of the same order as the SF density. For both
oxidation conditions, no precipitates were observed along the dislocations.
Contrast analysis at varying diffraction conditions revealed the
dislocations to be edge type with Burgers vectors of type $\frac{1}{2}$ <110>.

For the same duration of oxidation, more silicon is consumed to grow the
thicker oxide at the higher temperature. Hence the SFs can be expected to
grow longer and deeper into the silicon substrate for the 950°C oxidation
compared to the 900°C oxidation.

It is interesting to note that the SF density is lower and that there is a
considerable variation in SF length for 950°C oxidation compared to 900°C
oxidation. As seen in Figure 2(a) and (b) the SFs in (100) Si lie on {111}
planes in two orthogonal <110> directions. The Burgers vectors of these
SFs are identified as $^1/3$<111> type. As the SFs grow deeper and longer,
considerable interaction between the orthogonal SFs takes place when the SF
density is very high as seen here. This interaction is obviously
pronounced for SFs generated during 950°C oxidation. The SFs grow in size
due to the capture of interstitials by the partial dislocations (Leroy
1982). When two orthogonal SFs interact, especially at their partial
dislocations, the local concentration of interstitial silicon is shared
between the two SFs. This concentration of interstitials may not be able
to sustain the growth of two faults simultaneously; instead one of the
faults grows at the expense of the other. The stress generated by the
impingement of the partial dislocations also has some effect on the
annihilation of one of the faults. An excellent example of this unfaulting
mechanism has been reported by Ravi (1974). A similar effect is seen in
Figure 2(b). The marker A shows the interaction of two orthogonal faults,
where the two partial dislocations are visible whereas the main body of the
fault is almost extinct. Similarly, at the marker B one of the two
orthogonal faults is smaller than the other. The shapes of the partial
dislocations of the smaller fault are also changed to truncated Frank
loops. Further interaction would lead to the annihilation of the smaller

fault by the other. The consequences of such fault interactions are a
lower SF density and variation in SF length for the 950°C oxidation. In
addition, some of the implantation induced microdefects which act as the
nuclei for SF growth may not be stable at 950°C, so accounting for part of
the observed reduction in SF density.

3.3 Effect of BF_2^+ implantation dose on stacking faults

For BF_2^+ doses of 1×10^{14} cm^{-2} and above, an SF density of the order of 10^8
cm^{-2} was generated during the oxidation, whereas no faults were observed
for doses below 1×10^{14} cm^{-2}. It appears that a BF_2^+ dose between 5×10^{13}
and 1×10^{14} cm^{-2} is a threshold value for SF generation during oxidation.
In comparison the threshold value for boron ion implantation is higher
since no SFs were observed in oxidised wafers implanted with a boron dose
of 10^{14} cm^{-2}. A similar threshold dose between 5×10^{13} and 1×10^{14} cm^{-2} has
been reported for Ne implantation into silicon by Prussin et al (1977).

It was also found that the SF density was not linearly related to the
implantation dose. As in the case of Ne implantation into silicon, the
size and the density of SFs were found to be interrelated: the higher the
density, the smaller the fault size. For a given BF_2^+ dose higher than
10^{14} cm^{-2}, the SF density generated by the 900°C oxidation was higher than
that for 950°C oxidation, as reported above.

No stacking faults were generated by the oxidation of wafers implanted with
5×10^{13} BF_2^+ cm^{-2}; however, a dislocation etch pit density of about 2×10^2
cm^{-2} was revealed by dislocation etch studies.

3.4 EBIC studies of the defects induced by the oxidation of BF_2^+ implanted silicon

Figure 3(a) shows the EBIC image of the surface of a device implanted with
10^{14} BF_2^+ cm^{-2} and oxidised at 950°C. The dark regions in orthogonal
directions are due to stacking faults which contain recombination centres
for the minority charge carriers. The strain field variations surrounding
the dislocations may also be responsible for the contrast seen in the EBIC
image. It can be noticed the the dark regions are not all of the same
contrast which could be due to a variation in the level of decoration of
the stacking faults with impurities. The surface of the wafer after defect
etching is shown in the SEM photograph (Figure 3(b)) confirming the
presence of stacking faults corresponding to the electrical activity seen
in the EBIC image. The effect of pre-oxidation anneal on the implantation
induced defects was also studied using EBIC. Figure 3(c) is the EBIC image
of the surface of a device fabricated incorporating a 60 minutes pre-
oxidation anneal in a nitrogen ambient. There is no contrast variation in
the EBIC image corresponding to stacking faults. This is verified in
Figure 3(d) which is an optical micrograph of the same surface after defect
etching. This confirms that the pre-oxidation annealing is effective in
eliminating the BF_2^+ implantation induced defects.

The reverse bias leakage currents and breakdown voltages of completed
diodes were influenced by the presence of defects in the BF_2^+ implanted
field regions. It was found that typically an order of magnitude
improvement in leakage current and breakdown voltage was obtained for such
devices fabricated incorporating the pre-oxidation annealing step.

(a)   EBIC micrograph

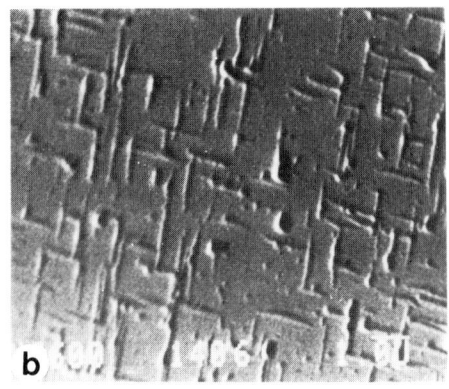

(b)   SEM of the device in (a)
       after Wright etching

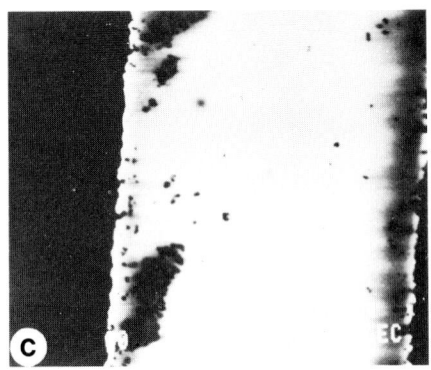

(c)   EBIC micrograph of a device
       with pre-oxidation anneal

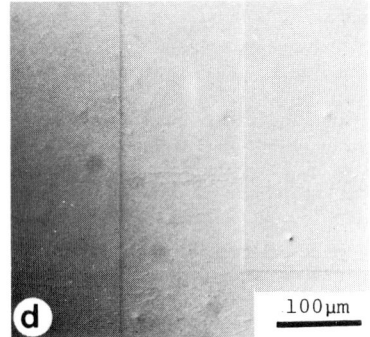

(d)   Optical micrograph of the
       device in (c) after Wright
       etching

Fig 3

4   Conclusions

Oxidation of BF_2^+ implanted silicon generates a high density of electrically active stacking faults. There is a critical dose between 5×10^{13} and 10^{14} BF_2^+ cm^{-2} for the SF generation. There is considerable interaction among the orthogonal faults at this high density. Anealing prior to the oxidation step is required to reduce the fault density; given this, BF_2^+ implantation is an attractive alternative to low energy boron implantation for auto-registered field doping in CMOS processes.

References

Gibbons J F 1972 Proc IEEE 60 1062
Jenkins M W 1977 J Electrochem Soc 124 757
Leroy B 1982 J Appl Phys 53 4779
MacIver B A and Greenstein E 1977 J Electrochem Soc 124 273
Prussin S, Li S P and Cockrum R H 1977 J Appl Phy 48 4613
Ravi K V 1974 Phil Mag 30 1081
Robinson M T and Torrens I M 1974 Phys Rev B 9 5008
Sansbury J D and Gibbons J F 1969 Appl Phys Lett 14 311
Tsai M Y and Streetman B G 1979 J Appl Phys 50 183

Inst. Phys. Conf. Ser. No. 67: Section 4
Paper presented at Microsc. Semicond. Mater. Conf., Oxford, 21–23 March 1983

Combined RBS, SEM and TEM studies of hydrogen and helium effects in silicon

+Susan Wood, +J Greggi, Jr, +J A Spitznagel, +N J Doyle, ++R B Irwin, ++J R Townsend and †W J Choyke

+ Westinghouse R&D Center, Pittsburgh, PA 15235
++University of Pittsburgh, Pittsburgh, PA 15260
† Westinghouse R&D Center and University of Pittsburgh

Abstract   A microstructural study in which single crystal silicon was implanted with hydrogen from 20 keV to 150 keV with fluences ranging from 1×10^{16} ions/cm^2 to 8×10^{17} ions/cm^2 at RT is reported.   RBS/ channeling data are correlated with cross-sectional TEM and show a progressive broadening of the damage zone, preferentially towards the implant surface.   An initial comparison with He implanted at 110 keV is included to differentiate between the effects of a chemically active (H) and inert (He) specie.

1.   Introduction

Interest in the behavior of hydrogen in silicon has grown considerably in recent years, in part, because, as discussed by Imura (1980), hydrogenated amorphous silicon has become a promising material for large scale solar cells.   Silicon is also a candidate for near plasma armor in fusion reactors where hydrogen effects are extremely important.   Since single crystal silicon is the most well characterized semiconductor, hydrogen-implanted single crystal Si was selected in the present study to assist in understanding the role which hydrogen plays in this material, particularly with respect to inducing microstructural changes.   Previous work by Chu et al (1977a,b) utilized Czochralski grown <100> Si and concentrated primarily on the effect of specimen temperature during ion bombardment. This paper reports a comprehensive transmission electron microscopy (TEM) study in which <111> and <100> float zone single crystal silicon has been implanted with hydrogen from 20 keV to 150 keV with fluences ranging from 1×10^{16} ions/cm^2 to 8×10^{17} ions/cm^2 at room temperature.   Supportive channeling results are also presented, in addition to comparative TEM data from an initial helium implantation experiment.   The intent was to distinguish between the effects of a chemically reactive specie (hydrogen) and of an inert specie (helium).

2.   Experimental

All samples used in these experiments were N-type float zone <111> or <100> Si with a conductivity of 30-60 Ω-cm.   Sample details were discussed by Choyke (1983).   Implants were made with a 200 keV implanter equipped with post acceleration magnetic mass separation and turbo-pumping to

---

*Supported in part by NSF Grant DMR-81-02968

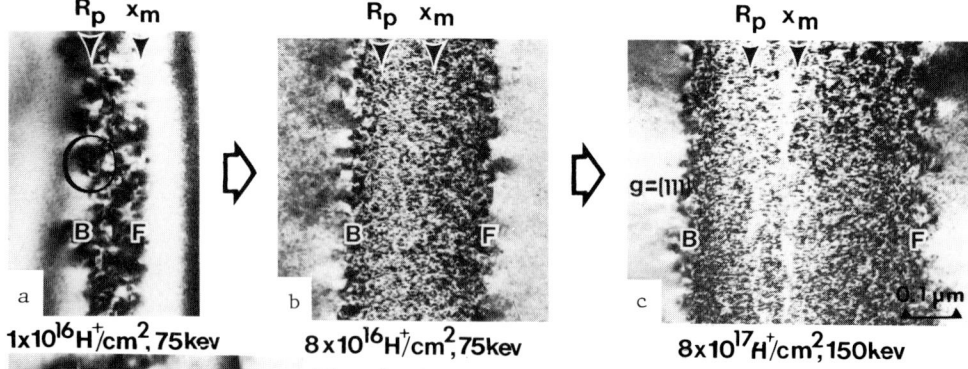

Fig. 1 Cross section micrographs of the damage layer produced in single crystal Si by increasing H^+ fluences. Inset shows 1×10^{16} H^+/cm^2 at higher magnification in <111> Si.

pressures of $\sim 10^{-7}$ Torr. Efforts were made to reduce hydrocarbon build-up to a minimum. To conduct the channeling experiments we used a 1.5 MeV He beam from a 2 MeV Van de Graaff. Beam currents are normally 15 nA and the spot size is 0.8 mm. The beam divergence is less than 0.03° and back-scattering measurements are made at 168°. Cross-section samples were prepared by first evaporating a ~ 100 nm thick layer of SiO_2 onto the implanted surface of 2 mm squares. The samples were then mounted in a low viscosity embedding medium, cured, and ground and polished to produce wafers ~ 0.1 mm thick. Final thinning to obtain electron transparent regions was achieved by ion milling with 6 keV Ar ions. Planar TEM specimens were also prepared using conventional chemical thinning from the back surface combined with ion milling of the front surface to select the desired section depth. A Philips 400 T electron microscope operating at 120 keV was used for the TEM studies.

3. <u>Results and Discussion</u>

3.1 Hydrogen-implanted Silicon

Figure 1 shows the progression of microstructural change in the damage layer, produced as a function of hydrogen fluence. F and B refer to the front and back limits of the damage layer, respectively. Although two different implantation energies were used here, additional data (not given) has shown that over the range of energies employed (20-150 keV) the micro-structure within the implanted layer appears to be energy independent. It was also not dependent upon crystal orientation. As the fluence of implanted H^+ increases, the number of displaced Si atoms and implanted hydrogen increases concomitantly. The disorder is expected to be at a maximum at a penetration depth, x_m, slightly smaller than the projected range R_p and to have a distribution skewed toward the implanted surface. Such a distribution $S_D (x)$ can be calculated using a modified EDEP-1 code of Manning and Mueller (1974) using the hydrogen stopping powers given by Anderson and Ziegler (1977). Calculated values of R_p and x_m are super-imposed on the micrographs shown in Figure 1. It is clear that, even at the lowest fluence of 1×10^{16} H^+/cm^2, the damage layer is displaced toward the front surface with respect to R_p. A preferential broadening toward the implant surface continues with increasing fluence in agreement with the backscattering data presented below.

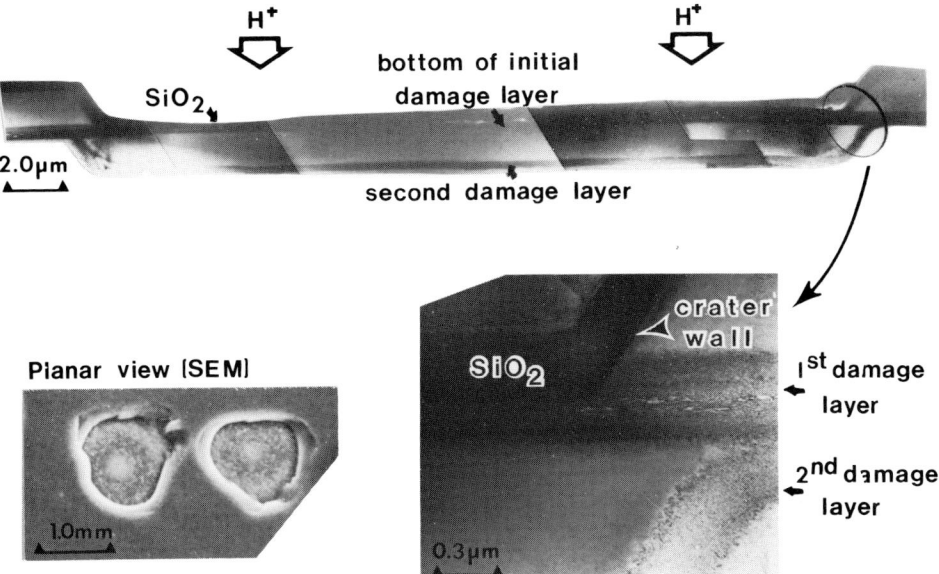

Fig. 2 Crater montage of <100> Si implanted with 8 x 10^{17} H^+/cm^2 at 150 keV and annealed 1.0 h at 450°C after SiO$_2$ deposition. Insets show planar view of crater and crater details at higher magnification

Precise interpretation of the microstructure is difficult. At the lowest fluence (Figure 1a), imaging with a 2-beam dynamical diffraction condition shows strain contrast around features which could be a second phase (such as a silicon hydride) or loops. Their projection is consistent with (111) planes and most are parallel to the original specimen surface. The nature of the black white contrast suggests that loops are more probable, which is supported by analysis of planar samples implanted to 6 x 10^{17} H^+/cm^2 and 8 x 10^{17} H^+/cm^2 and sectioned to a depth slightly greater than R_p. In addition, close inspection of the outer limits of the damage band at these higher fluences using 2-beam dynamical conditions reveals defect contrast identical to that imaged in Figure 1a. It is therefore deduced that the complex structures observed at fluences $\geq$8 x 10^{16} H^+/cm^2 are derived from loops, but some chemical effects from the hydrogen may also be occurring. Picraux, Vook and Stein (1979) report IR data on Si-H bonding in H-implanted Si consistent with SiH$_4$ formation and evidence that the Si atom to which H is bonded is in association with defects. Since each Si vacancy can accommodate 4 hydrogen atoms and a divacancy 6 atoms, we may be observing vacancy loops with trapped hydrogen.

The line of low contrast observed between x_m and R_p at 8 x 10^{16} H^+/cm^2 correlates with the appearance of the lenticular areas of lower atomic density (as suggested by absorption contrast) at higher fluences (Figure 1c) Microdiffraction has shown that, even at 8 x 10^{17} H^+/cm^2, the silicon retains its crystallinity within the damage layer, although some diffuseness of the diffraction spots is visible and HOLZ lines are absent. The region of the crystal between F and the implant surface reveals little damage.

Voids or bubbles have not been observed in specimens implanted at 25°C in agreement with Chu. However, the lenticular regions seen in Figure 1c appear in double rows at high fluences and seem to be linked to observed exfoliation phenomena (cratering). Figure 2 is a TEM montage of a crater cross-section produced at 8 x 10^{17} H+/cm^2, with insets showing a planar view (SEM image) and a higher magnification view of the crater bottom and side delineated by the SiO$_2$ layer. As observed by Chu (1977a), blisters and craters occur readily at these fluences and after annealing. Their formation mechanism, however, is not clear. The crater bottom shown in Figure 2 intersects the upper row of lenticular regions. Although this specimen was annealed, this particular exfoliation event took place during the implantation because the crater profile is further defined by the formation of second damage layer, in which regions of low density are already beginning to form. Small voids or bubbles are also present and are believed to have formed during the anneal. It is presumed that the implantation-induced stresses coupled with the weakening of the material in this layer produce local failure. This mechanism is being addressed more fully in a further paper.

As mentioned earlier, a broadening of the damage layer with fluence, preferentially towards the front surface is apparent in Figure 1. This is quantified in Figure 3 where the damage layer width measured at zero tilt with dynamical diffraction conditions is plotted as a function of fluence. No systematic variation with implantation energy was observed. The threshold for broadening is well below the fluence at which cratering begins ($\sim$4 x 10^{17} H/cm^2) and is probably linked to a "saturation" of the micro-structure observed in Figure 1a. It is further interesting to note that a similar broadening has been deduced from channeling spectra. Hydrogen atoms can make no direct contri-bution to the direct back-scattered He ions (at 168°) used for the

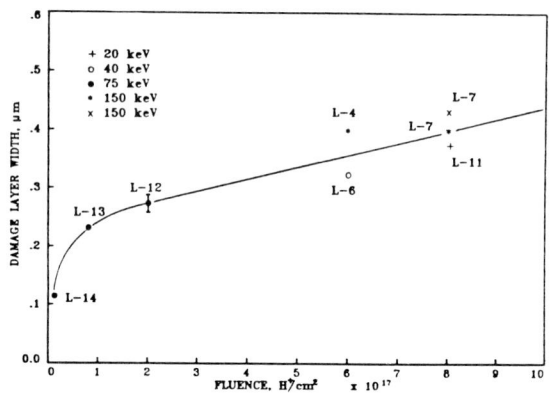

Fig. 3 Width of damage layer in H+ implanted Si measured from TEM data at 0° tilt

channeling measurements and therefore the observed backscattering is determined entirely by the displaced Si atoms. Following Rimini (1978) and as discussed further by Choyke (1983) we have determined N$_D$, the number of "equivalent displaced atoms" per unit volume from

$$\frac{N_D(x)}{N_{Si}} = \frac{\chi(x) - \chi_R(x)}{1 - \chi_R(x)}$$

where N$_D$(x) represents the equivalent density of Si atoms (as seen by the channeled He ions) required to produce the observed direct backscattering. N$_{Si}$ is the density of Si atoms. χ_R(x) is that fraction of the He channeling beam which has been dechanneled when the sample is positioned in an aligned, low index direction. χ(x) denotes the backscattered yield from the same crystal in a randomly oriented position. Figure 4 shows the resulting N_D (x) distributions for various fluences of 75 keV H^+ ions. As the fluence increases, the defect concentration N_D (x) reaches 90% of N_{Si} after a fluence of 8 x 10^{17} H^+/cm^2. At low fluences, the distributions are well-centered about x_m, but then the broadening of the peak becomes asymmetrically shifted towards the front surface. The widths of the distributions, determined at N_D/N_{Si} ≈ 0.1 are in qualitative agreement with the TEM data in Figure 3. Since

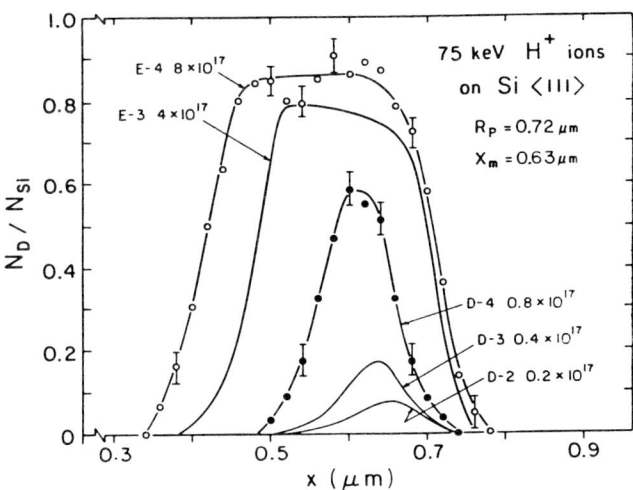

Fig. 4 N_D (x)/N_{Si} as a function of distance, x, from the implanted surface for various fluences of 75 keV H^+ ions

the backscattering is due primarily to single defects whereas the microscopy images defect clusters, the two techniques are complementary in terms of defining the damage zone. Thus, the agreement between the data indicates that most of the defects are located within the same zone defined by FB in Figure 1. The broadening observed in Figure 4 before N_D (x) reaches a near-saturation value is probably linked to the "saturation" of the initial microstructure depicted in Figure 1a and maybe linked to the broadening of the hydrogen concentration observed by Brice and Doyle (1981).

3.2 Helium-implanted Silicon

Comparison of H and He implanted single crystal Si was performed because the latter does not react chemically with the Si lattice. It was thus hoped that damage and chemistry components of the microstructure might be separated by matching fluences to yield the same dpa (displacement per atom) level at x_m. Fluences represented by the microstructures in Figure 5 are close to achieving this goal. Under a 2-beam dynamical diffraction condition at low magnification, the damage microstructure produced by He appears similar to that produced by H, although the layer width differs slightly. Under kinematical conditions, at higher magnification, however, voids or bubbles are readily imaged in He^+-implanted Si, but are not visible after H^+-bombardment. Through focus images were consistent with voids or bubbles which are best represented by the over-focussed image in Figure 5d, crystallinity within the He produced damage zone was preserved.

The line of low contrast which appears after H^+-implantation (Figure 1b) is not present after He^+ bombardment. Furthermore, the outer limits of the damage zone do not yield defect images like those observed with H^+. These differences probably reflect the lack of chemical interaction of He with the Si lattice. The presence of voids or bubbles obviously suggests that the He has sufficient mobility at RT to nucleate and grow stable clusters (diameters <15 to 70 Å). At this fluence (2 x 10^{16} He^+/cm^2), the

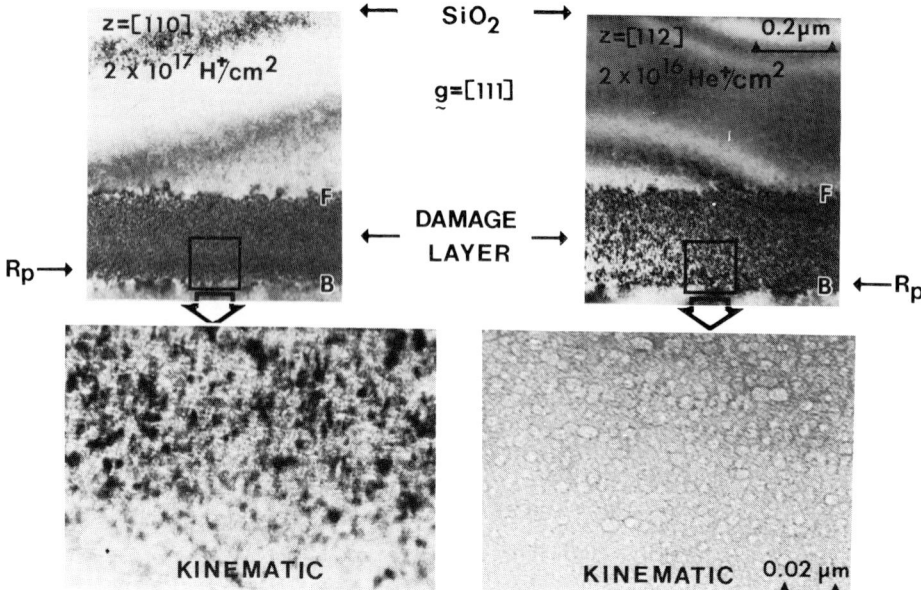

Fig. 5 Comparison of microstructures produced in single crystal Si at comparable peak damage levels of H⁺ and He⁺

details of the damage microstructure are too complex to interpret and lower fluence work is in progress. The data do indicate, however, that H interaction with the Si lattice does change the nature of the damage microstructure with respect to that formed by an inert specie.

4. Conclusions

H^+-implantation of single crystal Si produces a damage zone composed of loops which broadens preferentially towards the front surface with increasing fluence. Regions of low density are produced within the zone which are linked to the cratering phenomena observed at high fluence. Even at fluences of 8×10^{17} H^+/cm^2 the silicon was not driven amorphous by hydrogen implantation. Comparison of the damaged zones produced by RT implantation of hydrogen and helium at comparable peak dpa levels revealed differences, the most striking of which was that the helium formed bubbles.

References

Anderson H H and Ziegler J F 1977 'Hydrogen Stopping Powers and Ranges in all Elements' Vol 3 Pergamon Press

Brice D K and Doyle B L 1981 J Nucl Mat 103 and 104 503

Choyke W J, Irwin R B, McGruer J N, Townsend J R, Doyle N J, Hall B O, Spitznagel J A and Wood S 1983 Nucl Instr and Methods

Chu W K, Kastl R H, Lever R F, Mader S, Masters B J 1977a Physical Rev B 16 Number 9 3851

Ibid 1977b 'Ion Implantation in Semiconductors 1976' ed Chernow, Borders and Brice Plenum Press 483

Imura T, Kubota K, Ushita K and Hiraki A 1980 Jap J Appl Phys 19 Supplement 19-2 99

Manning I and Mueller G P 1974 Computer Physics Comm 7 85

Picraux S T, Vook F L and Stein H J 1979 Inst Phys Conf Ser No 46 31

Rimini E 1978 'Materials Characterization Using Ion Beams' ed J P Thomas and A Cachard Plenum Press

Inst. Phys. Conf. Ser. No. 67: Section 4
Paper presented at Microsc. Semicond. Mater. Conf., Oxford, 21–23 March 1983

The structure and energy of reconstructed {211} twins in silicon

R C Pond, D J Bacon and A M Bastaweesy

Department of Metallurgy and Materials Science, University of Liverpool, P.O. Box 147, Liverpool L69 3BX

Abstract   The structure and energy of {211} twins in silicon have been investigated using computer simulation. The simulation allows relaxation of atomic positions, and the contributions to the interfacial energy arising from both bond stretching and deviations of interbond angles from that in tetrahedral coordination are calculated. Two twin boundary structures were found having low interfacial energy; both configurations contain no dangling bonds. One structure exhibits mirror symmetry across the interfacial plane, whereas in the other structure the two crystals are rigidly translated to a relative position in excellent agreement with recent experimental measurements.

1. Introduction

In a previous paper (Vlachavas and Pond, 1981) an experimental investigation of the structure of {211} twin boundaries in silicon was described. The relative position of the adjacent crystals was determined accurately using transmission electron microscopy, and it was found that the crystals were relatively displaced away from an interfacial structure having high symmetry. Initial modelling of the boundary structure, using the experimentally determined relative displacement, indicated that bond reconstruction was also necessary in order to obtain a favourable structure (Pond 1982). The purpose of the present paper is to describe further modelling studies and analysis of favourable structures using computer simulation. In particular, modelling has shown that a favourable reconstructed boundary can be obtained without translation, and the energy of this structure is compared to that of the reconstructed configuration mentioned above.

2. Computer Simulation

The simulation package XLITE, developed at Liverpool from the DEVIL program of AERE Harwell, has been used. XLITE was developed for the study of defects in metals (Bacon and Martin 1981), for which ion-ion pairwise bonding dominates the structure-dependent energy. In solids such as silicon, the energy depends also on the angles between bonds formed by nearest-neighbour atoms, and its evaluation requires potentials which are functions of the relative positions of three or more atoms. For the present study, XLITE has been extended to handle both two- and three-body potentials and to generate efficiently the neighbour lists required for energy evaluation.

Empirical potential functions are not available for diamond-structure crystals, but the valence potentials (e.g. Musgrave and Pople 1962,

Keating 1966) used in lattice dynamics can be modified for use. They
describe crystal energy for small (harmonic) distortions by force constants
obtained from properties such as elastic constants and phonon frequencies.
For general distortions, functions with the same harmonic behaviour and
reasonable anharmonic form have been proposed (Sinclair 1975, Markland 1980,
Altmann et al. 1982). A simple modification of a potential suggested by
Sinclair has been used here. The total structure energy has two components:

$$E_T = E_r + E_\phi$$

$$= \frac{1}{2!} \sum_i \sum_j \frac{1}{2} F_r \left\{ [\Delta_r(r_{ij})]^2 - \frac{1}{n^2\beta^2} \right\} + \frac{1}{3!} \sum_i \sum_j \sum_k \frac{1}{2} F_\phi [\Delta_\phi(r_{ijk})]^2$$

where $\Delta_r(r_{ij}) = \begin{cases} [1-\beta^n(r_c-r_{ij})^n]/n\beta & \text{if } r_{ij}<r_c \\ 1/n\beta & \text{if } r_{ij}\geq r_c \end{cases}$

$$\Delta_\phi(r_{ijk}) = \frac{3r_o}{2\sqrt{2}} \left[\cos\phi_{ijk} + \frac{1}{3} \right] \frac{d\Delta_r(r_{ij})}{dr} \frac{d\Delta_r(r_{jk})}{dr}$$

Here, $r_{ij} = |x_i-x_j|$, where x_i and x_j are the positions of the ith and jth
atoms, and $\beta=(r_c-r_o)^{-1}$, so that the bond-stretching potential has a minimum at the perfect bond length r_o ($=\sqrt{3}a/4 = 0.235$ nm, where a is the lattice
parameter)and goes smoothly to zero at r_c, the cut-off distance. The
component E_ϕ arises from differences in angle ϕ_{ijk} from $\cos^{-1}(-1/3)$ between
bonds r_{ij}, r_{jk}, meeting at atom j. The force constants F_r and F_ϕ are 161
and 9.32 Nm^{-1}, respectively, and give elastic constant values $C_{11} = 167$,
$C_{12} = 65$, $C_{44} = 70$ GNm^{-2}. The parameter β is related to the sublimation
energy E_s by $\beta = \frac{1}{n}(F_r/E_s)^{\frac{1}{2}}$, and Sinclair (1975) used n = 3, so that r_c was
just larger than the second-neighbour spacing $a/\sqrt{2}$. This results in a
perfect lattice under hydrostatic stress. In the present study, a value of
n = 1.5 has been used, so that with $E_s = 0.746 \times 10^{-18}$J, r_c lies between r_o
and $a/\sqrt{2}$.

For studies of the {211} twin boundary, perfect crystallites were generated
with faces bounded by (011), (1$\bar{1}$1) and (21$\bar{1}$) planes. They were surrounded
by a mantle of atoms such that all atoms had a complete set of neighbours
as required by the potentials. During relaxation to minimise energy, atoms
outside the (21$\bar{1}$) faces were fixed and periodic boundary conditions were
adopted in the [011] and [1$\bar{1}$1] directions. Blocks containing 2 (or 4) (022)
planes, 3 (1$\bar{1}$1) planes and 18 (42$\bar{2}$) planes were employed. A symmetric twin
was introduced by shears of $\frac{1}{3}[\bar{1}1\bar{1}]$ on successive (42$\bar{2}$) planes, i.e. a simple
shear of magnitude $2\sqrt{2}$ was applied (figure 1). Other possible twin
structures were generated from this form by rigid-body and atomic displacements. Subsequent energy minimization of either E_T, or E_r and E_ϕ
separately, was achieved by the method of conjugate gradients. Since the
potential functions describe structure-energy with no restriction placed on
atomic coordination number, they can give misleading energies for some configurations . In the present work, they have been used to investigate
particular structures where the coordination number is limited to four.

3. Interfacial Structures

The structure of {211} twins in silicon has been the subject of several
investigations,Hornsta (1959), Kohn (1958), Möller(1981), and Fontaine
and Smith (1982). Figure 1 shows a schematic drawing of a configuration
with high symmetry. It is noted that atoms occupy coincidence site

positions at the interface but, since these atoms have three and five-fold coordination, there are two dangling bonds per period of the structure. This configuration resembles closely that proposed by Möller (1981), who suggested that the energy of this structure is relatively low despite the presence of dangling bonds on the grounds that the other bond lengths and bond angles are not distorted much compared to those in single crystals. The structures suggested by Hornstra (1959) and Kohn (1958) also have two dangling bonds per period; the former author envisages the boundary as an array of partial dislocations, and the latter as being facetted on a very fine scale. In the present paper, we demonstrate that it is possible to obtain, by bond reconstruction, two boundary structures without any dangling bonds and with only small distortions of tetrahedral bonding. One of these structures involves rigid translation of one crystal with respect to the other and is designated the translation twin. The second structure does not involve translation and is referred to as the reflection twin since the interface remains a plane of reflection; the two structures are considered separately below.

3.1 The reconstructed reflection twin

Consider first the symmetrical boundary shown in figure 1. As mentioned above, the left hand coincident atom, L, has only three nearest neighbours, and the right hand coincident atom, R, has five. It follows that there are two dangling bonds per period which will contribute about 0.9 J/m^2 to the boundary energy if the energy of a single dangling bond is taken to be about 1 eV. In addition, it is clear that the contribution E_ϕ to the total energy, E_T, will be substantial since some interbond angles are 90°. An accurate calculation of E_T has not been made, but an estimate based on the potentials described in section 2 suggests that it is considerably greater than the energies of reconstructed configurations to be discussed below.

The boundary structure depicted in figure 1 may be reconstructed in order to remove the dangling bonds as follows. If the atom R is displaced by about $\frac{1}{4}[0\bar{1}1]$ from its position in figure 1, the atom L will consequently have tetrahedral coordination. However, the atoms adjacent to R, marked T and S in the top and bottom crystals respectively, now have only three nearest neighbours which are approximately coplanar. However, if atom T is displaced by about $\frac{1}{10}[0\bar{1}1]$ and atom T', where T' is the atom originally $\frac{1}{2}[0\bar{1}1]$ above T in the unrelaxed structure, is displaced by about $\frac{1}{10}[01\bar{1}]$, these two atoms can bond together as a coupled pair. The corresponding atom pairs S,S' in the lower crystal can also bond in this way thereby saturating all bonds. The atom positions obtained by computer simulation of this reconstructed boundary are shown in figure 2. We note that although the boundary plane remains a mirror plane the primitive translation vector parallel to $[0\bar{1}1]$ is doubled in magnitude so that some atom pairs which were initially separated by $\frac{1}{2}[0\bar{1}1]$ in the unrelaxed structure no longer appear superimposed in the relaxed state. It can be seen that atom pairs S,S' and T,T' experience large displacements parallel to the (022) plane in addition to the displacements parallel to $[0\bar{1}1]$ which leave pair spacings close to r_0. Reference to the Table shows that the major contribution to the energy of this structure arises from E_ϕ, the bond stretching contribution, E_r, being small.

TABLE

Energy (in Jm^{-2}) of two reconstructed {211} twin boundaries in silicon. γ_r and $\gamma_\emptyset$ are the contributions to the total energy γ_T arising from the terms E_r and $E_\emptyset$ defined in section 2.

Boundary	γ_r	$\gamma_\emptyset$	γ_T
Reflection twin	0.21	0.98	1.20
Translation twin	0.62	0.81	1.43

3.2 The reconstructed translation twin

The atom positions obtained by computer simulation of the reconstructed translation twin are shown in figure 3. The reconstruction of this twin from the configuration of figure 1 can be envisaged in the following manner. The pair of coincident atoms, L and R, in figure 1 debond and one, say L, becomes associated with the upper crystal, and the other, R, becomes associated with the lower crystal. Atom L subsequently bonds to atoms P, T and T' in the upper crystal, and R bonds to Q, S and S' in the lower crystal, thereby making five-member rings. The lower crystal is rigidly displaced by the translation components 0.54 $[1\bar{1}1]$ and $\frac{1}{4}[0\bar{1}1]$. The unsaturated bonds on L and R can then bond across the interface to atoms J and K respectively as shown in figure 3. It is seen from the Table that γ_r and $\gamma_\emptyset$ make roughly equal contributions to γ_T, which is approximately 20% larger than the boundary energy of the reflection twin.

4. Discussion and Conclusions

The results reported here are the first simulations of twin boundary structures in silicon to be carried out using both bond stretching and inter-bond angle terms in the interatomic potential. These calculations provide a knowledge of both types of contribution to the boundary energy, and the relaxed atomic positions. Earlier estimates, such as that by Möller (1981), did not incorporate bond stretching and relaxation effects. For the reconstructed boundaries studied here, atomic coordination is such that the potential functions are evaluated in the ranges of r_{ij} and $\emptyset_{ijk}$ where they are most reliable. Contributions to the energy arising from interactions between distant atoms, such as those discussed by Altmann et al. (1982), are unlikely to be significant in twin structures.

In the course of this work two twin boundary structures have been found which have no dangling bonds and relatively low grain boundary energy. Both structures can be regarded as being comprised of sequences of 5, 6 and 7 member rings in the interfacial region. In the case of the reflection twin the boundary plane is a mirror plane, and the continuous traces of $(1\bar{1}1)$ planes have been indicated on figure 2 to emphasize this. The reconstruction involved leads to a doubling of the repeat distance parallel to $[0\bar{1}1]$, but retention of mirror symmetry parallel to (011); this dissymmetrisation implies that energetically degenerate domains separated by interfacial dislocations with Burgers vector $\frac{1}{2}[0\bar{1}1]$ can exist (Pond and Vlachavas 1983). In the case of the translation twin the adjacent crystals take up a relative position such that the boundary plane is not a mirror plane, but mirror symmetry is retained parallel to (011), and the periodicity of the interface is not modified as a result of reconstruction. The dissymmetrisation at the translation twin can lead to interfacial domains, and in this case the defects separating neighbouring domains are dislocations with Burgers vector either $0.08[1\bar{1}1]$ or $(\frac{1}{3}-0.08)[1\bar{1}1]$. The relative position of the adjacent crystals in the translation twin is in excellent agreement with experimental measurements (Vlachavas and Pond 1981). For example, the (022) planes were found to be continuous and the $(1\bar{1}1)$ planes were found to be offset by 0.38 interplanar spacings (modulo 1): the traces of these latter planes are indicated in figure 3 to illustrate this agreement. Furthermore, the simulated structure has expanded at the interface, as was detected experimentally. It is felt that the excellent correspondence between theory and experiment supports the conclusion that the structure observed experimentally is very similar to the calculated configuration in figure 3. In the light of this result it is surprising that the reflection-twin structure, which has a smaller calculated interfacial energy than the translation-twin structure, has not been observed experimentally. Electron microscopical and theoretical studies are continuing in this laboratory in order to investigate this matter further.

References

Altmann S L, Lapiccirella A, Lodge KW and Tomassini N 1982 J.Phys.C: Sol.St.Phys. 15 5581.

Bacon D J and Martin J W 1981 Phil. Mag. A43 883.

Fontaine C and Smith D A 1982 Appl. Phys. Lett. 40(2)153.

Hornstra J 1959 Physica 25 409.

Keating P N 1966 Phys. Rev. 145 637.

Kohn J A 1958 Am.Mineral. 43 263.

Marklund S 1980 Phys.Stat.Sol.(b) 100 77.

Möller H J 1981 Phil. Mag. A43 1045.

Musgrave M J P and Pople J A 1962 Proc.Roy.Soc. A268 472.

Pond R C 1982 J.de Phys. 43 Colloque Cl 51.

Pond R C and Vlachavas D 1983 Proc. Roy. Soc. Lond. A386 95.

Sinclair J E 1975 Phil. Mag. 31 647.

Vlachavas D S and Pond R C 1981 Inst. Phys. Conf. Ser. No. 60 159.

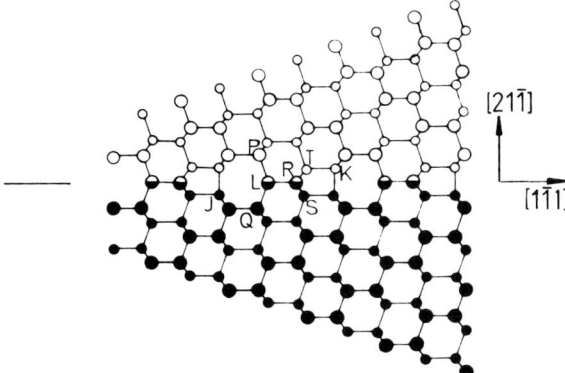

Fig. 1. Schematic illustration of a (211̄) twin boundary exhibiting mirror symmetry across the interface. The projection direction is [011̄], and the two atom symbols represent the ...ABAB... stacking along this direction.

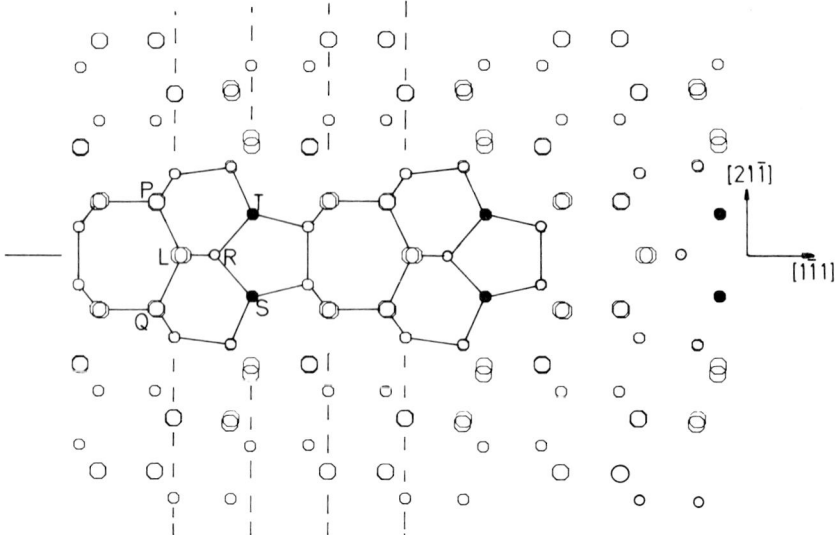

Fig. 2. Atomic positions at the reconstructed reflection-twin boundary obtained by minimising E_T Note that the symbols at S and T represent bonded atom pairs displaced from the (022) planes (see text). The vertical lines show the traces of the $(1\bar{1}1)$ planes.

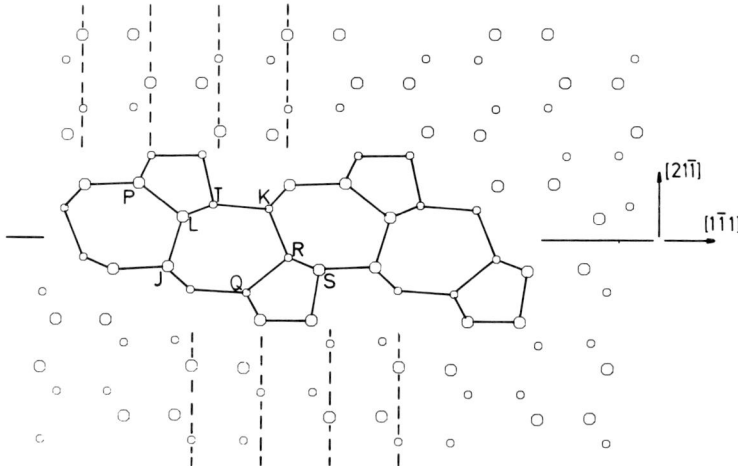

Fig. 3. Atomic positions at the reconstructed translation-twin boundary obtained by minimising E_T. The vertical lines show the traces of the $(1\bar{1}1)$ planes.

Inst. Phys. Conf. Ser. No. 67: Section 5
Paper presented at Microsc. Semicond. Mater. Conf., Oxford, 21–23 March 1983

The sources of defects in InP/InGaAsP emitters

S Mahajan

Bell Laboratories, Murray Hill, New Jersey, U.S.A.

Abstract   The following sources of defects in InP/InGaAsP emitters have
been delineated: (i) grown-in dislocations in underlying substrates,
(ii) epi-holes, dissolution pits and non-planar growth resulting from
melt-carry-over, (iii) phase separation in quaternary layers, and (iv)
dark spot defects resulting from the migration of gold. Ramifications
of these observations on device behavior are discussed.

1. Introduction

It is generally agreed that the performance, reliability, and degradation
behavior of opto-electronic devices are affected by the presence of disloca-
cations (Stringfellow 1972, Olsen 1975, Petroff and Hartman 1973, Hutchinson and Dobson 1975, Ishida, Kamejima and Matsui 1977, Mahajan, Johnston,
Jr., Pollack and Nahory 1979, Dutt, Mahajan, Roedel, Schwartz, Miller and
Derick 1981). Olsen (1975) has investigated the influence of interfacial
lattice mismatch on device performance. It is found that the increased
mismatch has deleterious effects on the electro-optical properties of
transmission photocathode and transmission secondary electron multipliers.
Furthermore, that dislocations play an important role in the degradation
behavior of (Ga, Al)As/GaAs DH lasers is well documented (Petroff and Hartman 1973, Hutchinson and Dobson 1975, Ishida, Kamejima and Matsui 1977).
A consensus has emerged that dark line defects (DLD's) observed in degraded
regions originate from existing dislocations. DLD's oriented along the
<110> directions may evolve by the glide of threading dislocations, whereas
the growth of <100> DLD's could occur by glide and climb. In addition, the
recent work on optically degraded InGaAsP epi-layers indicates that non-
luminescent regions, as revealed by spatially resolved photoluminescence,
contain dislocation networks (Mahajan, Johnston, Jr., Pollack and Nahory
1979). More recently, Dutt et. al. (1981) have shown that dislocations and
stacking faults have a marked effect on the performance of (Ga, Al)As/GaAs
DH LED's.

In recent years, there has been a tremendous surge of interest in InP,
InGaAs and InGaAsP materials. This stems primarily from their device applications as emitters and detectors in the current light-wave communication systems based on fused silica fibers. Since state-of-the art fibers
exhibit low attenuation (Horiguchi 1976) and minimum material dispersion
(Payne and Gambling 1975) in the 1.1-1.5-μm region, diode lasers and LED's
emitting in this wavelength regime constitute optimal light sources. Several investigators (Antypas and Moon 1973, Shen, Hsieh and Lind 1977,
Nakajima, Kusunoki, Akita and Kotani 1978, Dentai, Lee, Burrus and Buehler
1977, Pollak, Nahory, DeWinter and Ballman 1978) have recently demonstrated that emitters based on the InP/InGaAsP system, emitting at these

wavelengths, can be fabricated. One of the attractive features of this system is that, over a wide range of compositions, lattice-matched InGaAsP layers can be grown on InP substrates by liquid phase epitaxy (LPE). It is therefore feasible to tailor the band gap and thus the emission wavelength of optoelectronic devices to be compatible with the spectral properties of the fiber.

In this paper, some of the sources of "defects" in InP/InGaAsP emitters, grown on (001) InP substrates by LPE, have been delineated. In the context of the present paper, the term "defect" is quite broad and encompasses both macro-as well as micro-structural features which may affect luminescence properties of different epitaxial layers and device structures.

2. Substrate as a Source of Dislocations

It has been demonstrated previously that Huber etch (Huber and Linh 1975) reliably delineates the majority of material defects in InP (Mahajan and Chin 1981). Consequently, the macroscopic perfection of homoepitaxial layers grown on differently doped substrates has been assessed using this etchant. Illustrated in Fig. 1(a) is an etch pit pattern observed on a 2-μm thick epitaxial layer grown on the Sn-doped substrate. Fig. 1(b) shows the same area as in Fig. 1(a) except that the layer has been removed using

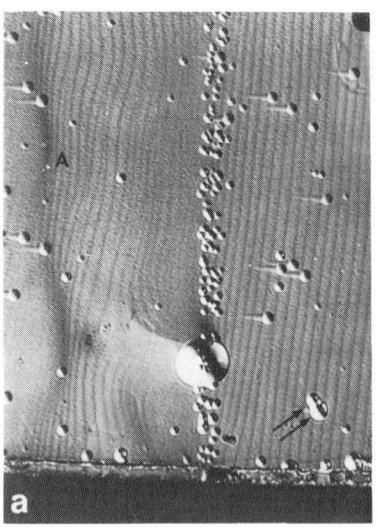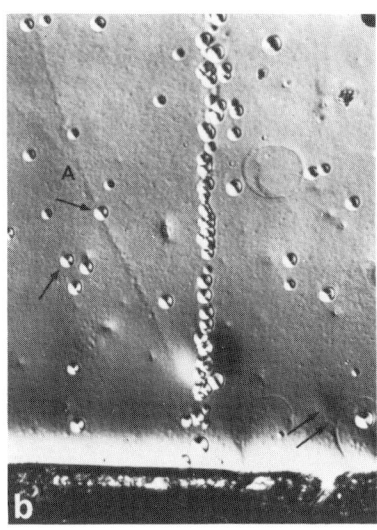

0.1 mm

Fig. 1 Etch pit pattern observed on (a) a 2-μm thick, epitaxial layer and (b) the underlying InP: Sn substrate. Single arrows denote some of the dislocations which do not have counterparts in the epitaxial layer. Double arrows show dislocation multiplication. Compare etch pit pattern around A in both micrographs.

a Br-methanol solution. Comparing the pattern observed in Figs. 1(a) and 1(b) (see the area marked A), it is clear that in the majority of cases there is one-to-one replication of imperfections during the epitaxial growth of a thin layer. Occasionally, some of the dislocations in the substrate do not have counterparts in the epitaxial layer, for example see singly arrowed etch pits in Fig. 1(b). Similar observations have been made by Kumar and Takagi (1977), Ishii et. al (1976) and Komiya et. al. (1976) on the replication of dislocations during the epitaxial growth on GaAs. The preceding result can be rationalized by assuming that some of the dislocations reorient during the epitaxial growth. In addition, dislocation multiplication has been observed in the epitaxial layer; compare the doubly arrowed regions in Figs. 1(a) and 1(b). Even though the etch pit patterns in Figs. 1(a) and 1(b) are essentially identical, the separation between the corresponding pits in these two micrographs is not the same. This effect must be due to the fact that dislocations in the substrate are not oriented normal to the (001) plane. Furthermore, similar results have been obtained for layers grown on the S- and Zn-doped substrates.

The cluster of etch pits, defining a broad band in the center and running vertically, results from the replication of dislocations associated with a scratch in the substrate. Since the cluster is oriented along the sliding direction, the scratch was very likely introduced when the substrate was slid after the melt-back, i.e., prior to the epitaxial growth.

The change in dislocation density of an epi-layer as a function of its thickness has been examined, and these results are listed in Table 1 (Mahajan, Keramidas, Chin, Bonner and Ballman 1981). It is clear that the defect density in the epitaxial layers is generally greater than that in the substrate. Moreover, the percentage increase in dislocation density is higher for the 10-μm-thick layers grown on the Zn- and Sn-doped substrates. These effects could arise from (i) the occurrence of double cross slip of dislocations induced by surface image forces, (ii) the nucleation of dislocation loops at surface ledges (Cherns and Stowell 1975), (iii) the climb of existing dislocations, and (iv) the condensation of point defects inherited by the epitaxial layer during its growth from the liquid.

TABLE 1 - Dislocation Densities Observed in Epitaxial Layers and Underlying Substrates by Etch Pitting

Type of Substrate	2-μm epi	Substrate	10-μm epi	Substrate
InP:Zn	1.3×10^3	1×10^3	2×10^4	5×10^3
InP:Sn	1.5×10^5	1.4×10^5	1.1×10^5	7.7×10^4
InP:S	6.1×10^4	5.1×10^4	7.9×10^3	6.8×10^3

defect density (cm^{-2})

A question may arise whether or not it is possible to reduce the density of threading dislocations in the epi-layer over that in the substrate. Several attempts have been made to achieve this objective in the past (Rozgonyi, Petroff and Panish 1974, Olsen, Abrahams, Buiocchi and Zamerowski 1974, Saul 1971). They can be broadly grouped into two categories: growth schemes involving either the imposition of a stress (Rozgonyi, Petroff and Panish 1974, Olsen, Abrahams, Buiocchi and Zamerowski 1974) or intermittent growth procedures (Saul 1971). The suitability of these procedures to the InP system has been examined (Chu, Mahajan, Strege and Johnston, Jr., 1981, Mahajan, Keramidas and Bonner 1982).

Fig. 2(a) shows an example of the etch pit distribution observed on the surface of an iso-epitaxial layer grown by vapor phase epitaxy and the corresponding area of the substrate is shown in Fig. 2(b). In addition, a cross-hatched pattern along the <110> directions is observed in Fig. 2(a). This is due to the presence of misfit dislocations and implies a lattice mismatch between the epi-layer and the substrate. The source of the misfit stress is not clear. It is conceivable that the starting layer of iso-epitaxy is nonstoichiometric.

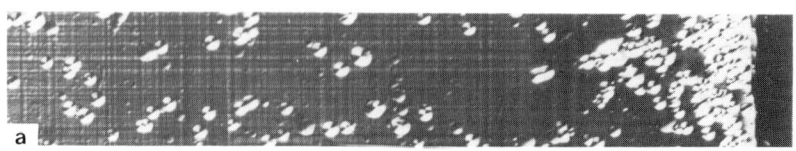

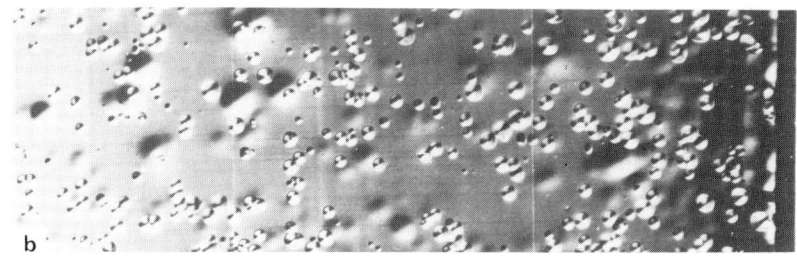

50 μm

Fig. 2 Etch pit pattern observed on (a) the iso-
epi-layer and (b) the substrate; this area is
close to the edge of the wafer. The general
reduction of threading dislocations and pileup
at the edge is shown in the epi-layer.

A comparison of the etch pit patterns in Figs. 2(a) and 2(b) shows that a large number of dislocations in the substrate do not have counterparts in the epi-layer. In addition, it is evident that near the epi-edge, the dislocation density in the epi-layer is considerably higher than that in the underlying substrate. This phenomenon is quite general and can be rationalized in terms of Matthews' hypothesis (Matthews 1966). It is visualized that during the epi-growth, the substrate dislocations undergo replication and misfit-stress-induced glide. As a result only those substrate dislocations which do not glide during growth have counterparts in the epi-layer. This effect can thus produce a general reduction of threading dislocations in the epi-layer. Furthermore, when the misfit stress becomes very small or vanishes, threading dislocations would stop gliding. This effect could cause the increase in dislocation density observed near and at the epi-edges.

In order to assess the influence of intermittent growth procedures on the reduction of threading dislocations in epi-layers, Mahajan et. al. (1982) have employed three different growth cycles. In cycle A, epi-growth was initiated at 700°C and the resulting homoepitaxial layer was isothermally aged for 2 hrs. at 660°C in contact with the melt from which it was grown. After aging, the layer was cooled, resulting in additional thickness. In cycle B, an epi-layer was partially back-melted by raising the temperature from 660° to 670°C and then an additional thickness was grown on the back-melted surface. In cycle C, A and B were combined with B preceding A. Furthermore, these layers were grown on highly dislocated substrates (dislocation density - $\sim$ 2-7 × 10^5 cm^{-2}).

It is observed that cycle A is only marginally effective in reducing the number of dislocations in iso-epitaxial layers, whereas cycles B and C are not effective at all. It is envisaged that the reduction occurs as a result of the formation of closed loops during the growth cycle which follows the isothermal hold. The formation of closed loops would be facilitated by glide and climb. Both of these processes would be affected by the attractive elastic interactions that occur between dislocation of opposite signs.

It is visualized that the incorporation of an isothermal hold during growth would allow more time for glide and climb to occur. In addition, as suggested by Saul (1971), the climb kinetics should be enhanced because of the influx of point defects from the melt into the epi-layer. This suggestion is borne out by the recent work of Small et. al. (1980) on the diffusion of Al into GaAs. They have observed that when GaAs epi-layers are in contact with a Ga-Al-As melt, Al diffuses considerably faster into the layers.

A dislocation density of $\sim$ 10^5 cm^{-2} in the substrate may represent a lower limit for cycle A to be effective. Assuming that these dislocations are homogeneously distributed, the interdislocation spacing would be $\sim$ 30μm. Since both glide and climb are assumed to be involved in the formation of closed loops and since the attractive force between dislocations of opposite Burgers vectors varies inversely as a function of the separation, a distance greater than $\sim$ 30μm would be difficult to close in a reasonable amount of time.

It is apparent from the preceding discussion that the above mentioned approaches for reducing dislocation density are not very practical from the technological viewpoint. If device applications require low dislocation densities, a more desirable approach would be to reduce the density of grown-in dislocations in InP substrates. As shown previously (Seki, Matsui and Watanabe 1976, Seki, Watanabe and Matsui 1978, Mahajan and Chin 1981), this can be achieved by doping heavily with S.

3. Macroscopic Defects Caused by Melt-Carry-Over

In general, device wafers used for the fabrication of InP/InGaAsP emitters consist of a Sn-doped InP buffer, a nominally undoped InGaAsP (Q) active layer, a Zn-doped InP confining layer and a Zn-doped quaternary (Q) contact layer. These structures are grown in a multiple well graphite boat involving multiple wipe-offs. It is relatively easy to visualize that if the wipe-off is incomplete, melt can be carried from one well to the next. This section addresses some of the situations which may result from melt-carry-over (MCO).

Shown in Figs. 3(a), (b), (c), (d) and (e) are morphological features observed on surfaces of Q-contact, p-InP confining, Q-active and n-InP buffer layers and the underlying highly S-doped substrate, respectively. In each micrograph X's identify corresponding areas in different layers and the substrate. The layers in Figs. 3(b) and 3(c) and the substrate, Fig. 3(e), have been Huber etched to reveal dislocations. It is clear that the growth morphology in Fig. 3(a) is rough and non-planar. Melt

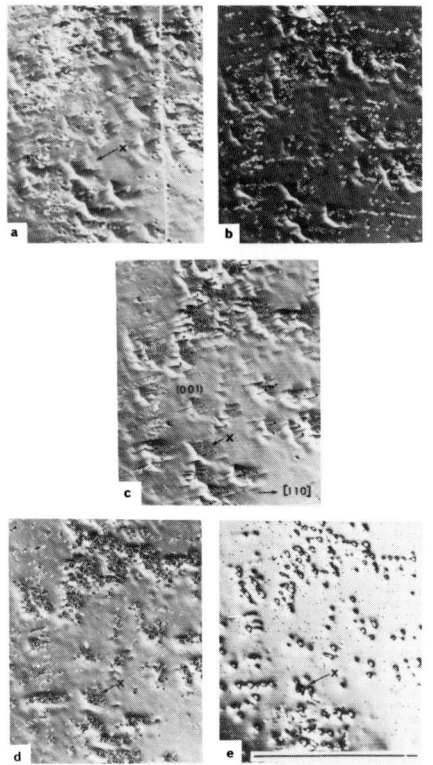

Fig. 3 Micrographs showing epi-holes
in different layers and the substrate;
(a) Q-contact layer, (b) Huber etched
p-InP confining layer, (c) Q-active
layer, (d) Huber etched n-InP buffer
layer, and (e) the underlying Huber
etched S-doped substrate. X's in
different micrographs identify corresponding areas in various layers
and the substrate. Marker represents
1 mm.

gets trapped in these ridged regions (see areas of high defect density
near left top and center right). Comparing Figs. 3(a), (b), (c) and (d),
it is apparent that the observed non-planar morphology may develop as a
result of poor buffer layer growth. This may, in turn, be related to
morphological features observed in the substrate. It was ascertained
experimentally that they represent depressions in the substrate.

It is clear from these micrographs that there exists a one to one correspondence between the rectangular shaped features in all the layers and
depressions in the substrate. However, their sizes are not the same in
all the layers because of lateral growth. This behavior is well delineated in Fig. 3(c). Combining the preceding observations, it is inferred
that these features are holes which penetrate all the epitaxial layers and
could be caused by the presence of depressions in the substrate. Further,

it is very likely that they contain In-rich alloy because during the wipe-off liquid will be trapped in them.

As shown earlier (Mahajan, Brasen, DiGiuseppe, Keramidas, Temkin, Zipfel, Bonner and Schwartz 1982), rectangular holes could form as a result of MCO after an In-melt-back, a necessary step in the growth sequence and is used to remove decomposition-induced damage from substrates. The carried-over In-melt produces dissolution pits (DP) in the substrate because it wants to dissolve P. Since melt could be trapped in these holes during a wipe-off, it can be argued that once DP's have formed in the substrate, holes penetrating all the epi-layers could result. Further, it is assumed in the preceding discussion that lateral growth over the holes or DP's is absent. It is conceivable that, depending upon the growth conditions, the lateral growth may cover the underlying DP's, thus leading to epi-holes which do not penetrate all the epi-layers.

Figure 4(a) shows various macroscopic features observed on a Huber etched surface of an InP buffer layer. In addition to rectangular holes (Hl's) and DP's, dislocation pits (D) are clearly evident. It is suggested that DP's arise as a result of MCO after the growth of the buffer layer. The situation which prevails following the MCO is quite complex. For its assessment, an interplay between the epi-layer, two-phase melt used to grow the buffer layer, and a Q-layer melt must be considered. The Q-layer melt would deplete P from the two-phase melt because the activity of P in the former may be lower than that in the latter. The amount of P depleted may be replenished by dissolving the underlying buffer layer which acts as

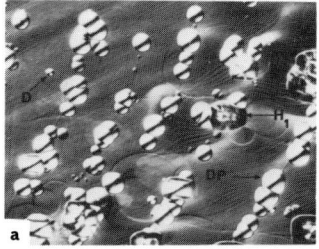

Fig. 4 Micrographs illustrating different manifestations of melt-carry-over observed in LPE growth of InP and InGaAsP epi-layers. (a) Huber etched n-InP buffer layer showing dislocation pits D, dissolution pits (DP) and holes (Hl). (b) Q-active layer showing two types of holes, Hl and H2, which result from the occurrence of melt-carry-over. (c) Huber etched surface of a p-InP confining layer showing dislocation pits (D) and dissolution pits (DP). Marker represents 0.1 mm.

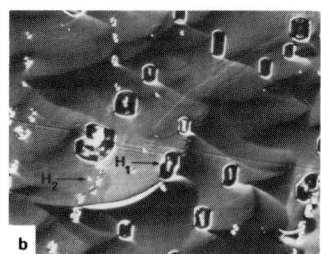

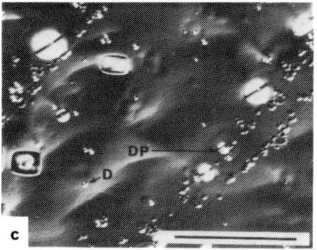

an infinite source for P. It is therefore envisaged that the occurrence
of MCO after the growth of a buffer layer may lead to the formation of DP's
as well as local variations in the composition of a Q-layer. The latter
assessment is consistent with the observations of Kopf and Sumski (1975)
and Ladany et al. (1981) regarding MCO-induced contamination of GaAs/
GaAlAs epitaxial layers. Furthermore, once DP's have formed in the buffer
layer, wipe-off after the growth of a Q-layer will be poor in these depres-
sed regions. Holes H2's in Fig. 4(b), which shows a Q-layer, could form in
this fashion.

Figure 4(c) shows a Huber etched surface of a p-InP confining layer and
exhibits some features whose characteristics are similar to those of DP's
shown in Fig. 4(a).

The shapes of holes H1 and H2 in Figs. 4(a) and 4(b) are different. It
could be that the pure In melt used for melt-back is more anisotropic in
its etching behavior than the P-enriched In melt used to grow the InP
buffer layer.

Recently, it has been shown that the formation of holes results in increased 'infant mortality' and electrically unstable devices (Mahajan, Temkin,
Zipfel, DiGiuseppe, Brasen, Bonner and Schwartz 1983). It is therefore
inferred that in order to effect a high yield of reliable devices in this
material system, complete wipe-off, i.e., the prevention of MCO, is a
critical part of the epitaxial process.

4. Phase Separation in Quaternary Layers

Based on the analysis of diffusion - limited growth of a solid solution
from a liquid, de Cremoux et. al. (1981) have predicted the existence of
a miscibility gap in quaternary InGaAsP alloys. Recently, this assessment
has been borne out experimentally (Henoc, Izrael, Quillec and Launois
1982). We have examined the microstructures of Q-layers spanning the
1.24-1.33µm range and evaluated their luminescence properties (Dutt,
Mahajan, Temkin, Bonner, Mayo and Tsakalakos 1983).

Reproduced in Figs. 5(a), (b), (c) and (d) are electron diffraction patterns obtained from Q-layers with λ = 1.24, 1.3, 1.33 and 1.33µm, respectively; an arrow on each pattern delineates the 400 spot. The last
two samples differ in their growth schemes for Q-layers. It is evident
that the 400 spots are elongated in the <100> direction. The diffuse
intensity is observed on the low angle as well as on the high angle side.

Typical microstructures seen in the four respective layers are shown in
Figs. 6(a), (b), (c) and (d); arrows in each micrograph mark the <100>
directions lying in the (001) plane. It is evident that the basic characteristics of the quasi periodic, fine scale structure observed along
the <100> directions are the same in four micrographs. The wavelength of
this structure is estimated to be ∿ 135-150Å. In view of the possibility
that the strain contrast could be a contributory factor to the size of the
observed images, the computed value is an over estimate. Based on the
preceding observations, it is inferred that these epitaxial layers could
undergo phase separation or spinodal decomposition.

The other distinguishing characteristics of the micrographs shown in Fig.
6 are rectilinear features aligned along the <100> directions, Fig. 6(c),
and the barely discernible basket-weave pattern, Fig. 6(d); the

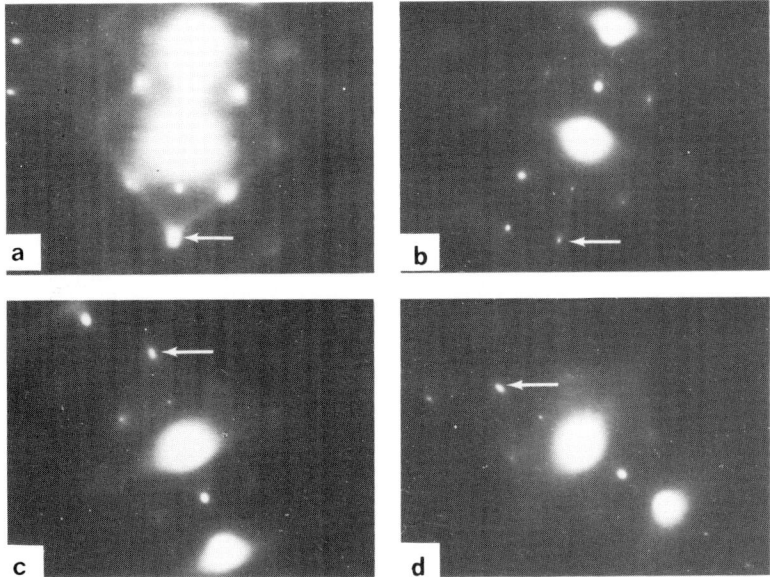

Fig. 5 Electron diffraction patterns obtained from different Q-layers: (a) λ = 1.24μm, (b) λ = 1.30μm, (c) λ = 1.33μm and (d) λ = 1.33μm. The last two samples differ in their growth schemes for Q-layers. The arrow in each micrograph delineates the 400 spot.

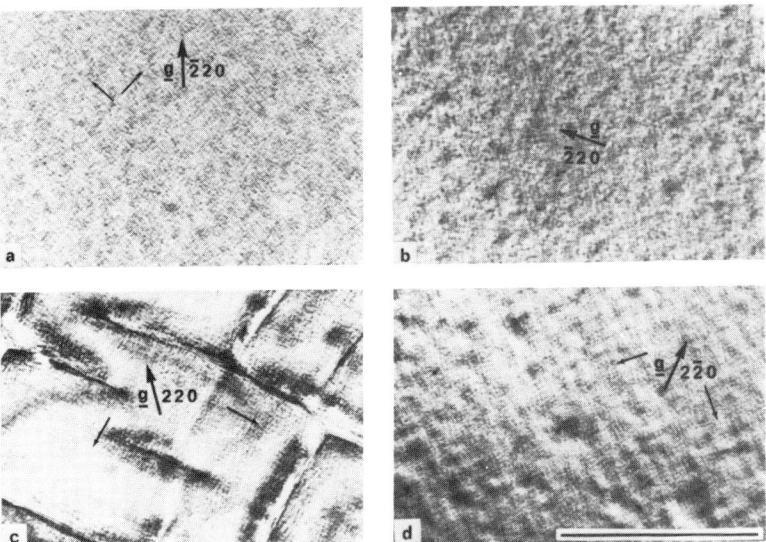

Fig. 6 Electron micrographs obtained from Q-layers of different compositions: (a) λ = 1.24μm, (b) λ = 1.30μm, (c) λ = 1.33μm and (d) λ = 1.33μm. Arrows in each micrograph mark the <100> directions lying in the (001) plane. Marker represents 1μm.

periodicity of this weave is of the order of ∿ 1250Å. Further, contrast experiments were carried out to ascertain the direction of strain associated with the boundaries. It was thus established that the principal strain is normal to the boundary.

Khachaturyan (1969) has considered the decomposition of an elastically anistropic homogeneous solid solution into two phases having the same structure as that of the original solution. Assuming that (i) the phases differ in their composition and the specific volume only, and (ii) the precipitating phases are coherent and have the same elastic moduli as that of the matrix, he has concluded based on the elastic energy considerations that three types of metastable periodic distributions are possible. In order of increasing elastic energy they are: (i) one-dimensional, (ii) two-dimensional, and (iii) three-dimensional distributions. It is envisaged that the microstructural features shown in Fig. 6(c) arise as a result of the transformation of a higher order distribution into a one-dimensional one. The rectilinear boundaries are walls separating regions differing in their <001> transformation direction. Further, the weakly developed, basket-weave pattern in Fig. 6(d) could be a consequence of the transformation of a three-dimensional distribution into a two-dimensional one. The observed large scale periodicity could be due to the interaction of strain fields of the two one-dimensional distributions.

It may be argued that the occurrence of spinodal decomposition has deleterious effects on luminescence properties of these epitaxial layers. We have evaluated this situation and find that these layers have very good luminescence (Dutt, Mahajan, Temkin, Bonner, Mayo and Tsakalakos 1983).

The occurrence of spinodal decomposition in Q-layers has ramifications in long term reliability of InP/InGaAsP emitters. As discussed earlier, the evolution of DLD's in degraded GaAs/(Ga, Al)As diode lasers could involve both glide and climb. Since spinodal decomposition is known to strengthen the matrix (Cahn 1963, Dahlgren 1977), it will be difficult for dislocations to glide and multiply in InGaAsP epitaxial layers. Furthermore, climb will also be difficult in a spinodally decomposed matrix because its occurrence would create surfaces within the two phases across which energetically favorable atom pairing could be destroyed. It is therefore inferred that if growth-induced defects are eliminated, this material system is capable of yielding highly reliable light emitting devices.

5. Dark Spot Defects in Aged LEDs

Chin et al. (1982) have evaluated the formation of dark spot defects (DSD's) in aged InP/InGaAsP LEDs by cathodoluminescence imaging and energy dispersive x-ray spectroscopy. DSD's are shown to be located in either the p-contact layer, the p-InP confining layer or the Q-active layer. Results strongly suggest that the migration of gold from the p contact during device processing and aging could result in the formation of DSD's. We have extended this study and examined microstructural features of DSD's. (Mahajan, Chin, Zipfel, Chin, Nakahara and Tung 1983).

Fig. 7 shows the situation observed in the p-contact layer of an LED aged at a junction temperature of 200°C for 1800 hours and a current density of 40KA/cm². The diffraction patterns associated with different regions are also included. The semicircular feature delineates the boundary between the metallized region to the right and the unmetallized area to the left. Dark, oblate features are very likely an intermetallic which could not be

thinned using Br-methanol solution. Further, the diffraction pattern
obtained from the center of the contact is quite complex and could be
indexed in terms of $AuIn_2$ and Au_9In_4.

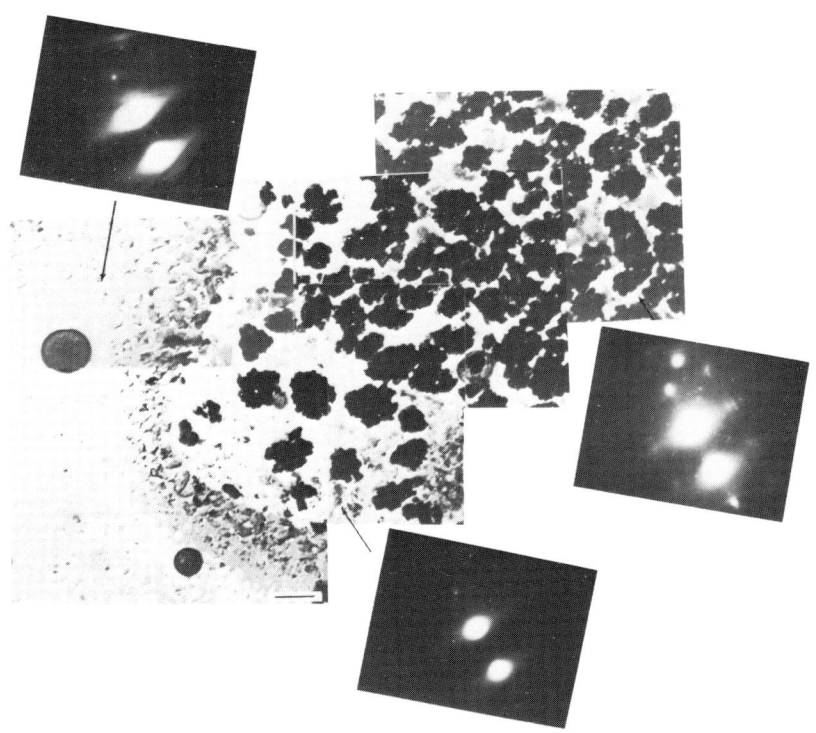

Fig. 7 Electron micrograph obtained from the p-contact layer of
an aged LED. Diffraction patterns associated with different
regions are also included. Marker represents 1μm.

Shown in Figs. 8(a) and (b) are the microstructural features observed in
the p-confining layer of an aged LED. In this micrograph, AB and CD represent the projections of the [101], [10$\bar{1}$] and [011], [0$\bar{1}\bar{1}$] vectors onto the
(001) plane. Several inclusions, $I_1...I_7$, are seen and they are associated with multiple faults and dislocations. The faults are out of contrast
in Fig. 8(b); whereas most of the ½ <101> dislocations are in contrast in
both micrographs. This observation rules out ½[110] and ½[1$\bar{1}$0] Burgers
vectors. Further, comparing the projections of various dislocations with
AB and CD, it is inferred that the majority of the dislocations are in
screw orientation. Wakita et. al. (1982) have reported similar observations in InP/InGaAsP double heterostructure lasers which have undergone
accelerated aging.

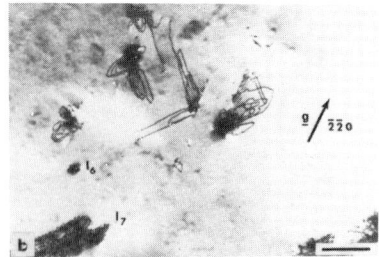

Fig. 8 Microstructural features observed in the p-confining layer of an aged LED. Inclusions, $I_1 \ldots I_7$, are associated with faults and dislocations. AB and CD represent the projections of the [101], [10$\bar{1}$] and [011], [01$\bar{1}$] vectors onto the (001) plane. Marker represents 1µm.

It appears from the preceding results that gold migrates from the p-contact during accelerated aging and produces inclusions within the p-confining layer; similar features could be present in the active layer. These inclusions are associated with faults and dislocation clusters which appear to evolve by glide. Since inclusions are likely to be good electrical conductors, injected current would tend to concentrate in these areas. This assessment is consistent with the observations of Fukuda et al. (1981) that temperature is high in the DSD regions. It is envisaged that this effect together with non-radiative recombination occurring at similar features in the active layer could produce the dark spot contrast observed in electroluminescent images.

6. Conclusions

(i) The grown-in dislocations in the underlying substrate is the principal source of threading dislocations in the epi-layer. Since it is possible to replicate the quality of the substrate in thin epi-layers, InP/InGaAsP emitters should be grown on substrates having low dislocation density. Further, it is not practical to reduce the dislocation density either by the imposition of a stress or by the incorporation of intermittent growth procedures.

(ii) Melt-carry-over, resulting from incomplete wipe-off during liquid phase epitaxial growth of multilayer structures, is the most serious problem.

(iii) Phase separation or spinodal decomposition is observed in quaternary layers. It is argued that its occurrence should make these emitters more degradation-resistant.

(iv) The migration of gold during high temperature reliability testing of LED's produces dark spot defects in the emitting region.

Acknowledgments

The author gratefully acknowledges the sustained collaboration of his colleagues W. A. Bonner, D. Brasen, A. K. Chin, S. N. G. Chu, M. A. DiGiuseppe, B. V. Dutt, V. G. Keramidas, H. Temkin and C. L. Zipfel in this work, and is thankful to T. Boone for his expert technical assistance.

References

Antypas G A and Moon R L 1973 J. Electrochem. Soc. 120 1574
Cahn J W 1963 Acta Met. 11 1275
Cherns D and Stowell M J 1975 Thin Solid Films 29 107
Chin A K, Zipfel C L, Mahajan S, Ermanis F and DiGiuseppe M A 1982
  Appl. Phys. Lett. 41 555
Chu S N G, Mahajan S, Strege K E and Johnston, Jr. W D 1981 Appl. Phys.
  Lett. 38 766
Dahlgren S D 1977 Met. Trans. A 8 347
DeCremoux, Hirtz P and Ricciardi J 1981 GaAs and Related Compounds ed
  H W Thim (London: Inst. of Phys.)
Dentai A G, Lee T P, Burrus C A and Buehler E 1977 Electron. Lett. 13
  484
Dutt B V, Mahajan S, Roedel R J, Schwartz G P, Miller D C and Derick L
  1981 J. Electrochem. Soc. 128 1573
Dutt B V, Mahajan S, Temkin H, Bonner W A, Mayo W E and Tsakalakos T
  1983 submitted for publication to Mat. Letters
Fukuda M, Wakita K and Iwane G 1981 Jap. J. Appl. Phys. 20 L87
Henoc P, Izrael A, Quillec M and Launois H 1982 Appl. Phys. Lett. 40
  963
Horiguchi M 1976 Electron. Lett. 12 310
Huber A and Linh N T 1975 J. Crystal Growth 29 80
Hutchinson P W and Dobson P S 1975 Phil. Mag. 32 745
Ishida K, Kamejima T and Matsui J 1977 Appl. Phys. Lett. 31 397
Ishii M, Hirano R, Kan H and Ito A 1976 Jpn J. Appl. Phys. 15 645
Khachaturyan A G 1969 Phys. Stat. Sol. 35 119
Komiya S, Hirano R, Nishitani Y, Isozumi S and Kotani T 1976 J. Appl.
  Phys. 47 3367
Kopf L and Sumski S 1975 J. Crystal Growth 28 365
Kumar K and Takagi M 1977 Jpn J. Appl. Phys. 16 957
Ladany I, Smith R T and Magee C W 1981 J. Appl. Phys. 52 6064
Mahajan S, Brasen D, DiGiuseppe M A, Keramidas V G, Temkin H, Zipfel C L,
  Bonner W A and Schwartz G P 1982 Appl. Phys. Lett. 41 266
Mahajan S and Chin A K 1981 J. Crystal Growth 54 138
Mahajan S, Chin A K, Zipfel C L, Chin B H, Nakahara S and Tung R T 1983
  submitted for publication to Materials Letters
Mahajan S, Johnston, Jr. W D, Pollack M A and Nahory R E 1979 Appl. Phys.
  Lett. 34 717
Mahajan S, Keramidas V G and Bonner W A 1982 J. Electrochem. Soc. 129 1556
Mahajan S, Keramidas V G, Chin A K, Bonner W A and Ballman A A 1981 Appl.
  Phys. Lett. 38 255
Mahajan S, Temkin H, Zipfel C L, DiGiuseppe M A, Brasen D, Bonner W A and
  Schwartz G P 1983 submitted for publication to J. Electrochem. Soc.
Matthews J W 1966 Phil. Mag. 13 1207
Nakajima N, Kusunoki T, Akita K and Kotani T 1978 J. Electrochem. Soc.
  125 123
Olsen G H 1975 J. Cryst. Growth 31 223
Olsen G H, Abrahams M S, Buiocchi C J and Zamerowski T J 1974 J. Appl. Phys
  46 1243
Payne D N and Gambling W A 1975 Electron. Lett. 11 176
Petroff P M and Hartman R L 1973 Appl. Phys. Lett. 23 469
Pollack M A, Nahory R E, DeWinter J C and Ballman A A 1978 Appl. Phys. Lett
  33 314
Rozgonyi G A, Petroff P M and Panish M B 1974 Appl. Phys. Lett. 24 251
Saul R H 1971 J. Electrochem. Soc. 118 793
Seki Y, Matsui J and Watanabe H 1976 J. Appl. Phys. 47 3374

Seki Y, Watanabe H and Matsui J 1978 J. Appl. Phys. 49 822
Shen C C, Hsieh J J and Lind T A 1977 Appl. Phys. Lett. 30 353
Small M B, Ghez R, Potemski R and Reuter W 1980 J. Electrochem. Soc. 127
   1177
Stringfellow G S 1972 J. Appl. Phys. 43 3455
Wakita K, Takaoka M, Seki M and Fukuda M 1982 Appl. Phys. Lett. 40 525

Inst. Phys. Conf. Ser. No. 67: Section 5
Paper presented at Microsc. Semicond. Mater. Conf., Oxford, 21–23 March 1983

273

Emission and transmission cathodoluminescence analysis of InGaAsP/InP LPE double heterostructures emitting at 1.3 and 1.6 microns

M Cocito, C Papuzza, F Taiariol

CSELT - Centro Studi e Laboratori Telecomunicazioni S.p.A.
Via G. Reiss Romoli, 274 - 10148 TORINO - Italy

Abstract   An investigation of the crystallographic defects and the surface morphology of InGaAsP/InP LPE double heterostructures (DH) is reported. A combination of emission cathodoluminescence (ECL), transmission cathodoluminescence (TCL), Si and Ge detectors at various beam energies allowed the identification of crystallographic defects on different layers in complete DHs with a simple and non destructive technique.   Preliminary work on InP substrates and on single layer structures enabled us to understand the main cathodoluminescence (CL) contrast mechanisms and to emphasize some basic rules for the interpretation of the complete DH images.

1. Introduction

LPE InGaAsP/InP double heterostructure (DH) LEDs and LASERs emitting at 1.3 and 1.6 μm are very interesting devices for optical fiber communications. In this paper two factors affecting device   yield and reliability are   investigated: the surface morphology and the crystallographic defects. SEM CL techniques have been used to detect electrically active defects and to understand the origin of surface hillocks.

CL examinations were performed with a JEOL JSM-U3 Scanning Electron Microscope equipped in order to detect both the upward emitted radiation (emission cathodoluminescence: ECL) and the downward emitted radiation (transmission cathodoluminescence: TCL). The experimental setup for ECL - TCL analysis consists of a solid state detector (Vale and Greer 1977), a low noise preamplifier and an external amplifier. A pair of photovoltaic detectors with large active areas (0.2 cm^2), low noise, high bandwidth and multidecade output linearity were used in   order to cover the wide emission range of our samples. The Si detector used (U.D.T., PIN-6 DP) covers the 0.4 to 1.06 μm range while the Ge detector (Judson Infrared, J16 8) covers the 0.5 to 1.8 μm range. The large active area and the close specimen mounting result in a very high collection efficiency. The specimens were mounted on metal slides with a central hole or directly on the detector window. In the case of InGaAsP and InP the material transparency to its own CL radiation allowed analysis of the actual structures without any problems.

The TCL image contrast is determined by local variations of the generated or transmitted CL intensity. The first mechanism is common to both   TCL and ECL, the second one is   TCL exclusive (Chin et al 1978, Cocito et al 1982). However local variations of the transmitted CL intensity seldom occur and produce very low contrast. Normally ECL and TCL images show the same contrast.

A particular feature of the work was the combination of ECL, TCL, Si and Ge detectors at various beam energies on the same specimen area. Consequently, it was possible to determine for the first time the defect distribution of different layers in complete DHs with a non destructive technique.

2. InP substrates and single layer structure analysis

CL studies were performed on InP substrates (as received as well as after various technological steps) and single layer structures in order to understand the main contrast mechanisms and to obtain a basic outline for the interpretation of the complete DH images.

Our InP substrates were n-doped (Sn and S) (100) oriented, produced by different manifacturers: MCP, Sumitomo and Lasertron. Relying on a previous work (De Padova and Stano 1982) which showed that on ECL images of Huber etched samples dark spots are associated with nearly all of the pits, we have used ECL or TCL to measure the defect density on as-received substrates. Defects consisted of three main types, which are: residual damage, single and/or multiple dislocation clusters. Dopant striations were also detected.

The residual damage, due to incomplete removal of surface layers damaged during wafer polishing, appeared on CL images as dark lines. The various substrates differed mainly from one another in dark line intensity and density; in some cases induced dislocations starting from the dark lines were also detected (Fig. 1 and 2). Combining CL analysis and mechanical-chemical polishing (Br_2; CH_3OH) we found a residual damage depth of 70 to 80 μm on the worst substrates.

After Al-Jassim et al (1981) we consider the randomly distributed small dark spots present on the CL images as single dislocations. On some substrates we found large (15 to 20 μm) dark spots showing an internal structure with tetragonal symmetry aligned along the < 011 > direction (Fig. 3). This observation is in good agreement with the work of Al-Jassim et al (1981) who demonstrated that the large dark spots are dislocation clusters.

Table 1 summarizes our results on defect type and density for the various InP substrates. Among the various technological steps of DH LED and LASER production we investigated in more detail the etching process at the beginning of LPE growth. MCP (first delivery) InP (100) Sn doped substrates were etched with both pure In and un-saturated InP melt ($\Delta T = 20$°C) for 5 and 15 seconds. The un-saturated InP melt etch failed to remove the thermal damage caused by heating and showed different etching rates in the two < 011 > and < 01$\bar{1}$ > directions. The best surfaces were obtained after 15 sec of pure In etch corresponding to about 7 μm of material removed.

In all cases 25 to 35 μm diameter hillocks were present on the surface. A small melt inclusion was always detected on the hillocks. CL analysis showed a close correspondence between hillocks and dislocation clusters; i.e. under each hillock a dislocation cluster was found. However not all clusters produce hillocks (Fig. 4a) and b)). Single dislocations seem to have no influence at all on surface morphology.

Small melt droplets(1 to 2 μm diameter) were found on the surface with higher density at the steepest regions. Their origin is probably related to the InP thermal dissociation during the last thermal cycle.

The various InP or InGaAsP (not intentionally doped) single layers of the DHs were grown on the same type of InP substrates. The hillocks and the melt droplets on the surface were also observed in this case and the same relationship with dislocation clusters was shown to exist.

Fig. 5a) and b), and c) show respectively surface morphology, quaternary active layer defects, (TCL, Ge detector) and InP substrate defects (TCL Si detector).

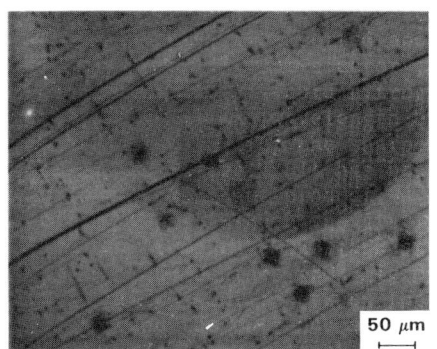

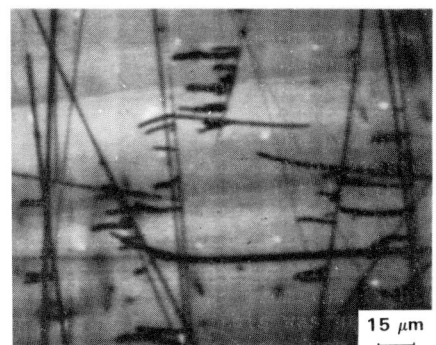

Fig. 1 - *MCP InP substrate (Sn; 1.3x10^{18} cm^{-3}) showing residual damage, dislocations and clusters. ECL-Si 25 KeV*

Fig. 2 - *LASERTRON InP substrate (Sn; 1.5x10^{18} cm^{-3}) showing residual damage, native dislocations, induced dislocations and doping striations. TCL-Si 25 KeV*

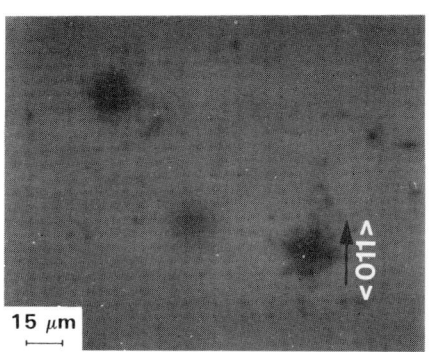

Fig. 3 - *Dislocation cluster on MCP InP substrate (Sn; 1.3x10^{18} cm^{-3}). ECL-Si 25 KeV*

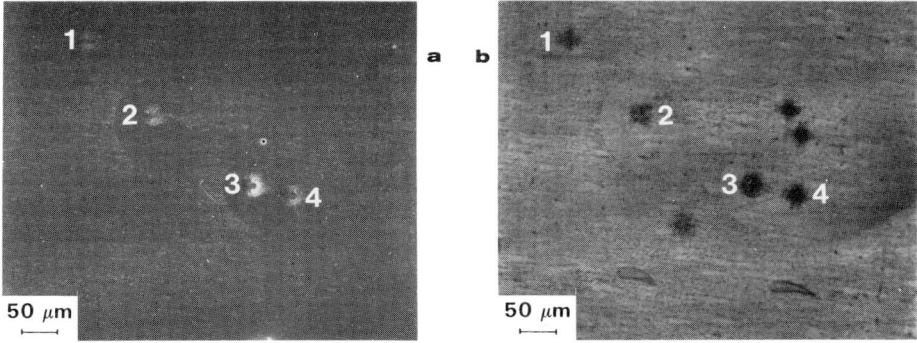

Fig. 4 - *Hillocks on MCP InP (Sn; 1.3x10^{18} cm^{-3}) substrate after 15 sec pure In etch. a) morphology and b) ECL-Si 25 KeV on the same area*

DEFECTS	SUBSTRATES				
	MCP FIRST DELIVERY Sn 1.6×10^{18} cm^{-3} epd $\sim 10^4$ cm^{-2}	MCP SECOND DELIVERY Sn 1.2×10^{18} cm^{-3} epd 2.3×10^4 cm^{-2}	MCP S 6×10^{17} cm^{-3} epd 5×10^3 cm^{-2}	SUMITOMO Sn 1.9×10^{18} cm^{-3} epd $< 10^4$ cm^{-2}	LASERTRON Sn 1.5×10^{18} cm^{-3} epd $< 3 \times 10^4$ cm^{-2}
SINGLE DISLOCATIONS (cm^{-2})	5.2×10^4	2.1×10^4	2.8×10^3	2.4×10^4	1.8×10^5
DISLOCATION CLUSTERS (cm^{-2})	2.4×10^3	VERY FEW	NO	NO	NO
RESIDUAL DAMAGE	VERY STRONG	NO	NO	VERY WEAK	VERY STRONG
INDUCED DISLOCATIONS	NO	NO	NO	NO	YES
DOPANT STRIATIONS	YES	YES	YES	YES	NO

Table I - Defects and inhomogeneities of different InP substrates

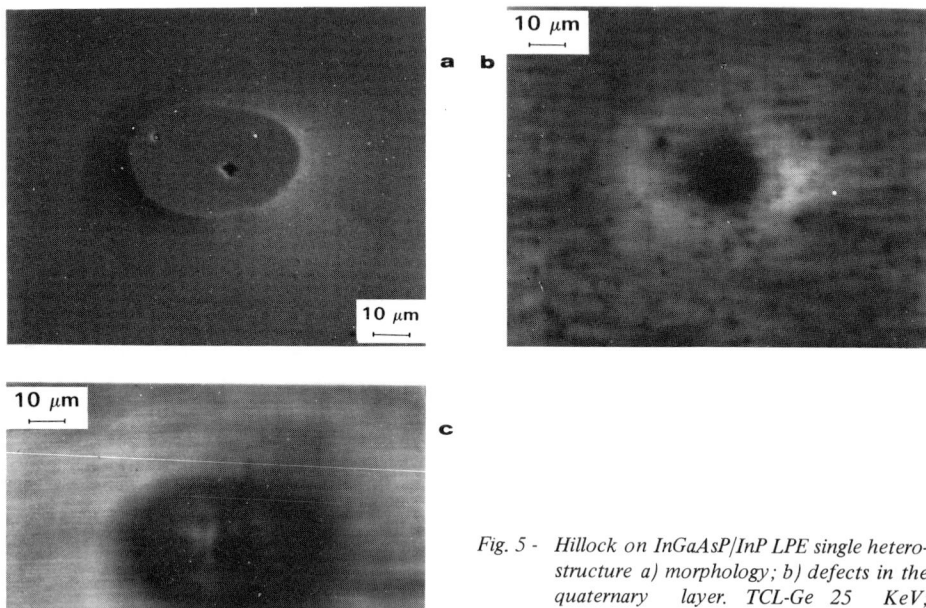

Fig. 5 - Hillock on InGaAsP/InP LPE single hetero-structure a) morphology; b) defects in the quaternary layer. TCL-Ge 25 KeV; c) defects in the substrate. TCL-Si 35 KeV

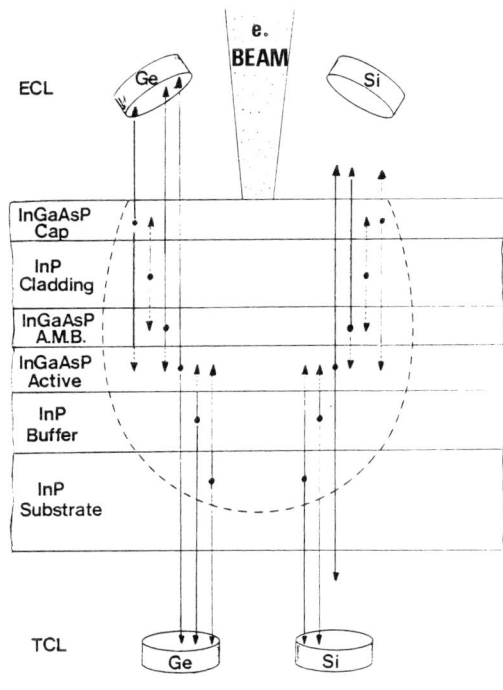

Fig. 6 - *Outline of ECL - TCL analysis of InGaAsP/InP DH semiconductor LASER (λ=1.6 μm). Dashed arrows mean absorption while short arrows mean wavelength outside the spectral response of the detector*

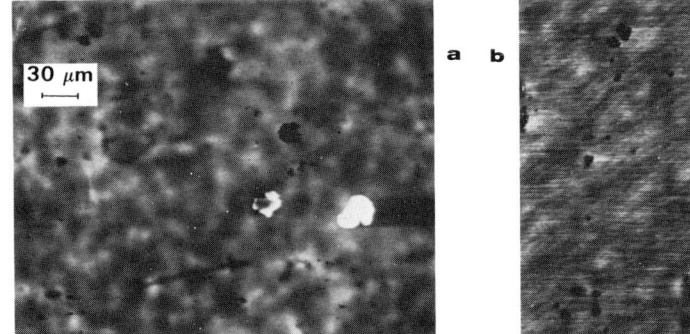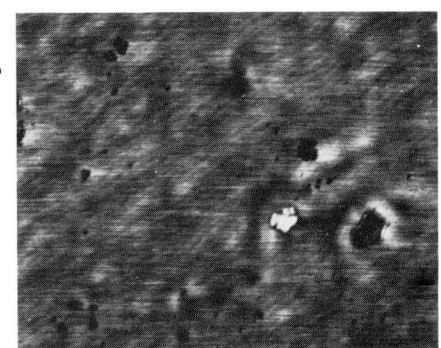

Fig. 7 - *Defect propagation at the buffer-active layer interface on InGaAsP/InP LPE DH LASER (λ = 1.6 μm) a) TCL-Si 45 KeV; b) TCL-Ge 25 KeV*

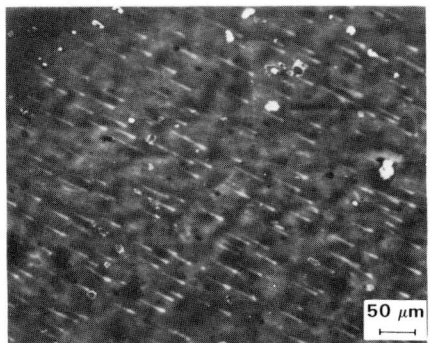

Fig. 8 - *Thermal etching at the substrate buffer layer interface on InGaAsP/InP LPE DH LASER (λ = 1.3 μm)*

3. Double Heterostructure Analysis

Complete DHs were grown on MCP (first delivery) InP Sn doped substrates. The hillocks in this case appeared very smooth due to the covering effect of the grown layers. The hillocks actually entirely disappeared using cluster free substrates (MCP S doped).

The basic outline for the interpretation of the CL contrast mechanisms in complete DH images is shown in Fig. 6. It was possible to evaluate the number of the propagated, stopped and generated dislocations at the InP buffer - quaternary active layer interface by comparing a series of TCL images of the same area.

Using TCL images similar to that of Fig. 7a) and b) from a DH emitting at 1.6 μm ($\Delta a/a\%$ < 0.03) we found that the 70% (2.1×10^4 cm^{-2}) of the dislocations present in the buffer layer are stopped at the interface and the 30% (9.0×10^3 cm^{-2}) transmitted. The density of the new dislocation generated in the quaternary layer was 1.07×10^4 cm^{-2}.

The pattern shown in Fig. 8 is often observed in the TCL images of DHs. The white lines are due to thermal etching of the substrate not completely removed and filled up by the In buffer layer (more doped).

4. Conclusions

The combination of ECL, TCL, Si and Ge detectors at various beam energies is a powerful analytical tool for the characterization of LED and LASER semiconductor multistructures. The main results achieved on complete DHs were:
- obtaining the defect density of the various layers;
- following defect propagation at the various interfaces;
- detecting the thermal etching at the substrate-first layer interface;
- evidencing that the grown hillocks are always due to dislocation clusters originating in the substrate.

This technique will be used to investigate other technological problems relevant to device yield degradation mechanisms.

5. Acknowledgements

The authors wish to thank Mr. Gorgellino for the assistance in CL analyses and Mr. Fornuto for providing the LPE structures.

References

Chin A.K., Temkin H. and Roedel R. 1978 Appl. Phys. Lett. 34, 477

Cocito M., Gorgellino F. and Troia A. 1982, 10th International Congress on Electron Microscopy Hamburg 2, 417

De Padova M. and Stano A. 1982, CSELT Internal Report 82.09.312

Al-Jassim M.M., Warwick C.A. and Booker G.R. 1981, Inst. Phys. Conf. Ser. No. 60, 357

Vale B.H. and Greer R.T. 1977, Scanning Electron Microscopy 1977, 1, 241

Inst. Phys. Conf. Ser. No. 67: Section 5
Paper presented at Microsc. Semicond. Mater. Conf., Oxford, 21–23 March 1983

Optically induced catastrophic degradation in InGaAsP laser structures

H Temkin, S Mahajan, R A Logan

Bell Laboratories, Murray Hill, New Jersey 07974

ABSTRACT

A detailed study of optically induced catastrophic damage in layers of InGaAsP is carried out. The catastrophic dark line defects are found to propagate preferentially in the $<110>$ direction common to the $\{111\}_{In,Ga}$ and (001) planes. Results of transmission electron microscopy studies indicate that such defects are generated by localized melting and solidification in the InGaAsP layer. The directionality of defect propagation is discussed.

1. Introduction

Degradation of InGaAsP light emitting devices is of considerable technological and scientific interest. Optically induced degradation of as-grown wafers can be used to separate the intrinsic recombination induced effects from other degradation modes present in fully processed devices. Slow degradation under moderate optical excitation levels has been investigated by Mahajan et al. and more recently by Komiya et al. These experiments have evaluated the influence of optical excitation on the glide of inclusion-generated dislocations and of misfit dislocations. We have previously investigated (Temkin et al) the features of catastrophic degradation under very high excitation level and compared them, on a macroscopic scale, to these observed in GaAlAs layers by Henry et al. This work has now been extended to include the orientation dependence of CDL propagation. In addition, substructural features associated with CDLs have been evaluated by transmission electron microscopy.

2. Experimental Details

The experimental arrangement used to generate and evaluate catastrophic damage has been described previously. A beam of $\lambda=1.06\mu m$ light from a Q-switched Nd:Yag laser was directed towards the sample mounted on a translation stage. The double heterostructure wafers of $1.3\mu m$ InGaAsP/InP were grown by the conventional liquid phase epitaxy on (001) oriented InP substrates. Active layer thickness of $0.2\mu m$, typical of the semiconductor laser structures, was used. The active layer doping density was $n \approx 1\times10^{17}cm^{-3}$ in unintentionally doped layers; layers doped with Zn to $p \approx 5\times10^{17}cm^{-3}$ and Sn to $n \approx 1\times10^{18}cm^3$ were also examined. After the growth samples were Br-methanol polished to a thickness of $\sim100\mu m$ and cleaved into $254\mu m$ long (where length refers to the facet-to-facet distance) and up to 5mm wide bars. As the incident power was increased by adjusting a variable neutral density filter, the lasing threshold was observed by a sharp increase in the 1.3μm light generated at the mirror facets. In the samples with undoped active layer threshold occurred at the incident peak power density of $\sim 2 \times 10^4$ Watt/cm^2, subject to an experimental uncertainty of a factor of two. At the ambient temperature of 20°C catastrophic dark line defects (CDL) form at a pump power level 7 to 15 times greater than that required for lasing. No dependence on the dopant type or concentration has been observed. Analysis of the power levels involved shows that CDLs are generated by the 1.3μm light confined to the active layer and not directly by the pump laser itself. The radiative flux required for such damage in InGaAsP layers is estimated to be well in excess of 100 MW/cm^2. This is at least a factor of 10 higher than the CDL threshold in the comparable GaAlAs structures. Since CDLs in InGaAsP originate at the material defects, rather than the mirror facets as is the case of GaAlAs structures, degradation can be induced in a wafer without cleaving it into bars.

After the power level necessary for CDL generation is reached, it is possible to translate the sample, always keeping the end of the CDL within the pump laser beam, and propagate the CDL continuously through the sample over the distance of many milimeters. However, this can be done only in the <110> direction which is common to the {111}$_{In,Ga}$ and (001) planes. This procedure leads to a very heavy degradation described previously. The tips of CDLs "drawn" in such fashion appear very bright in the real-time luminescence images, consistent with the scattering or reflection of light at small regions of molten semiconductor. The presence of such melting was

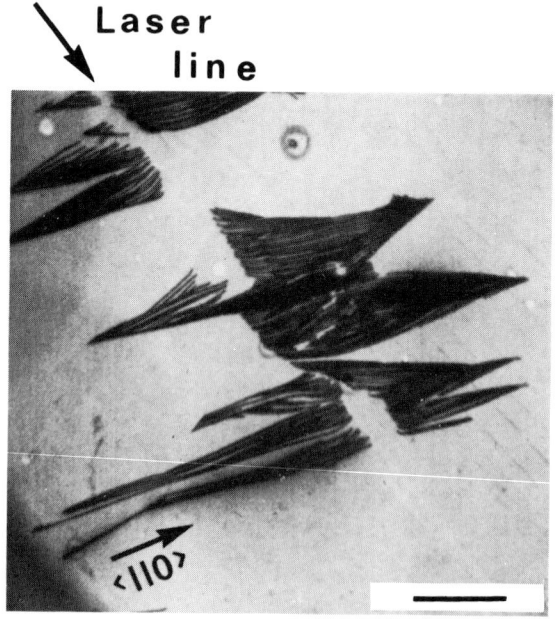

Figure 1.   Transmission cathodoluminescence image of a typical CDL pattern. Direction of the laser beam motion is indicated by an arrow, black bar represents 100μm.

confirmed by the optical Nomarski microscopy of the exposed active layers in which CDLs have been produced.

The macroscopic structure of CDLs was further investigated using transmission cathodoluminescence. A lightly degraded region showing the characteristic damage pattern is shown in Figure 1. The laser beam was moved, by translating the sample, in the <110> direction common to the $\{111\}_{As,P}$ and (001) planes, as indicated by the arrow on Figure 1. The CDLs did not propagate continuously in this direction. Instead a number of CDLs originating at random outside the laser illuminated area and propagating in the orthogonal <110> direction towards the laser beam was observed. Once these CDLs converged towards the laser beam, the radiative efficiency of the material dropped down and the degradation process was self-terminated. While the CDL patterns of Figure 1 show a general crystallographic directionality, their fan-like shapes and detailed structure give a distinctly non-crystallographic impression. This appears different from the CDL behavior in GaAlAs where a clear <110> direction has been observed by Henry et al. Furthermore, the CDL propagation velocity in InGaAsP is of the order of 200-400μm/s, considerably slower than that in GaAlAs.

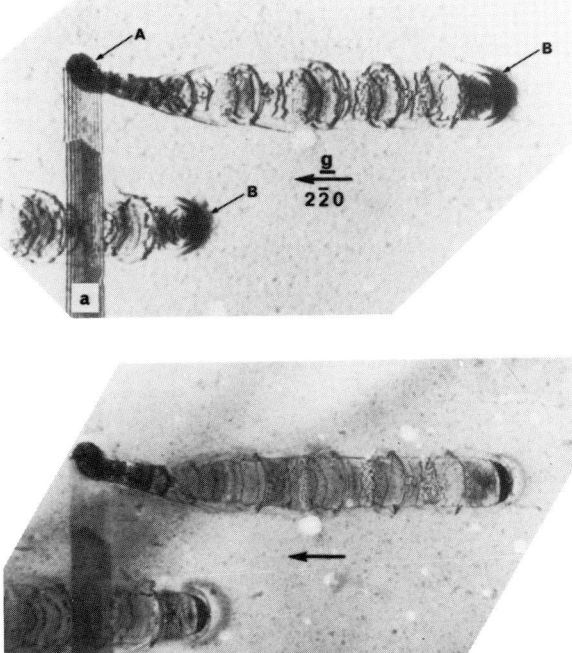

Figure 2.   Transmission electron micrograph showing (a) two short CDLs oriented along the [1$\bar{1}$0] direction and overlapping stacking faults. Origin and termination points are labeled A and B respectively. (b) same region under quasi-kinematic conditions. Black bar denotes 1μm.

Similar results have been obtained over a range of doping densities and active layer compositions ranging from 1.1μm to the 1.66μm In$_{0.53}$Ga$_{0.47}$As ternary layers. Since the CDL damage occurs through light generated within the semiconductor, it can occur only in very high quality, high luminescence efficiency material.

Some of the microstructural features observed in optically degraded samples by transmission electron microscopy are reproduced in Figures 2-4. Figure 2(a) shows two short CDLs oriented along the [110] direction and overlapping stacking faults. The upper defect line originates from a region marked A and terminates in the form of a hemispherical cap at B; the lower CDL terminates in a similar fashion. The propagation rate in the [110] direction is considerably higher than their widening rate in the [110] direction. The repetitive pattern shown on a microscopic scale by all the defect lines is attributed to periodic melting and solidification due to the pulsed nature of the laser beam. This substructural pattern consists of colonies of loops and bowed out dislocations with a repeat distance of ~1μm. In addition, when the crystal is tilted away from the strong two-beam diffraction situation and examined under quasi-kinematical condition, features similar to the ones shown in Figure 2(b) are observed. This suggests that the dislocations and loops are heavily decorated with impurities. The absorption effects are quite strong at regions A and B of Fig 2. The presence of black-greyish contrast at B implies that strain is associated with the CDL terminations.

The structure of a fully developed, wide, CDL shown in Figure 3 is quite complex. Three main structural components can be identified:

(i)   features identified by C; aligned along the [1$\bar{1}$0] direction.

(ii)   features identified by D exhibit light grey contrast and are aligned along the [110] direction; they do not exhibit lines-of-no contrast;

(iii)   jagged dislocation dipoles, identified by E, and loops aligned along the [110] direction.

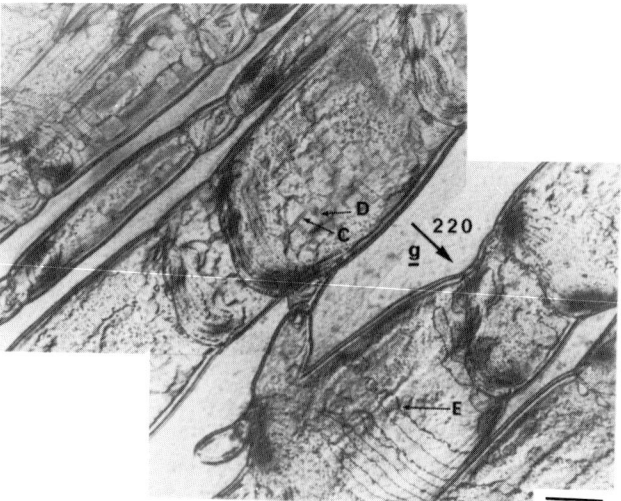

Figure 3.   Transmission electron micrograph of a fully developed CDL.

The contrast of features C and D indicate that their composition differs from the adjoining matrix, consistent with the melting-regrowth propagation mechanism. Again, as in the case of CDLs in Figure 2, these CDLs are parallel to the (001) plane.

In majority of cases, CDLs were observed to propagate along only one of the <110> directions which lie in the (001) plane. However, occasionally dark lines oriented along the [010] direction were observed to originate from the <110> oriented defect lines. A typical example is shown in Figure 4. Again, these CDLs show a repetitive substructural pattern whose repeat distance is ∼1.5μm.

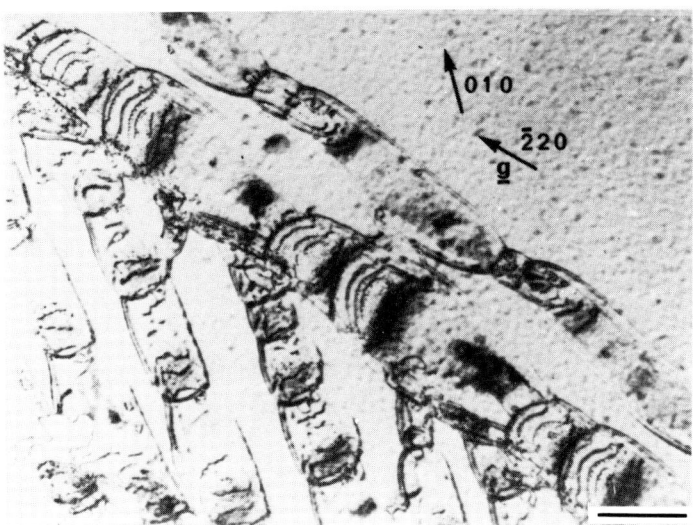

Figure 4.   Short segment of CDLs oriented along te [010] direction. These lines are rare and invariably originate at the <110> type CDLs. Black bar denotes 1μm.

3. Discussion

It is envisioned that these microstructures result from the migration of liquid droplets through the active InGaAsP layer. This appears similar to the temperature-gradient zone melting first proposed by Pfann and investigated in detail by Wernick and Cline and Anthony. The liquid droplets form because at the very high radiative fluxes generated by optical pumping localized melting may occur at defect regions, similar to the one shown at region A of Figure 2(a). The exact nature of such defects is not understood. It is known however that these must be regions of high non-radiative recombination or highly absorptive to the 1.3μm light. These could be regions of slightly altered composition perhaps related to the small In-inclusions and the associated strain fields present in the epitaxial layers. The migration of the droplets is accompanied by the dissolution of the layer and is followed by very rapid solidification with the resulting high defect density in the regrown material. Such droplets would tend to propagate along the <110> direction common to the $\{111\}_{In,Ga}$ and (001) planes since the dissolution rate of the $\{1\bar{1}1\}_{P,As}$ planes is higher than that of the equivalent In,Ga planes. This also results in the well known elongated thermal decomposition pits on InP reported by Lum and Clawson. Under the influence of the temperature gradient set-up in the process of optical pumping these droplets propagate through the InGaAsP layer towards the high temperature region. i.e. the area of the laser beam.

References

Cline H. E. and Anthony, T. R. 1976 J. Appl. Phys. *47* 2325.

Henry C. H., Petroff P. M., Logan R. A., and Merritt F. R. 1979 J. Appl. Phys. *50* 3721

Lum W. Y. and Clawson A. R. 1979 J. Appl Phys. *50* 5296.

Komiya S., Yamaguchi A., Umebu I., Kotani T., 1983 J. Appl. Phys., *54* 1058.

Mahajan S., Johnston Jr. W. D., Pollack M. A., and Nahory R. E. 1979 Appl. Phys. Lett., *34* 717.

Pfann, W. G. 1955 Trans. AIME, *203* 961.

Temkin H., Mahajan S., DiGiuseppe M. A., Dentai A. G. 1982 Appl. Phys. Lett., *40* 562.

Inst. Phys. Conf. Ser. No. 67: Section 5
Paper presented at Microsc. Semicond. Mater. Conf., Oxford, 21–23 March 1983

Microscopic examination of the deep donor EL2 in undoped semi-insulating GaAs

D J Stirland*, I Grant**, M R Brozel*** and R M Ware**

*   Plessey Research (Caswell) Limited, Allen Clark Research Centre,
        Caswell, Towcester, Northants NN12 8EQ, U.K.

**   Cambridge Instruments, Rustat Road, Cambridge CB1 3QH, U.K.

***   Dept. of Electrical and Electronic Engineering, Trent Polytechnic,
        Burton Street, Nottingham NG1 4BV, U.K.

Abstract   A high spatial resolution infra-red absorption technique
has been used to examine distributions of the principal deep donor
EL2 in undoped semi-insulating GaAs. Fine structure has been found
superimposed on characteristic "W" and "U" shaped variations in [EL2]
across <110>diameters of (001) specimens, and in {110} sections using
a CCTV vidicon system to detect transmitted infra-red radiation. The
fine structure has been correlated with bands of localised dislocations. The origin of these dislocation arrays is considered, and
reasons for correlation between [EL2] and dislocations is discussed.

1. Introduction

Crucial requirements for the current developments of GaAs integrated circuit
(IC) technology are an understanding and control of the properties of the
liquid encapsulated Czochralski (LEC) grown semi-insulating substrate material used for the ICs. Recently, it has been demonstrated (Holmes et al
1982) that the semi-insulating behaviour of undoped GaAs grown from polycrystalline boron nitride (PBN) crucibles is determined by the presence of
deep level defects, particularly the deep donor EL2, which compensates the
principal residual carbon (acceptor) impurity. Further, it has been shown
that the concentration of this principal deep donor,[EL2],can be controlled
by adjustments to the melt stoichiometry (Holmes et al 1982, Hobgood et al
1982). These advances are based on the important earlier studies carried
out by Martin and others (Martin 1981, 1982 and Martin et al 1977, 1981) on
the characterisation and estimation of EL2 in GaAs. Essentially two different methods were used. In low resistivity (n ~ 10^{16} cm^{-3}) material DLTS was
used to detect EL2 and determine concentrations in both bulk grown and VPE
GaAs. Subsequently, Martin (1981) showed that the EL2 level gave rise to a
specific absorption band in the near infra-red (~ 1 μm) in undoped semi-
insulating GaAs. By use of a capacitance technique (Huber et al 1979) to
measure [EL2] in low resistivity undoped GaAs Martin (1981) was able to cali-
brate the absorption coefficient α at different wavelengths as a function
of [EL2]. Spatial distributions of EL2 across substrates could then be
measured, and Martin et al (1981) demonstrated that variations in [EL2]
across <110>diameters of (001) wafers exhibited "W" or "U" shapes, which
matched variations in dislocation densities across the same regions.

We have developed (Brozel et al 1983) a direct observation technique based on the absorption property of EL2 at 1 μm, which has demonstrated that fine structure fluctuations occur in [EL2] variations across wafers, and are superimposed on the characteristic "W" and "U" shapes. The use of etching procedures has shown that these fine structures can be directly related to similar fluctuations in dislocation density distributions.

2. Experimental

(001) and (110) slices of 3 mm thickness cut from near the seed end of 2" diameter undoped LEC semi-insulating GaAs ingots supplied by Cambridge Instruments and Wacker-Chemitronic were prepared with double polished surfaces. The location of (110) specimens is shown in Fig.1. Preliminary examinations had shown that 3 mm thick specimens were most suitable for the infra-red absorption measurements. The infra-red absorption scanning system employed has been outlined previously (Brozel et al 1983). Briefly, it employs a Pye-Unicam SP700 double beam spectrophotometer, which produces from its monochromator a narrow illuminated area 5 mm high by $\sim$ 100 μm wide (for $\lambda \sim$ 1 μm) at the specimen surface. Absorption scans were performed by mounting specimens normal to the beam, through which they could be moved horizontally at $\sim$ 1cm min^{-1} on a simple trolley mechanism. Different scan paths could be selected by a specimen height adjustment, using visible wavelength light at the illuminated area. Absorption curves were recorded on a chart recorder, at λ = 1 μm and 2 μm. Semi-insulating GaAs is transparent at 2 μm, so that this scan provided a base-line for absorption coefficient measurements at 1 μm. (Scratches and superficial blemishes give 'absorption' peaks on this base-line, which can be subtracted from the 1 μm trace) [EL2] values were then calculated using the Martin et al (1981) calibration.

In addition to this quantitative assessment of [EL2] variations across different regions of the ingots, inhomogeneities (fine structure) in the near infra-red absorption were studied by infra-red microscopy. Transmission images of the GaAs specimens were taken using an infra-red sensitive silicon vidicon camera equipped with a 75 mm f/1.8 lens and displayed on a monitor. Light in the wavelength range from the GaAs bandgap of 0.88 μm to the vidicon cutoff at 1.1 μm was detected, but insertion of a narrow band pass (1065 ± 25nm) multilayer filter between the infra-red source and the sample gave the same infra-red images, thus demonstrating that they are qualitatively comparable with the 1 μm absorption scans.

Some specimen surfaces were etched to reveal dislocation distributions. Although the molten KOH etch (Grabmaier and Watson 1969) can be used for (001) GaAs it does not attack (110). However the Abrahams and Buiocchi (1965) A/B etch attacks both (001) and (110) surfaces in a similar manner, producing ridges at dislocation lines as previously described (Stirland 1977).

3. Results

We have previously described (Brozel et al 1983) the occurrence of fine structure, consisting of rapid amplitude fluctuations, superimposed on the characteristic "W" shaped variation of [EL2] across a <110> diameter of an (001) specimen when examined by the absorption of a scanned 1 μm wavelength infra-red beam. Similar fine structure has also been observed in scans of (110) surface specimens, across [$\overline{1}$10] directions normal to [001]. This orientation has the advantage that variations across <110> diameters can be made at different positions down the [001] axis of the ingot, under otherwise identical conditions.

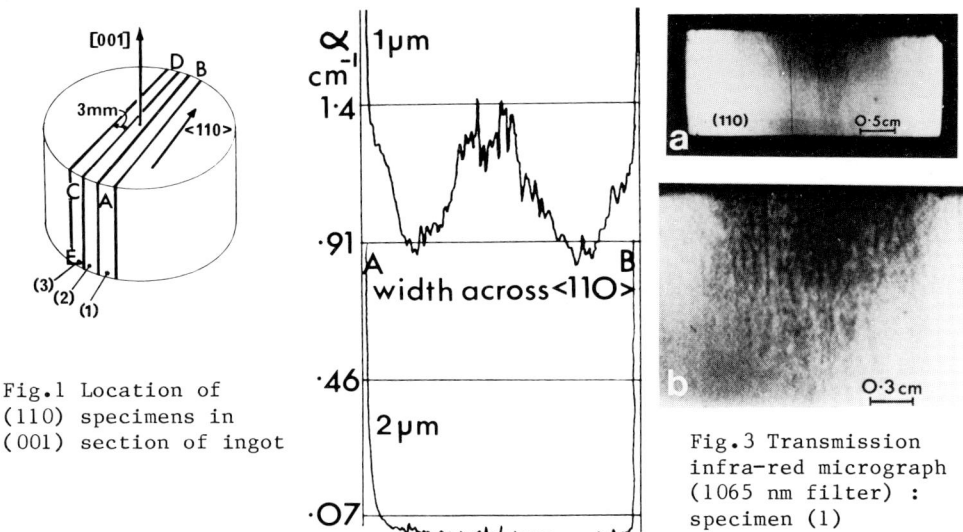

Fig.1 Location of
(110) specimens in
(001) section of ingot

Fig.2 Infra-red absorption scans at
1 μm and 2 μm : specimen (1)

Fig.3 Transmission
infra-red micrograph
(1065 nm filter) :
specimen (1)

The 1 μm scan in Fig.2 represents the variations in absorption across a
5 mm wide strip on the upper part (AB) of specimen (1), Fig.1. The scan at
2 μm is the base-line from which the absorption coefficient α was determined and [EL2] calculated. Strictly, [EL2] is given by the difference
between 1 μm and 2 μm values, but the error introduced by omitting this
correction is negligible, and on subsequent scans the base-line is not
shown. Fig.3(a) shows specimen (1) as viewed by the silicon vidicon camera.
It is evident that the central upper region exhibits greater absorption
than the remainder, and at higher resolution Fig.3(b) the region is resolved
into "streamers" of increased absorption. The 1 μm absorption scan of
Fig.2 reveals considerable fine structure in the equivalent region. Fig.4
shows 1 μm absorption scans, converted to [EL2] using the Martin (1981)

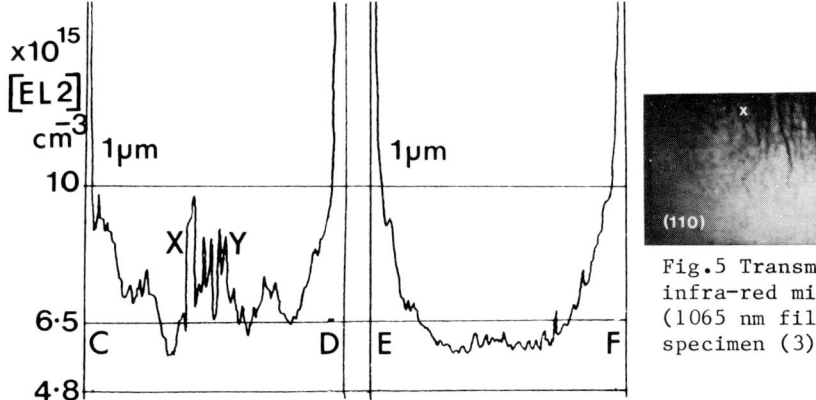

Fig.4 1 μm absorption scans across upper (CD)
and lower (EF) regions of specimen (3)

Fig.5 Transmission
infra-red micrograph
(1065 nm filter) :
specimen (3)

calibration, for upper and lower regions of specimen (3). These scans illustrate the characteristic "W" and "U" shaped distributions previously reported by Martin et al (1981) and Martin (1982) for different specimens, and it is believed that this is the first reported observation in which the same specimen exhibits the two types of distribution. The infra-red micrograph of Fig.5 indicates the reason for the difference between upper and lower regions. The highly absorbing streamers are only present in the upper region.

In order to confirm the clear indication that absorption peaks on 1 μm scans correspond with the streamers on infra-red micrographs, densitometer scans across the upper region of a negative transparency of Fig.5 were made. This specimen was more suitable than specimen (1), because of the smaller number of streamers. Measurements of peak positions are given in Table 1 (see below).

Finally, specimen (3) was A/B etched to delineate the dislocation distribution at the (110) surface. It was found that the pre-etch chemical polish (3:1:1; H_2SO_4: H_2O_2 : H_2O, Iida and Ito 1971) often employed to remove work damage from (001) GaAs does not attack (110) GaAs. Thus some background scratches are evident on the mosaic micrograph of Fig.6, which corresponds to the region marked XY on Figs.4 and 5.

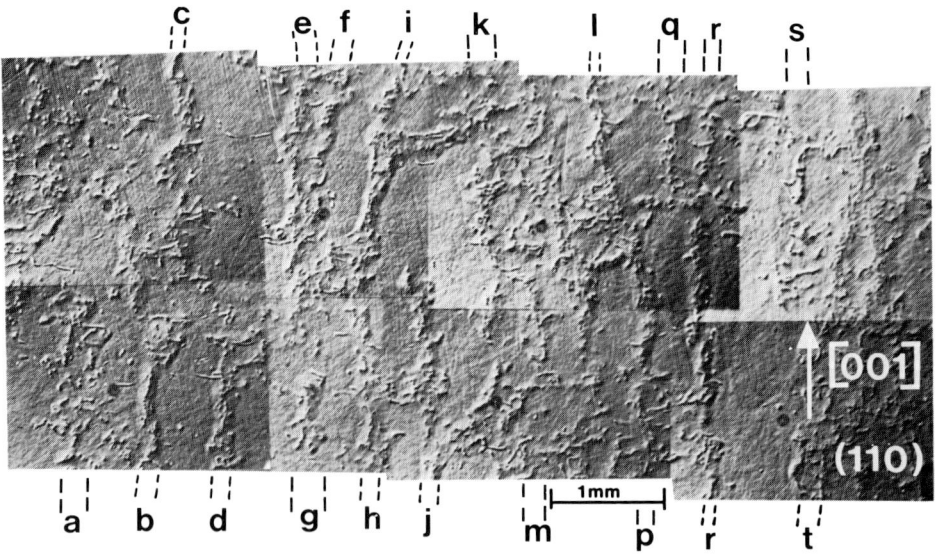

Fig.6   A/B etched surface; region of XY on Figs. 4 and 5. The locations of the lettered dislocation groups are shown in Table 1.

It is characterised by the irregular bunching of dislocations into streamers lying approximately along <001>, of width ~ 100 μm separated by ~ 500 μm wide dislocation - free areas. Measurements of the distances of the principal streamers from a marker (accidental scratch) which could also be seen on the transmission infra-red micrograph allowed comparison of the positions of infra-red streamers with dislocation streamers, as shown in Table 1. The agreement is excellent. The area of Fig.6 shows only the region between XY; across the remainder of the (110) surface the appearance of the dislocation distribution was similar to the cellular structure previously described for dislocation arrays on (001) A/B etched surfaces (Clark and Stirland 1981). Figure 7 shows an example of cellular structure on a 2" diameter (001) wafer after a molten KOH etch.

4. Discussion

It is somewhat surprising that such close agreement between the 1 μm absorption peaks and the dislocation bunches should exist, because the former is produced by a 3 mm thick specimen, and the latter by ~ 10 μm of surface. The implication is that the majority of streamers extend continuously in < 110> directions through at least several millimetres. This view is supported by comparing the infra-red micrographs of Figs. 3 and 5 which still show qualitative agreement between the streamer positions, even though the specimens were separated by 3 mm (Fig.1).

What are the streamers? We believe that these features have already been detected in (001) A/B and KOH etched specimens, such as that shown in Fig.7. In addition to the overall cellular structure, closely spaced etch pits form continuous lines several millimetres long oriented along <110> . Similar lines have recently been described by Chen and Holmes (1983) and interpreted as arising from polygonisation of dislocations into walls defined by {110} planes normal to {111} slip planes, which intersect (001) along <110> . This explanation does not account for the streamers seen in Figs.3, 5 and 6 which lie essentially in {110} sheets normal to the (001) ingot surface. However, the orientation of individual dislocations within the streamers has not yet been established, and it is clear that TEM examinations are required to determine the types and orientations of these dislocations. This work is currently in progress.

TABLE 1 - Correlation between infra-red absorption peaks and dislocation cluster positions. Numbers are distances in millimetres from a marker.

*D peaks	1 μm peaks	Dislocation streamers	
17.0	17.0	17.0	a
	16.5	16.5	b
15.9	16.1	16.0	c
15.4	15.4	15.6	d
15.0	15.0	15.0	e
		14.8	g
	14.4	14.6	f
	14.2	14.3	h
		14.1	i
13.6	13.6	13.7	j
13.1	13.2	13.1	k
	12.8	12.7	n
12.4		12.3	m
	11.9	12.1	l
11.7	11.7	11.8	p
11.6		11.5	q
11.3	11.2	11.2	r
10.8	10.8		-
10.2	10.2	10.3	s
	9.8	10.0	t

*D = Densitometer

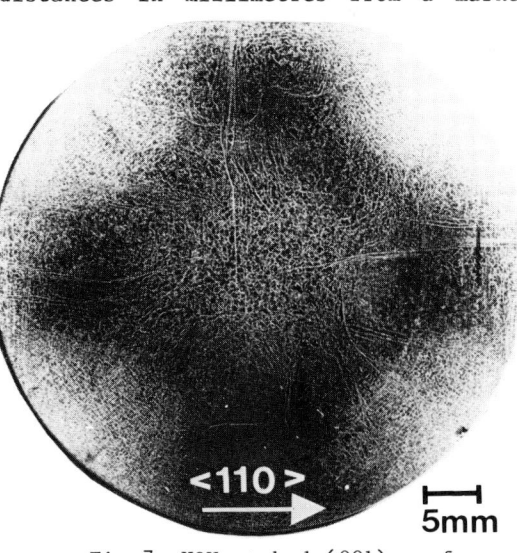

Fig.7 KOH etched (001) surface

The overall association between [EL2] variations and dislocation density variations has been reported by Martin et al (1981), and the present work has extended the correlation down to a microscopic level. Martin et al (1981) also found that the absolute values of [EL2] could not be correlated with dislocation densities, and that [EL2] might either increase, decrease or remain constant from the seed to the tail end of ingot. Thus any explanation of the correlation based on the attraction of the strain fields surrounding dislocations for impurities in the matrix seems unlikely, since the EL2 centre does not appear to behave as a simple substitutional impurity. An alternative explanation is that the EL2 centre is generated in the vicinity of dislocations. A mechanism has been proposed by Weber et al (1982) and independently by Stirland (1982) by which As_{Ga} anti-site defects could be generated at dislocations by climb. The climb process is considered to occur in the presence of a supersaturation of As, and negative climb results in the creation of a V_{Ga} for each absorbed As interstitial. Further As interstitials then interact with V_{Ga} to form As_{Ga} anti-site defects. Support for this model is provided by the observation that As precipitates occur on line dislocations in bulk grown GaAs (Cullis et al 1980). Thus if EL2 can be equated with As_{Ga} (Lagowski et al 1982) this might provide a plausible explanation for the association between [EL2] and dislocation density, and also the results of Martin et al (1981) since the concentration of EL2 might be controlled in some such manner as :

[EL2] $\propto$ [dislocation density] x [amount of dislocation climb]

However this speculative conclusion must await further experimental evidence, including a definitive identification of EL2 with As_{Ga}.

Acknowledgement. This work has been carried out with the partial support of Procurement Executive, Ministry of Defence, sponsored by DCVD.

References

Abrahams M S and Buiocchi C J 1965 J. Appl. Phys. 36 2855
Brozel M R, Grant I, Ware R M and Stirland D J 1983 Appl. Phys. Lett to be published
Chen R T and Holmes D E 1983 J. Crystal Growth 61 111
Clark S and Stirland D J 1981 Micros. Semicond. Mater. Conf. Oxford 1981 Inst. Phys. Conf. Ser. 60 p339
Cullis A G, Augustus P D and Stirland D J 1980 J. Appl. Phys. 51 2556
Grabmaier J B and Watson C B 1969 phys. stat. sol. 32 K13
Hobgood H M, Ta L A, Rohatgi A, Eldridge G W and Thomas R N 1982 Semi-insulating III-V Materials, Evian (Shiva Publishing Ltd) p28
Holmes D E, Elliott K R, Chen R T and Kirkpatrick C G 1982 Semi-insulating III-V Materials, Evian (Shiva Publishing Ltd) p19
Huber A M, Linh N T, Valladon M, Debrun J L, Martin G M, Mitonneau A and Mircea A 1979 J. Appl. Phys. 50 4022
Iida S and Ito K 1971 J. Electrochem. Soc. 118 768
Lagowski J, Gatos H C, Parsey J M, Wada K, Kaminska M and Walukiewicz W 1982. Appl. Phys. Lett.
Martin G M, Mitonneau A and Mircea A 1977 Elect. Lett. 13 191
Martin G M, Jacob G, Poiblaud G, Goltzene A, Schwab C 1981 Defects and Radiation Effects in Semiconductors, Oiso 1980. Inst. Phys. Conf. Ser.59 p281
Martin G M 1981 Appl. Phys. Lett. 39 747
Martin G M 1982 Physica Scripta T1 38
Stirland D J 1977 GaAs and Related Compounds Edinburgh 1976 Inst. Conf. Ser. 33a p150
Stirland D J 1982 Presented at BACG Conf. Sept. 1982 Oxford (unpublished)
Weber E R, Ennen H, Kaufmann U, Windschief J, Schneider J and Wosinski T 1982 J. Appl. Phys. 53 6140.

Inst. Phys. Conf. Ser. No. 67: Section 5
Paper presented at Microsc. Semicond. Mater. Conf., Oxford, 21–23 March 1983

291

Chemical and structural TEM of defects in doped GaAs

Y Kouh and C B Carter

Dept. of Materials Science and Engineering, Bard Hall,
Cornell University, Ithaca, NY 14853

Abstract   Two different types of defect have been observed in this
study of heavily-doped MBE-grown GaAs. The first of these consists
of dislocation clusters of varying dislocation density which appear
to originate at growth defects in the MBE layer. Such defects are
similar in appearance to those observed in material grown by other
techniques. The second defect-type is a second-phase particle which
appears to have been present as a liquid during the growth process
and would significantly influence the incorporation of dopant in the
epitactic layer.

1. Introduction

The incorporation of high concentrations of n-type dopants such as Si, Ge,
and Sn in GaAs has been found to be limited for MBE-grown material by the
formation of defect clusters (Carter et al 1982a,b) or by segregation of
dopant to the surface (Harris et al 1982). Surface segregation has also
been found to occur for p-type Mn-doped material; in the case of Mn-doping,
the critical concentration appears to be somewhat lower ($\lesssim 5 \times 10^{18} cm^{-3}$) than
for n-type doping and internal defects appear to be rare. The composition
of the new second-phase is Ga-rich for Mn-doped material, but appears to
be As-rich or pure Sn, for Sn-doped material. Observations on the composition of these new phases are important not only because they explain the
failure to grow heavily-doped material, but also because surface enrichment, particularly of Mn, can occur during the heat-treatment of substrate
material which is necessary for the epitactic growth process (Palmateer
1982, Woodall 1982). Such process-induced surface enrichment may also lead
to unintentional impurity incorporation in the epitactic layer. The formation of defect clusters which can be associated with highly-localized,
strain centers is clearly detrimental to any subsequent use of that material. The clusters discussed in this paper are strikingly similar to those
which have been observed in a variety of doped III-V compounds. The origin
of these clusters has been attributed either to surface damage caused by
handling (Kotani et al 1977) or to the presence of second-phase particles
in the material which may punch out the dislocations (Wagner 1981, Mahajan
and Chin 1981, Augustus and Stirland 1982). There has previously been no
direct identification of the particles responsible for the defect clusters
although evidence for their presence has been reported (eg Mahajan and
Chin 1981).

2. Experimental Details

The material examined in this study was grown using a Varian MBE 360

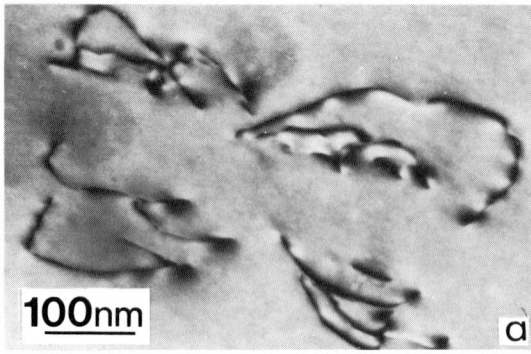

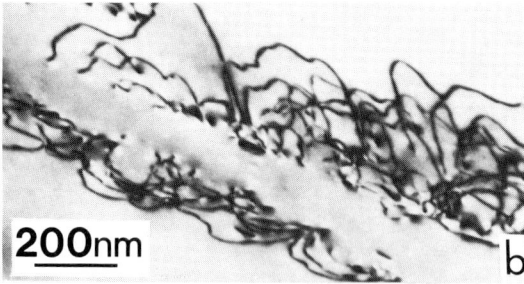

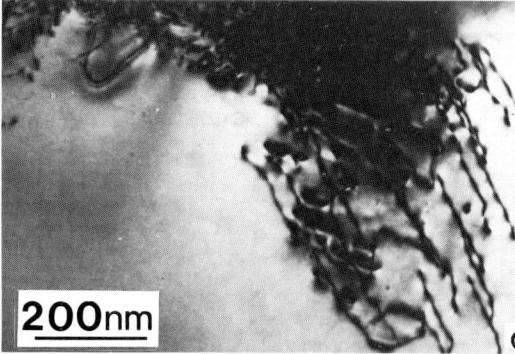

Fig 1 Defect clusters in n-type GaAs

system. The molecular-beam sources were Knudsen effusion cells in resistance-wound Ta-shielded furnaces. The source material was held in graphite or pyrolytic boron nitride crucibles; the latter was necessary for the more reactive Mn in order to reduce incorporation of oxygen and carbon in the layers. The substrate temperature was optimized for the layer in the range 520°C to 620°C. The initial surface of the GaAs substrate was cleaned to remove the oxide surface layer by heating at or above ∿550°C under an arsenic flux. Details of the MBE process are given in the review by Wood (1980) and the electrical characterization of the layers is also discussed elsewhere (DeSimone 1981).

Samples were prepared for examination in the transmission electron microscope (TEM) by etching with a bromine-methanol solution in the usual manner. Jet-polishing was carried out from one side only so that the final growth surface could be examined. The structure analysis was performed using a Siemens 102TEM operating at 125kV while the chemical analysis was carried out using a JEM200CX Temscan operating at 200kV and equipped with a Tracor-Northern 70°- take-off angle detector for energy dispersive x-ray spectroscopy (EDS).

3. Observations and Discussion

The appearance of the defect clusters in the Si-, Ge-, and Sn-doped material was essentially the same and ranged from small groups of dislocations such as shown in fig 1a, to more complex arrays (fig 1b) and finally to the type of tangle shown in fig 1c. The process giving rise to the defect clusters in each of these materials is, therefore, presumably the same. Lattice dislocations were rare except in the vicinity of these defect clusters.

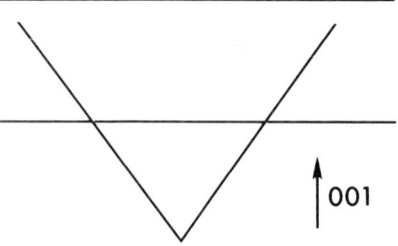

Fig 2 Diagram showing origin of clusters in figs 1a and b

Defect clusters similar to those shown in figs 1a and 1b have been
observed in LPE-grown material and might at first be interpreted as ariṣing
from surface damage (Kotani et al 1977). The actual cause in the present
study can, however, be appreciated from the schematic shown in fig 2: the
two sets of dislocations which bow out in opposite directions originated
at the same source which has been removed by the chemical polish. By pro-
jecting the traces back into the removed material, it is found that in
this case the source would have been a distance of ∿70nm below the lower
surface; i.e. the source was an inclusion, or group of inclusions in the
interior of the epilayer: it was neither at the final surface nor at the
original substrate surface. Such observations thus support the interpretation of Wagner (1981) that the source is a second-phase particle.

Several additional features of interest are present in fig 1: the bowing
dislocations are apparently not decorated, but their high curvature indicates relatively large pinning stresses. The dislocations close to regions
such as that shown in fig 1c actually move readily under the influence of
the electron beam (Carter et al 1982b). Fig 1b shows particularly clearly
that several sources must have been present during growth and that these
sources were: i) very nearly at the same depth, and ii) aligned along a
<110> direction. A possible example of such a source is shown in fig 3.
The individual defects at first appear similar to dislocations, but it is
soon apparent that they are actually highly localized strain centers.
However, in this example the defects lie approximately along a <100>
direction. The defect cluster shown in fig 1c is similar to those observed in Sn-doped InP ingots by Augustus and Stirland (1982). These authors

associate these defect regions with the
'grappes' which have been detected in
other InP substrate material. Augustus
and Stirland report that the dislocations
emitted from such grappes in InP have the
form of dislocation loops predominantly
on {110} planes. The dislocations
observed in fig 1 lie predominantly on
{111} planes.

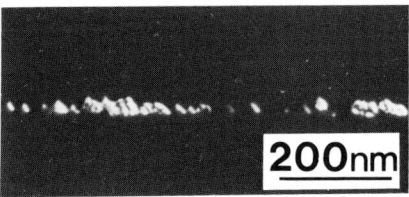

Fig 3 Row of defects in Si-doped
MBE-GaAs

Figure 4 illustrates the appearance of
the second-phase particles on the surface
of Mn-doped GaAs. The image is clearly
similar to that reported by Harris et al
(1982) (see their fig 1) for Sn-doped
material. The term Sn-doped is used here
to include material where the 'dopant'
has been deposited on top of a GaAs substrate. Figure 5a shows one of the particles in Mn-doped material which partly
overlaps the GaAs substrate. The geometry
of the particle is as shown schematically
in fig 5b: the particle is partly inside
the GaAs not simply on top of the substrate
as deduced for the Sn-doped material. The
nature of these second-phase particles has
been examined extensively using EDS,
electron diffraction and high-resolution
TEM. The results of this analysis will
now be briefly illustrated and full
details will appear elsewhere. In all

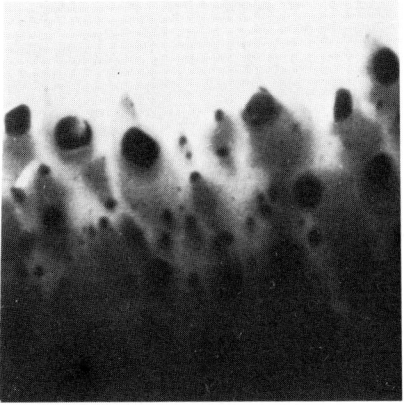

Fig 4 Distribution of surface
particles in Mn-doped MBE-GaAs

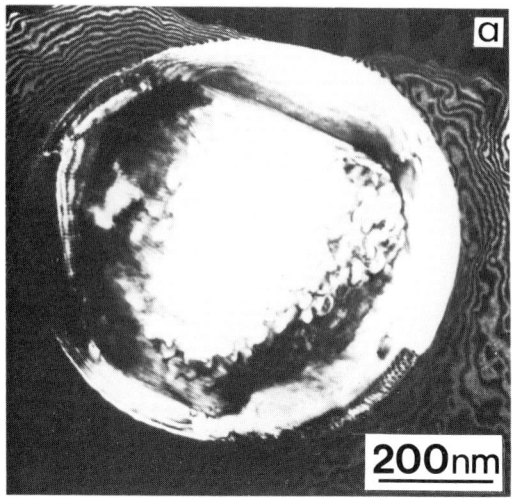

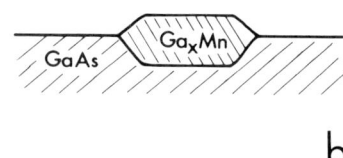

Fig 5 a) Image and b) side-on schematic of a second-phase particle on the surface of Mn-doped MBE-GaAs

cases where material which did not overlap the substrate has been examined it has been found to contain only Ga and Mn. Further work is in progress to test for the presence of light elements such as oxygen and carbon, but these are not thought to be present in significant quantity (DeSimone 1981). EDS spectra taken from two different particles are shown in fig 6 and clearly show a difference in composition. The K-factors relating Ga, As, and Mn have been both calculated and determined experimentally using MnS, As_2S_3, and GaAs as standards. Absorption effects have also been investigated and although they do affect the relative peak counts the effect is small for the geometry and specimen thickness in the present study.

The phase diagram is not well-defined for the Ga-Mn system although certain Ga-rich phases have been reported. In the present study, several phases

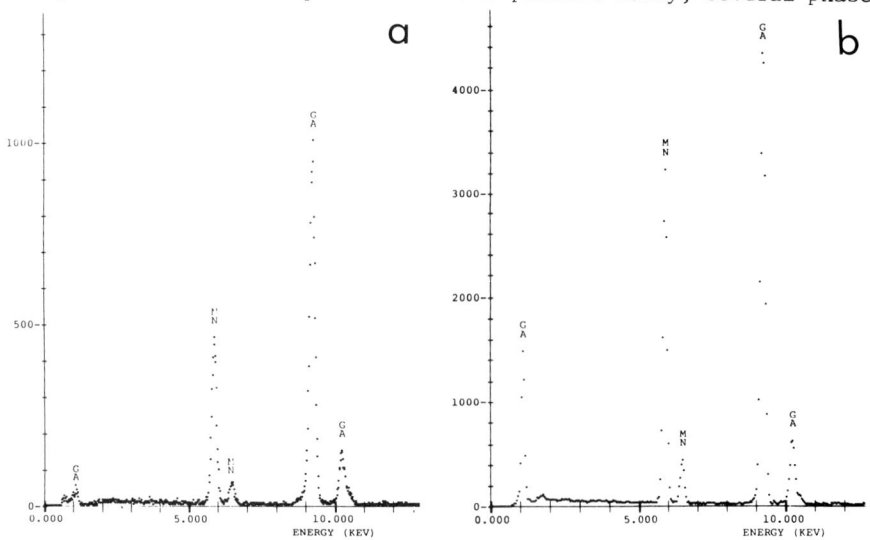

Fig 6 EDS spectra from a) the GaAs matrix and b) a second-phase particle

have been identified including the tetragonal phase Mn_2Ga_5. In many cases there is a well-defined topotactic relationship between the second-phase particle and the GaAs substrate and an example of a diffraction pattern from such a particle is shown in fig 7; the spot pattern of the new phase shows a 4-fold symmetry and does not appear to correspond to a previously identified phase. The diffraction pattern and high resolution image shown in fig 8 are taken from a particle which has one particularly large d-spacing, but also does not correspond to a known phase. The structure as revealed by the high resolution image appears to contain twin planes which are parallel to the 111 planes of the GaAs epilayer.

The growth surface of this Mn-doped material was not flat, but did not show features which would suggest the occurrence of Ga-spitting during MBE growth. Examination of the phase diagram suggests that the Ga-rich compounds will be low melting-point eutectics and it is thus proposed that once a Ga-Mn phase has nucleated, it will continue to exist as a liquid during growth and will, in fact, continue to grow. The separation and size of the islands in fig 4 are also presumably directly associated with the nucleation and growth process. When growth is concluded, the liquid will cool and solidify, i.e. the solidification begins at the GaAs-Ga_xMn interface; a topotactic relationship which minimizes interfacial stresses will be favored.

Fig 7 Diffraction pattern from topotactically aligned particle

4. Conclusions

The phases containing Ga and Mn which are found on the surface at MBE-grown GaAs which was heavily-doped with Mn are

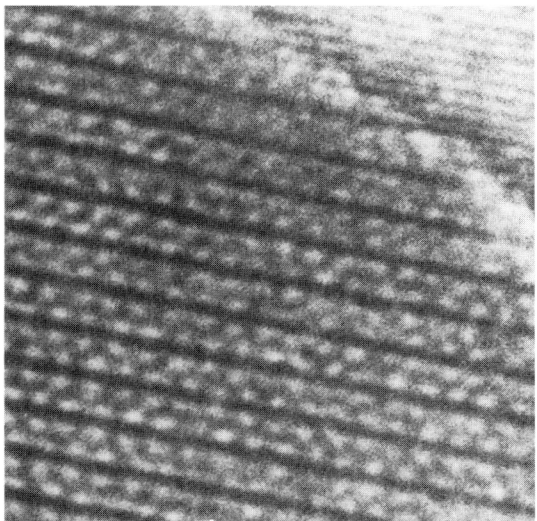

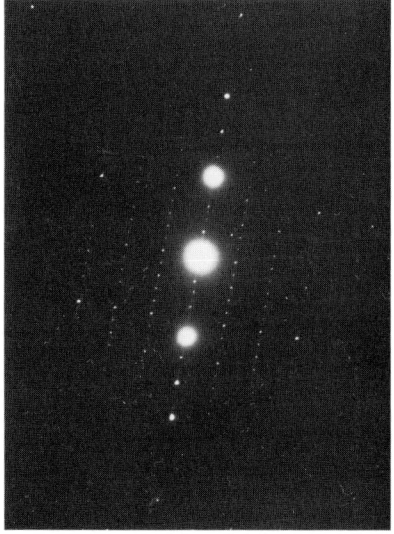

Fig 8 High-resolution image and diffraction pattern from Ga_xMn

proposed to occur because of the accidental nucleation and subsequent
growth of islands of liquid Ga_xMn eutectics which float on the surface during growth of the layer.   It is also proposed that the explanation advanced
by Wagner (1981) for the formation of defect   clusters in Ge-doped LPE-
grown GaAs and $Al_{0.4}Ga_{0.6}As$ also holds for MBE-grown GaAs doped with Si,
Sn, and Ge.   That is, that second-phase particles occur by accident during
growth and that, since these are higher melting-point (i.e. solid)
eutectics, they are then trapped inside the growing layer and become
highly-localized strain centers as the layer cools after growth.   The
observation that these strain centers may be aligned along <110> and
<100> directions suggest that they occur at ledges during MBE growth.
The case of Sn-doping is particularly interesting since the eutectics SnAs
and Sn_2As_3 have melting points which are close to the MBE growth temperatures and may thus give rise to either type of defect depending on the
actual temperature used.

Acknowledgments

The authors would like to thank Professor D.M. DeSimone and Dr. C.E.C.
Wood for providing the MBE material and for many discussions.   They also
thank H.T. Griem for specimen preparation, B.C. DeCooman and J. Hunt
for assistance with the EDS analysis and R. Coles for maintaining the
microscopes.   This research is supported by the U.S. Army Research Office.

References

Augustus P D and Stirland D J 1982 J. Electrochem Soc. 129, pp 614-621
Carter C B, DeSimone D M, Griem H T and Wood C E C 1982a Proc. 40th Ann.
   EMSA Meeting (Washington) pp 442-445
Carter C B, DeSimone D M, Griem H T and Wood C E C 1982b Proc. defects
   in Semiconductors, MRS Symposium (North-Holland) in press
DeSimone D M 1981 Ph.D. Thesis, Cornell University
Harris J J, Joyce B A, Gowers J P and Neave J H 1982 Appl. Phys. A28,
   pp 63-71
Kotani T, Ueda O, Akita K, Nishitani Y, Kusunoki T, and Ryuzan O 1977
   J. Cryst. Growth 38, pp 85-92
Mahajan S and Chin A K 1981 J. Cryst. Growth 54, pp 138-149
Palmateer S C 1982 M.S. Thesis, Cornell University
Wagner W R 1981 J. Electrochem Soc. 128, pp 2641-2644
Wood C E C 1980 Phys. Thin Films 11, p 35
Woodall J 1982 Ph.D. Thesis, Cornell University

Inst. Phys. Conf. Ser. No. 67: Section 5
Paper presented at Microsc. Semicond. Mater. Conf., Oxford, 21–23 March 1983

The structure of displacement cascades in III–V semiconductors

T J Chandler and M L Jenkins

Department of Metallurgy & Science of Materials,
University of Oxford, Parks Road, Oxford OX1 3PH

Abstract   The structures of displacement cascade regions in GaP and
GaAs created by bombardment with various heavy ions of energy 100kV and
ranging in mass from Ga (M=70) to W(M=184) have been investigated using
transmission electron microscopy. At ion doses $< 10^{16}$ ions/m^2, where
little cascade overlap occurs, heavier ions produce amorphous zones
within individual cascade regions whilst lighter ions produce no visible
damage. The amorphous zone sizes are relatively independent of ion mass.
No evidence is found for the production of regions rich in antisite
defects. Preliminary observations on the damage at higher doses are
also reported.

1. Introduction

The implantation of energetic heavy ions into both elemental and compound
semiconductors causes considerable damage and at high doses leads to the
amorphization of a region at or close to the ion entry surface. The problems caused by amorphization and the annealing processes required to restore crystallinity have attracted a great deal of attention but the microstructure of the primary radiation damage and the mechanism of development
of amorphous layers have been less widely studied. The work reported in
the literature   has concentrated largely on the elemental semiconductors
and particularly silicon. Howe et al (1980,1981) reported the observation
that in silicon ions heavier than about As (M=75) create small amorphous
zones with high efficiency and little dependence of zone size on ion mass.
Lighter ions produced little or no visible damage. Similar observations
have been made by other authors (Ruault and Jäger 1980, Pasemann and Werner
1979). It has generally been considered that these amorphous zones are
produced by a process of 'direct impact amorphization' within individual
displacement cascades (see section 4.2). No detailed studies have been
made of any III-V compound semiconductors.

Although GaP and GaAs have essentially the same crystal structure as, and
similar bonding to, Si there are factors which may lead to differences in
the primary radiation damage. The compound semiconductors have higher
average atomic masses than Si and the two component atoms have different
masses and possibly different displacement energies. There are six types
of primitive defects (Group III and V interstitials, Group III and V
vacancies, and two types of antisite defect) compared to just two in elemental semiconductors (interstitials and vacancies). Regions of high
density of antisite defects may be investigated in principle in the transmission electron microscope using reflections such as {200} which are forbidden in the Si structure. These reflections have structure factors

proportional to the difference in atomic structure factors between the Group III and Group V components and are allowed because the two components are ordered on distinct sublattices. This is analogous to the occurence of superlattice reflections in ordered alloys such as Cu_3Au, where it has been shown that displacement cascade regions may be investigated very directly by imaging 'disordered zones' created at cascade sites (Jenkins and English 1982).

This paper is primarily concerned with a transmission electron microscopy investigation of the structures of individual cascade regions in GaP and GaAs. We also report some preliminary results concerned with the development of amorphous layers at higher doses.

2. Experimental

<001> oriented specimens of GaP and GaAs were implanted with 100keV heavy ions at AERE Harwell. The ions used included Ga^+, As^+, Kr^+, Xe^+ and W^+. Low doses (10^{14}-10^{16} ions/m^2) were used for investigations of individual cascade regions; higher doses (> 10^{16} ions/m^2) are being used to investigate cascade overlap and the formation of amorphous layers. Specimens were thinned for transmission electron microscopy in a solution of chlorine in methanol, and were examined in a JEOL JEM 100B microscope operating at 100kV.

3. Results

3.1 Doses $\leqslant 10^{16}$ ions/m^2

At doses $\leqslant 10^{16}$ ions/m^2 little cascade overlap is expected to occur and any damage seen is assuumed to result from individual ion impacts. Only ions of mass greater than some threshold value - different for GaP and GaAs caused visible damage (see Table 1).

	Ion				
Target	Ga	As	Kr	Xe	W
GaP	x	✓	✓	✓	✓
GaAs	-	x	x	✓	✓

Table 1: Incidence of damage visible in the TEM for low doses of 100keV ions.

Where visible damage was found detailed contrast experiments were carried out with the following results.
   (i)   Defects imaged in dark field under dynamical two beam diffraction conditions using the fundamental reflections g = {220} and g={400} showed dark dot contrast on the thinner sides of thickness fringes and white dot contrast on the thicker sides. The contrast reverses in bright field. No pronounced sub-cascade structure has been seen. A typical dark field micrograph taken using g = 400 is shown in figure 1.
   (ii)   Some defects situated near the centre of thickness fringes showed weak black-white contrast with black-white vector l̲ antiparallel to g̲ in dark field. Some examples are arrowed in figure 1.
   (iii)   The defect yield (defined as the number of visible defects per

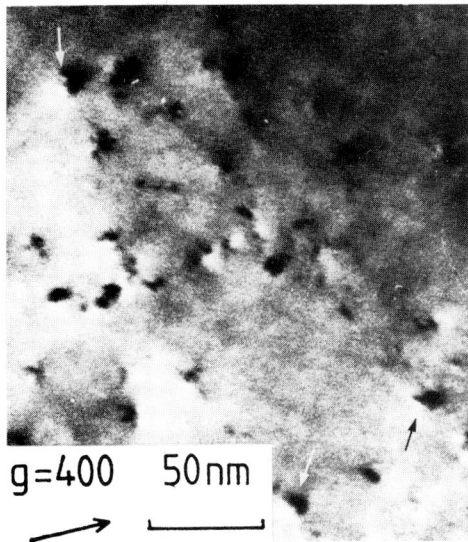

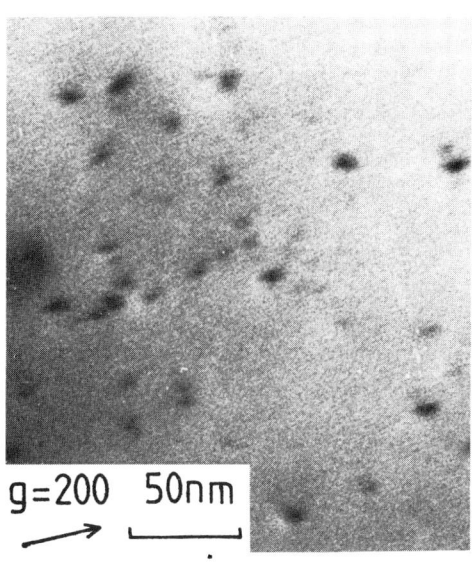

Figure 1: Typical cascades in GaP imaged in $\underline{g}$ = 400 dark field. The arrows indicate defects showing black-white contrast.
(Dose: 10^{15} 100keV Kr^+ ions/m^2)

Figure 2: The same area as figure 1, imaged in $\underline{g}$ = 200 dark field. Note that the same defects are visible.

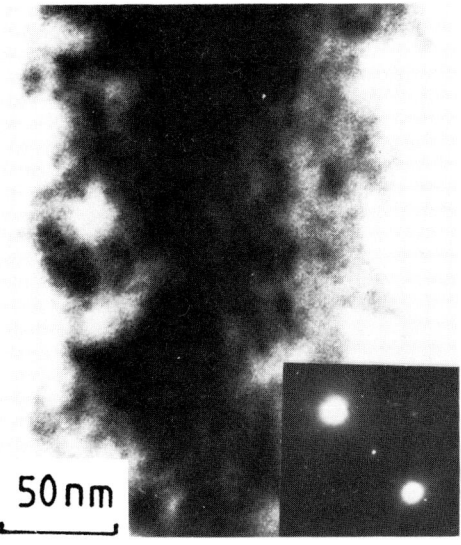

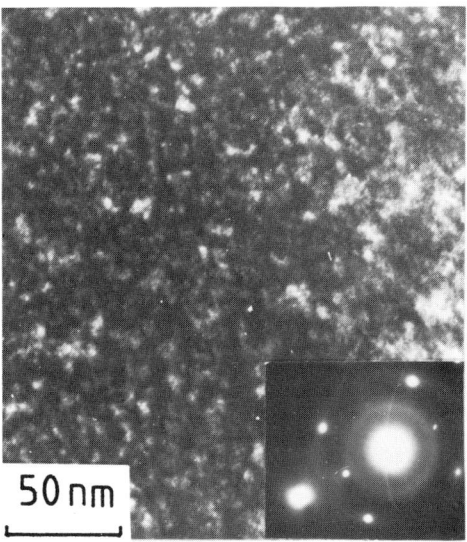

Figure 3: GaP irradiated with 10^{17} 100keV Kr^+ ions/m^2. Only weak damage contrast is visible with no effects in the diffraction pattern. ($\underline{g}$ = 400 dark field)

Figure 4: GaP irradiated with 10^{18} 100keV Kr^+ ions/m^2. Strong contrast is visible, and diffuse rings are present in the diffraction pattern. ($\underline{g}$ = 400 dark field)

incident ion) was close to unity for all ions producing visible damage.

(iv)   There was no marked variation of defect size with ion mass. Defect sizes ranged from 2nm to 8nm, averaging about 4nm.

(v)   Defects imaged in the superlattice type {200} reflections showed rather weak dark dot contrast in both dark- and bright-field images. No contrast features were found extra to those present in the fundamental reflections. Individual defects showed little change in size and shape between fundamental and superlattice reflections. Figure 2 shows the same region as in figure 1 imaged in $\underline{g}$ = 200.

(vi)   No diffraction pattern effects were found.

The defect contrast is consistent with small amorphous zones, as discussed in section 4.1 below.

3.2 Doses > 10^{16} ions/m²

At doses > 10^{16} ions/m² cascade overlap is expected to occur and further damage due to interactions between separate cascades might be expected. At very high doses, an amorphous layer will be formed in the specimen. Preliminary investigations of 100keV Kr⁺ irradiations of GaP and GaAs at 10^{17} and 10^{18} ions/m² show the following features.

(i)   After a dose of 10^{17} ions/m² the material is still essentially crystalline with rather weak damage contrast (see figure 3).

(ii)   After a dose of 10^{18} ions/m² amorphous rings have appeared in the diffraction pattern, with strong contrast present in all reflections (see figure 4).

4. Discussion

4.1 Interpretation of contrast observations

The contrast observed at the lower doses is consistent with the production of small amorphous zones at individual displacement cascades. The contrast is dominated by structure factor differences between the amorphous zones and the crystalline matrix. Thus a column of material containing an amorphous zone has a smaller effective thickness for all reflections than an adjacent column of perfect crystal. The magnitude of the contrast is expected to vary with the rate of change of intensity in the diffracted beam with foil thickness. This leads to contrast maxima on the flanks of thickness fringes and minima at the centres. This is consistent with our observations. The weak black-white contrast occasionally observed is consistent with an outwardly directed strain field probably arising from a small expansion of the cascade region as a consequence of the crystalline-amorphous transition. The lack of extra contrast features in {200} reflections indicates that regions of high density of antisite defects are probably not formed. This view is supported by the work of Werner et al (1982) who calculated from electron paramagnetic resonance measurements on neutron irradiated GaAs that only ∿ 1% of the primary defects are antisite defects. The non-reversal of contrast in {200} dark- and bright-field images is not fully understood but may arise from the small structure factor for {200} reflections compared with fundamental {400} reflections. Even with $\underline{g}$ = 200 fully excited (s_{200} = 0) the (400) reflection is of comparable intensity. Such effective three beam conditions have been found to produce anomalous contrast in Nb₃Sn (Jenkins et al 1982).

4.2 The mechanism of production of amorphous zones in cascades

Our results for GaP and GaAs bombarded with low doses of heavy ions are similar to those obtained in similar investigations in silicon (Howe et al 1980,1981). These authors have characterized the damage in terms of the average deposited energy density $\bar{\theta}_\nu$ as calculated from the WSS analytical theory of cascades (Winterbon 1970). The experiments reported here cover a range of $\bar{\theta}_\nu$ from 0.05eV/atom to 0.26eV/atom. The contrast observed agrees with the classification given by Howe et al in that we observe single contrast features with little or no subcascade structure. However, $\bar{\theta}_\nu$ does not appear to be a good predictor of the appearance of visible damage, nor does it lead to any particular mechanism for amorphous zone formation. The moderate to high energy density cascades such as we are concerned with here are complicated events which are not fully understood (Thompson 1981). Cascade effects are however frequently interpreted in one of two ways: as being due to 'thermal spikes' or 'displacement spikes'. The thermal spike mechanism depends on local retention of the energy deposited in the cascade for periods long enough for thermal or quasi-thermal effects to occur. The displacement spike mechanism - as applied to the formation of amorphous zones in semiconductors - is dependent on the concept of 'critical defect density', the point defect density at which the lattice becomes unstable and collapses to an amorphous state. This second concept has proved to be more popular in interpreting amorphization of semiconductors probably because it can be extended to include the damage caused by light ions which do not create amorphous zones directly, but which do produce amorphous layers at high doses. The best example of interpretation along these lines is probably the work of Carter et al (1979), although these authors do not set a value for the critical defect density. Several attempts have however been made to estimate the critical defect density, the most recent being by Christel et al (1981) who obtained a value of 10% for boron implantation into silicon. For individual cascades the average defect density will increase with increasing ion mass at constant ion energy. Assuming a critical defect density of 10%, we estimate from WSS theory and a modified Kinchin-Pease equation (Sigmund 1969) that cascades dense enough to produce amorphous zones $\gtrsim$ 4nm diameter cannot be produced by ions of mass $\leq$ 70 in GaP and GaAs. This agrees with our results in GaP although it is only an approximate figure and subject to uncertainty in the values of displacement energies. It is clear, however, that our results do support the hypothesis that amorphization occurs when a critical defect density is exceeded in a cascade, although it is not possible to rule out thermal spike effects.

The difference between GaP and GaAs - visible damage being produced by As and Kr ions in GaP but not in GaAs - was somewhat unexpected and is rather difficult to understand. Since P is lighter than As, cascades are expected to be larger and less dense in GaP than in GaAs for any particular ion. Calculations based on WSS theory however show the differences to be small. More critical in determining the average defect densities are the displacement energies which are not well established in these materials. Our results are consistent with higher displacement energies in GaAs than in GaP, assuming similar critical defect densities in the two materials. It should be noted, however, that room temperature annealing is known to occur for GaAs and may also occur in GaP; such effects cannot be ruled out in our experiments. Further experiments with low temperature irradiations are planned to explore this.

Another interesting aspect of our results is the lack of dependence of amorphous zone size on ion mass. A similar result was found for silicon (Howe 1980), and a possibly related effect was reported in studies of disordered zones produced in high energy density cascades in Cu_3Au by Sb atomic and molecular ions. Jenkins and English (1982) found that the

disordered zone sizes were determined by the total ion energy and were ide-
pendent of whether the ion was atomic or molecular. Experiments to explore
this effect by varying the energy of the incident ions are planned.

Finally it should be stated that whilst our results have been discussed
within the generally accepted framework of 'direct impact amorphization',
some recent observations of Ruault et al (1983) cast some doubt on this
interpretation. Ruault et al reported observations of defects in silicon
produced by in-situ Bi$^+$ ion bombardment in a microscope linked to a heavy
ion accelerator. Defects were observed which gave very similar contrast to
those reported here and those seen in silicon in other experiments. These
defects were, however, reported to develop from a 'damage track' over a
period of about 10 seconds. An effect on this timescale cannot be simply
related to cascade mechanisms. At present this effect remains unverified
and not understood although it is possible that it may be promoted by the
electron beam. We plan in-situ experiments to see if such an effect is
present in GaP and GaAs.

5. Conclusions

GaP and GaAs bombarded with sufficiently heavy ions at doses $\lesssim 10^{16}$ ions/m^2
show small damaged regions in the transmission electron microscope. The
contrast is consistent with these damaged regions being amorphous. Lighter
ions produce no damage visible in the transmission electron microscope.
There is no evidence of regions with high densities of antisite defects
being formed.

References

Carter G, Webb R 1979 Rad. Eff. Lett. 43 19
Christel L A, Gibbons J F and Sigmon T W 1981 J. Appl. Phys. 52 7143
Howe L M, Rainville M H, Haugen H K and Thompson D A 1980 Nucl. Inst. and
    Meth. 170 419
Howe L M and Rainville M H 1981 Nucl. Inst. and Meth. 182/183 143
Jenkins M L and English C A 1982 J. Nucl. Mat. 108/109 46
Jenkins M L, Roller G, Katerbau K-H and Wilkens M 1982 J. Nucl. Mat.
    108/109 603
Pasemann M and Werner P 1979 Phys. Stat. Sol. A 54 179
Ruault M O and Jäger W 1980 J. Micros. 118 67
Ruault M O, Chaumont J and Bernas H 1983 (in press)
Sigmund P 1969 Appl. Phys. Lett. 14 114
Thompson D A 1981 Rad. Eff. 56 105
Winterbon K B, Sigmund P and Sanders J B 1970 Mat. Fys. Medd. Dsn. Vid.
    Selsk. 37 No 14.
Werner R, Kaufman V and Schneider J 1982 Appl. Phys. Lett. 40 141

Inst. Phys. Conf. Ser. No. 67: Section 6
Paper presented at Microsc. Semicond. Mater. Conf., Oxford, 21–23 March 1983

Characterisation of semiconducting materials and devices by EBIC and CL techniques

K Löhnert and E Kubalek

Universität Duisburg Fachgebiet Werkstoffe der Elektrotechnik
Prof.Dr.-Ing. Erich Kubalek, Kommandatenstr. 60, 4100 Duisburg 1 F.R.G.

Abstract Electron beam induced current (EBIC) and cathodoluminescence
(CL) investigations in the SEM (or STEM) enable the evaluation of electronic material properties on a micron scale, which makes them extremely
useful analytical tools in semiconductor research. The theoretical basis
of quantitative signal analysis is treated for both techniques and recent results on the CL contrast of dislocations are reported. The experimental requirements are briefly discussed and a survey is given on the
types of information to be obtained. The significance of such information for a detailed understanding of device-related problems is demonstrated by illustrative examples of application. Advances in extending
the CL technique to the wavelength region above 1.1/um, as required for
the study of low bandgap semiconductors or deep impurity levels, are
reported.

1. Introduction

The physical basis of both EBIC and CL is given by the three processes of
generation, motion and recombination of excess charge carriers. Consequently all material and device properties which have an influence on these
processes can in principle be investigated. While in many cases already a
qualitative interpretation of the experimental results enables the understanding of the investigated physical phenomenon, a thorough theoretical
analysis is generally indispensible to obtain quantitative material and/or
device parameters.

Therefore this paper deals first with the treatment of the EBIC and CL signal formation under stationary conditions and gives a survey of the solved
problems and the underlying approximations. Dynamical signals, which occur
less frequently in practice, are not covered. They have been treated by
Kuiken (1976), Jacubowicz (1980) and Kamm (1977) with regard to EBIC while
for CL general recombination theories apply (see e.g.Blakemore 1962).Based
on Donolato's (1978/79) approach for the evaluation of the EBIC contrast at
defects of enhanced recombination, first calculations for the case of CL are
presented showing good agreement with experimental results on dislocations
of simple dark spot contrast.
After a short discussion of the experimental requirements, the most important types of application will be briefly mentioned and two recent studies
from our own laboratory on the degradation mechanism in GaP green light
emitting diodes and on grain boundary effects in ZnO varistor ceramics will be
used to exemplarily illustrate the potential also of qualitative EBIC and
CL studies for rendering an understanding of microphysical phenomena in
semiconductor materials and devices.
Finally new application possibilities for the CL technique in the field of

low bandgap materials and of deep radiative energy levels with an upper wavelength limit of about 1.9/um are reported and demonstrated by CL results on
silicon.

2. Signal formation

The three fundamental processes that establish and/or influence the signal
both in the case of EBIC and CL are the generation, the recombination and
the motion of excess electrons and holes. For a continuous electron beam at
a fixed position $\underline{r}$ the action of these processes results in the formation
of a stationary excess carrier distribution $\delta p(\underline{r}',\underline{r})$(for n-type semiconductor), which in the simplest case can be evaluated from the 3-dimensional
continuity equation

$$D \; \nabla^2_{\underline{r}'} \; \delta p(\underline{r}',\underline{r}) - \frac{1}{\tau(\underline{r}')} \quad \delta p(\underline{r}',\underline{r}) + g(\underline{r}',\underline{r}) = 0 \qquad (1)$$

and from the additional boundary conditions at the specimen surface and/or
at internal interfaces. D and τ are the diffusion coefficient and lifetime
of minority carriers respectively and g represents the generation function
of electron-hole pairs as defined by the electron beam parameters and the
energy dissipation process. Eq. (1) is valid under the following two conditions:
i) The carrier lifetime τ is independent of δp. This condition is satisfied quite well as long as δp is small compared to the majority carrier
density n_0 (or p_0) (Blakemore 1962), which can in experiment generally be
adjusted for highly doped material and short recombination lifetime τ. Care
has to be taken however in the interpretation of experimental data, where
n_0 (or p_0) is low.
ii) The second assumption included in eq. (1) is that the motion of excess
electrons and holes is purely diffusive as is valid in sample regions without electrical fields. For a wide variety of applications with p-n junctions
or Schottky barriers in highly doped materials ($\geqslant 10^{16} cm^{-3}$) this approach
is quite useful since here the field regions (depletion regions) are narrow
and can be treated as internal boundaries using appropriate boundary conditions.
Once δp has been determined from eq. (1) and the associated boundary conditions both the EBIC and the CL signal can be calculated. If there are no
recombination losses inside the field region of the electrical barrier used
to collect the minority carriers the EBIC signal I equals the total diffusion current entering the boundary plane F of the field region (Berz and
Kuiken 1976):

$$I \; (\underline{r}) = \quad D \int_F \nabla_{\underline{r}'} \; \delta p(\underline{r}',\underline{r}) \; d^2\underline{r}' \qquad (2)$$

The CL intensity I_{CL} under the widely applicable assumption of a linear
dependence on δp can be written:

$$I_{CL} \; (\underline{r}) = \int_V A \; R \; \eta(\underline{r}') \; \delta p(\underline{r}',\underline{r}) \; d^3\underline{r}' \qquad (3)$$

where A and R are correction factors for reabsorption losses inside the material (A) and reflection losses at the surface (R), and η is the internal
quantum efficiency. A, R and η are dependent on wavelength so that for the
integral (undispersed) CL they have to be regarded as average values. In
principle A and R may also depend on the coordinate $\underline{r}$. This can be neglected in most cases, however.
To calculate I or I_{CL} from eq. (1), (3) we require the solution $\delta p(\underline{r}'\underline{r})$
of eq. (1), which represents the main problem. Exact solutions are

feasible only for homogeneous material, i.e. $\tau(\underline{r}') = \tau$ = const. and for
simple boundary conditions. The boundary occurring in all problems is the
irradiated surface z=0, which in terms of the surface recombination velocity v_S introduces the condition

$$D \frac{\partial}{\partial z'} \; \delta p(\underline{r}'\underline{,r}) \; = v_S \; \delta p(\underline{r}'\underline{,r}) \tag{4}$$

The solution of eq.(1) + (4) for arbitrary values of v_S and for generation
by a point source has been given by van Roosbroeck (1955). In particular
an analytical solution is obtained for the important case $v_S=\infty$, which applies to many compound semiconductors (e.g. GaAs, GaP) with v_S-values large
compared to the diffusion velocity $v_D = \sqrt{D/\tau}$ and for Schottky barriers or
flat p-n junctions parallel to the irradiated surface with the width W_d of
the depletion layer small compared to the primary electron range R_e. Further boundary conditions as inferred by a planar p-n junction at a depth
z=h below the surface (Possin 1977) or perpendicular to the surface (Berz
and Kuiken 1976, van Opdorp 1977) have been treated with respect to the
calculation of EBIC signals. The corresponding results provide the basis
for the determination of the diffusion length $L = \sqrt{D\tau}$ and/or v_S from experimental EBIC data.

The main lack of all existing results is the use of the point source generation (or sphere of constant generation), which can be regarded valid only
for distances from the boundaries large compared to R_e and/or L. The solution for arbitrary generation functions g $(\underline{r}',\underline{r})$ can in principle be obtained by linear superposition of the point source solutions, but due to
the enormous computational efforts required this has not been undertaken
yet and presents a challenging field for further work.

Other largely uncovered fields are the analysis of wide depletion layers,
where drift motion of the excess carriers due to the electrical field and
even field variations associated with doping gradients have to be regarded,
and the analysis of imperfect semiconductors with lifetime variations $\tau'<\tau$
in the vicinity of defects.
The former problem has been recently attacked by Marten and Hildebrand
(1982), who presented as a first step calculations of EBIC linescans perpendicular to the depletion layer in a completely one-dimensional approach
and used the results to reconstruct the dopant profile.
The latter problem was treated first by Donolato (1978/79), who introduced
an iterative solution scheme for δp to calculate the EBIC contrast of
localized defects.

In contrast to the numerous efforts on EBIC signal analysis practically no
attempts have been made to quantitatively treat the spatial signal dependence in the case of CL, except for the simplified approach of van Opdorp
(1977) to CL linescans across p-n junctions with hypothetical profiles of
the quantum efficiency.
Detailed calculations of the CL intensity profile at dislocations perpendicular to the surface have been presented for the first time by Löhnert
and Kubalek (1982) on the basis of Donolato's (1978/79) iterative procedure for δp, which to a first approximation is given by

$$\delta p(\underline{r}',\underline{r}) = \delta p_0(\underline{r}',\underline{r}) - \int_{V_d} \gamma(\underline{r}'') \; \delta p_0(\underline{r}'',\underline{r}) \; G \; (\underline{r}',\underline{r}'') \; d^3\underline{r}'' \tag{5}$$

δp_0 is the solution for the homogeneous semiconductor without a defect,
$\gamma = (1/\tau' - 1/\tau)/D$ describes the recombination activity of the defect and
G is the point source solution of eq.(1) obeying the same boundary conditions as δp.
Isoconcentration lines of δp for a dislocation at right angle to the surface

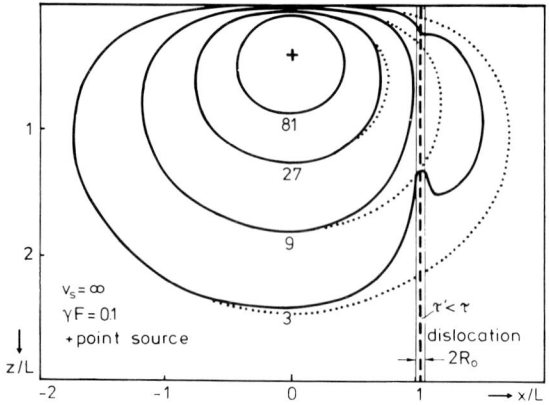
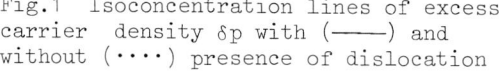

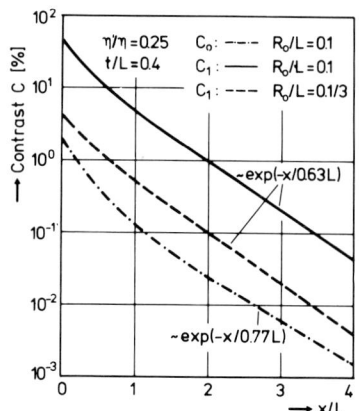

Fig.1 Isoconcentration lines of excess carrier density δp with (———) and without ($\cdots$) presence of dislocation

Fig.2 Calculated CL contrast of dislocation as depicted in fig. 1

in a distance L from the electron beam are shown in fig.1 and compared to those of δp_0 (dotted). As indicated, a point source and $v_s=\infty$ are assumed and the dislocation is modelled as a cylinder of cross-section $F=\pi R_0^2$ with constant lifetime $\tau' < \tau$ inside. Calculated CL contrast profiles for the situation of fig.1 are presented in fig.2 assuming $\tau'=0.25\tau$ or $\eta'=0.25\eta$ respectively, which implies constant radiative lifetime τ_r, since $\eta=\tau/\tau_r$. The dash-dotted curve corresponds to the hypothetical case $\tau'=\tau$ and $\tau_r' < \tau_r$, which applies if the only effect of the dislocation is a lowering of the radiative recombination rate and if the latter is negligible against the non-radiative rate. As revealed in fig.2 the contrast turns into an exponential decrease at sufficient distance from the dislocation, which was found to be in good agreement with experimental linescans across dislocations of plane dark spot contrast in e.g. GaP and InP. According to the theoretical result the decay constant equals 0.63 L, suggesting the interesting possibility of evaluating the diffusion length L from CL linescans across localized non-radiative defects.

3. Experimental

The experimental requirements for EBIC and CL are characterized by two contrary tendencies:
i) EBIC signals are usually big enough to be measured without serious difficulties using conventional amplifiers or picoammeters and there is no need for any detector, because the sample itself is the detector. Therefore little effort has been put in tayloring special EBIC equipment except for some recent work by Holt and coworkers (1981).
At very low signal levels or when a bias voltage has to be applied, which cannot be sufficiently suppressed by ac coupling, lock-in amplification and pulsing of the electron beam may be required (Balk et al 1975). This setup is particularly useful for electron beam induced conductivity measurements, where large dc currents may be present with only little modulation by the electron beam.
If a fast beam chopping system is available time-resolved studies down to the subnanosecond region can be performed by gated amplification with a boxcar integrator (Balk et al 1980). Gated amplification in combination with a pulsed beam can also be used to improve the image resolution in EBIC (and CL) micrographs (Spivak et al 1977) by setting the gate to the rising

slope of the pulse, where only carriers with short diffusion paths contribute to the signal.

If such special amplification techniques are not required the main prerequisite for EBIC investigation of unprocessed semiconductors is the preparation of a uniform, electron-transparent Schottky contact, which requires some experience, however, at least as far as uniformity is concerned.

ii) The situation is just opposite for CL measurements, where any polished or cleaved semiconductor can be investigated without further sample preparation, but where substantial additional equipment for collecting, analyzing and sensitively detecting the emitted radiation has to be provided for most applications. Detailed information thereabout can be obtained from a recent publication (Löhnert et al 1981), on the SEM based CL measurement system developped in our laboratory, which includes light collection by an ellipsoidal mirror adjustable under vacuum, different rapidly interchangeable detector types for sequential or simultaneous detection of both stationary and time-dependent signals, automated operation under control of a desk-top computer and the capability of temperature-dependent measurements between room temperature and liquid helium temperatures. Another very efficient system has been introduced at the same time by Davidson and co-workers (1981), who for the first time presented spatially resolved CL investigations on silicon (Cumberbatch et al 1981).

4. Applications

EBIC and CL techniques have been applied to a wide variety of research problems in semiconductor physics and technology and recent reviews were given by Leamy (1982) on EBIC, by Holt and Datta (1980) on CL, and by Balk et al (1980), Davidson and Dimitriadis (1980) and Booker (1981) on both EBIC and CL. Roughly speaking two categories of application can be distinguished: I) Quantitative determination of material and/or device parameters and II) qualitative studies with the aim of elucidating differences in material or device properties dependent on e.g. growth technique, processing steps or device operation.

In the case of EBIC the most exploited example of category I is the measurement of the diffusion length L using the type of signal analysis described in section 2. The most convenient method is a linescan across a Schottky barrier or p-n junction perpendicular to the scanned surface as analysed in detail by Berz and Kuiken (1976) and van Opdorp (1977). In structures where the electrical barrier does not protrude to a surface accessible to the electron beam the variation of the EBIC signal with primary electron energy, i.e. with penetration range, can be used to evaluate L (Possin and Kirkpatrick 1979).

Typical device type applications are the measurement of the barrier hight of a p-n junction (Balk et al 1975), or its depletion layer width W_d versus reverse bias (Oelgart and Wagner 1980), which provides a means of obtaining the contributions to W_d and the corresponding doping profile on both the n- and the p-side of the junction and which therefore can be superior to capacitance-voltage measurements. One has to be aware, however, that this works only if the doping is low and the depletion width and its variation with reverse bias is sufficiently large (>1 /um) to be accurately determined from EBIC linescans. Further parameters, such as minority carrier lifetime and diffusion velocity can be obtained by time-resolved or phase-shift measurements, which has been applied by various workers(e.g. Balk et al 1980, Kamm 1977).

Time-resolved EBIC may also be used to study the emission of trapped carriers after the electron beam is switched off, which has been first intro-

duced by Petroff and Lang (1977) and which in combination with the tempera-
ture-dependence of results yields the energy level of the associated traps.
Due to the generally too low signal levels this technique is only applicable in special cases, however.

In the case of CL quantitative data have been deduced rather making use of
its spectral properties or its decay behaviour than of spatial intensity
variations. Well-known examples are the determination of local doping level
(Balk et al 1976) or of local temperature in operated devices (Jones et al
1974). Carrier lifetime and its spatial variation at individual dislocations
have been determined from CL decay measurements by several workers(Dimitriadis et al 1978, Hastenrath et al 1979, Steckenborn et al 1981). A recent
review on time-resolved CL has been given by Hastenrath and Kubalek (1982).

For many practical problems truly quantitative data are not required fortu-
nately, but the information of interest can be drawn from a qualitative interpretation or from comparative studies and sometimes even from imaging
alone. The possibilities for this type of work are almost unlimited (see
e.g. the bibliographies of Leedy 1977 for EBIC and Bröcker and Pfefferkorn
1978 for CL), the main applications for characterization of semiconducting
materials and devices up to now being:
a) Imaging of dopant and point defect inhomogeneities (de Kock et al 1977,
Kyser and Wittry 1966).
b) Imaging of extended defects like dislocations, stacking faults or grain
boundaries and characterization of their recombination activity in terms of
the associated image contrast (Titchmarsh et al 1977, Petroff et al 1980,
Ourmazd et al 1981).
c) Investigation of dopant redistribution around dislocations by spectral-
ly resolved CL (Balk et al 1976, Rasul and Davidson 1977).
d) Failure analysis of devices, including e.g. pipes in transistor structures (Schick 1981) or degradation induced effects in laser diodes (Shaw
and Thornton 1970, Wakefield and Robertson 1981).
e) Correlation of the performance of light emitting diodes(LED's) with the
CL efficiency in the material region involved (Calverley and Wight 1969).
f) Investigation of process-induced damage such as formation of dislocations and stacking faults during high temperature steps (Varker and Ravi
1974) or ion implantation damage (Norris et al 1973, Cocito et al 1981).

In the following two applications recently persued in our own laboratory
will be discussed exemplarily to illustrate the potential of EBIC and CL
techniques for the evaluation of micro-physical phenomena in semiconductor
materials and devices more closely:

A. Degradation mechanism in green emitting GaP:N:Zn LED's

Valuable insight into the processes responsible for degradation has been
obtained both by EBIC and CL measurements (Löhnert 1982) and can be briefly summarized as follows:
EBIC micrographs and linescans rendered the result that extended non-radiative defects were not generated during degradation and that the recombination activity of the previously existing grown-in dislocations, visible as
dark spots in the EBIC micrograph of fig. 3, did not increase during degradation. Dislocations could therefore be excluded as the source of degradation.
On the other hand CL micrographs of cleaved diodes revealed that the luminescence efficency on the n-side of degraded devices was lowered while the
p-side did not seem to be affected appreciably (see fig.4).
From the doping levels of $2 \cdot 10^{18} cm^{-3}$Zn on the p-side and $5-10 \cdot 10^{16} cm^{-3}$ S

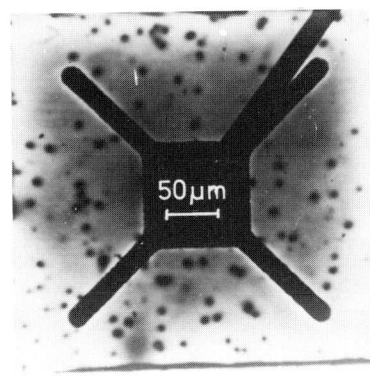

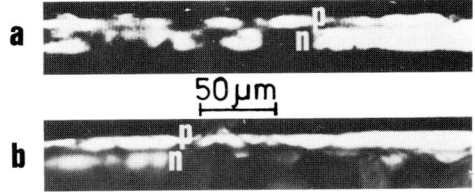

Fig.4 CL micrographs of cleaved GaP LED's, (a) undegraded and (b) degraded device

Fig.3 EBIC micrograph of GaP LED (top surface)

on the n-side hole injection was expected to strongly dominate the total injection of minority carriers, leading to the conclusion that the electro-luminescence (EL) should be produced mainly on the n-side of the junction and that therefore the observed decrease in luminescence efficiency on the n-side might be the cause of LED degradation. This assumption could indeed be confirmed by a comparison of the EL spectrum with the CL spectrum of the n- and p-side (fig.5), which shows that the EL spectrum is largely identical with the CL spectrum on the n-side but differs from the more structureless one on the p-side. To obtain clearly resolved details the spectra in fig. 5 have been taken at low temperature (T = 85 K), where additional luminescence processes visible at room temperature are quenched, however. At room temperature and forward currents ≤ 1mA the EL spectrum in addition to the desired green luminescence at 560 nm exhibits a weak red luminescence peaking near 680 nm, which is due to Zn-O centers formed through the zinc doping and a slight unintentional contamination with oxygen. The intensity of Zn-O luminescence in the EL spectrum increases during degradation as can be seen from fig.6. CL linescans at the peak wavelength (680 nm) of the Zn-O luminescence showed a broadened profile towards the n-side of the junction in degraded devices (fig.7), revealing that during degradation some zinc diffuses across the junction. Part of this zinc forms Zn-O centers with the pre-existing oxygen leading to the increase of the Zn-O luminescence in the EL spectrum.

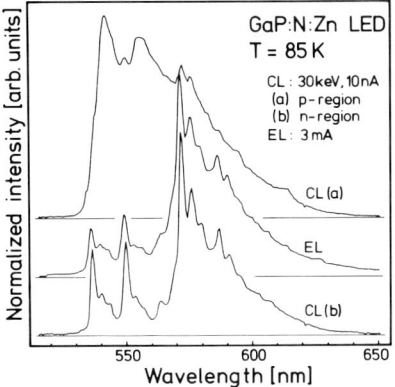

Fig.5 CL and EL spectra of GaP LED at T = 85 K.

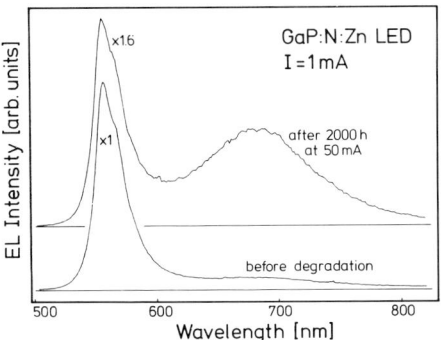

Fig.6 Room-temperature EL spectrum of same GaP LED before and after degradation

Based on the above experimental observations the degradation was attributed to the formation of non-radiative centers on the n-side of the junction, associated with the zinc diffused in. By a theoretical treatment of this process under simplifying

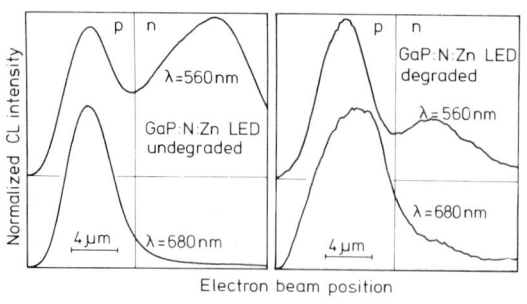
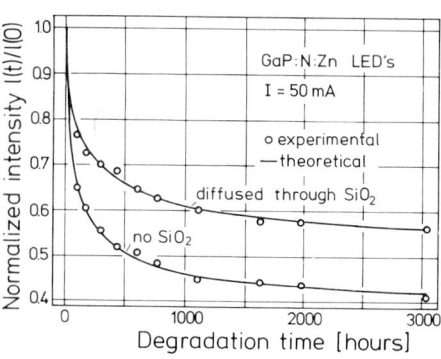

Fig.7 Monochromatic CL linescan across
p-n junction of undegraded and degraded
GaP LED.

Fig.8 Degradation at constant
current I=50mA of GaP LED's

approximations an equation for the degradation of the EL intensity at constant current could be deduced (Löhnert 1982), which excellently fits the
experimental degradation data (fig.8) and thus provides a good confirmation
for the proposed degradation mechanism.

B. Grain boundary effects in ZnO varistor ceramics

A completely different field of application is offered by polycrystalline
materials, which attain increasing technological interest and where again
both EBIC and CL techniques can supply important information. So far we
have been concerned mainly with zinc oxide ceramics, which by addition of
various metal oxides obtain a strongly non-linear current-voltage characteristics, similar to two diodes back to back, and which therefore are of
great practical interest as varistor devices. The non-linear electrical
properties are caused by the grain boundaries, and since different added
metal oxides are able to produce such behaviour it has been proposed (Einzinger 1978) that not the metal component but the oxygen is the responsible
factor. The idea is that oxygen is incorporated near the grain boundaries
and there makes the material intrinsic, which is otherwise always n-type
due to a stoichiometric excess of zinc. Thus two adjacent grains would form
a n-i-n structure and the associated potential barrier would be responsible
for the observed current-voltage characteristic.

To reinforce this proposal we have performed CL measurements starting out
for a comparison with pure air-sintered ZnO ceramics without metal oxide
additives. It turned out that there are specimen regions with strongly
enhanced oxygen content near the grain boundaries (Löhnert and Kubalek 1981)

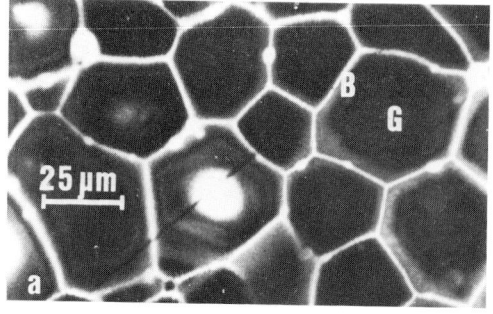
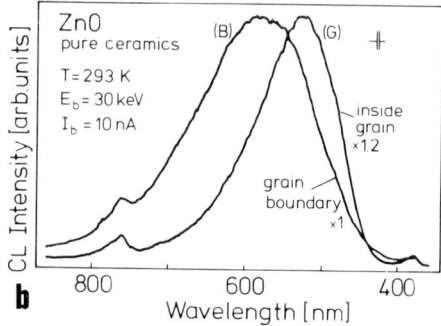

Fig.9 Integral CL micrograph (a) and CL spectra (b) from pure ZnO ceramic

as can be learnt from fig.9, which shows an integral CL micrograph of such
a region and the normalized CL spectrum measured inside the grain (G) and
in the bright grain boundary region (B) respectively. Spectrum (B) is found
in material sintered in pure oxygen everywhere in the grains while spectrum
(G) is typical of material sintered in pure nitrogen, which leads to the
above conclusion. Despite the enhanced content of oxygen at the grain boundaries it has not been possible, however, to observe a notable non-linearity
in the electrical properties. Therefore it must be concluded that in the
varistor type ceramics the influence of oxygen, if really responsible for
the non-linearity, is strongly enhanced compared to the pure ceramics.

Unfortunately the above CL results cannot be transferred to varistor type
ceramics because the luminescence in the latter ones is dominated by centers associated with the metal oxide additives (fig.10). From CL micrographs of the cross-section area in var-
istor devices it can be seen that these
centers are enhanced at the grain boundary regions (fig.11a), which indicates
that they also might be responsible for
the electrical properties. An EBIC micrograph (fig.11b) of the same area reveals,
that not all of the grain boundaries visible in the CL micrograph contribute to the
electrical characteristics of the varistor.
Roughly there can be seen five barriers
constituted from grain boundary sections, which run mainly parallel to the two
contacted surfaces, i.e. perpendicular to
the applied field. Without applied field

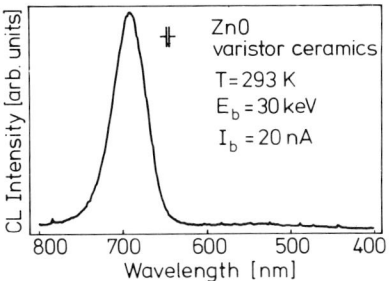

Fig.10 CL spectrum of
ZnO varistor ceramic

or bias voltage respectively no EBIC signal is obtained. The type of signal
involved in fig.11b therefore should be more precisely designated as electron beam induced conductivity. From the current voltage characteristics of
the investigated varistor a threshold voltage of 17 V for superlinear current increase was determined. Taking
into account the five barrier regions delineated in fig.11b one obtains a potential drop of 3.4 V per
barrier, which is in excellent agreement with point contact measurements between isolated grains (Einzinger 1979). The grain boundaries
not imaged by EBIC are shunted by

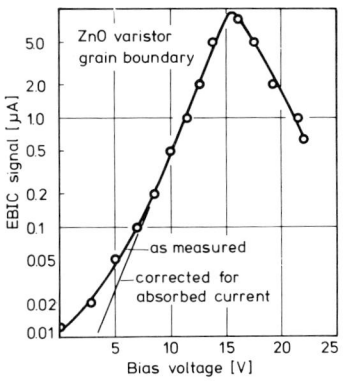

Fig.12
EBIC sig-
nal at
grain
boundary
in ZnO
varistor
versus
bias
voltage

Fig.11 CL micrograph (a) and corresponding EBIC micrograph (b) of ZnO
varistor

parallel running current paths and are therefore also not active in varistor operation.

Further information is supplied by the dependence of the grain boundary EBIC signal on bias voltage (fig.12). The signal first increases exponentially and after passing through a maximum decreases again. The voltage where this occurs corresponds roughly to the varistor threshold voltage, where the dc bias current increases strongly. Above this voltage the resistance of the grain boundaries drops rapidly and thus the effect of the conductivity contribution by the electron beam becomes smaller. The exponential signal increase at lower bias voltages can be explained assuming Schottky emission of majority carriers across the grain boundary barriers (Card 1981, Bernds et al 1983).

5. Advances in infrared CL at wavelengths above 1.1/um

CL microanalysis in the infrared spectral range above 1.1/um is of great practical interest since i) radiative recombination levels deep in the bandgap can be studied and ii) silicon as the most important semiconductor of today becomes accessible. Unfortunately the sensitivity of photodetectors useful in this spectral range is orders of magnitude lower than that of photomultipliers, which are limited, however, to a maximum wavelength of about 1.1/um (S-1 cathode). Additionally the luminescence yield of silicon, which will be focussed upon, is extremely low. Therefore spatially resolved CL studies have not been undertaken up to the recent pioneering work by Davidson et al (1981). The beam currents of >3/uA required in their experiments have been unsatifactory high, however, resulting in a temperature rise of up to 140 K at room temperature according to their own estimation.

Meanwhile first experiments have been done also in our laboratory using a S-1 type photomultiplier for imaging and a high efficieny germanium photodetector (North Coast EO-817) with a response limit of about 1.9/um for spectral analysis. Integral CL micrographs of Wacker solar grade poly-silicon (fig.13) have been feasible with beam currents $I_b \simeq 1$/uA at 30 kV accelerating voltage and T = 80 K. CL spectra of the near band-edge luminescence in device grade silicon (see fig.14 right part) were obtained down to $I_b \simeq 200$-300nA at a bandwidth of 6 nm. From the specified detector response of roughly $7 \cdot 10^9$ V/W the maximum signal in fig.14 has been determined about 10^{-14} W or $5.8 \cdot 10^4$ photon/s of energy 1.1 eV. The CL peak intensity was found to vary as I_b^2 in the useful range 0.2/uA $< I_b < 3.2$/uA, which is characteristic of high excitation (Kyser and Wittry 1966). Probably this effect is also responsible for the fast initial CL decay observed by Davidson et al (1981). While this situation is in principle undesirable it is advantageous on the other hand to enhance the spatial resolution of CL micrographs because most of the signal is produced in the center of the excita-

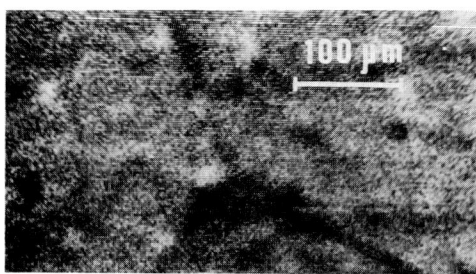

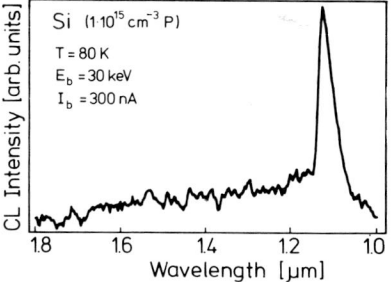

Fig.13 Integral CL micrograph of solar grade poly-silicon

Fig.14 CL spectrum of device grade silicon

tion volume where the density is highest.
The broad background in fig.14 extending up to about 1.7/um (0.73 eV), indicates the presence of deep radiative states in the bandgap. Such deep
states are also supposed to be associated with dislocations in silicon since
by plastic deformation at elevated temperatures a characteristic luminescence near a wavelength of 1.5/um can be generated (Gwinner 1982). Fig.15
shows CL spectra of such a sample with a

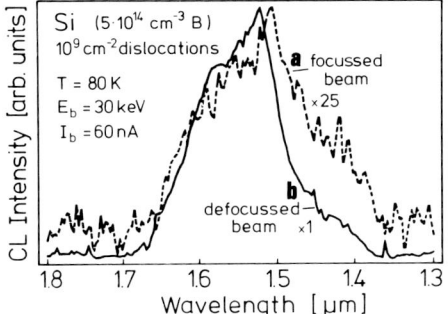

dislocation density of $10^9 cm^{-2}$ recorded
at 80 K with a focussed (a) and a strongly defocussed (b) electron beam at the
same beam current of 60 nA. Due to a sub-
linear increase of this luminescence at
high excitation densities the intensity
obtained with a focussed beam is about
25 times lower than with completely defocussed beam. Additionally the weak
shoulder near 1.4/um grows in intensity
causing a slight peak shift in comparison to spectrum (b) which agrees well
with corresponding photoluminescence
spectra (Gwinner 1982). Except for these
differences the same features are dis-

Fig.15 CL spectra of plasti-
cally deformed silicon

closed, however, in both spectra showing that relevant information is obtained also under conditions of focussed beam and thus the investigation
of individual dislocations at densities $<10^6 cm^{-2}$ is likely to be successful.

Parts of the reported work have been supported by the Ministry of Science
and Research of North Rhine-Westfalia and the German Research Society. The
EBIC studies on ZnO varistors were done by Mr. A. Bernds. Dr. D. Gwinner
kindly supplied the silicon sample of fig. 15.

References

Balk L J, Kubalek E and Menzel E 1975 IEEE Trans.on Electr.Dev. ED-22
    pp 707-11
Balk L J, Kubalek E and Menzel E 1976 SEM/I pp 257-64
Balk L J, Kubalek E and Menzel E 1980 Proc.8th Int. Congr.on X-ray Optics
    and Microanalysis, ed D R Beaman et al (Midland, Michigan: Pendell)
    pp 613-24
Bernds A, Löhnert K and Kubalek E 1983 to be published at 10th Int. Congr.
    on X-ray Optics and Microanalysis, Toulouse
Berz F and Kuiken H K 1976 Solid-State Electr. 19 pp 437-45
Blakemore J S 1962 Semiconductor Statistics (Oxford : Pergamon)
Booker G R 1981 Inst.Phys. Conf. Ser. No. 60 pp 203-14
Bröcker W and Pfefferkorn G E 1978 SEM/I pp 333-51
Calverley A and Wight D R 1969 Solid-State Electr. 13 pp 382-5
Card H C 1981 J.Appl.Phys. 52 pp 3671-3
Cocito M, Dal Pozzo L and Gorgellino F 1981 Inst.Phys.Conf.Ser.No.60
    pp 345-50
Cumberbatch T J, Davidson S M and Myhajlenko S 1981 Inst.Phys.Conf.
    Ser. No.60 pp 197-202
Davidson S M and Dimitriadis C A 1980 J.Microscopy 118 pp 275-290
Davidson S M, Cumberbatch T J, Huang E and Myhajlenko S 1981 Inst.Phys.
    Conf. Ser.No. 60 pp 191-6
de Kock A J R, Ferris S D, Kimerling L C and Leamy H J 1977 J. Appl.Phys.
    48 pp 301-7
Dimitriadis C A, Huang E and Davidson S M 1978 Solid-State Electr.21
    pp 1419-23

Donolato C 1973/79 Optic 52 pp 19-36
Einzinger R 1978 Appl. of Surf. Sci. 1 pp 329-40
Einzinger R 1979 Appl. of Surf. Sci. 3 pp 390-408
Gwinner D 1982 Ph.D Thesis, Clausthal University FRG
Hastenrath M, Löhnert K and Kubalek E 1979 BEDO 12/1 pp 163-75
Hastenrath M and Kubalek E 1982 SEM/I pp 157-73
Holt D B and Datta S 1980 SEM/I pp 259-78
Holt D B 1981 Inst. Phys. Conf. Ser.No. 60 pp 165-78
Jakubowicz A 1980 Solid-State Electr. 23 pp 635-9
Jones G A C, Nag B R and Gopinath A 1974 J. Phys. D7 pp 183-93
Kamm J D 1977 Semiconductor Silicon, ed H R Huff and E Sirtl (Princeton N J:
    Electrochem. Soc.) pp 491-501
Kuiken H K 1976 Solid-State Electr. 19 pp 447-50
Kyser D F and Wittry D B 1966 Proc. of the Electron Microprobe Symp., ed
    T D Mc Kinley et al (New York : Wiley) pp 691-74
Leamy H J 1982 J. Appl. Phys. 53 pp R 51-80
Leedy K O 1977 Solid State Technol. pp. 45-8
Löhnert K, Hastenrath M, Balk L J and Kubalek E 1981 Inst. Phys. Conf.
    Ser. 60 pp 179-84
Löhnert K and Kubalek E 1981 BEDO 14 pp 147-52
Löhnert K and Kubalek E 1982 Proc.10th Int. Congr. on Electron Microscopy
    Vol. 2 pp 419-20
Löhnert K 1982 Ph.D. Thesis, Duisburg University FRG
Marten H W and Hildebrand O 1982 BEDO 15 pp 121-32
Norris C B, Barnes C E and Beezhold W 1973 J.Appl.Phys. 7 pp 3209-21
Oelgart G and Wagner G 1980 phys.stat.sol. (a) 59 pp 43-51
Ourmazd A, Weber E, Gottschalk H, Booker G R and Alexander H 1981 Inst.
    Phys. Conf. Ser. No. 60 pp 63-3
Petroff P M and Lang D V 1977 Appl. Phys. Lett 31 pp 60-2
Petroff P M, Logan R A and Savage A 1980 Phys.Rev.Lett. 44 pp 287-91
Possin G E 1977 J. Appl.Phys. 48 5245
Possin G E and Kirkpatrick C G 1979 SEM/I pp 245-56
Rasul A and Davidson S M 1977 SEM/I pp 225-31
Schick J D 1981 SEM/I pp 295-304
Shaw D A and Thornton P R 1970 Solid-State Electr.13 pp 919-24
Spivak G V, Saparin G V and Komolova L F 197 SEM/I pp 191-9
Steckenborn A, Münzel H and Bimberg D 1981 Inst.Phys.Conf. Ser. No.60
    pp 185-90
Titchmarsh J M, Booker G R, Harding W and Wight D R 1977 J.Mat. Science
    12 341
van Opdorp C 1977 Philips Res. Repts. 32 pp 192 - 249
van Roosbroeck W 1955 J.Appl.Phys. 26 pp 380-91
Varker C J and Ravi K V 1974 J.Appl.Phys. 45 pp 272-86
Wakefield B and Robertson M J 1981 Inst.Phys.Conf. Ser.No 60 pp 447-51

SEM 1973-77 : IIT Res. Inst., Chicago IL 60616 U.S.A.
SEM 1978-82 : SEM Inc., AMF O'Hare IL 60666 U.S.A.
BEDO   : Beitr. Elektronenmikroskop.Direktabb. und Analyse von Ober-
              flächen, ed. G. Pfefferkorn, Münster, W.-Germany

Inst. Phys. Conf. Ser. No. 67: Section 6
Paper presented at Microsc. Semicond. Mater. Conf., Oxford, 21–23 March 1983

315

Low temperature SEM cathodoluminescence of life-tested GaAlAs/GaAs double heterostructure lasers

B Wakefield

British Telecom Research Laboratories, Martlesham Heath, Ipswich, Suffolk, IP5 7RE, UK

Abstract   The cathodoluminescence emitted at 4K from different regions of long lived GaAlAs DH stripe lasers has been measured in an SEM. Changes have been found in the spectra emitted from the active layer under the stripe as a result of operating the devices for several thousand hours at 70°C. These changes are explained in terms of operation induced, strain enhanced native defect migration.

1. Introduction

Long lived GaAlAs/GaAs oxide isolated, stripe geometry, double heterostructure lasers (Fig.1), which have been lifetested for several thousands of hours at elevated temperatures, have previously been examined in the scanning electron microscope using its cathodoluminescence (CL) and electron beam induced current (EBIC) modes, and have frequently been found to contain regions of enhanced and reduced carrier recombination (Wakefield and Robertson 1981). These regions reveal themselves in the SEM as regions of bright and dark CL or EBIC contrast (Fig.2). Two main types of degradation features are observed. The CL and EBIC images of all those devices studied which have been aged at elevated temperatures (ie between 55°C and 90°C) have to a greater or lesser extent been found to display bright and dark bands running parallel to their stripes. These degradation related bands were shown to be related to the strain fields introduced into the devices by cutting a window in a layer of SiO_2 on the p-side of the devices to form the laser stripe and are thought to be caused by the migration of native defects in the laser material towards or away from positions of maximum or minimum volume dilation (Wakefield and Robertson 1981), the driving force being non-radiative recombination enhanced defect migration (Lang and Kimerling 1974). A less common degradation feature has been the appearance of <100> dark lines apparently radiating from dark spots on the laser stripe. A closer examination of these dark lines has shown that they are in fact arrays of dark spots aligned in the <100> direction, being thus different from the dark line defects which were responsible for the premature death of early GaAlAs lasers. Skeats et al (1983), using electron probe X-ray microanalysis, detected copper at just those positions on the laser stripe from which the dark spot arrays were radiating, but not on the dark spots outside the stripe. The copper is thought to have diffused from the heatsinks, through the p-side metallisation and into the device stripe, but why the copper should produce <100> dark spot arrays outside the stripe is not understood.

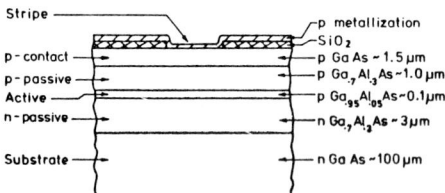

Fig.1 Schematic cross-section of a typical GaAlAs DH laser.

Fig.2   Cathodoluminescence micrograph of a laser lifetested at 70° C.

In order to attempt to obtain further information about the degradation mechanisms occurring here, the light emitted by these devices under CL excitation in the SEM has been measured spectroscopically. Earlier measurements by Hatch and Robertson (1981) using photoluminescence excitation at room temperature had indicated that some changes had occurred to the emission spectrum during the operation of a device. They found that the active layer light emitted from a 50μm diameter area across the stripe appeared to be at a slightly shorter wavelength than that emitted from a similar area well away from the stripe, which presumably represented undegraded material. By repeating the measurements using CL excitation in the SEM and by working at near liquid helium temperatures, it was hoped to achieve not only a much better spatial resolution but also a higher spectral resolution. It might then be possible to relate the spectral changes to specific degradation features in the cathodoluminescence images.

2. Experimental

The cathodoluminescence measurements were made on a Cambridge Instruments S180 scanning electron microscope, fitted with a liquid helium cooled cryostage to enable measurements to be made at any temperature between room temperature and about 4K. The cryostage and CL light collection system is based on a design described by Holt (1981) and manufactured by Oxford Instruments. It differs from the original system developed by Holt and co-workers in that the plane mirror used to collect and direct the emitted luminescence onto the output lightpipe, has been replaced with an ellipsoidal mirror. The specimen is placed at one focus of the ellipsoid. A small convex lens forms an image of the second focal point on the input of a fibre optic bundle, which conducts the light outside the SEM chamber. This modification increases the light collection efficiency of the system. The output of the fibre optic bundle can be either coupled directly to a detector or to the input of a Spex "Minimate" monochromator. For this work the detector used was a photomultiplier with either a GaAs photocathode, or one with an S1 photocathode.

The specimens to be examined were prepared as described before (Wakefield and Robertson 1981). Electron energies between 20 and 30keV and beam currents of about 100nA were used to excite the CL in the specimens. These electron energies are necessary to give a sufficiently large cathodoluminescence output, but it does mean that the primary electron beam is able to excite the substrate as well as the active layer of the lasers. Cathodoluminescence images of the devices were first formed in order to locate the particular areas of interest. A stationary electron beam was then focused to a spot on the specimen at a chosen place and cathodoluminescence spectra measured. Areas of interest included points on the bright and dark bands, on and off dark spots, and on and off the stripe region. The electron beam was chopped at a frequency of 200Hz, enabling phase sensitive detection to be used. Measurements were made at both room temperature and near 4K. The slit width on the monochromator was set at 0.5mm, giving a spectral bandwidth of 4nm.

3. Results

In figure 3(a) is shown the spectrum, which was measured at 4K, of the cathodoluminescence emitted from the centre of the stripe of the same laser as that shown in figure 2. The emitted light is seen to contain two contributions. The peak at 836nm is the light emitted by the substrate, while that at 800nm is from the active layer. (The active layer light is at a shorter wavelength than that of the GaAs substrate because the active layer, being GaAlAs containing about 4% AlAs, has a larger bandgap than GaAs). In other devices the active layer emission was at slightly different wavelengths, reflecting small variations in the active layer composition from device to device. The active layer CL peak is about 50nm wide, which is quite broad for 4K. This presumably is caused by the high doping level of the material. (The active layer is p-type and doped with Si to a carrier concentration of about 5×10^{17} cm^{-3}.).

No significant differences in the shape of the spectra were found on moving the incident electron beam from the centre of the stripe to the bright or dark bands bordering the stripe. Only the intensity of the emitted light was seen to vary. Very little light at all was emitted from the dark spots on the stripe. However when the electron beam was moved to a point well away from the stripe, into what should represent undegraded material, quite different spectra were recorded. An example is shown in Figure 3(b). Clearly there is an additional emission peak present, between the

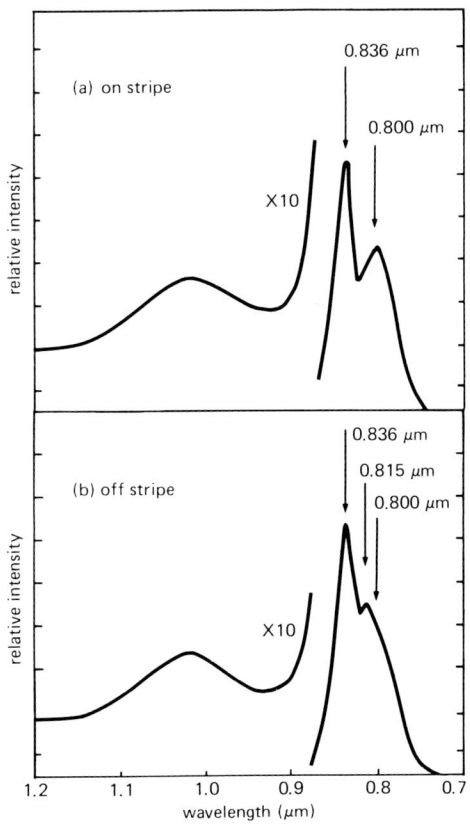

Fig.3 Cathodoluminescence spectra measured at 4K from a lifetested laser:- (a) on stripe, (b) off stripe.

active layer peak at 800nm and the substrate peak at 836nm. The wavelength of this additional peak is about 815nm in the device illustrated here. As well as these spectral differences, it was noticed that the emission intensities from the centre of the stripe was on average slightly different from those in the undegraded region. The intensity of the 800nm (active layer) peak was on average slightly greater on the stripe than off it, while the substrate peak intensity (at 836nm) was on average slightly lower on than off the stripe. All the above effects were found in each of the lifetested devices examined in this way. A laser that had not been lifetested was also studied. In this device no differences were found in the cathodoluminescence spectra measured on or off the stripe. Also no differences in the emitted spectra were found from point to point on a device which had been held at 70°C for 5000 hours, but with no current passing through it. This therefore shows that the effects reported here for the lifetested devices must have occurred as a result of the operation of the devices.

4. Discussion

On examining lifetested GaAlAs/GaAs double heterostructure lasers in the SEM using cathodoluminescence spectroscopy at liquid helium temperatures the following observations were made:-

(i) The spectrum emitted from parts of the active layer well outside the stripe consisted of two broad CL peaks, one at about 800nm and one at about 815nm.

(ii) The spectrum of the light emitted from the active layer inside the stripe consisted of only one broad peak, at about 800nm.

(iii) The active layer spectrum from an undegraded device consisted of two broad peaks (800nm and 815nm), both inside and outside the stripe.

(iv) In degraded lasers, the active layer CL intensity was slightly greater under the stripe than outside it.

(v) In degraded lasers, the substrate CL intensity was slightly lower under the stripe than outside it.

The most surprising of these observations was the disappearance of the 815nm peak from the active layer spectrum under the stripe region of the degraded devices. Using the measurements made at room temperature the room temperature bandgap and hence the composition of the active layer could be deduced. Then knowing the composition, the bandgap at 4K could be calculated. This allowed one to assign the 800nm CL peak to the band to acceptor transition, while that at 815nm is at an energy of about 30meV less. The implication of the disappearance of the 815nm peak must be that the optical emission properties of the material under the stripe have improved. It is consistent though with the EBIC work on these lasers reported earlier (Wakefield and Robertson 1981), where it was found that when the devices were imaged using the EBIC mode of the SEM, the centre of the stripe region appeared brighter than the regions well away from the stripe, implying a lower carrier recombination rate in the bright areas. It was concluded in this work that the bright areas had been partially cleared of native defects, brought about by non-radiative recombination enhanced defect migration (Lang and Kimerling 1974), the direction of migration being determined by the strain gradients within the laser structure. The defects tend to move towards positions of maximum or minimum volume dilation, depending on whether they are vacancies or interstitials, with interstitials moving to positions of maximum dilation and vacancies to where the crystal lattice is being compressed most (see Fig.4). Combining this with the CL work, it is reasonable to account for the 815nm CL peak seen in the undegraded material as being related to the presence of the same defects as those which are gradually cleared from the stripe during operation, and from the direction in which the defects migrate it seems as if they are a vacancy type. In addition, the small increase in the active layer CL intensity found in the stripe region of the degraded lasers, lends support to this conclusion; viz that of the removal of one source of recombination centre.

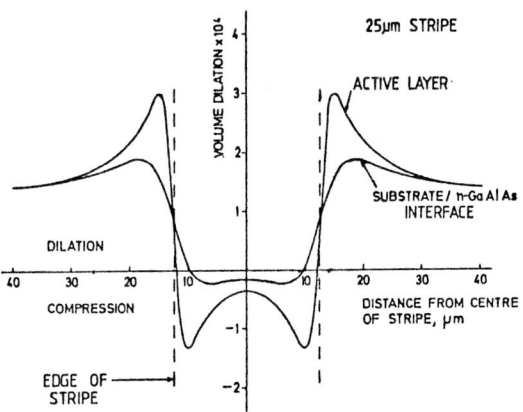

Fig.4   Volume dilation calculated for the active layer and substrate of a 25μm wide stripe laser.

If native defects are indeed moving out of the stripe area, then it is reasonable to relate the disappearance of the 815nm CL peak to the removal of the defects. These defects in the undegraded material give rise to a radiative recombination centre which leads to a radiative transition (at 4K) of lower energy than the normal band to acceptor transition. Removal of the defects removes the recombination centres and hence the radiative transition associated with them. Where defects accumulate to produce the dark EBIC and CL bands, they are known to coalesce into small dislocation loops (Robertson at al 1981). These loops are expected to be non-radiative recombination centres, and so would explain why no extra CL peaks are seen at the position of the dark bands.

The elimination or reduction of one of the recombination routes will increase the probability and hence the intensity of the normal near band radiative transition, and this is just what is observed. The radiative strength transfers from the peak near 815nm to the one near 800nm.

In the case of the substrate CL there has been a slight reduction in the CL efficiency under the stripe. The centre of the stripe at the substrate-epilayer interface is at a maximum of compression (Fig.4), so one would expect that the mobile defects would accumulate there. This might then cause an increase in non-radiative recombination, and as a result a reduction in CL intensity. Such a reduction is observed.

It is not obvious how these results relate to the actual performance degradation undergone by the lasers. During the lifetests at 70°C the threshold currents of these lasers had risen by, on average, about 3% per thousand hours. Yet on the face of it, the results presented here, and those reported earlier (Wakefield and Robertson 1981), appear to indicate that the material in the active stripe of the devices improves during operation. However, the "improvement" is by no means uniform along the length of the stripe. When examined using EBIC (Stevenson et al 1979, and Wakefield and Robertson 1981) the stripe region is found to have a mottled appearance, indicating that the recombination properties of the stripe material are not uniform. If this is the case then the optical wave passing up and down the laser cavity will be travelling through a medium with a fluctuating gain, which as a result could affect the overall gain of the laser. The longer the device is operated, the more non-uniform the gain profile becomes, leading therefore to a gradually increasing threshold current.

5. Conclusion

Cathodoluminescence spectroscopy at 4K on GaAlAs double heterostructure lasers has shown that changes occur in the emission properties of the active layer during extensive operation at elevated temperatures. Initially it is found that the active layer material contains a defect radiative recombination centre, which gives rise to a CL peak at about 815nm, depending on the precise composition of the active layer. This peak is excited at the expense of the band to acceptor peak. After lifetesting at 70°C, the 815nm peak is absent, or reduced in intensity, in those parts of the active layer underneath the stripe of the lasers. This effect has been explained by the removal of the defect centre from these parts of the devices by non-radiative recombination enhanced defect migration in the strain field within the devices. The removal of the defects does not lead to improved device performance because the clearing of defects is not uniform along the stripe. However their elimination from the material from which the devices are made should lead to an improvement in performance of these already very reliable lasers.

6. Acknowledgements

The author thanks Dr M Robertson for providing the lifetested lasers for this study. Acknowledgement is also given to the Director of the British Telecom Research Laboratories for permission to publish this paper.

7. References

Hatch C and Robertson M J, 1981, Private communication.
Holt D B, 1981, Inst Phys Conf Ser No.60 165-178
Lang D V and Kimerling L C, 1974, Phys Rev Lett. 33 489
Robertson M J, Wakefield B and Hutchinson P, 1981, J Appl Phys 52 4462
Skeats A P, Wakefield B and Robertson M J, 1983, J Appl Phys 54
    (to be published)
Stevenson J L, Skeats A P and Heckingbottom R, 1979, J Microscopy 118 321
Wakefield B and Robertson M J, 1981, Inst Phys Conf Ser No.60 447-451

Inst. Phys. Conf. Ser. No. 67: Section 6
Paper presented at Microsc. Semicond. Mater. Conf., Oxford, 21–23 March 1983

Use of SEM CL spectra to determine local variations in Ge doping concentration in LEC InP ingots

C A Warwick[*] and G R Booker

Department of Metallurgy and Science of Materials, University of Oxford, Parks Road, Oxford, OX1 3PH, UK

*Present Address: RSRE, St Andrews Road, Malvern, Worcs. WR14 3PS, UK.

Abstract   Scanning electron microscope (SEM) cathodoluminescence (CL) spectra were used to determine the local variations in Ge doping concentration in LEC InP ingots by measuring the Moss–Burstein peak shift and the impurity broadening. Total light SEM CL micrographs showed doping striations as bright and dark bands spaced ~ 25 µm apart. CL spectra were recorded from local areas at 88K. The band-to-band peak was found to overlap with a "ghost" peak. Modifications to the standard CL collection system eliminated this peak and enabled accurate determinations of the inhomogeneity of the doping concentration to be made.

1. Introduction

The advent of LEC n^+-InP as a substrate for InGaAsP/InP lasers and InP microwave devices has led to an interest in the perfection and uniformity of the substrate material. Dislocation free material can be obtained by heavy (~ 1 x 10^{19} cm^{-3}) Ge doping (Brown, Cockayne and MacEwan 1981). The aim of this work is to assess the dopant inhomogeneity of Ge doped LEC InP ingots grown under different conditions in order to understand the causes of the non-uniformity and to minimize the effect.

Local SEM CL spectra were recorded in order to use the Moss–Burstein shift and the impurity broadening to determine accurately the doping level with a high spatial resolution (~ 3 µm). Two overlapping peaks were found in each spectrum. The higher energy peak corresponds to the Moss–Burstein shifted band-to-band transition expected in degenerately doped material. The peak width was impurity broadened. The behaviour of the peak energy of the lower energy peak cast doubt on its origin, which was initially thought to be degenerate-band-to-Zn-acceptor recombination. Subsequent experiments, involving modifications to the CL collection system, showed the peak actually to be a "ghost" peak (Carr and Biard 1964). The presence of the ghost peak impairs the accuracy of measuring the doping level from the width or energy of the main band-to-band peak since the two peaks overlap. Elimination of the ghost peak by a modification of the CL collection system enabled increased accuracy to be obtained over our previous work. These results allowed conclusions to be drawn about the mechanism of striation formation and its dependence on average doping level, crystal growth conditions, radial position in the ingot and solidification mechanism i.e. facet and non facet growth (Warwick et al. 1983).

2. Experimental

Single crystal ingots of InP doped with Ge in the range 4 x 10^{16} to
2 x 10^{19} cm^{-3} were grown by the liquid encapsulated Czochralski (LEC)
method. Each crystal was sectioned longitudinally to reveal the ($\bar{1}$10)
plane and chemically polished flat with a 2% v/v Br$_2$/CH$_3$OH solution. Part
of each slice was used for Hall effect measurements using the Van der Pauw
technique. An adjacent part was used for the CL measurements, which were
made using 30 keV electrons at a probe current of 15 nA and probe diameter
of 0.2 μm. The generation volume in InP at 30 keV is about 3 μm in diameter. The specimens were cooled to 88K during examination and the CL
collected using one of the two systems illustrated in figures 1(a) and
1(c). Figure 1(a) is the standard system where light emitted from any-
where on the specimen surface within a circle of diameter 1.4 mm (centred
on the electron optic axis) is collected. Figure 1(b) is a detailed view
of the specimen in figure 1(a) showing rays A and B emerging at and away
from the electron probe and entering the collection system. Figure 1(c)

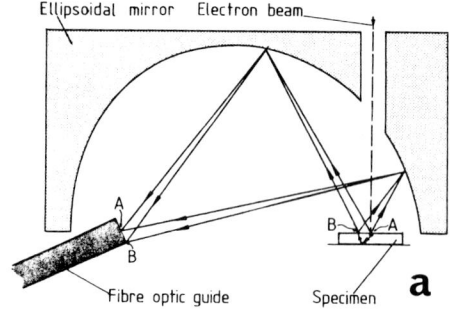

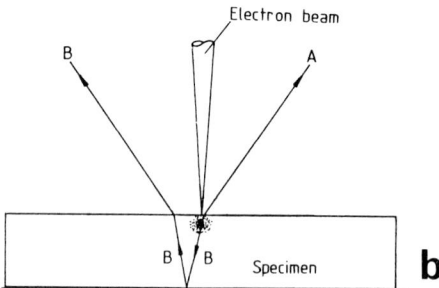

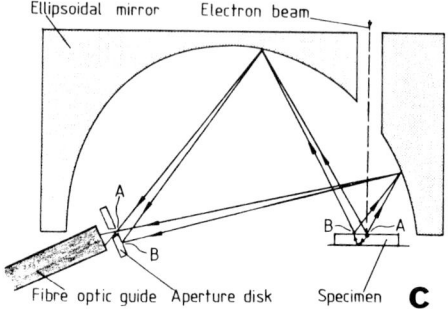

Fig. 1 CL collection systems
(a) standard system (b) as (a)
but showing the rays leaving
the specimen (c) modified
system with restricted field
of view

shows a modified system where an aperture of diameter 250 μm is placed at
the second focus of the mirror. The fibre optic is placed behind the
aperture. This has the effect of excluding ray B from entering the fibre
optic. The field of view is reduced to 100 μm. In both cases the fibre
optic was coupled to a f = 600 mm f4 grating monochromator and the spectra
were recorded using an S1 response photomultiplier (EMI type 9808B),
cooled to 200K.

3.   Results and Discussion

Figure 2 shows a total light micrograph of the seed end of crystal MR796.
The curved bands are the dopant striations. Edge facets (E) and dislocation dark spots (D) as well as some scratches (S) are also visible. For
each specimen CL spectra from light and dark striation bands were recorded
using the standard system (figure 1(a)). Figure 3(a) shows the pair of
spectra for crystal L911 and figure 3(b) those for L979. The mean carrier
concentration of L911 was 1.2 x 10^{18} cm^{-3} and that of L979 was 8.2 x 10^{18}
cm^{-3} (as measured by the Van der Pauw method). The curve labelled H corresponds to the heavily doped part of the striation and L to the lightly
doped part. Two overlapping peaks are seen in each spectrum; the band-to-
band peak (shorter wavelength) and the ghost peak (longer wavelength).
The expected dependence of peak wavelength versus carrier concentration
for a peak is as follows; the peak wavelength should shorten as the
carrier concentration is increased, whether by changing the specimen to a
more highly doped one or by moving the electron probe from a lightly doped
striation to a heavily doped one. The wavelength shifts are referred to
the peak wavelength in undoped material i.e. 883 nm for the band-to-band
and 900 nm for the ghost peak (see figure 6, curve S). This dependence is
seen for the band-to-band peak as shown in figure 3 i.e. shift A<B<C<D.
For the ghost peak this is not seen since shift W = X < Y = Z. The fact
that the ghost peak does not shift on exciting different striations but
does shift on changing the specimen suggests that the peak is characteristic of the bulk of the specimen. The following hypothesis is proposed
to explain this. The light emerging from the specimen is from two paths
(labelled ray A and ray B in figure 1(b)). Only ray A is normally considered, but since this forms only ~ 1.5% of the light generated (Gooch
1973) and because of the broadness of the spectrum and the position of the
absorption edge, ray B must also be taken into account.

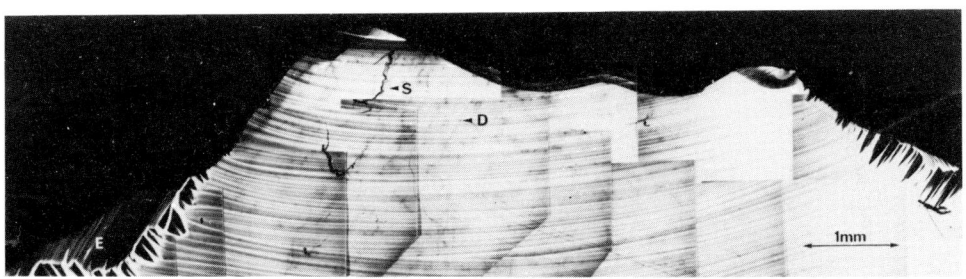

Fig. 2   SEM CL micrograph.   Crystal MR796 seed end showing striation bands
N_e = 4.7 x 10^{18} cm^{-3}

98.5% of the light generated cannot leave the specimen directly along path
A due to either total internal reflection or being directed away from the
top surface of the specimen. The absorption edge is at an energy such
that most of the shorter wavelengths of the band-to-band peak are absorbed
but the absorption coefficient at the long wavelength tail of the band-to-

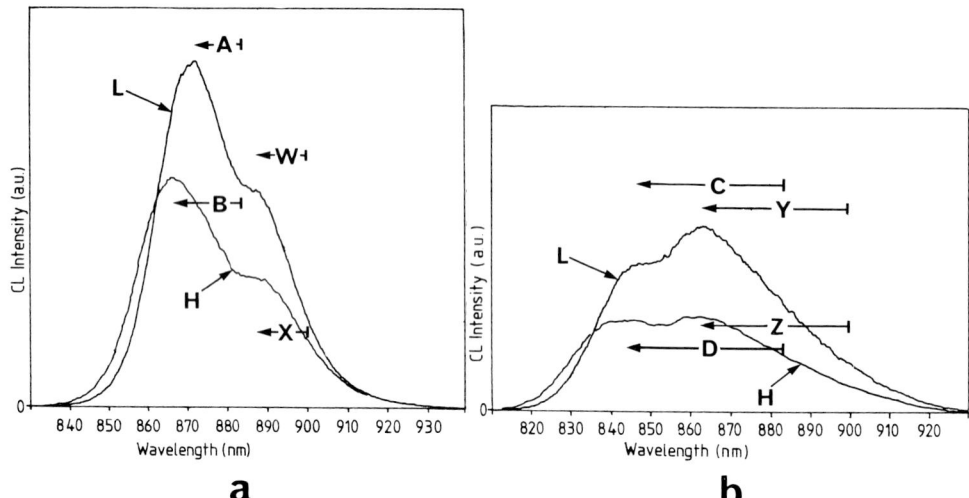

Fig. 3   Local CL spectra with 1.4 mm field of view showing band-to-band and ghost peak.   (a) L911: $N_e = 1.2 \times 10^{18}$ cm^{-3} (b) L979: $N_e = 8.2 \times 10^{18}$ cm^{-3}.

band peak is small enough to ensure that a large fraction of the 98.5% of the ray B light can be scattered off the back surface and leave the front face of the specimen away from the position of the electron probe.   The spectral properties of the light will be determined by the bulk of the specimen, not by the emission properties of the excited volume.   To test this hypothesis the second light collection system in figure 1(c) was used to record spectra from the same specimens.   Figure 4(a) shows the spectra from the heavily (H) and lightly (L) doped part of crystal L911 and figure 4(b) those for L979.   By comparison with figures 3(a) and 3(b) it can be seen that the ghost peak has been removed since the collection system used (figure 1(c)) excluded ray B light from the fibre optic.   This shows that the ghost peak light does indeed travel by the path labelled B on figure 1(b).   This result was confirmed with the following experiment using the standard collection system (figure 1(a)) on two specimens of un-doped InP.   One was a standard specimen (eg figure 1(b)) but the other had a 0.25 μm thick 750 μm diameter aluminium dot evaporated on it.   The dot was electron transparent but light opaque (see figure 5).   The spectrum recorded when the electron beam was positioned at the centre of the dot is labelled D in figure 6.   Only ray B light enters the collection system so only the ghost peak appears (900 nm).   Also shown on figure 6 is curve S which is the spectrum for undoped InP using the standard specimen.   The band-to-band (883 nm) and ghost peaks (900 nm) appear.   It was found that in undoped InP (where the absorption edge is much steeper than in doped InP), the ghost peak could not be completely removed using the modified system of figure 1(c), a smaller field of view was required.   However, when a smaller aperture was used, aberrations in the mirror prevented a sufficiently small field of view being obtained.   The system in figure 7 was adopted.   Here an aperture disc is glued directly on to the specimen. Figure 8(a) shows the spectra from a 50 μm diameter aperture and figure 8(b) from a 10 μm diameter aperture.   The specimen in both cases is the same undoped InP.   Only the 10 μm diameter aperture gives the correct spectrum.   In the case of undoped InP, the longer wavelength tail of the

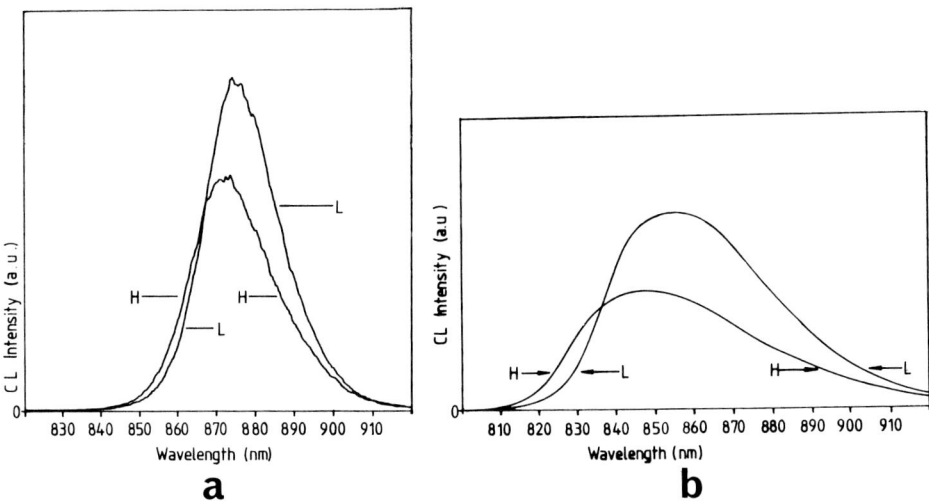

a　　**b**

Fig. 4　Local CL spectra with 100 µm field of view showing band-to-band peak.　(a) L911　(b) L979.

band-to-band peak (which is narrow compared to doped material) is small in intensity compared to the acceptor recombination due to Zn impurities, which account for the small shoulder in figure 8(b) at 900 nm.

4.　Conclusions

The use of CL spectra at 88K to measure doping levels can be made quantitative only if the field of view of the system excludes ghost peaks.　A

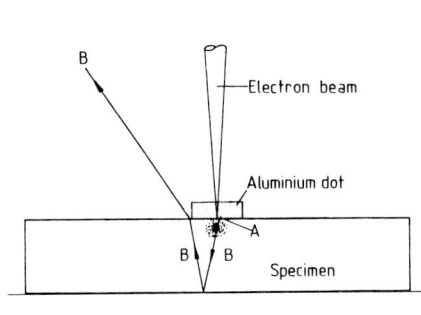

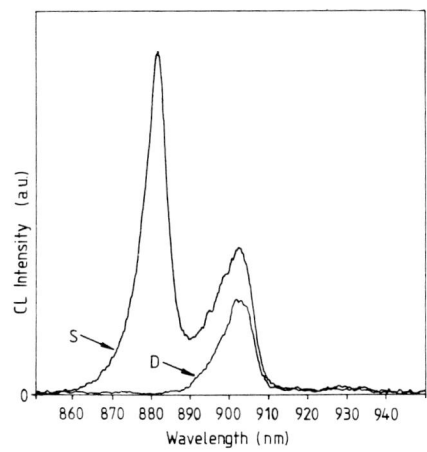

Fig. 5　Modified specimen for ghost peak enhancement.　Compare with figure 1(b).

Fig. 6　Local CL spectra of InP:Un. S-standard specimen (eg figure 1(b)) Curve D - aluminium dot specimen (eg figure 5)

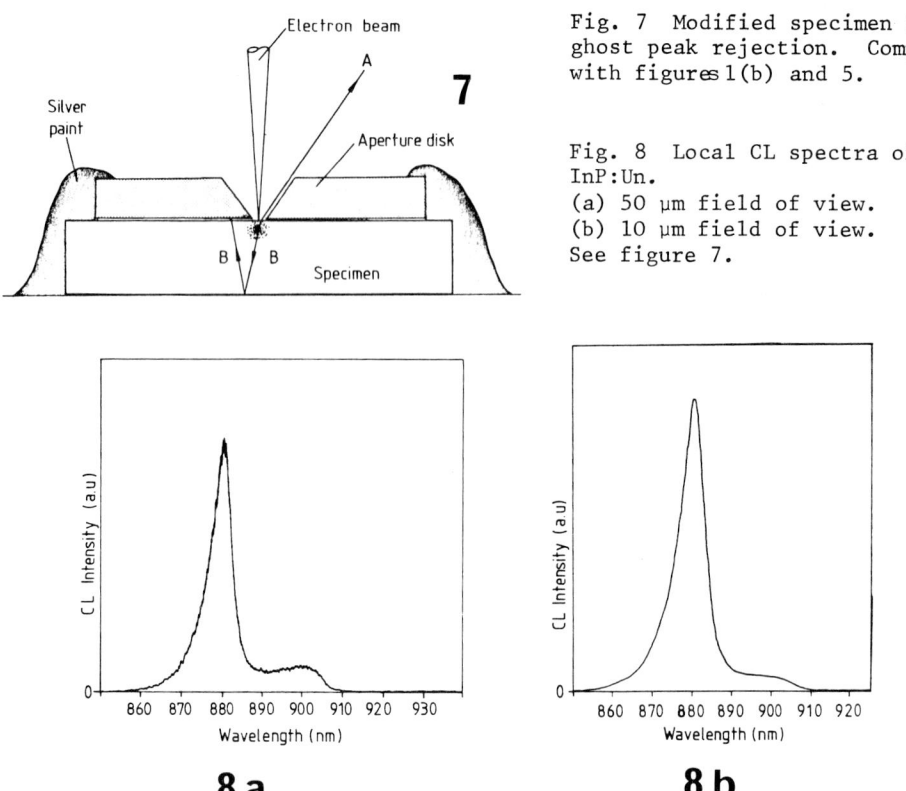

Fig. 7 Modified specimen for ghost peak rejection. Compare with figures 1(b) and 5.

Fig. 8 Local CL spectra of InP:Un.
(a) 50 μm field of view.
(b) 10 μm field of view.
See figure 7.

field of view of 100 μm is small enough for doped InP but 10 to 50 μm is necessary in the purest material. The method of conversion of the peak wavelength and width into doping concentration and the results obtained from different ingots and growth conditions are presented elsewhere (Warwick et al. 1983).

Acknowledgements

This work was carried out under SERC(UK) support through a CASE award with RSRE Malvern. Specimens and Hall measurements provided by RSRE Malvern and Cambridge Instruments.

References

Brown G T, Cockayne B and MacEwan W R 1981 J. Cryst. Growth 51 369.
Carr W N and Biard J R 1964 J. Appl. Phys. 35 2776.
Gooch C H 1973 Injection Electroluminescent Devices (London: John Wiley & Sons) pp 58-9.
Warwick C A, Brown G T, Booker G R and Cockayne B 1983 J. Cryst. Growth Special Edition "2nd NATO Workshop on Materials Aspect of InP" (in press)

Inst. Phys. Conf. Ser. No. 67: Section 6
Paper presented at Microsc. Semicond. Mater. Conf., Oxford, 21–23 March 1983

327

SEM CL assessment of minority carrier lifetime in silicon

S Myhajlenko*, S M Davidson+ and B Hamilton

Department of Electrical Engineering and Electronics,UMIST, Sackville St, Manchester M60 1QD.

* Now at H H Wills Physics Lab, University of Bristol.

+ GEC Hirst Research Centre, Wembley, Middx HA9 7PP

Abstract   The potential value of cathodoluminescence assessment of silicon is considered. The excitation requirements for silicon work and the experimental set-up used are described. Results are presented of imaging, spectra and decay measurements performed on a variety of silicon samples. In particular, a quantitative description is given of deep level, Auger and surface recombination effects observed in cathodoluminescence decays.

1. Introduction

The cathodoluminescence (CL) mode of the scanning electron microscope (SEM) potentially offers a non-destructive, contactless method of material/device assessment with reasonable spatial resolution (micron regime). In the first instance, the luminescence efficiency of a semiconductor can often provide a preliminary indication of the electrical quality. In such cases the band edge CL efficiency is sensitive to the presence of non-radiative point defect recombination centres. Spatially extended defects, such as dislocations, grain boundaries, etc., may also give rise to non-radiative regions. Such features can be conveniently monitored by CL imaging. This is important if the contrast is associated with a local reduction in the lifetime, which implies electrical activity. More quantitative information may be derived from CL spectra and decay measurements. The latter is significant if the decay data yield a minority carrier lifetime which correlates with the presence of a particular defect state in the material. Such studies in the past on GaAs and GaP have made important contributions to our understanding of defect states in these materials and hence have contributed to significant improvements in material processing and device performance. Similar studies on silicon would be useful in the light of recent developments in processing and devices. However, silicon has hitherto been excluded from routine CL studies due to its low internal radiative efficiency. In addition, given the relatively poor performance of detectors in the spectral region of interest (1.15 µm for band edge), progress in silicon CL work has been slow. Nevertheless, accounts of CL imaging (Cumberbatch et al 1981) and CL spectra studies of silicon (Vouk and Lightowlers 1977) have appeared in the literature.

2. Experiment

The typical experimental set-up used in this work is shown in Fig. 1.

The SEM based system has been specifically built and optimized for CL studies of semiconductors. Further details concerning the design features of the instrument can be found elsewhere (Davidson et al 1981).

For CL imaging work, the integrated CL route is normally used, where detection is by means of a S1 photomultiplier. For CL spectra work, conventional phase sensitive detection is used, where a grating spectrometer and an InSb detector are used in conjunction with the necessary instrumentation. For CL decay measurements, a variation of the delayed coincidence technique is used. The arrival of the first photon (pulse) following beam cut off initiates (START) timing circuits in a Le Croy 3001 multichannel analyser (MCA). A suitably delayed STOP pulse derived from the beam blanking signal terminates the timing measurement. The time difference between the START and STOP pulses is then stored in an appropriate channel of the MCA memory. Provided that the photon detection probability per blanking cycle is low (< 10%), then the resulting histogram of the time distribution of photon arrivals will represent the CL decay. Also, the START/STOP pulses have been transposed relative to convention in order to overcome the excessive MCA dead time which would otherwise result as a consequence of unconsummated STARTS. As a result of this transposition, the recorded CL decays appear inverted.

3. Excitation Conditions

On the assumption that the CL system detection efficiency is a fixed and optimized quantity, then the observed CL signal strength will be primarily determined by the excitation and the recombination physics. Hence for silicon, the excitation used tends to be higher than that used for III-V material evaulation. In this work, beam voltages in the range 30-35 kV, beam currents of 2 - 5 µA, with spot sizes between 1 - 5 µm have been used. This excitation regime is comparable to that used for deep level luminescence measurements on III-V materials. This higher excitation has two undesirable effects, (1) high excess carrier generation rates and (2) beam heating effects.

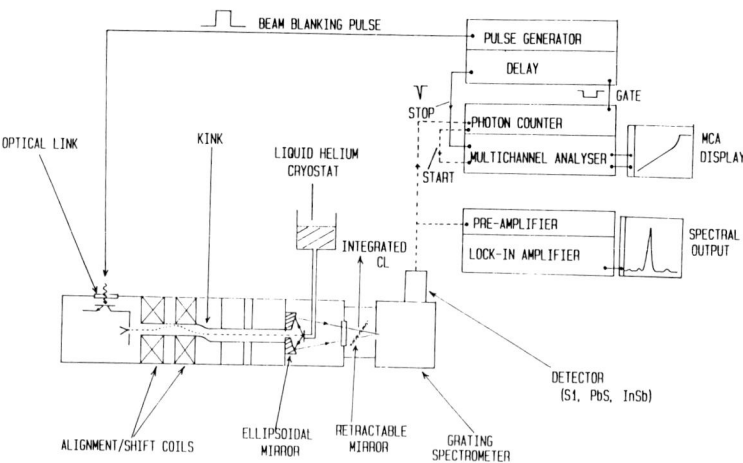

Fig. 1. The SEM based experimental set-up.

The excess carrier densities generated by the above excitation can be calculated by solving the relevant carrier diffusion equation with appropriate boundary conditions. From Davidson and Dimitriadis (1980), the peak density which occurs within the generation volume, is expected to be in the range 10^{17} - 10^{19} cm^{-3} for the above excitation regime. In comparison, the mean carrier density which is highly sensitive to the minority carrier properties, can be many orders of magnitude lower than the peak value.

Beam heating of the semiconductor is potentially a more serious effect. This time the heat conduction equation must be solved in order to determine the temperature rise. For this purpose, a modified form of the Vine and Einstein (1964) model is used which allows for the temperature dependence of the thermal conductivity. The predicted maximum steady state temperature rise for a 30 kV beam (3 μm spot diameter) as a function of the substrate temperature (T_S) is shown in Fig. 2. The results indicate that substrate cooling can be quite effective in limiting the temperature rise, i.e. for T_S < 100 K.

4. CL Spectra Results

Nuclear transmutation doped (NTD) silicon (100 Ω cm) has been used in order to establish the origin of the CL emission. A single broad emission band at 275 K (peak $\simeq$ 1.08 eV, FWHM = 110 meV) is resolved into four components at 17 K, see Fig. 3. An effective increase in the CL efficiency is observed with the decrease in temperature as the radiative route becomes dominated by excitonic decay processes. This alleviates the excitation requirements (predicted beam heating at 17 K: < 0.5 K). The CL spectrum at 17 K is similar to that reported by Haynes et al (1960) using photoluminescence. The I_{TO}(FE) (1.099 eV) and I_{TA}(FE) (1.138 eV) bands are due to free exciton decay with transverse optical and tranverse acoustical phonon emission, respectively. Band A can be attributed to a two phonon assisted process. Band B on the other hand does not conform with the selection rules for such a process. It is tempting to ascribe this emission to some residual 'nuclear damage' since it is also observed at 75 K. Further, this emission is not observed in any other silicon sample assessed in the same temperature regime. The slight structure (C) is probably associated with the unresolved phosphorus lines.

5. CL Imaging Results

Fig. 4 shows defects of unknown nature/origin observed in argon implantation damaged silicon. Of interest is the resolution, where the linewidth is similar to the expected spatial extent of the generation volume (9 μm at 30 kV). The imaging facility in principle allows further quantitative high resolution CL spectra/decay measurements. Fig. 5 shows CL decays recorded at line features (grain boundaries) in solar development grade ribbon silicon. The difference in lifetimes implies that some boundaries are electrically more active - a significant observation, since the solar cell efficiency will be reduced by the presence of the 'dark line' features.

6. CL Decay Results

The minority carrier lifetime in silicon is known to be sensitive to the material growth and subsequent device processing, e.g. introduction of deep levels. The CL decay behaviour of silicon containing intentionally introduced deep level gold impurities is reported here. This has both

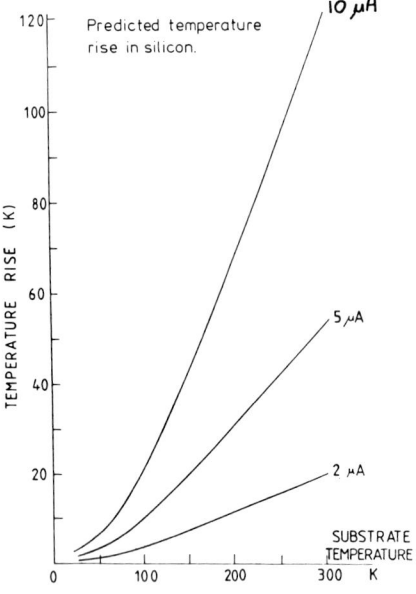

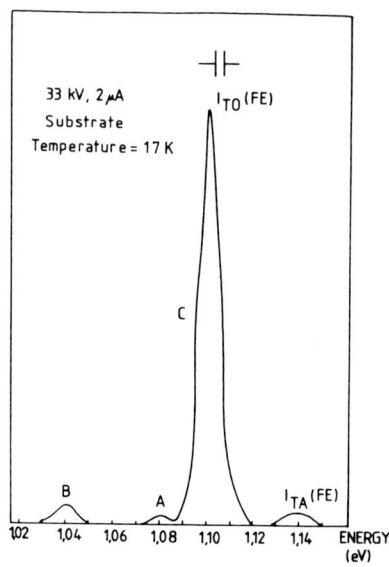

Fig. 2.   Beam Heating in silicon
for 30 kV beam.

Fig. 3.   CL spectrum from NTD
silicon.

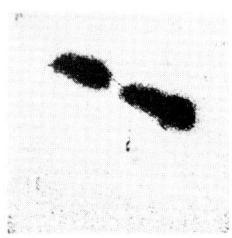

⊢—⊣ 20 μm

Fig. 4.   Defects in
argon-implantation
damaged silicon.

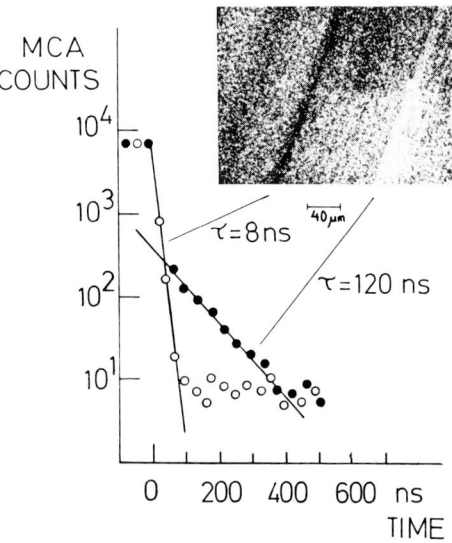

Fig. 5.   Decays from line features in
ribbon silicon.

technological (switching/power devices) and fundamental relevance. Namely that the behaviour of the gold centre (donor/acceptor levels) in silicon is now no longer thought to be associated with the simple substitutional impurity, but with gold related defects, i.e. gold complexes with host defects (Van Vecten and Thurmond 1976). Fig. 6 shows typical CL decays (log-linear) recorded over four orders in signal from n-type (2 x 10^{15}cm^{-3}) phosphorus) Czochralski silicon at 300 K containing various amounts of gold: (a) 10^{13} cm^{-3}, CL bulk lifetime τ = 1.62 μs (b) 10^{14} cm^{-3}, τ = 98ns and (c) 3 x 10^{14} cm^{-3}, τ = 40 ns, where the control sample starting CL lifetime is 2.25 μs. The initial non-exponential phase of the decay is normally attributed to surface recombination effects. For silicon, this process is expected to be limited by carrier diffusion, i.e. assume in-finite surface recombination velocity. The observed enhancement of the initial decay component in samples with increasing τ is consistent with such an interpretation. After about one lifetime the CL transient settles into a single mode exponential decay which should relate to the bulk excess carrier recombination process, i.e. yield the bulk lifetime.

Schottky barrier deep level transient spectroscopy (DLTS) measurements have also been performed on these samples. A correlation is observed between the CL bulk lifetime in n-type silicon and the electrically active gold acceptor concentration N_T, see Fig. 7, where the fit to the predicted behaviour is quite good (dashed line), i.e. τ = $(C_p N_T)^{-1}$ the Hall-Shockley-Read (1952) low level lifetime, where C_p is the hole capture coefficient of the gold related acceptor level. The capture cross section data of Wu et al (1981) has been used for this purpose. No CL lifetime-deep level correlation
is observed with gold diffused p-type (boron
doped) Czochralski silicon, where in contrast
to the 'clean' DLTS spectra observed in the
n-type samples, the p-type DLTS spectra contain multiple deep levels (including the gold
donor level). The lifetime behaviour
(typically < 10 ns) is not consistent with
that expected from the donor or combined
donor/acceptor properties. This is thought
to be due to the formation of boron related
complexes in the diffusion temperature regime
used (800 - 950 °C) (Gerson et al 1979).
Further details of this work will appear
elsewhere (Myhajlenko).

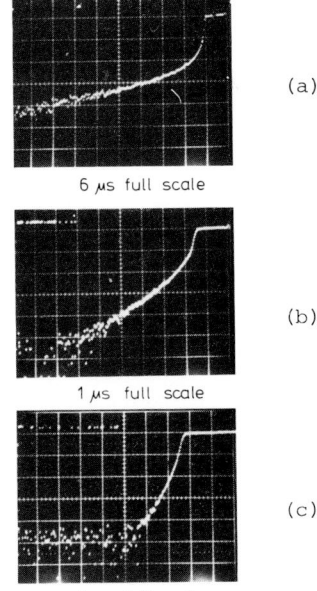

(a)

6 μs full scale

(b)

1 μs full scale

(c)

1 μs full scale

Fig. 6. CL decays from gold diffused silicon.

The CL decay behaviour of n-type antimony
doped (2 x 10^{18} cm^{-3}) Czochralski silicon
(not gold diffused) is found to be Auger recombination controlled. For silicon, a
phonon assisted Auger process should yield
the following lifetime dependence $\tau_A = (Cn^2)^{-1}$,
where C is the Auger coefficient and n is the
electron concentration. The CL decay data
yield a value of C = 3 x 10^{-31}cm^6 s^{-1} (300K)
which is consistent with reported values
(1.7 - 3.4 x 10^{-31} cm^6 s^{-1}). Further, the
CL lifetime is found to be independent of
the temperature between 100 - 300 K
(τ_A = 825 $\pm$ 15 ns) and decreases between
300 - 400 K (τ_A(400 K) = 720 ns), which again

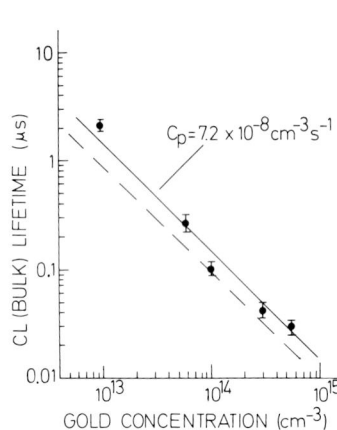

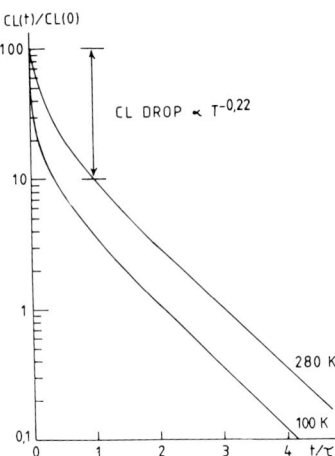

Fig. 7. CL lifetime-deep level correlation.

Fig. 8.   Temperature dependence of CL decays.

is consistent with the observations of Dziewor and Schmid (1977) who used photoluminescence decays. Also, in the constant CL (bulk) Auger lifetime regime, the initial CL drop is found to be temperature dependent, see normalized CL decays in Fig. 8. One would expect that the magnitude of the initial CL drop after one lifetime, representing the transition from surface to bulk dominated excess carrier recombination, to have the temperature dependence of the bulk diffusion length, namely $T^{-0.25}$ for lattice scattering, i.e. mobility α $T^{-3/2}$ and via the Einstein relation. Experimentally, a $T^{-0.22}$ dependence is observed which adds support to the surface recombination argument for silicon.

References

Cumberbatch T J, Davidson S M and Myhajlenko S 1981 Inst.Phys.Conf.Ser.No. 60 pp 197-202.
Davidson S M and Dimitriadis C 1980 J.Microsc. 118 275.
Davidson S M, Cumberbatch T J, Huang E and Myhajlenko S 1981 Inst.Phys. Conf.Ser.No. 60 pp 191-6.
Dziewior J and Schmid W 1977 Appl.Phys.Lett. 31 346.
Gerson J D, Cheng L J and Corbett J W 1979 J.A.P. 48 4821.
Hall R N, Shockley W and Read W T 1952 Phys.Rev. 87 p387 and p835.
Haynes J R, Lax M and Flood W F 1961 Int.Conf.Phys.Sem. Prague p423.
Myhajlenko S (to be published).
Van Vecten J A and Thurmond C D 1976 Phys.Rev. B14 3539.
Vine F and Einstein P A 1964 Proc. IEEE 111 921.
Vouk M A and Lightowlers E C 1977 J.Phys.C 10 3689.
Wu R H, Myhajlenko S, Hamilton B and Peaker A R 1981 3rd Lund Deep Level meeting Abstracts p104.

Acknowledgements

Financial support by the SERC and the Paul Institute Fund is gratefully acknowledged. Thanks are due to Dr. W. Ke for useful comments and Dr. R. Musil for conceptual support.

Inst. Phys. Conf. Ser. No. 67: Section 6
Paper presented at Microsc. Semicond. Mater. Conf., Oxford, 21–23 March 1983

A low-temperature, high-resolution, computer-controlled SEM CL mode detection system

F M Saba and D B Holt

Department of Metallurgy and Materials Science, Imperial College of Science and Technology, London, SW7 2BP.

Abstract   The use of a liquid helium stage makes phenomena visible in CL emission spectra that cannot be seen at higher temperatures. Examples are quoted related to GaAs and a III-V alloy. To take full advantage of the sharpening of some emission bands at lower temperatures, a mono-chromator with a higher resolution-throughput product and a computer-controlled stepper motor drive were introduced. The effectiveness of the signal averaging this makes possible is illustrated.

1. Introduction

The development of CL mode detection systems has been pursued in this labor-atory for some time, Steyn et al (1976) reported an efficient spectroscopic detection system for CL mode studies of specimens at room temperature in an SEM.   Holt (1981) gave a preliminary account of a liquid helium stage for SEM CL mode work based on an Oxford Instruments continuous flow liquid helium cryostat system.   This stage was designed for maximum thermal efficiency without regard to the efficiency of light collection and transmission to the monochromator and photomultiplier.   To illustrate the value of low temperatures in the CL mode, two recent observations will be reported.

2. CL From GaAs Epitaxial Multilayer Material

Two samples grown at the GEC Hirst Research Centre were examined.   Each had three epitaxial layers grown on GaAs liquid-encapsulated Czochralski (LEC) substrates.   A buffer layer 2μm thick was grown first, then an undoped layer about 3.5μm thick and finally a top n-type layer.   In one specimen, this was 0.7μm thick whereas in the other, the top n-layer was only 0.45μm thick.   At room temperature using a beam voltage giving a penetration depth of a couple of μms both specimens gave similar fundamental band CL emission at $h\nu \simeq E_g$, the gap energy.

At liquid helium temperatures (8K) however they gave quite different spectra. The specimen with the thicker top n-layer gave a spectrum in which the fundamental emission band dominated, with its peak shifted to a photon energy corresponding to the lower-temperature wider energy gap value. There appeared also to be a shallow extrinsic emission and a hint of a deeper recombination level band as shown in Figure 1.   The specimen with the shallower n-type top layer exhibited only apparent evidence of a low intensity, noisy deep-level CL band.   This observation could be of interest in relation to the contaminants in LEC GaAs and the distances they can diffuse during epitaxial layer growth.   However, the poor resolution and the low signal to noise ratio in these spectra showed that improved detection was needed.   It would be premature to place any reliance on the details of these spectra.

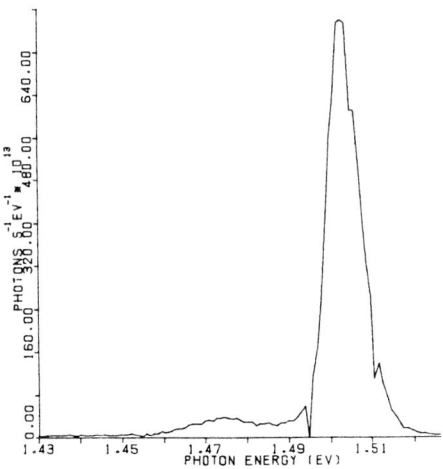 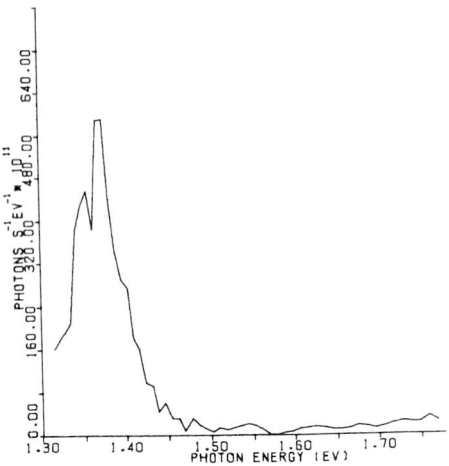

Figure 1. CL emission spectrum
obtained from a triple epitaxial
layer GaAs specimen at 8.2K with
a low-resolution detection system.

Figure 2. CL emission spectrum of
a $Ga_xIn_{1-x}As_yP_{1-y}$ alloy at 6.8K
recorded with a low-resolution
detection system.

3.   Low Temperature CL from a III-V Alloy

Previous observations on samples of $Ga_xIn_{1-x}P$ alloy material showed that the
single fundamental CL band seen at higher temperatures is joined below about
8K by smaller bands.   One falls on either side of that corresponding to the
energy gap of the average composition of the alloy.   It was thought possibl
that this was indicative of spinodal decomposition of the alloys (Holt,1981)

More recent observations on a quaternary $Ga_xIn_{1-x}As_yP_{1-y}$ alloy gave a
similar result as shown in Figure 2.   There are emissions at photon energie
higher than that of the fundamental band of the average composition at the
left.   No localized level in the band gap would produce such emission.
This again suggests significant local variations from the average compositio
i.e. some form of phase separation or "clustering".   TEM observations on
this specimen showed contrast characteristic of spinodal decomposition.
Similar TEM observations were explained in terms of clustering by Gowers
(1983).   The spectrum of Figure 2 also illustrates the need for improving
the detection system.

The problem is compounded by the fact that practical interest is increasingl
concentrated on materials luminescing in the wavelength range from 800 nm
upwards.   Hence S1 photocathode photomultipliers have to be used.   The
quantum efficiency of this type of long-wavelength photomultiplier is
typically only about 0.3%.   Hence not only must the monochromator throughput be maximized but techniques for improving the signal to noise ratio such
as signal averaging should be adopted.   This technique is well-known in
other fields but appears to be new to the CL mode field.   Hence, this paper
presents some details of a system employing it, and gives an example
illustrating the effectiveness of signal averaging.

4. The Modified CL Detection System

Experience like that reported above confirmed that the monochromator that
had previously been satisfactory for use on the thermally-broadened CL
spectra observed at liquid nitrogen temperatures and above had neither the
spectral resolution (wavelength bandwidth) nor the throughput (transmissive
efficiency) needed for the sharper liquid helium spectra of the prototype
liquid-helium stage. It has been replaced by a monochromator of greater
resolution-throughput product, a Bentham Instruments M3OOE fitted with order-
sorting filters. Both the grating and the filter wheel are driven by a
Bentham Instruments SMD3B/IEEE stepper motor drive unit which is controlled
by a Research Machines 380Z microcomputer via an IEEE 488 interface. A
lens system was built between the fibre-optic light guide from the light
collector and the entrance slit of the monochromator. This was necessary
to reduce the loss of light due to the divergence of the beam from the
light guide. The system is shown in Figure 3.

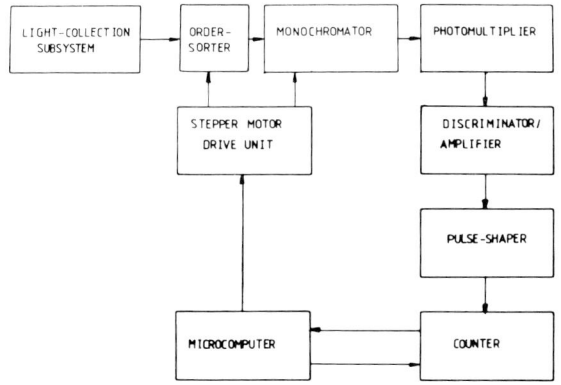

Figure 3. Schematic diagram of the computer-controlled CL detection system.

The multi-channel scaler (MCS) previously used for data logging was replaced
by a 24-bit synchronous counter built from standard TTL integrated circuits
and interfaced to the microcomputer. The counter can be stopped, cleared,
started and read under the control of the computer. A program was written
to record CL spectra and store them as files on floppy discs. The program
obtains its operating parameters from the user through a question and
answer sequence. These parameters include the passband (spectral
resolution, which is selected manually on the monochromator), the wavelength
interval between spectral data points, the dwell time (photon counting
time for each wavelength recorded) and the number of spectral scans to be
carried out successively in the increasing and decreasing wavelength directions. The required initial and final wavelengths of the spectrum are also
entered. It then drives the monochromator to the required wavelength,
clears and starts the counter and waits for the dwell time to expire. The
data are transferred to memory and the cycle is repeated until the final
wavelength of the last spectral scan has been recorded. The data are then
plotted on the computer's visual display unit using high resolution graphics.
Dwell times can be set in the range from 20μs to 1.3s. Far more spectral
wavelength points can be recorded than the 1024 to which the MCS was limited.
The multiple scanning of CL spectra averages out the random background noise.
This is known as signal averaging.

5. Signal Averaging of CL Spectra

The system is working but has not yet been employed in low-temperature observations. However, Figure 4 illustrates the improvement in signal to noise ratio obtained by averaging over 40 scans rather than 20 scans across the spectral range. The noise in Figure 4(a) appears greater than in Figures 1 and 2 partly because very many more closely spaced wavelengths were recorded with the new higher-resolution system.

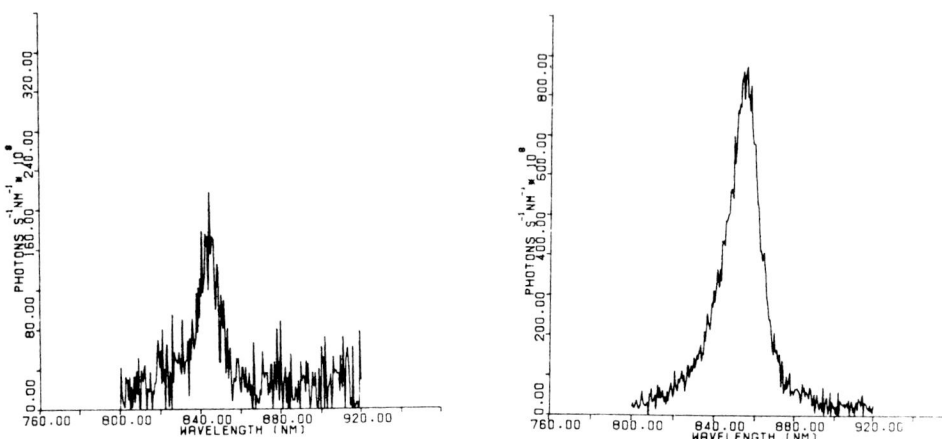

Figure 4. CL spectra of a triple epitaxial layer GaAs specimen at -40°C with the signal averaged over (a) 20 scans across the spectrum and (b) 40 scans.

Up to 90 scans, taking an hour have been recorded with continuing improvements in signal to noise ratio. A limit must ultimately be set by various forms of instrumental drift but it is not yet known what that limit is.

Computer control also makes possible selective data logging and feedback control of the system operating parameters. These techniques are being explored.

Acknowledgements

Thanks are due to Dr. D. Wickenden of the GEC Hirst Research Centre for supplying the GaAs samples and to Dr. J.P. Gowers of the Philips Research Laboratory for the supply of the GaInAsP alloy specimen.

References

Gowers, J.P., 1983, Appl. Phys. to be published.
Holt, D.B., 1981, Microscopy of Semiconducting Materials, 1981, ed. A.G. Cullis and D.C. Joy, Conf. Series No. 60 (Bristol: Inst. Phys.) pp.165-178.
Steyn, J.B., Giles, P. and Holt, D.B., 1976, J. Microscopy, <u>107</u>, 107-128.

Inst. Phys. Conf. Ser. No. 67: Section 6
Paper presented at Microsc. Semicond. Mater. Conf., Oxford, 21–23 March 1983

337

Spatial distribution of donors in MOCVD ZnSe

S H Roberts Oxford Instruments Ltd., Osney Mead, Oxford OX2 0DX
M Grande, J L Batstone, J W Steeds, University of Bristol
P J Dean, P J Wright, B Cockayne, RSRE, St. Andrews Rd, Malvern.

Abstract We report an investigation of the spatial and spectral features of
the cathodoluminescence (CL) in zinc selenide, undertaken on the CL system at
Bristol University. At specimen temperatures of ~35K the CL emission due to
a number of recombination processes is resolved. False colour monochromatic
scanning topographs were obtained using a computer to store and process
single photon photomultiplier pulses. A number of interesting features are
seen in these images, indicating inhomogeneities in the donor concentration
in the specimens investigated.

1. Introduction

Examination of the spatial and spectral features of the cathodoluminescence(CL)
excited in a sample by the beam of an electron microscope provides a powerful
analytic tool. When combined with the other microanalytic facilities available in
this environment (TEM, EBIC, EDX etc) it is especially appropriate for investigating small scale defects in semiconductor devices. It is of particular value in
the non-destructive examination of thin layers to be used in optical applications.

In comparison with photoluminescence (PL), CL lacks the potential for selective
excitation obtained by monochromatic illumination, and for the very low temperatures accessible in PL techniques. However, the very fine probe size and high
energy density attainable with an electron beam means that the emission features
can be imaged with a spatial resolution much less than the wavelength of the
emitted light, in this case ~ 0.2μ (Roberts and Steeds 1982). Due to the non-selec-
tive nature of electron excitation, CL spectra may be complex and difficult to
interpret. By using samples already well characterised by PL techniques, it is
possible to interpret many of the spectral characteristics of the CL emission.

Signal intensities in CL microscopy can be very low because of the small volume
of material excited at high magnification and the narrow wavelength bandpass for
spectral analysis. To exploit the full advantages of the technique it is essential to
use an efficient light collection system.

Moreover, this limited amount of data must be stored accurately and comprehen-
sively. For this reason we have interfaced the CL detection system implemented
on the Bristol University Philips EM400 TEM to a dedicated Link Systems 860
system II computer. This provides interactive control of the monochromator
grating position and the microscope scanning coils, as well as data recording
and image processing facilities.

The CL system is being used to investigate a wide range of materials including II-VI and III-V compound semiconductors, at temperatures between 35K and 300K using a liquid helium cooled specimen holder (Eades 1982). Here we report an investigation of zinc selenide samples grown for applications in optical devices and previously characterised by PL. The results given show the value of computer recording and processing the signals.

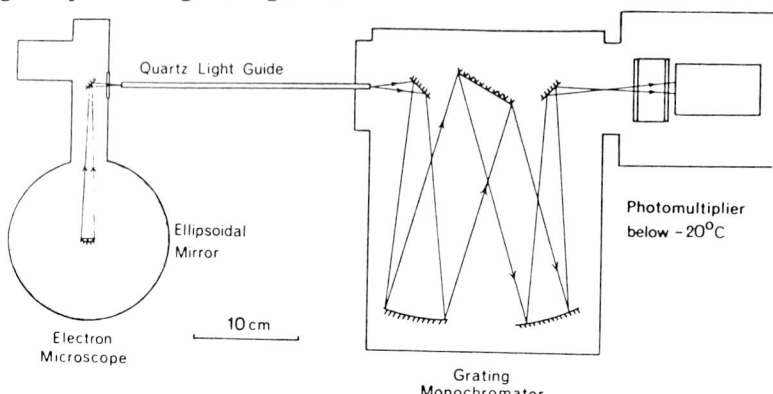

Fig. 1 Plan view of optical path for CL collection and analysis.

2. Experimental

Fig. 1 shows a schematic diagram of the light collection optics (Roberts 1981) employed in this experiment. The CL intensity is measured by an RCA 31034A-02 photomultiplier tube with a single photon collection capability and a spectral range from 200 to 900nm. These pulses can be acquired and stored in the computer in two modes, either as a spectrum or a monochromatic scanning topograph. The two acquisition modes interact very fruitfully; thus spectra may be acquired at points found to be of interest by monochromatic imaging, while images can be obtained at wavelengths corresponding to interesting spectral features. Using the framestore it is possible to image very weak spectral features, by integrating over times of 1000 seconds or more; the total time is limited only by specimen stability. The image processing facilities include spatial and temporal filtering, background subtraction and the addition, subtraction, multiplication and division of images.

3. Spectral Features of Samples

In this communication we report the investigation of four ZnSe samples, three grown by the organometallic chemical vapour deposition (MOCVD) technique (Stutius 1978, Wright and Cockayne 1982) onto a $<100>$ GaAs substrate, and the fourth a polycrystalline sample grown by the CVD technique (Current Raytheon) which exhibits luminescence with a range of unusual characteristics. The samples were initially examined as grown.

At low ($\sim$35K) specimen temperatures the CL spectrum is resolved into a number of recombination processes. Fig. 2 shows the CL emission spectrum of an undoped sample (PW267) of high optical quality, indicating at least five separate processes and their associated series of phonon replicas, identified by comparison with low temperature PL spectra. The small discrepancy between the peak energies presented here and the accepted values from PL may be ascribed to specimen temperature effects, higher excitation levels and a possible offset in the monochromator calibration.

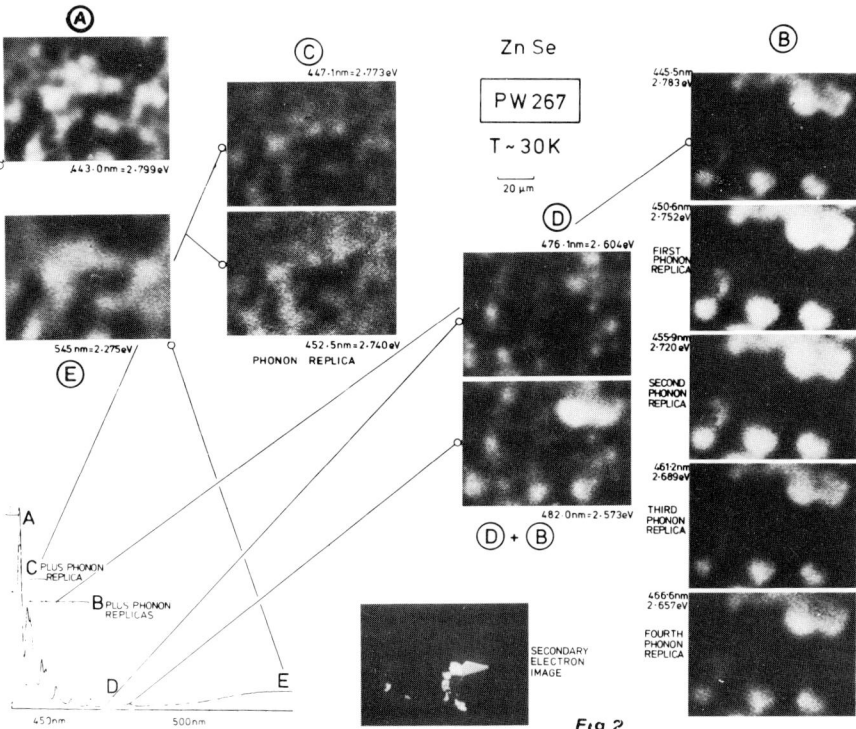

Fig 2

Peak A at 2.799eV is attributed to the I_{20}^{Cl} transition due to radiative recombination of an exciton bound (D^OX, BE) to a neutral chlorine donor. Peak B (2.783eV) and its phonon replicas are attributed to inelastic scattering of free excitons from neutral donors (FE-(1s-2s)). The alternative interpretation of this feature as I^{deep} BE recombination at an acceptor (presumably copper) (Bhargava 1982, Dean et al. 1981), down-shifted by strain in the heteroepitaxial layer, is rejected due to the weak coupling exhibited to LO phonons (Dean, Wright and Cockayne 1983). Peak C (2.773eV) is identified with "two electron" satellites of the D^OX, BE in which the donor is left in a 2s or 2p excited state.

Peak D (2.604eV) is the Y_O feature (Dean et al. 1982). This feature was tentatively attributed to FE recombination coupled with inelastic scattering at a transition metal impurity, accounting for the very weak LO phonon coupling and strong dependence on excitation density. However, more recent detailed studies (Dean et al. 1983) have failed to substantiate the change of spectral lineshape under selective excitation through the FE polariton region, which are exhibited by the 2.78eV feature. The identification of this unusual spectral feature, a striking characteristic of the heteroepitaxial MOCVD ZnSe, is the subject of further study. The continuum E centered on 2.27eV is the familiar copper green luminescence.

4. Results and Discussion

Fig. 2 also shows monochromatic scanning topographs, over the same area, of the various peaks. Of particular note is the strong   correlation between peaks A and C, and their anticorrelation with peak B. The obvious interpretation is that the regions of high B luminosity, typically of $\sim 20\mu$ scale, represent areas of high chlorine donor concentration. Here excitons are not bound to a particular donor because of concentration-induced delocalisation of the bound state, and hence the FE process dominates; in regions of lower doping excitons are not free to diffuse and hence BE processes contribute most of the CL intensity. In this sample as well as the other two MOCVD samples (PW184 and PW272), highly localised areas with emission at 2.720eV were observed. These are almost certainly the Q_o process, attributed to the D-Li shallow DAP luminescence (Merz et al. 1973) and hence representing areas strongly contaminated by lithium. An example in sample PW272 is shown in the lower right quadrant of Fig. 3 in which three such features are superimposed on the approximately isoenergetic third phonon replica of B. Fig. 3 also shows the results of image processing on the topographs, in which the monochromatic topographs have been normalised to the sum of the emissions in four peaks (Grande 1982). A strong anticorrelation is observed between the B_o and Y_o emission features, such as might be generally consistent with competitive recombination between deep states responsible for Y_o, thus far unidentified, and shallow donor-related recombinations. In an Al doped sample (PW184) the emission in all the lines appeared to be correlated, but was inhomogeneous on a scale of $\sim 1\mu$.

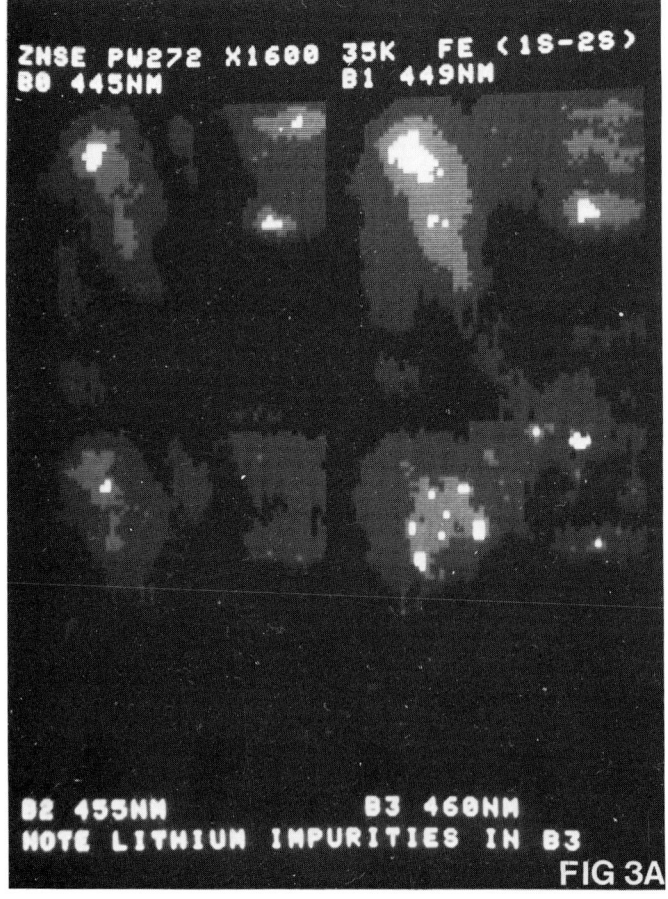

FIG 3A

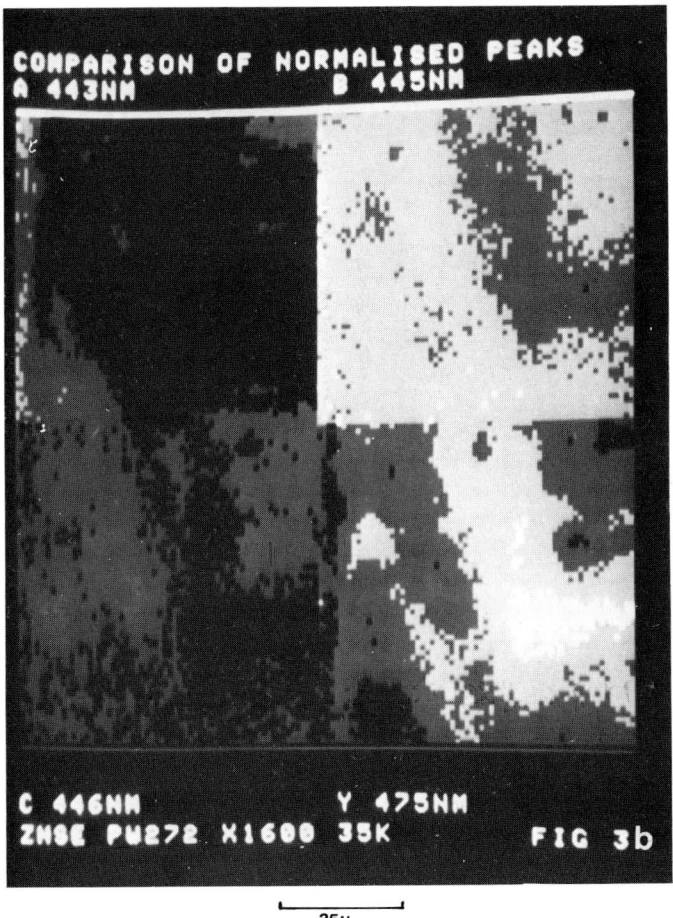

25μ

Figs. 4a and 4b show a comparison of the spatial distribution of emission features
observed in the CVD sample, in which a further spectral feature, the so-called
Z process peaking at 2.445eV is observed (Dean 1980). No physical explanation
for this emission is available, although it may be associated with defects introduced during specimen cutting. This explanation is supported by the spatial distributions shown in Figs. 4a and 4b which show the Z luminescence concentrated
around the edges of regions of bright overall luminescence, apparently individual
grains from the SEM morphology. Areas of high Z emission were chemically
analysed by EDX techniques but no contamination was observed.

5. Conclusion

Investigations on the samples discussed above and others continue, with a view to
further clarification of the processes involved, using TEM and CL studies on thin
samples to correlate emission features with morphology. In conclusion, the
technique of scanning monochromatic CL topography, performed in situ in a TEM
with a range of supporting microanalytical techniques available is a powerful tool
in investigating the microstructure of compound semiconductors.

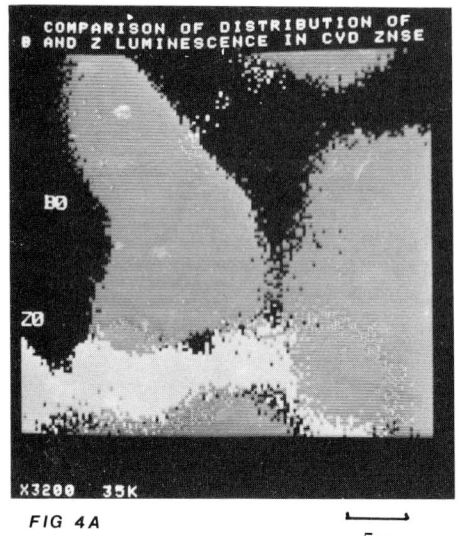

FIG 4A

5μ

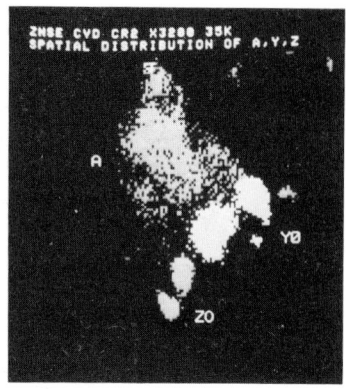

FIG 4B

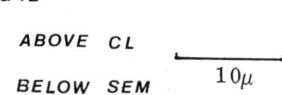

ABOVE CL

BELOW SEM   10μ

Acknowledgements

Three of the authors (MG, JLB, SHR) wish to thank the SERC for financial support. We are most grateful to Keith Lewis of RSRE for supplying the CVD sample.

References

Bhargava R M 1982 J. Cryst. Growth. 59 15
Dean P J 1980 Phys. Stat. Sol. (b) 98 439
Dean P J, Herbert D C, Werkhoven C J, Fitzpatrick B J and Bhargava R N 1981 Phys. Rev. 23 4888
Dean P J, Pitt A D, Skolnick M S, Wright P J and Cockayne B 1982 J. Cryst. Growth. 59 301
Dean P J, Wright P J and Cockayne B 1983 J. Phys. C. to be published
Eades J A 1982 J. Phys. E. 15 184
Grande M 1982 late presentation I.O.P. Benefits of digital imaging
Merz J L, Nassau K and Shiever J W 1973 Phys. Rev. B. 8 1444
Roberts S H 1981 I.O.P. Conf. Ser. 60.7 377
Roberts S H and Steeds J W 1982 J. Cryst. Growth. 59 312
Stutius W 1978 J. App. Phys. 33 656
Wright P J and Cockayne B 1982 J. Cryst. Growth. 59 148

Inst. Phys. Conf. Ser. No. 67: Section 6
Paper presented at Microsc. Semicond. Mater. Conf., Oxford, 21–23 March 1983

Cathodoluminescence spectra and images of crystalline ZnSe in the scanning electron microscope

T N Chin

FSL, ARRADCOM, Dover, New Jersey 07801, USA.

Abstract   A cathodoluminescence (CL) study of ZnSe compound semiconductor in the scanning electron microscope (SEM) is reported.
With the polycrystalline samples, deep-level emissions prevail over
the band-edge emission in the room-temperature spectrum. As the
sample temperature is lowered, the edge emission increases but deep-
level emissions decrease. In the case of monocrystalline ZnSe
samples, the band-edge emission is dominant over deep-level luminescence even at room temperature. However, nonuniform CL is still
observable in the high-quality monocrystalline ZnSe layers grown on
GaAs substrates.

1. Introduction

For scientific and technological purposes, considerable effort has been
devoted to the study of ZnSe compound semiconductor. Difficulties in
the present understanding of II-VI semiconductors preparation have yet
to be overcome. By examining the characteristics at microscopic areas,
one may obtain some insight into the basic problems in further
development of compound semiconductors.

1.1 Electron Penetration in Solids

As illustrated by Liljequist (1978), the Bethe range R has a nearly
linear dependence with the transport mean-free-path for electron
excitations of medium energy (10 < E < 60 keV). With this empirical
relationship, the Bethe range can be calculated from

$$R = 9.14(10)^{-4} \frac{A\ E^2}{\rho Z\ \ln\ (0.108EZ^{-2/3})} \quad \text{in °A}$$

where A is the atomic mass in gm, ρ is the density of the solid in
gm/cm^2, E is the average energy of the electrons in eV, and Z is the
atomic number.

1.2 Electron Energy-Loss Distribution

In the diffusion model (Kanaya and Ono 1978), the electrons are
considered to spread from a diffusion center X_D with a spherical
symmetry. Since electrons penetrate into the solid from the surface,
the maximum energy dissipation occurs at depth X_E before they reach the
diffusion center depending upon Z. By using the average atomic values
of ZnSe as approximation, one finds that

$X_D/R = 0.35$ and $X_E/R = 0.24$.
The diagram in Fig. 1 indicates the
conditions appropriate for crystalline
ZnSe.

1.3 Electron Excitation in Crystalline ZnSe

The cubic ZnSe has the same zincblende
structure as that of GaAs crystals
(Blakemore 1982) but with increased
ionic bonding. Since the information
on GaAs seems more complete at
present, it is natural to compare the
observed properties of ZnSe with the
known values for GaAs. The
commonly used value for the energy
required in pair production in GaAs
is 4.3 eV which is essentially 3 times
the band-gap energy. Perhaps, it is
reasonable to select 2.5 times the bandgap energy, or 6.75 eV, for the average
energy required for ZnSe semiconductor.

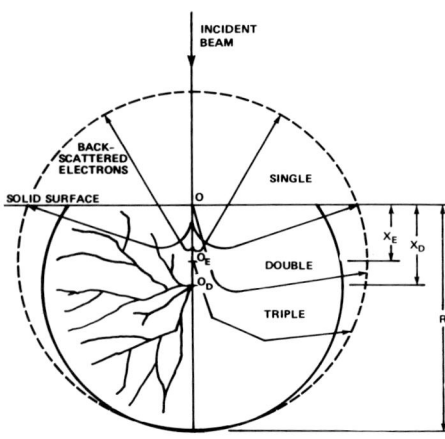

Fig. 1   A diffusion model of electron
beam penetration into a homogeneous
solid. Relative values of X_E, X_D and R
are shown as appropriate for ZnSe.

2. Experimental

The polycrystalline ZnSe flats (Donadio et al 1974) used in this investigation were obtained from the Raytheon Research Division. At the
deposition temperature of 750 °C, the average grain size is approximately 40 μm. Samples were selected after   heat treatment in
high-purity molten Zn at 800 °C for 10 hours.

Recently, a high-quality ZnSe layer has been successfully grown
(Stutius 1982) on (100) GaAs substrates by using organometallic
chemical vapor desposition (OM-CVD).
Since these layers were prepared at a
substrate temperature of 340-350 °C,
it is expected to have reduced several
well-known problems, such as contamination, interdiffusion, charge-compensa-
tion, and native vacancies. The
dominance of band edge emission in
the photoluminescence spectra has
been used to demonstrate a high-
degree of crystal perfection, while
the intensity   of deep level cathodoluminescence (CL) (Chin 1983) is
interpreted by the presence of
impurities and native defects.

All the samples were examined in an
AMR 1000 SEM. An experimental
arrangement of the SEM implemented
with optical coupling to a monochromator is shown in Fig. 2.

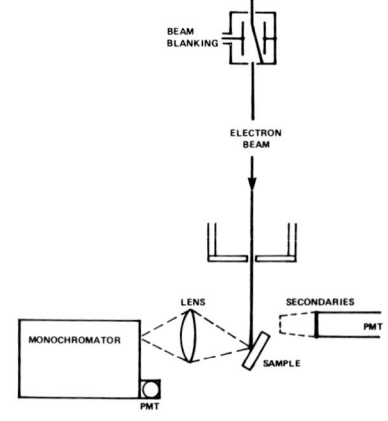

Fig. 2   An experimental arrangement
of the scanning electron microscope
implemented with optical coupling to
a monochromator.

3. Results

On the basis of the simplified model, the domain of electron interactions in ZnSe compound semiconductor may be qualitatively visualized with three parameters, i.e., Bethe range, diffusion-center depth and the depth of

maximum energy-loss. The following table indicates these calculated values under electron beams of three different voltages, namely, 10, 20, and 30 kV.

Micrometers	10 kV	20 kV	30 kV
Bethe range	0.814	2.837	5.934
Depth of diffusion-center	0.285	0.993	2.077
Depth of maximum energy-loss	0.195	0.681	1.424
Numbers of pairs per incident electron	1480	2960	4440

The estimated number of electron-hole pairs generated per incident electron is also given for electrons at the respective beam voltages, 10, 20, and 30 kV. However, the lifetimes of these generated carriers are not known and the pair density at any time is difficult to estimate.

3.1 CVD Polycrystalline ZnSe

Micrographs in Fig. 3A and Fig. 3B show the surface texture of an annealed CVD ZnSe sample. A collection of surface grains is shown in Fig. 3C. The CL images of the same sample under the same magnifications are presented in Fig. 4. Several geometrical features in Fig. 3 may be observed and used to identify the CL of the same area in Fig. 4.

Although the resolution of the CL images is not as sharp as that of the SEM micrographs, there is clear indication that a group of grains has bright luminescence among many non-radiative grains. It is possible to select a strongly luminescent area of a grain for examination at a higher magnification. A

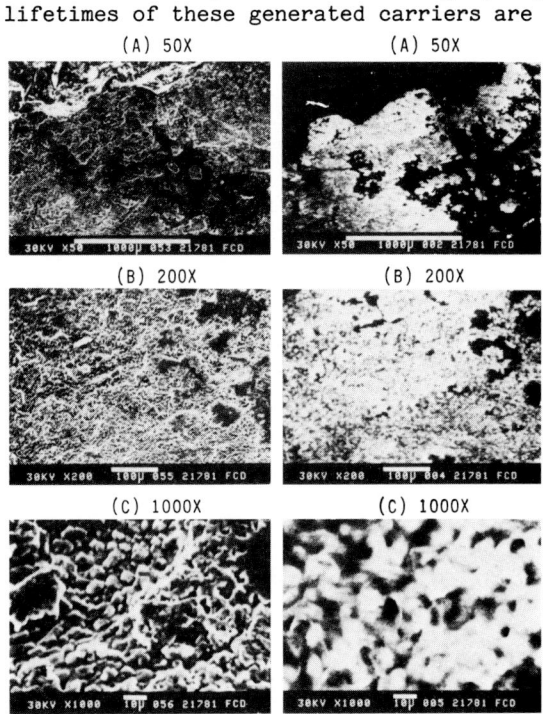

(A) 50X   (A) 50X

(B) 200X   (B) 200X

(C) 1000X   (C) 1000X

Fig. 3 Surface textural images of an annealed CVD polycrystalline ZnSe sample examined in SEM mode.

Fig. 4 Cathodoluminescence images of the same areas as those shown in Fig. 3.

typical CL spectrum from a ZnSe
sample at room temperature
taken from a scan area of
4 X 6 μm² is shown in Fig. 5.
Here the CL spectrum from the
annealed polycrystalline ZnSe
sample shows that the deep-
center luminescence prevails
over the edge emissions in that
area.

When the polycrystalline ZnSe
sample is cooled to lower temperatures, the intensity of edge
emission is greatly enhanced but
the deep-level luminescence is
substantially decreased. The
spectrum shown in Fig. 6 was
taken with reduced slit width
and at 200 K approximately. The
peak intensity of the edge
emission in the spectrum shown
in Fig. 7 is not increased very
much when the temperature is
further lowered from 200 K to 100
K. However, there are striking

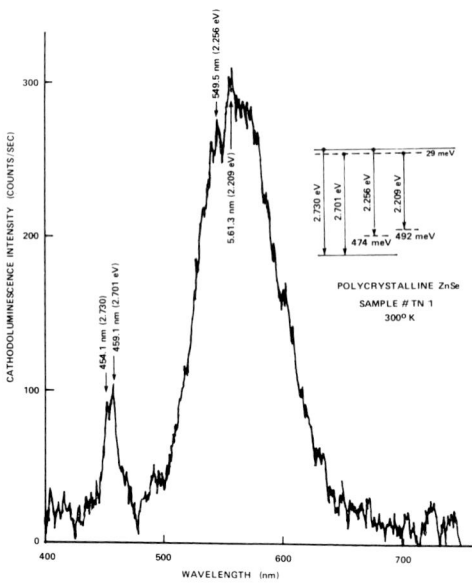

Fig. 5   A cathodoluminescence spectrum from
annealed CVD polycrystalline ZnSe sample at
room temperature.

changes in the deep-center luminescence: the intensities are much
reduced and two luminescent maxima are revealed instead of one peak.

3.2   OM-CVD Monocrystalline ZnSe

The ZnSe layer on (100) GaAs substrate was prepared without intentional
dopant.   This film, under 10-mW He:Cd laser radiation, produced strong

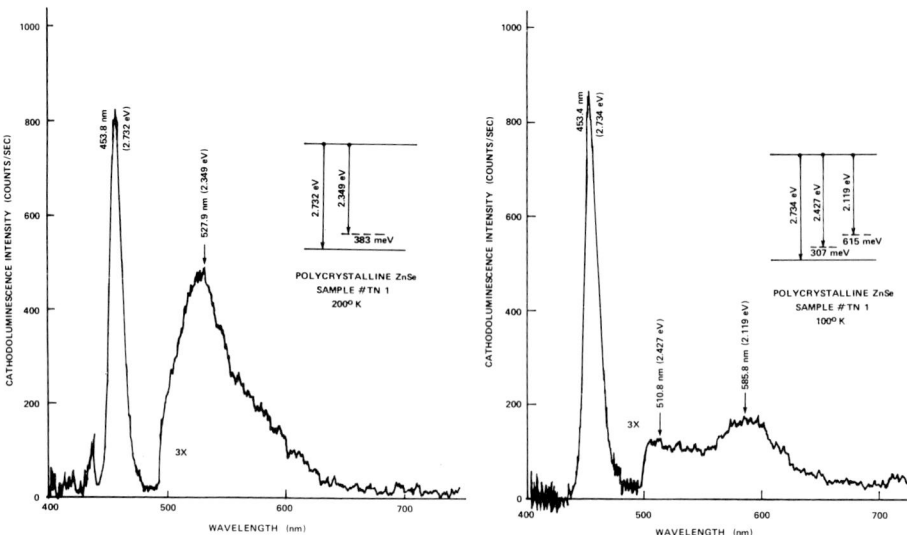

Fig. 6 A cathodoluminescence spectrum from annealed CVD polycrystalline ZnSe sample at 200K.

Fig. 7 A cathodoluminescence spectrum from annealed CVD polycrystalline ZnSe sample at 100K.

near band-gap photoluminescence.
Fig. 8A shows the micrograph of a
typical surface area of the ZnSe
layer grown on (100) GaAs. Although
these ridged hillocks are irregular,
they appear to orient nearly parallel
in one direction with an average
spacing of 4600 A. From one cleaved
edge of the ZnSe layer, the thickness
of the grown ZnSe is determined to be
2.87 μm. Although the cleaved surface does appear smooth, structures
can be clearly observed in the micrograph in Fig. 8B. The first interface
layer is especially thick, perhaps 0.4
μm in thickness. Futhermore, the ZnSe
layer seems to have the structure of
14 sub-layers, each sub-layer being
0.19μm thick. Stacking faults are
known to occur frequently in II-VI
crystals but the identity of these
sub-layers has not been determined.
No attempt has been made to examine
the interdiffusion problem.

Based upon the Pseudo-Kikuchi
patterns obtained from the hillock surface, the crystallinity of
this ZnSe layer seemed quite good.
As shown in Fig. 9A, the nonuni-
formity in the light emission can still
be readily observed. The micrograph
with a larger magnification shown in
Fig. 9B indicates the outlines of a
few bright CL spots in this undoped
ZnSe layer. It may be noted here that
the underlying light emission domains
cannot be correlated with the moro-
phology at the top surface. A typical
CL spectrum for room temperature taken
in this area is shown in Fig. 10. In
contrast to that of the polycrystalline
ZnSe, the band-edge emission from the
monocrystalline ZnSe is dominant in the
spectrum and the CL peak near the bandgap energy is increased by a factor of
70 approximately as compared to the
polycrystalline sample. In addition,
several deep-level luminescent peaks
are also observable even at room temperature conditions.

4. Conclusion

CL is an important approach in the
investigation of semiconducting materials.
It is especially useful in characterizing
the luminescent crystals with high yields.

(A) Surface morphology of ZnSe

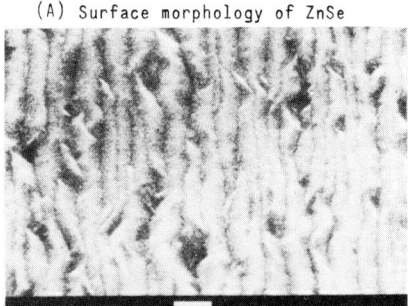

(B) Cross-sectional view of ZnSe

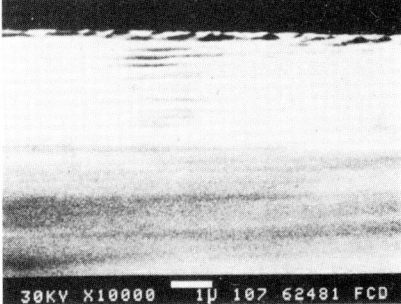

Fig. 8 Micrographs of a monocrystalline
ZnSe layer grown on (100) GaAs substrate.

(A)

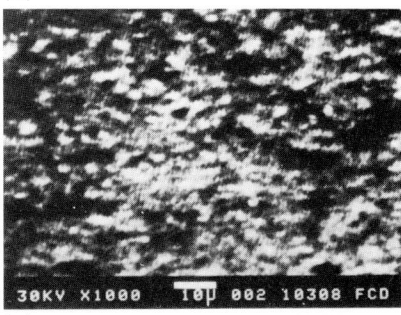

(B)

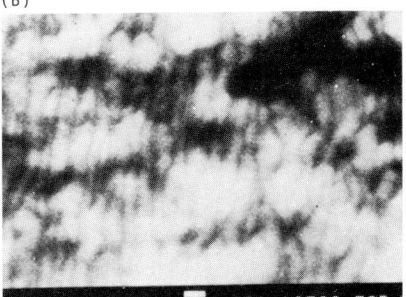

Fig. 9 Cathodoluminescence images of a
high-purity ZnSe layer grown on (100)
GaAs substrate.

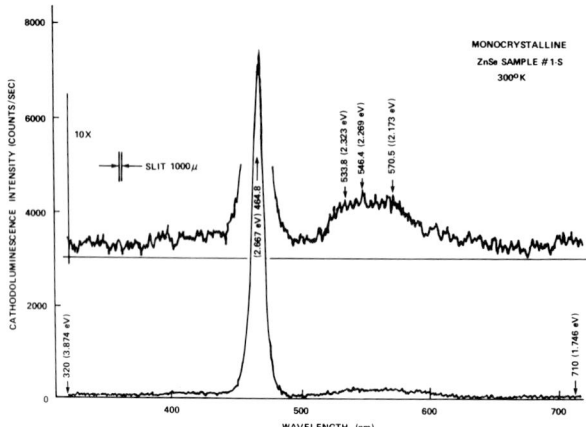

Fig. 10 A room-temperature cathodoluminescence spectrum from
OM-CVD monocrystalline ZnSe layer grown on (100) GaAs substrate.

The CL images and spectra obtained from high-purity ZnSe semiconductor
demonstrate their sensitivities. The donor and deep-center levels
indicated in Fig. 5-7 are qualitative interpretations based upon the
spectra obtained at microscopic areas. As implemented in SEM, CL may
become a powerful tool if correlations can be made with material
processing or other experimental findings. With the increased knowledge
of the electron loss mechanisms in crystals, an improved understanding of
the semiconducting materials may be acquired through CL investigations.

Acknowledgment

The author is grateful to Wolfgang Stutius of Xerox Palo Alto Research
Center for providing the OM-CVD ZnSe samples used in this investigation.

References

Blakemore J S 1982 J. Appl. Phys. 53 R125
Chin T N 1983 J. Vac. Sci. & Tech. A, Apr/Jun 831
Donadio R N, Swanson A W and Pappis J 1974 Proc. Ann. Conf. on Infrared
Laser Windows, (Air Force Materials Lab., WPAFB/OH) 494-509
Kanaya K and Ono S 1978 J. Phys. D: Appl. Phys. 24 1495
Liljequist D 1978 J. Phys. D: Appl. Phys. 11 839
Stutius W 1982 J. Crystal Growth 59 1

Inst. Phys. Conf. Ser. No. 67: Section 7
Paper presented at Microsc. Semicond. Mater. Conf., Oxford, 21–23 March 1983

349

Determination of atom locations on surfaces with x-ray standing waves

J R Patel and J A Golovchenko

Bell Laboratories, Murray Hill, New Jersey, 07974

Abstract   Standing waves of x-rays have been utilized to locate adsorbed atom positions on crystal surfaces. In addition to positions normal to the surface the component relative to a plane inclined to the surface has been determined. This suffices to establish the registration of the adsorbed surface atoms to the crystal lattice below.

1. Introduction

In a classical experiment performed over ninety years ago Wiener (1890) demonstrated the existence of optical standing waves formed by light reflected from the surface of a silver mirror. In establishing their existence he also proved that the E field of the electromagnetic wave was responsible for the observed optical effect (i.e. the darkening of a photographic plate placed at a small angle to the mirror surface). We will be concerned in this paper with the extension of Wiener's discovery to the x-ray region of the electromagnetic spectrum.

The existence of standing waves of x-rays inside perfect crystals was shown by Batterman (1964) some twenty years ago. Our interest in x-ray standing waves arises from the exciting possibility that they can be used as an *Angstrom Gauge* to characterize structure on an atomic scale. There lies the potential not only of characterizing the interior of solids but also of adsorbed layers on crystal surfaces. In what follows we discuss the generation of standing waves in a crystal and near its surface. We proceed then to describe recent experiments on the location and registration of adsorbed surface atoms on crystals.

Replacing Wiener's silver metallic reflector with a crystal for the x-ray experiment confers a significant advantage. For the metallic reflection case the phase of the standing wave is fixed by the condition that the tangential component of the electric vector at the surface vanishes. In the x-ray region under strong Bragg reflection, the conditions that determine the standing wave phase are considerably more complex but have been thoroughly treated by the dynamical theory. For strong Bragg scattering dynamical theory predicts a *Bragg bandgap* region in which the intensity of the incident and diffracted beams are very nearly equal. The coupling between these two coherent beams in the crystal gives rise to a standing wavefield with nodal planes parallel to and having the periodicity of the Miller planes. A unique feature distinguishes the metallic reflection from the Bragg x-ray case. For the former the phase of the standing wave is fixed while for the latter the phase of the reflected beam is a strong function of angle and changes by π from one side of the Bragg bandgap region to the other. This feature allows us to tune the phase of the standing waves as the crystal is scanned through the Bragg reflection. The nodal planes move continuously from coincidence with planes of maximum to planes of minimum electronic charge density for the relevant Fourier component of the charge density associated with the Bragg reflection.

In demonstrating the existence of x-ray standing waves Batterman (1964) showed that the standing wavefield can be used as a probe to excite host atom or impurity fluorescence. The angular dependence of the excited fluorescence will depend on the electric field intensity in the standing wave field at the host atom or impurity site. However as Golovchenko, Brown and Batterman (1974) have emphasized an undesirable strong angular dependent extinction signal accompanies the characteristic fluorescent signal especially in the bandgap region. The extinction signal dominates the impurity fluorescence making accurate atom location studies difficult. A clever solution to the problem is to maintain the depth from which the fluorescent signal originates small compared to the characteristic minimum extinction depth. Golovchenko et al (1974) and Anderson et al (1976) have shown that for conditions where the above criterion applies extinction effects are negligible. One can in effect treat this situation as if all the impurities are at the surface. Obviously for impurities adsorbed at the surface which is the condition maintained and described in this paper extinction effects are irrelevant.

2. Theory

The detailed theory of the nature of the wavefield in thick nonabsorbing crystals have been treated by James (1963) and in a review by Batterman and Cole (1964). Here the bare outline of the development is presented to enable the reader to qualitatively understand and interpret the experimental results. Consider two plane waves in a crystal in the primary and diffracted beam directions, the total electric field in the crystal is given by the sum of two coherent plane waves

$$\overline{E} = \overline{E}_o \exp(2\pi i \overline{k}_o \cdot \overline{r}) + \overline{E}_H \exp(2\pi i \overline{k}_H \cdot \overline{r}) \qquad (1)$$

where $\overline{E}_o$ and $\overline{E}_H$ are the amplitudes and $\overline{k}_o$ and $\overline{k}_H$ the wave vectors in the incident and diffracted beam directions respectively. When the Bragg condition is satisfied $\overline{k}_o + \overline{H} = \overline{k}_H$ where $\overline{H}$ is the relevant reciprocal lattice vector. The normalized field intensity in the crystal can then be expressed as

$$\left|\frac{\overline{E}}{\overline{E}_o}\right|^2 = 1 + \left|\frac{\overline{E}_H}{\overline{E}_o}\right|^2 + 2P\left|\frac{\overline{E}_H}{\overline{E}_o}\right| \cos(2\pi\overline{H}\cdot\overline{r}+\phi) \qquad (2)$$

where P is the polarization factor and ϕ a phase factor. Equation (2) states that there are planes of constant intensity (the standing wave pattern) in the crystal for constant $\overline{H}\cdot\overline{r}$ where $\overline{r}$ is the position vector. Since $\overline{H} = |1/d|$ and is perpendicular to the Bragg reflection planes the standing wave field is parallel to and has the periodicity of the lattice planes. The quantity $\left[\overline{E}_H/\overline{E}_o\right]$ in (2) is the ratio of the field amplitude in the diffracted to the incident beam direction. From dynamical theory its variation with angle can be shown to be

$$\left|\frac{E_H}{E_o}\right|^2 = \eta \pm \sqrt{\eta^2-1} \qquad (3)$$

where η is a dimensionless parameter proportional to $\Delta\theta$ the deviation from the Bragg angle. Since the fluorescent yield is proportional to the standing wave field intensity, evaluation of equation (2) should give the fluorescent intensity for various positions $\overline{H}\cdot\overline{r}$ in the crystal. Figure 1. shows the angular dependence of the fluorescent signal calculated using equation (2). Because equation (2) contains no correction for extinction effects we shall refer to the fluorescent yield calculated from (2) as the *surface yield*. The dashed curve in Fig. 1. is the Bragg reflectivity for a 220 reflection of $MoK\alpha_1$ x-rays from silicon. The two surface yield curves are displaced $\pm 0.1\ d_{220}$ from the exact 220 Miller planes. The large difference in fluorescent yield between the A and B positions for planes separated by $0.2\ d_{220}$ underscores the high sensitivity to atomic position inherent in the standing wave method.

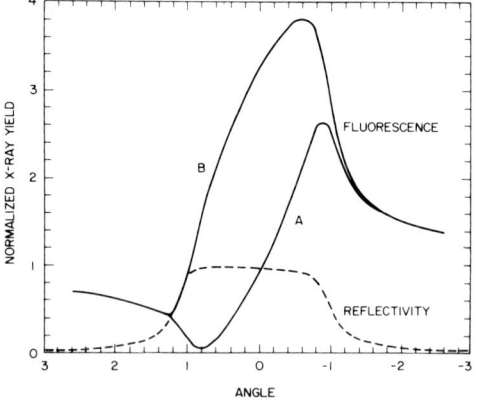

FIG. 1. Theoretical reflectivity and surface fluorescent yield for $MoK\alpha_1$ reflection from silicon 220. Curves A & B are for planes displaced $\pm 0.1d$ from the 220 Miller planes.

A further complication in interpreting position information from surface yield arises when all the atoms are not necessarily at coherent lattice sites. Some fraction of the atoms of interest may lie in random positions.

A concrete example could be a thin liquid like layer at the crystal surface where the impurity of interest could occupy random positions. To analyze the fluorescent yield from a crystal under these conditions it is necessary to consider the fraction of atoms that occupy coherent lattice sites at the crystal surface. The problem can be treated as follows. In Fig. 2. we again show fluorescent surface yield and reflectivity curves for a 220 reflection from silicon for MoKα_1 x-rays curve A refers to atoms at the exact 220 planes of the crystal lattice. If the atoms were not exactly at lattice sites but distributed at random between consecutive 220 plane positions the relevant yield curve is B which is an average of the yield curves for all positions between two neighboring 220 planes. Finally if only half of the atoms are at coherent lattice sites we obtain curve C which is the weighted average of the totally coherent curve A and the totally random yield curve B. In what follows we will see many examples of curves such as C in Fig. 2., since our surface treatments invariably resulted in some fraction of the adsorbed atoms at random positions. It is possible to get contributions to fluorescent yield similar to that obtained from incoherent atoms if there are two or more unique atomic positions between the Miller planes. Ambiguities in data interpretation can be removed in such instances by considering additional reflections.

Finally before leaving this discussion we include for the sake of completeness a full set of surface fluorescent yield curves Fig. 3. for the positions indicated in the inset. The curves were generated for MoKα_1 x-ray incident on silicon (220). The marked dependence of surface fluorescent yield on atom position is amply evident from this set of curves.

3. X–ray Standing Waves at Crystal Surfaces

Some years ago Andersen, Golovchenko and Mair (1976) predicted that fluorescent scattering from x-ray standing waves could provide a method for accurately determining impurity atom positions on crystal surfaces. This prediction evolved naturally from pioneering observations by Batterman (1964) referred to earlier. Cowan, Golovchenko and Robbins (1980) recently took the first step in realizing the prediction of Andersen et al (1976) by using x-ray standing waves to demonstrate that submonolayer amounts of bromine atoms could be adsorbed into surface lattice sites strongly correlated to the crystal substrate (i.e. coherent atomic positions).

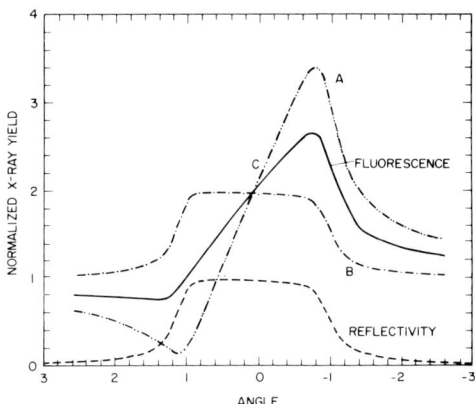

FIG. 2. Angular dependence of reflectivity and surface fluorescent yield for MoKα_1 reflected from silicon 220. Curve A for atoms located at the exact 220 Miller planes. Curve B for atoms at random positions. Curve C is for 50% of atoms at coherent lattice sites with the rest random.

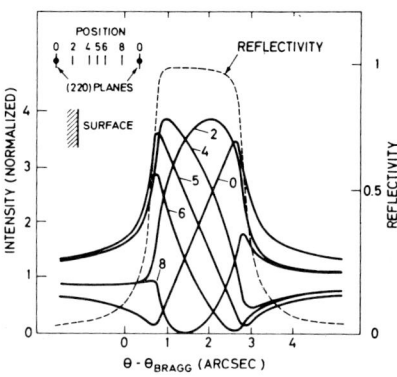

FIG. 3. Theoretical reflectivity and electric-field strength versus incidence angle for several atomic positions between (220) planes. The integers labeling each curve give the atomic position in units of the planar spacing upon division by 10. The curves are normalized to unity far off the Bragg condition.

The bromine atomic position measurements normal to the (110) crystal surface are described below.

The conventional approach to the preparation of a crystal surface with adatoms involves sputtering, annealing, and impurity atom exposure followed by a number of standard measurements such as LEED, Auger spectroscopy, etc. all done in an ultra high vacuum environment. While this route is pursued in some of the surface x-ray interference experiments, a different kind of surface preparation is reported here. The strong penetrating power of both incident and fluorescence x-rays is employed to demonstrate the existence of surface standing waves at a crystal-liquid boundary. This represents one of the few means of obtaining microscopic atomic location information at this kind of interface. A single crystal of silicon was cut and syton polished with the (220) planes parallel to the surface.

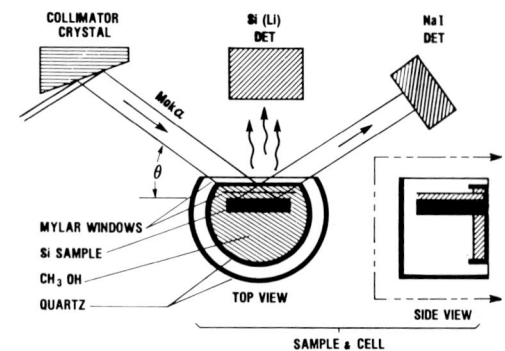

FIG. 4. Schematic layout of experimental apparatus including sample cell detail.

Oxide was removed with an HF etch followed by a quick methanol rinse. The sample was then immediately transferred to a solution of 50 ml methyl alcohol to which a single drop of liquid bromine was added, the goal being to deposit bromine atoms on the clean silicon surface. Without allowing the sample surface to come in contact with the atmosphere it was transferred successively to six clean methanol baths. A 0.0005" mylar window wet with methanol was placed in contact with the sample and the combination of surface wetting and tension forces trapped a thin liquid layer (several microns thick). The sample was then placed in a quartz cell (see Fig. 4.) that continually resupplied methanol to the layer via capillary action replacing liquid lost due to evaporation from the sample edges. This procedure is necessary to minimize Compton and diffuse scattering off the liquid from entering our fluorescence detector.

The cell described above was then placed in an x-ray apparatus illustrated in Fig. 4. A fine focus molybdenum x-ray tube served as the source of $MoK\alpha_1$ x-rays. The beam was then collimated-monochromated by a 220 reflection from an asymmetrically cut silicon crystal. The asymmetry of the collimator crystal produced an outgoing beam having a theoretical angular width 1/7 the natural width of the (220) reflection for a symmetric crystal. This broad highly collimated beam then fell upon our sample in solution after passing through the thin mylar window. The sample and its holder were placed on a piezoelectrically controlled and long term stabilized goniometer capable of stabilities of order .01 arc sec over many days. The goniometer was programmed to sweep back and forth through the region of total reflection associated with the 220 Bragg angle (10.6°). Details of the electronic feed back and stabilization circuits are given by Miller et al (1979). A lithium drifted silicon x-ray detector monitored the inelastic x-rays scattered from the sample. In particular the fluorescent x-ray yield from the surface bromine atoms was monitored as a function of crystal angle relative to the beam.

Bromine surface coverage could be evaluated immediately by comparing "off Bragg" bromine fluorescence yields with another similar sample that had $10^{15}/cm^2$ bromine atoms ion implanted to a depth of ~300Å instead of the chemical treatment. A surface coverage of 4×10^{14} bromine/cm^2 was thus obtained for the chemically prepared sample. Evidence that the chemically deposited bromine was actually on the surface came from observing that a brief dip in hydrofluoric acid completely removed it. (Attempts were made to establish bromine coverage by Rutherford backscattering and electron induced Auger yields, but bromine appears to be rapidly desorbed by these techniques.)

The closed circle data points in Fig. 5. which were acquired over a 14 hour period represent the experimental angular yield of bromine Kα fluorescence around the total reflection region. The yield is clearly asymmetric with respect to the center of the reflectivity curve which is shown as curve A in the figure. Had there been no standing waves at the surface or no coherent bromine atom positions this yield would have been simply an incoherent sum of fluorescent yields from the incident and reflected beams (curve *B*).

Consider now the coherent aspects of the bromine fluorescent signal. First note that in moving across the region of total reflectivity the aforementioned phase change of the reflected beam results in the x-ray antinodes shifting from being midway between, to being in step with the crystal 220 planes. Thus, for an angle of incidence at the left of the total reflection region the first external x-ray antinode lies one half a 220 planar spacing above the surface. For an angle of incidence at the right of the total reflection region the antinode lies one whole planar spacing above the surface (1.92Å). Therefore the presence of the fluorescence maximum on the right side of the total reflectivity region indicates a significant fraction of the adsorbed bromine lies approximately 1.9Å above the crystal surface. A calculation using equation (2) above and standard silicon parameters in a 2 beam dynamical theory is also shown in Fig. 5. (curve *C*). The fit assumes that 30% of the bromine lies 1.73Å above the crystal surface and 70% is incoherent relative to the crystal lattice below. The theory for completely coherent (curve *D*) and completely incoherent (curve *B*) results is also indicated. Fig. 5b. shows the expected results for completely incoherent, completely coherent and a mixture using a different atomic position than we have observed. Our experimental value of 1.73 ± .07Å corresponds closely with that expected from a Si-Br covalent bond in volatile molecules. The value for the latter given by Gutmann (1967) is 1.76 ± .16Å with geometrical factors taken into account. Here we assume the angular orientation of the Si-Br bond is identical to that expected for the equivalent Si-Si bond.

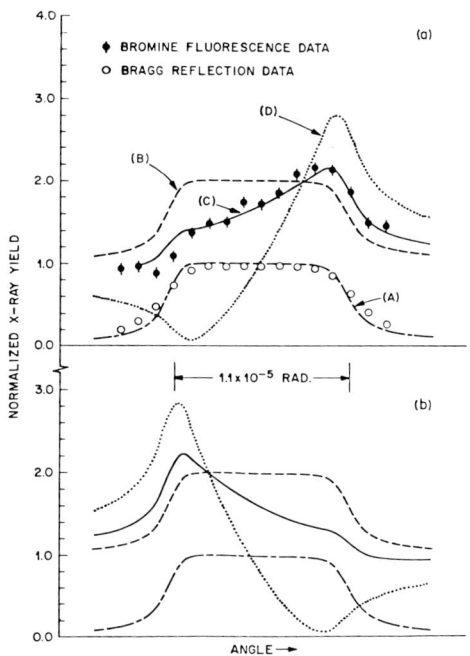

FIG. 5. (a) Normalized x-ray yields from both fluorescence and reflected-beam detectors. Bromine fluorescence, solid circles; Bragg reflectivity, open circles. Curve *A*, Bragg reflectivity theory; curve *B*, standing-wave theory with coherent fraction 0%; curve *C*, standing-wave theory with coherent fraction 30% at $\delta = 1.72$Å; curve *D*, standing-wave theory with coherent fraction 100% at $\delta = 1.72$Å. δ is distance above surface plane. (b) Bragg reflectivity theory, dot-dashed curve; standing-wave theory with coherent fraction 0%, dashed curve; standing-wave theory with coherent fraction 30% at $\delta = 1.15$Å, solid curve; standing-wave theory with coherent fraction 100% at $\delta = 1.15$Å, dotted curve.

It has been shown previously Anderson et al (1976) that the incident and diffracted beams are highly coherent even at distances of ~1000Å above the single crystal region which produces the standing waves. The 70% incoherent signal is believed to be actually connected with some bromine atom sites being essentially uncorrelated with the crystal lattice below. Thus, while some of the deposited Br atoms are able to situate themselves at preferred sites relative to the crystal substrate, most of the Br present may be caught in some more complex structure. Surface impurities from the air or the methanol such as oxygen or hydrocarbons which are too light to give an observable fluorescence signal may be responsible for the complex environment. On the other hand, while it is not appropriate to go into great detail here, we just mention that samples prepared and measured in a UHV environment and samples allowed to age in air after the bromine treatment and samples prepared in a CF_3Br plasma all gave rather similar results with regard to bromine incoherent fraction (and position). Attempts to get any coherent fraction from bromine in the bulk (i.e. ion implantation followed by thermal or laser annealing) however have all failed. This leaves open the possibility that some bromine may actually penetrate into the solid and take up incoherent positions or that a combination of locations is populated that simulates an incoherent result in the fluorescence data.

The statistical accuracy of the data in Fig. 5. is rather poor compared to data from standing wave experiments on bulk impurities by Anderson et al (1976). This is due partly to the low coherent fraction ($\sim 10^{14}$ Br/cm^2) and partly to the limited times a well defined liquid solid interface was maintained. It must also be realized that one cannot from the single measurement described above exclude the possibility that the in-planar structure of the 30% coherent Br atoms is random or liquid-like. The question of the *registration* of Br to the silicon surface below must be resolved by further measurements to be described in the section below.

4. Solution to the Surface Registration Problem using X—ray Standing Waves

In the following we describe two significant developments in the evolution of the application of x-ray standing waves to surface studies Golovchenko et al (1982). First we demonstrate that these methods can solve the *registration problem* of surface physics. This is the problem of determining the absolute position of the periodic impurity sites on a crystal surface particularly with regard to translation of the surface impurity lattice vectors in the surface plane. Secondly we extend the original measurements of Cowan et al. (1980) described above to the silicon (111) surface which seems to be of more interest because it is a surface with which one must more commonly deal in applications.

Until now lattice location studies by the standing wave method have utilized standing waves with nodal planes that lie parallel to the crystal surface. Consequently, only position information perpendicular to the surface has been obtained. If, however, the reciprocal lattice vector of the reflection does not lie perpendicular to the surface the impurity can be located with respect to a plane that intercepts the surface at an angle. Such a measurement together with the distance measured normal to the surface constitutes a solution to the registration problem since by *triangulation* the impurity positions along the surface in the direction of the reciprocal lattice projection is completely determined. It should be pointed out that this type of determination of the bromine position utilizes phase information which is not normally available from elastic scattering measurements (x-ray diffraction, leed) because the diffracted intensities do not contain absolute phase information.

We have studied the fluorescent signals from standing waves excited from both (111) and (220) reflections for the same crystal. The first reflection provides the surface normal distance which is designated as distance "A" in the schematic diagram of the 110 projection in Fig. 6. The (220) reflection data provides the distance labeled "B" in that figure. The complexities caused by an asymmetrical reflection in this case are avoided by having the plane formed by incident and diffracted beams lie perpendicular to Fig. 6. with small angular divergence along the diffraction cone see Fig. 6. inset.

The sample was prepared with a syton polish mirror finish. Oxide was stripped off in a HF etch after which the surface was rinsed in anhydrous ethyl alcohol. The surface was then rubbed on a pad saturated with an $\sim.01\%$ bromine in ethanol solution under uv illumination for five minutes. This treatment is similar to one recently reported by Aspnes and Studna (1981). Little or no loss of coherent or incoherent bromine atoms from a (111) surface prepared in this way has been noted over a period of 3 days. This is in contrast to recent observations on (220) surfaces where substantial loss over approximately 12 hours was observed by Bedzyk et al (1982).

The experimental arrangement is similar to that shown previously except that each of the reflections reported here required its own collimating crystal. Also we have found that bromine on the (111) surface is so stable that the sample need not be immersed in alcohol during the measurement. All measurements were thus performed in open air.

Figure 7. shows an angular scan of both the intensity of the reflecting beam and the bromine fluorescence in the vicinity of the (111) reflection. Here the standing waves move normal to the surface and hence the bromine position along this direction can be deduced. The data were acquired over a period of three days using a rotating anode molybdenum x-ray tube whose output beam was collimated with an asymmetrically cut (111) silicon crystal. A beam flux of 10^6 to 10^7 photons per sec was delivered to the sample in an angular divergence of $\sim 1/6$ the natural (111) reflection width. The bromine fluorescence signal shows a clear minimum inside the region of total reflection which corresponds to the nodal plane of the exciting standing wave passing through the bromine atom positions. The theoretical fit to the data also shown in the figure takes into account both the position of the bromine atoms and the fraction that actually occupies a coherent lattice site.

The deduced position is 2.56 ± .03 angstroms above the extrapolated last maximum in the 111 Fourier component of the silicon charge density at the surface (which the reader should know does not correspond to a physical silicon atomic plane see figure (6)). This position is indicated as distance "A" in figure (6). Within the experimental error this is just where covalently bonded bromine atoms would reside if they terminated the silicon lattice on the (111) plane that would yield only one terminating bond per surface Si atom. Therefore we see directly that the last silicon layers are the closely spaced physical (111) atom planes as drawn in figure (6). Furthermore to within .03 angstroms no surface relaxation can be allowed without changing the Si-Br covalent bondlength (2.17Å) by a corresponding amount. A perusal of tetrahedrally bonded silicon halide bondlengths from electron scattering in vapors Sutton (1958) excludes this possibility. The coherent part of the bromine signal in figure (2) amounts to 67% of the total coverage. The latter was $\sim 10^{14}/cm^2$.

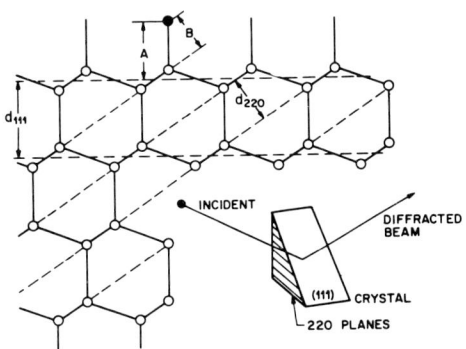

FIG. 6. Schematic illustration of a silicon (111) surface viewed edge-on along a (110) projection. Distances A and B indicate bromine-atom positions above surface. Silicon and bromine atoms are represented by open and closed circles, respectively. The position of the relevant (111) and (220) Fourier components of the charge density are indicated by dashed lines. Inset shows reflection from the 220 planes for Br adsorbed on (111) silicon.

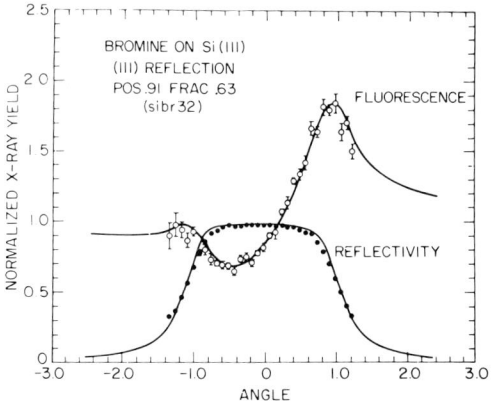

FIG. 7. Bromine fluorescence and reflectivity angular yields for (111) Bragg diffraction on a silicon (111) surface. Angular scale is in reduced units where rocking curve width is 2.

The above discussion of the bromine atom location is based upon a single distance measurement and bonding arguments from physical chemistry. The x-ray standing wave data by itself could support a model where bromine atoms lay a distance "A" above the surface and form a two dimensional liquid along the surface uncorrelated to the crystal below. Alternatively bromine atoms could occupy sites correlated to the crystal below but not directly above the top-most silicon atoms as required by the discussion of the previous paragraph. This matter can be resolved by performing standing wave experiments using diffraction planes that do not lie parallel to the surface. In this geometry a liquid layer would show no coherent fluorescence signal. Observation of a coherent signal here shows not only the presence of atoms in sites correlated to the crystal but the absolute transverse position can be extracted from the details of the fluorescence angular yield data.

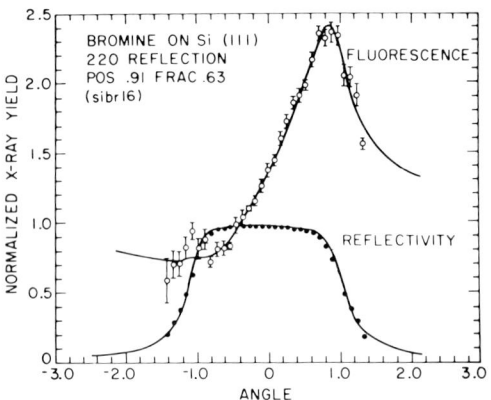

FIG. 8. Bromine fluorescence and reflectivity angular yields for (220) Bragg diffraction on a silicon (111) surface. Angular scale is in reduced units where rocking curve width is 2.

Figure 8. shows the results of a standing wave experiment using one of the 220 planes for diffraction which intersects the 111 surface at an angle of 35.26 degrees (see inset Fig. 6.). A clear coherent signal is observed. Its magnitude is 63% of the total signal which is similar to the (111) case. The theoretical fit in the figure corresponds to a bromine position $1.75 \pm .02$ angstroms normal to the 220 plane of silicon atoms pointing away from the surface. This is the value for distance "B" in figure (1). Together with the analogous results from the two other (220) planes that intersect the surface at the same angle, we uniquely deduce that the bromine atoms sit directly above the top-most silicon surface atoms to an accuracy of $\sim.04$ angstroms. Coverage of bromine and coherent fraction of atoms are as in the (111) reflection study. As a consequence of more refined instrumentation (improved lock-in amplifiers and electronic circuitry) the statistical accuracy of the fluorescent data Br on silicon (111) in Fig. 7. and 8. is much improved over the corresponding earlier work on (220) surfaces in Fig. 5.

5. Conclusions

There are several important points to be addressed concerning the interpretation of these atom location results. Except for the coherent bromine atom locations we are completely ignorant of the state of the surface. Thus 90% of the surface Si atoms are terminated either by bonding to oxygen or other surface impurities or are reconstructed in some undetermined way. The simple result we deduce both with regard to bromine position and surface relaxation seem remarkable in view of the conceivable complications our uncharacterized surface could induce. To our knowledge no one has yet been equal to the challenge of performing an x-ray standing wave measurement in a controlled ultra-high vacuum environment. The relationship of our results on "practical" surfaces to the more controlled surfaces that should thereby be obtained awaits such an achievement. We are particularly interested in the possibility of terminating all 111 surface atoms in covalent Si-Br bonds. Such a surface would be well suited as a base for studies of further physically adsorbed layers by standing wave measurements. For example melting and crystallization of two dimensional inert atom layers physically adsorbed on such a substrate could be characterized in great detail by the methods demonstrated above.

References

Anderson S K, Golovchenko J A and Mair G (1976) Phys. Rev. Lett. *37* 1141.
Aspnes D E and Studna A A (1981) Appl. Phys. Lett. *39* 316.
Batterman B W (1964) Phys. Rev. *133* A759.
Batterman B W (1969) Phys. Rev. Lett. *22* 703.
Batterman B W and Cole H (1964) Rev. Mod. Phys. *36* 681.
Bedzyk M J, Gibson W M and Golovchenko J A (1982) J. Vac. Sci. Technol. *20* 634.
Cowan P L, Golovchenko J A and Robbins M F (1980) Phys. Rev. Lett. *44* 1680.
Golovchenko J A, Batterman B W and Brown W L (1974) Phys. Rev. *B10* 4239.
Golovchenko J A, Patel J R, Kaplan D R, Cowan P L and Bedzyk M J (1982) Phys. Rev. Lett. *49* 560.
Gutmann V (1967) ed. Halogen Chemistry Vol. 2 (Academic Press, New York).
James R W (1963) Solid State Physics *15* 55 (Academic Press, New York, 1963).
Sutton L E (1958) ed. Tables of Interatomic Distances and Configurations in Molecules and Ions Chemical Soc. Lond.
Wiener O (1890) Am. Phys. Chem. *40* 203.

Inst. Phys. Conf. Ser. No. 67: Section 7
Paper presented at Microsc. Semicond. Mater. Conf., Oxford, 21–23 March 1983

Impurity drag on climbing misfit dislocations in phosphorus-implanted (001) silicon

M P A Viegers, C W T Bulle-Lieuwma and W J Bartels

Philips Research Laboratories, P.O. Box 80.000, 5600 JA Eindhoven, the Netherlands

Abstract   Misfit dislocations in phosphorus-implanted (001) silicon were studied by TEM and X-ray double-crystal diffractometry. The misfit strain was found to be overcompensated by excess dislocations. This shows that the dislocations climbed in consequence of a self-interstitial supersaturation. The excessive climb was not observed after intentional contamination with gold. It is argued that this has to be due to Cottrell pinning.

1. Introduction

Misfit dislocations in silicon can be generated during high concentration phosphorus-diffusion owing to solute lattice contraction by undersized phosphorus atoms. For chemical source diffusion there is a critical misfit beyond which a stable dislocation network is formed (Fair 1978). For diffusion of ion-implanted phosphorus it requires a dose of 10^{16} cm^{-2} to obtain a network of misfit dislocations (Tamura 1977). It is generally considered that the misfit dislocations may move by climb involving the absorption or emission of intrinsic point defects (vacancies or interstitials). Therefore their movement may be related to a number of anomalous diffusion effects in silicon, reviewed by Willoughby (1977) and more recently by Gösele and Frank (1981). Hence dislocation movement may offer clues to diffusion mechanisms, provided information is available on forces acting on the dislocations.

Little attention has been paid to measurement of the state of strain of diffused layers containing climbing misfit dislocations. The state of strain gives information on those forces acting on the dislocation network which are due to internal stress. In addition it may permit an estimation of forces associated with non-equilibrium concentrations of intrinsic point defects.

In the present work we studied the effect of phosphorus diffused in silicon from a chemical source (POCl$_3$) and from ion-implantation. The state of strain of the lattice was derived from double-crystal rocking curves. TEM was used to characterize the dislocations. Our first results (Viegers et al. 1983) revealed an unexpected lattice distortion, corresponding to compressive stress in the diffused layer, which was attributed to excess dislocations. Here we present the results of further experiments on silicon wafers that were intentionally contaminated with gold. It has been reported that gold is gettered at substitutional sites in phosphorus-diffused regions (e.g. Meek et al. 1975). Our goal has been to observe the effect of impurity gettering on dislocation climb.

2. Experimental

Experiments were performed on 3" diameter p-type Si wafers (20-30 Ωcm), (001) oriented, and grown by the floating zone method. The polished front-side of the wafers was either implanted with P^+ at 80 keV (the projected range is 0.1 /um) and a dose of 2.5×10^{16} cm^{-2} followed by a diffusion at 950°C for 20 minutes in dry nitrogen, or diffused with phosphorus by thermal deposition from POCl$_3$ at 950°C for 15 minutes. In both cases the resulting phosphorus concentration was 8×10^{20} cm^{-3} in a layer of 0.5 /um as determined by secondary ion mass spectrometry (SIMS).

Prior to the phosphorus treatment some wafers were contaminated with gold. It was implanted in the non-polished rear-side of the wafers and diffused at 950°C during 5 hours in dry nitrogen. Thereafter about 5 /um silicon was etched from the surfaces to remove gold accumulated in these regions. Then the wafers were thoroughly cleaned. The resulting gold concentration was 10^{15} cm^{-3} as determined by neutron activation analysis.

TEM specimens were prepared by thinning the wafers from the rear-side. The etchant was composed of 5:1:1 HNO$_3$-HF-CH$_3$COOH. TEM observations were made with a Philips EM400 microscope. Common methods were employed to charac-terize dislocations (Loretto and Smallman 1975). The depth of a dislo-cation was determined using the oscillatory behaviour of its contrast.

The |115| diffraction peaks were recorded with an X-ray double-crystal diffractometer in the parallel (+,-) setting as described by Bartels and Nijman (1978). The in-depth concentra-tion profile of P is flat enough to be represented by a model in which a uni-formly doped layer (L) is fitted on the substrate (S) (fig. 1). Two parameters define the lattice mismatch of L with respect to S, parallel $(\Delta a/a)_{//}$ and per-pendicular $(\Delta a/a)_\perp$ to the wafer surface. The relaxed lattice constant of the layer $(\Delta a/a)_{relax}$ was obtained after the application of a correction factor for the elastic deformation according to Hornstra and Bartels (1978). Strains in the substrate were neglected, because the thickness of the layer is much smaller than the dimensions of the substrate.

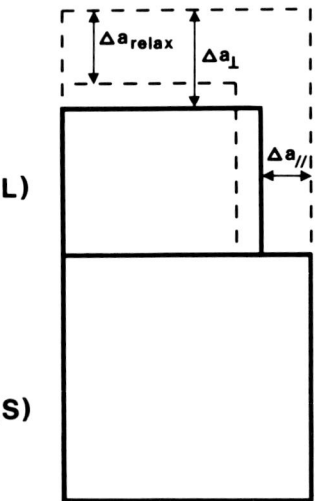

3. Results and discussion

Lattice mismatches are given in table 1 for layers diffused from POCl$_3$, and from P^+ implantation with a dose of 2.5×10^{16} cm^{-2}. The implantation was also carried out after intentional contamination with gold to a level of 10^{15} cm^{-3}.

Fig. 1 Schematic representation of deformed (solid line) and un-deformed (dashed line) lattice of an epitaxial layer L on a sub-strate S.

Table 1. Lattice mismatches

	$(\Delta a/a)_{//}$	$(\Delta a/a)_{\perp}$	$(\Delta a/a)_{relax}$
diffused from POCl$_3$ (uncontaminated)	$< 0.5 \times 10^{-4}$	15.3×10^{-4}	8.9×10^{-4}
ion-implanted and diffused (uncontaminated)	11.8×10^{-4}	7.0×10^{-4}	9.1×10^{-4}
ion-implanted and diffused after contamination with gold	10.3×10^{-4}	10.3×10^{-4}	10.3×10^{-4}

For the uncontaminated cases the relaxed mismatches are almost equal. This
is to be expected since the phosphorus concentrations are about the same
(8×10^{20} cm^{-3}). The absolute value of the relaxed mismatch in the case of
contamination with gold is 13% larger. The influence of gold on the
lattice mismatch is expected to be small because of its low concentration,
and of opposite sign as compared to phosphorus. Thus the larger relaxed
mismatch has to be attributed to a higher concentration of phosphorus.
Since the same dose had been implanted, the higher concentration might indicate that the diffusion of phosphorus has been retarded by the presence
of gold. However, minor changes in annealing temperature and time may
produce a similar effect. Therefore these numbers are not yet conclusive.

The other parameters of table 1 show significant differences. For the
layer diffused from POCl$_3$, $(\Delta a/a)_{//} \approx 0$. This means that the layer resides
coherently on the substrate, i.e. without misfit dislocations. It agrees
with our TEM observations where neither dislocation lines or loops nor
precipitates were found. The tetragonal distortion which follows from
$(\Delta a/a)_{//} < (\Delta a/a)_{\perp}$ has to be attributed to elastic strains. The stress in
the layer is tensile, as would be expected for a coherent lattice contai-
ning undersized phosphorus atoms.

In the layer that was implanted and diffused without intentional contamination we found a sessile network of edge type dislocations (fig. 2) at
a depth of about 0.5 μm, i.e. at the slope of the phosphorus profile.
The network resides on the (001)-plane with Burgers vectors a/2 [110] and
a/2 [1$\bar{1}$0]. The extra half planes have been inserted from the top of the
layer. In this case $(\Delta a/a)_{//}$ is much larger than $(\Delta a/a)_{\perp}$. The distortion
has a direction corresponding to compressive stress in the layer, which
is the opposite of the stress in the layer diffused from POCl$_3$.

Apparently the misfit strain has been overcompensated by too many dislocations. It is known that a misfit network develops in the very early
stages of annealing (Tamura et al. 1975). The P concentration is then high
and the misfit strain correspondingly large, providing an opportunity for
the formation of a relatively high density of misfit dislocations. During
the drive-in under dry nitrogen the local concentration of phosphorus, and
thus the lattice mismatch, decreases. After that the internal stress acts
in such a manner as to move excess dislocations out of the layer. Nevertheless too many dislocations have climbed in depth. It shows, as suggested by Willoughby (1978), that climb is not the origin of the supersaturation of point defects. Instead, the observed climb probes the presence
of a non-equilibrium concentration of point defects. There may be an
undersaturation of vacancies or a supersaturation of interstitials.

It has been argued (Gösele and Strunk 1979) that anomalous phenomena,

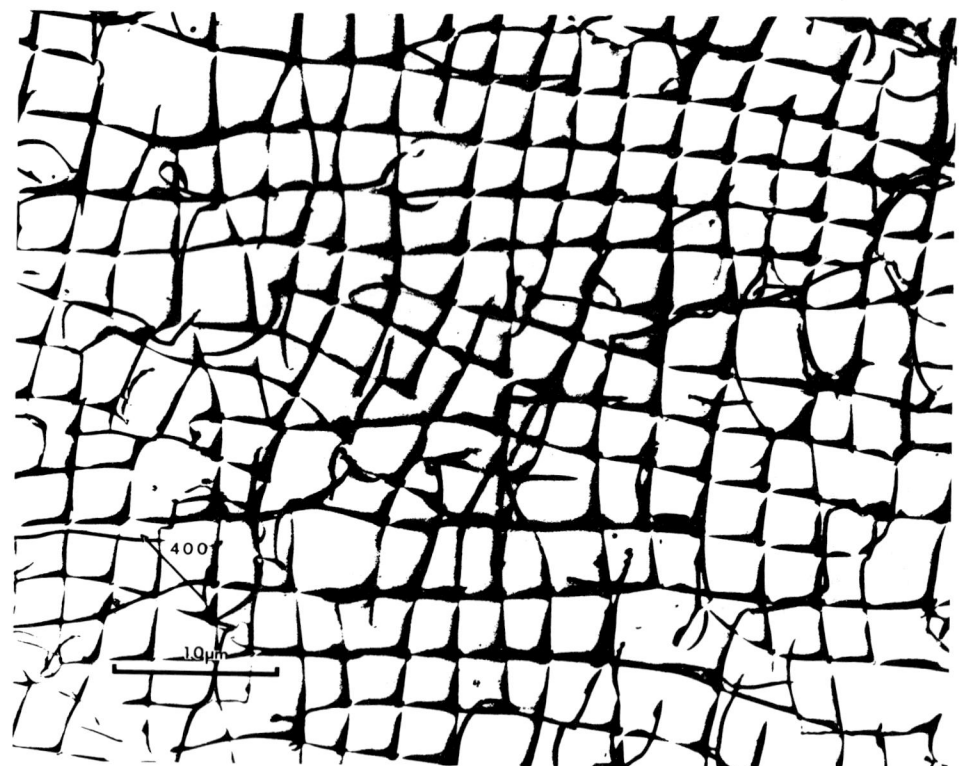

Fig. 2 Misfit network in Si implanted with 2.5×10^{16} P^+ cm^{-2} at 80 keV and diffused at $950^\circ C$ for 20 minutes in dry nitrogen.

generally associated with the diffusion of phosphorus, require a supersaturation of intrinsic point defects and cannot be explained in terms of an undersaturation. Accordingly, the observed climb must be due to a supersaturation of silicon self-interstitials, which agrees with the climb of screw dislocations into helices observed by Strunk et al. (1979) and the climb of oxidation-induced stacking faults, studied by Claeys et al.(1978).

In the presence of gold the dislocation network has broken into relatively short segments (fig. 3) as compared to the uncontaminated case. It would appear that the development of a more complete network has been impeded. No evidence could be obtained for impurity precipitation on the dislocations using weak beam imaging. The misfit strain, however, is not over-compensated but just relaxed, since now $(\Delta a/a)_{\!/\!/} \approx (\Delta a/a)_\perp$. This could be explained if the supersaturation of silicon self-interstitials had been strongly reduced by the presence of gold, but existing evidence indicates the contrary. It has been shown (Meek et al. 1975) that gold is gettered in phosphorus-diffused regions essentially at substitutional sites. This is attributed to the formation of P^+Au^- complexes. Dash (1960) showed that gold diffusion into silicon caused left-hand screw dislocations to form right-hand helices. In this way the dislocations either generate vacancies or absorb interstitials, which means that the effect of gold arriving at a

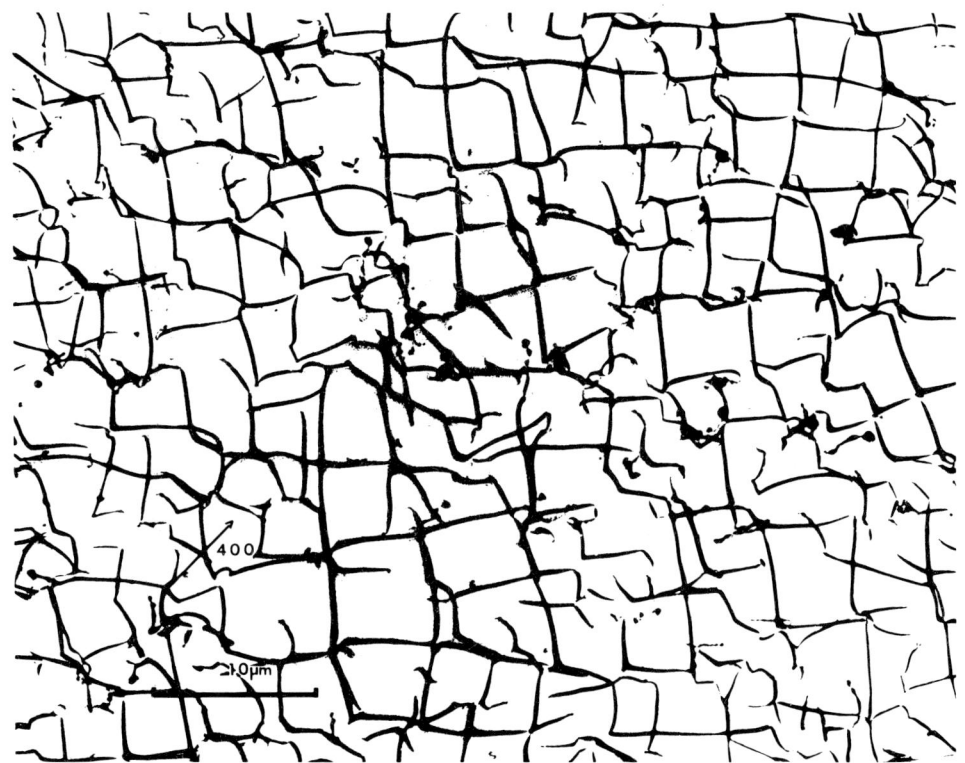

Fig. 3 Misfit network consisting of small segments after intentional contamination with gold.

substitutional site is to deplete the lattice of vacancies or to generate
interstitials. Today there exist strong experimental indications, recently
reviewed by Gösele and Tan (1982), that in the diffusion of gold there is a
kick-out mechanism that generates interstitials. In conclusion the effect
of gold is not expected to cause a reduction but rather an enhancement of
the supersaturation of self-interstitials. Nevertheless the dislocation
motion has been restricted, probably by Cottrell pinning. As a result
only small segments of the dislocations could climb to a depth of 0.5 μm.
The ends could not break away from their atmosphere of impurities and
remained pinned near the surface. Consequently a complete network had not
developed. The extra half-planes extended only partially to the slope of
the phosphorus profile, so that the misfit strain was just relaxed.

4. Conclusions

Combining the results of X-ray double-crystal diffractometry and TEM we
observed an overcompensated misfit strain caused by excess dislocations.
This shows that dislocation climb in phosphorus-diffused silicon is not
the origin of supersaturated self-interstitials but rather a consequence
of it. Excessive climb of dislocations was hampered by the presence of
gold as an additional impurity. This can be attributed to Cottrell pinning.

It possibly contributes to the getter performance. On the other hand absorption of silicon self-interstitials by climbing dislocations is impeded at the same time. It follows as a matter of course that the supersaturation of interstitials increases, and this in turn lowers the substitutional solubility of gold, therefore reducing the getter performance. Altogether there might be no net effect of Cottrell pinning.

Acknowledgements

We wish to thank A.J. Linssen for making the clean and contaminated phosphorus-diffused wafers available to us, H.A.M. de Grefte for SIMS measurements and P. Bruisten for neutron activation analysis.

References

Bartels W J and Nijman W 1978 J. Cryst. Growth 44 518
Claeys C L , Declerck G J and Van Overstraeten R J 1978
    Semiconductor Characterization Techniques, eds. Barnes P A and
    Rozgonyi G A (Princeton : Electrochem. Soc.) pp 366
Dash W C 1960 J. Appl. Phys. 31 2275
Fair R B 1978 J. Electrochem. Soc. 125 923
Gösele U and Strunk H 1979 Appl. Phys. 20 265
Gösele U and Frank W 1981 Defects in Semiconductors, eds. Narayan and Tan
    (Amsterdam : North Holland) pp 55
Gösele U and Tan T 1982 Proc. Satellite Symposium on Aggregation
    Phenomena of Point Defects in Silicon, Munich Sept. (to be published)
Hornstra J and Bartels W J 1978 J. Cryst. Growth 44 513
Loretto M H and Smallman R E 1975 Defect Analysis in Electron Microscopy
    (London : Chapman and Hall)
Meek R L, Seidel T E and Cullis A G 1975 J. Electrochem. Soc. 122 787
Strunk H, Gösele U and Kolbesen B O 1979 Appl. Phys. Lett. 34 530
Tamura M, Joshihiro N and Ikeda T 1975 Appl. Phys. Lett. 27 427
Tamura M 1977 Phil. Mag. 35 663
Viegers M P A, Bulle-Lieuwma C W T and Bartels W J 1983 Physica 116B 612
Willoughby A F W 1977 J. Phys. D.: Appl. Phys. 10 455
Willoughby A F W 1978 Rep. Progr. Phys. 41 1665

Inst. Phys. Conf. Ser. No. 67: Section 7
Paper presented at Microsc. Semicond. Mater. Conf., Oxford, 21–23 March 1983

Measurement of grading in heteroepitaxial layers

M A G Halliwell, J Juler and A G Norman[+]

British Telecom Research Laboratories, Martlesham Heath, Ipswich, IP5 7RE.
[+] Department of Metallurgy and Science of Materials, University of Oxford, Parks Road, OXFORD OX1 3PH.

Abstract Heteroepitaxial layers of III-V semiconductors can show compositional grading with thickness, arising from variations in the conditions during layer growth. In this paper the Taupin theory for the diffraction of x-rays in distorted crystals has been used to predict x-ray rocking curve shapes for linearly graded layers.
The variation in the indium content of gallium indium arsenide layers on indium phosphide substrates has been measured using electron probe x-ray microanalysis. Rocking curves have been calculated from the elemental variation observed and compared to those recorded with the double crystal diffractometer.

1. Introduction

The optical communications devices currently being developed at BTRL are fabricated from heteroepitaxial layers. The new generation of detectors are made from gallium indium arsenide layers on indium phosphide substrates. In order to produce satisfactory devices the composition of the layers must be controlled such that the lattice parameter difference between layer and substrate is small. The procedure for measuring lattice parameter differences between epitaxial layers and substrates, using the double crystal diffractometer is well established (Halliwell, 1981). In the diffractometer the x-ray beam is diffracted by a slice of substrate material followed by diffraction from the sample. The sample is rotated through the Bragg condition and the resulting variation of intensity with angle is known as the "rocking curve". Detailed analysis of peak shapes of the rocking curves recorded from heteoepitaxial layers has not yet been developed. In this paper we present a preliminary attempt to derive theoretical rocking curves for graded layers of gallium indium arsenide on indium phosphide. The layers are assumed to have a low density of crystalline defects such as dislocations and stacking faults. A variation of composition within the layer will be accompanied by a variation in lattice parameter and we would expect the layer peak to be broadened. In order to calculate the rocking curves in the presence of a lattice parameter gradient, we represented the layer by a series of laminae each with a different constant lattice parameter. In this preliminary report the layers were subdivided into laminae of equal thickness with equal lattice parameter differences between them to model linear grading with depth.
The calculations were performed using Taupin's (1964) theory of diffraction by non-uniform crystals. This theory has already been used to

predict rocking curves for diffused and ion-implanted layers in silicon (Burgeat and Taupin 1968, Fukuhara and Takano 1977 and Larson and Barhorst 1980) and a reasonable correlation between theory and experiment has.been established. In this paper we compare theoretical and experimental rocking curves for three samples of gallium indium arsenide on indium phosphide. All three gallium indium arsenide layers were found to show a linear compositional grading using electron probe microanalysis.

2.Calculation of Rocking Curves

The structure factors for indium phosphide and gallium indium arsenide were calculated from the atomic scattering factors given in the International Tables (1974). The values for the ternary were expressed as a function of m, the mismatch between the layer and the substrate. Table 1 lists all the numerical values used.

Table 1 Numerical values used in rocking curve calculations

Wavelength	$CuK\alpha_1 = 1.540562$Å	
Lattice Parameters: InP	= 5.8688 Å	
$Ga_{(1-y)}In_yAs$	= 5.6535(1-y) + 6.0585y Å	

Structure factors:

Material	000Reflection	004Reflection
InP	256.00+21.92i	161.03+21.92i
(Ga,In)As	(287.45+.0011m)+(16.22+.0003m)i	(184.72+.0008m)+(16.22+.0003m)i

m is the mismatch (=the fractional difference in lattice parameter)

The following relationship for ratio of incident to diffracted amplitude (X) as a function of depth (Z) below the sample surface was used:

$$dX/dZ = (i\pi/\lambda \sin\theta_o)(Y_H X^2 + (2Y_o - \alpha_H(Z))X + Y_H) \ldots\ldots (1)$$

where θ_o is the Bragg angle, Y is related to the structure factor F by the formula $Y = -(\lambda^2/V)r_e F$ where V is the unit cell volume, λ is the wavelength and r_e is the classical electron radius. Subscript H refers to the 004 reflection and subscript o refers to the 000 reflection. $\alpha_H(Z) = -2\lambda(\theta - \theta_o) \cos\theta_o/d$, where θ is the angle of incidence.

The d spacing used for the layer, assumed tetragonal distortion occured in the layer (Halliwell,1981). This is essentially equivalent to that used by previous workers on silicon but in this case we had to consider a variation of structure factor with depth as well as a variation of θ_o. At first glance it would seem reasonable to find X(0), the ratio of the amplitudes at the surface of the crystal, by numerically solving equation (1) by a standard technique. This was not possible as X oscillates rapidly. Instead, we must consider the sample as made up of a series of thin laminae of constant lattice parameter and use the value X obtained for the top surface of one layer to be the value of X at the bottom surface of the adjacent layer. For a uniform layer equation (1) can be written in the form :

$$dX/dZ = iC(A(X^2 + 1) + 2BX)$$

where A, and C are constants and B depends on ($\theta - \theta_o$), the deviation from the Bragg condition. As a boundary condition we assume X = 0 at a point z deep inside the crystal. The solution obtained is:

$$X(Z) = iA \tan(CS(Z - z))/(S - iB\tan(CS(Z - z))) \ldots\ldots\ldots\ldots(2)$$

where $S = (B^2 - A^2)^{\frac{1}{2}}$. The value for X at the surface of a thick uniform crystal, such as the substrate, is the limiting value of equation (2) when (Z - z) is large and negative.

$$X(0) = (- B + Ssign(Im(S)))/A \dots\dots\dots\dots\dots\dots\dots\dots (3)$$

To obtain the reflectivity as a function of the angle of incidence, equation (3) must be evaluated for different values of B.
We next consider the appropriate boundary conditions for a number of different crystalline layers each of uniform composition. If, for instance, the boundary condition was X(w) = K, equation (2) would be replaced by

$$X(Z) = (KS + i(A +KB)\tan(SC(Z - w)))/(S(B + AK) \tan(SC(Z - w))) \dots (4)$$

Thus to find a value of the amplitude ratio at the surface of a substrate with a single uniform epitaxial layer, we first use equation (3) to give a value for the ratio at the surface of the substrate. Then using different values of A, B and C appropriate to the layer, we calculate the ratio at the surface of the layer using equation (4). For graded layers the epitaxial layer first has to be sub-divided into a number of laminae, each assumed to be of constant lattice parameter. Equation (4) is then evaluated repeatedly to obtain the value of the amplitude ratio at the top surface. The reflectivity(R) is given by the square of the modulus of the complex amplitude ratio at the surface of the crystal.

The reflectivity for the first crystal (indium phosphide substrate) is convoluted with the reflectivity for the sample to give the final rocking curve in the manner described by James (1965).

A series of computations was performed in order to establish the optimum lamina thickness required to calculate rocking curves. The approach was to repeat the calculations with increasingly large numbers of laminae until the rocking curve reached a stable shape. For this purpose all layers were assumed to have lattice parameter differences (m) of 400 parts per million (ppm) at the layer/substrate interface, increasing to 800 ppm at the top surface. Inspection of curves calculated for a layer of total thickness 2 μm indicated that the rocking curve reached a stable shape when the layer had been divided into at least eight laminae, ie lamina of thickness 0.25 μm. Similar tests on layers up to 5 μm thick again indicated laminae 0.25 μm thick were adequate to calculate the rocking curve.

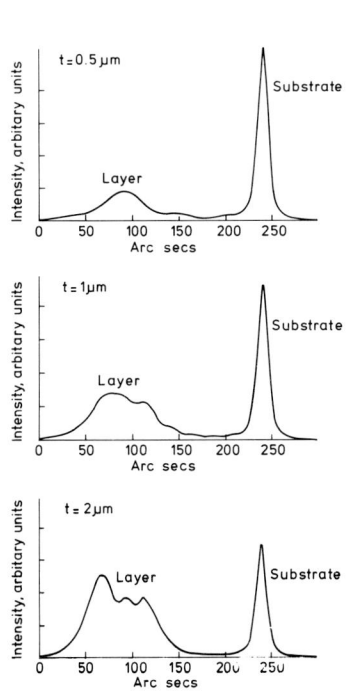
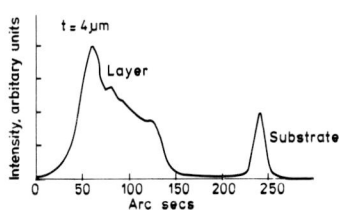

Fig. 1 Calculated x-ray rock-ing curves for layers with m varying from 400 to 800 ppm.

Eight laminae were used for all layers less than 2 μm in thickness. Fig 1 shows calculated rocking curves for graded layers with thicknesses (t) of 0.5, 1, 2, and 4 μm. Since we have calculated the rocking curve by dividing the layer into a series of perfect laminae we might have expected to obtain a wedge shaped layer peak corresponding to the sum of a series of equally spaced peaks of identical half width, and of increasing intensity as we consider laminae closer to the layer surface. The computed peak for a thicker layer is approximately wedge shaped, but it does also show some unexpected fine structure. At 1 μm and below the overall shape no longer approximates to a wedge.

3. Microprobe Determination of Grading

Layers of gallium indium arsenide on indium phosphide substrates were grown at BTRL by a vapour phase technique employing a mixed gallium/indium source (Chatterjee et al 1982). In this technique a monotonic variation in layer composition can occur during growth. In order to determine the magnitude of the compositional grading, electron probe X-ray microanalysis (XRM) was carried out on angle lapped (approx 2° angle) and polished specimens. The specimens were polished using a special jig and Syton on a glass plate. Polished specimens were then cleaned thoroughly and mounted using silver DAG on a brass holder for microprobe work. Bevel angles and layer thicknesses were measured using optical interferometry techniques. The XRM was carried out on a Camebax microprobe linked to a microcomputer for automatic stage control and on line data processing (including ZAF corrections). An accelerating voltage of 10kV was used to achieve good spatial resolution. Wavelength dispersive crystal spectrometers and count times of 50 seconds were used to optimise the accuracy of the analyses and the elements were analysed using In Lα, Ga Lβ and As Lα peaks to avoid problems of peak overlapping.

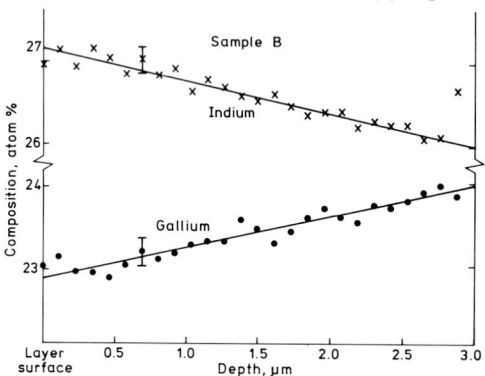

Fig. 2 Variations in In and Ga concentrations with depth for sample B

To maximise the sensitivity to small variations in layer composition, for each analysis the surface of the layer itself was used as the standard, the composition of the surface being taken as the value determined by the rocking curve method. Although this procedure does not give independent absolute values of composition, it should give accurate relative values of composition and hence of grading rates.

A line of point analyses at intervals of 4 or 5 μm was then made automatically down the bevel, perpendicular to the interface, to measure the grading in composition with depth in the layer. Results of analyses were then plotted versus the depth in the layer as in Fig.2. For the layers analysed the results indicated a uniform grading, and so a straight line, least squares fit was made, and the slope used to give the mean grading rate. The results of several runs were averaged to give the results quoted in Table 2. In all three samples the indium content was lowest at the interface with the substrate. A further account of this XRM method and results will be published later (A.G.Norman and G.R.Booker). Some analogous results for grading in LPE layers were recently reported

(Darby, Boyes and Booker 1982).

Table 2 XRM data

Sample	Layer Thickness	Bevel angle	Gradient-In% per μm
A	3.4 μm	1.88°	0.24%
B	3.4	1.65	0.32%
C	3.2	2.02	0.25%

4. Comparison of Theoretical and Experimental Rocking Curves

The total changes in m, the mismatch, measured by XRM were 1060, 1500 and 1035 ppm for samples A, B and C respectively. X-ray rocking curves were calculated using the thicknesses and mismatch ranges measured by XRM. The absolute value of the mismatch at the layer surface was adjusted so that the calculated rocking curve gave the same separation between the layer and substrate peaks as that recorded experimentally from the sample. The results are shown in Fig.3.

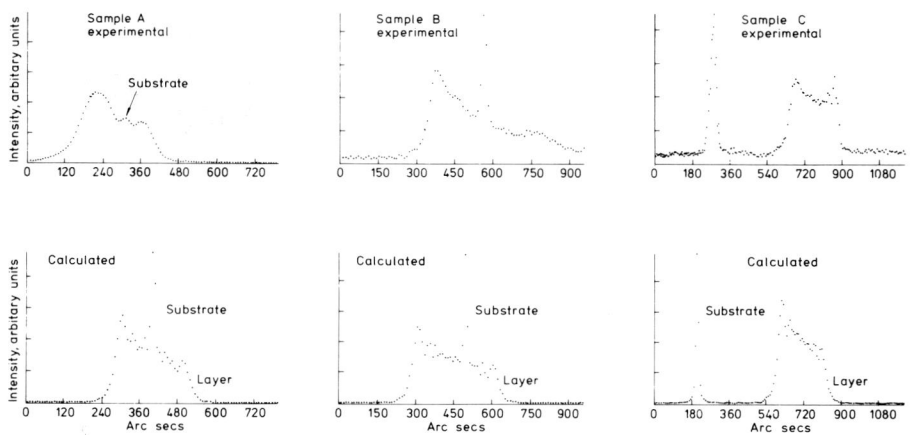

Fig 3 Comparison between theoretical and experimental x-ray rocking curves for linearly graded gallium indium arsenide layers on indium phosphide substrates. (CuKα 004 reflection)

In all three cases the overall breadth of the calculated and experimental rocking curves are similar. This confirms that the peak width of the x-ray rocking curve is a useful non-destructive method of determining the compositional variation in epitaxial layers.

More detailed comparison between the calculated and experimental data does reveal discrepancies. This is not unexpected since our preliminary calculations have made a number of simplifying assumptions. For instance we have assumed both layer and substrate are of good crystalline quality. Reduced perfection would be expected to give a broadening of the features which probably accounts for the more rounded appearance of the experimental curve for sample A. Samples with epitaxial layers with m not equal to zero will normally become curved. This has not been taken into account and would be expected to lead to peak broadening. This accounts for the increased substrate peak breadth in the experimental curve recorded from sample C. In addition we have only considered one of the polarization states of the incident x-ray beam; inclusion of the second polarization state will give small changes in the peak shape. Further discrepancies could be due to the compositional variations not being exactly linear. Although it was possible to interpret the XRM results as indicating linear grading, a small deviation from linearity would not have been detected. In particular the XRM method could not measure right up to the interface because the excitation volume included substrate material when the probe was positioned within about 0.7 μm of the interface. The detailed shape of the right hand side of the layer peak for sample C may well be due to a change in the In gradient near the interface. Work is in hand to extend the calculations to include both polarization states, non-linear grading and sample curvature.

5. Conclusions

A method has been developed for calculating x-ray rocking curves for heteroepitaxial layers showing compositional variations with thickness. In this preliminary report linear variations in lattice parameter are considered. The calculated rocking curves showed more fine structure than was expected from simple predictions.

The indium variations in three layers of gallium indium arsenide on indium phosphide were measured using electron probe microanalysis. The values obtained were used to calculate x-ray rocking curves which were compared with those obtained experimentally. The agreement between calculated and experimental curves was considered acceptable in view of the simplifying assumptions made. In particular the width of the layer peaks were very similar in the calculated and experimental curves. Thus we have demonstrated that rocking curve measurements offer a rapid non destructive method of detecting and quantifying linear grading in epitaxial layers. The method will be further developed for more complex compositional variations.

Acknowledgements

Acknowledgement is made to the Director of Research, British Telecom for permission to publish this work. We would like to thank C.R.Elliott and M.H.Lyons for samples, and B.Wakefield, T.Ambridge and M.R.Taylor (BTRL), and G.R.Booker (Oxford) for helpful discussions.

References

Burgeat J and Taupin D 1968 Acta Cryst. $\underline{A24}$ 99
Chatterjee A K, Faktor M M, Lyons M H and Moss R H 1982 J.Cryst.Growth $\underline{56}$ 591
Darby D B, Boyes E D and Booker G R 1982,10th International Congress on Electron Microscopy, Hamburg, $\underline{1}$ 687
Fukuhara A and Takano Y 1977 Acta Cryst. $\underline{A33}$ 137
Halliwell M A G 1981 Inst Phys Conf Ser No60: Microsc Semicond Mater Conf Oxford, April 1981, 271
International Tables for X-ray Crystallography, 1974 Vol 4, Kynoch Press (Birmingham)
James R W 1965 The Optical Principles of the Diffraction of X-rays, Bell(London) Chap 6
Larson B C and Barhorst J F 1980 J.Appl.Phys. $\underline{51}$ 3181
Taupin D 1964 Bull.Soc.Fr.Miner.Crist. $\underline{87}$ 469

Inst. Phys. Conf. Ser. No. 67: Section 7
Paper presented at Microsc. Semicond. Mater. Conf., Oxford, 21–23 March 1983

An examination of dislocations in Si-doped LEC GaAs by double crystal x-ray topography, SEM cathodoluminescence and chemical etching

G T Brown, C A Warwick[+*], I M Young and G R Booker[+]

Royal Signals and Radar Establishment, St Andrews Road, Malvern, Worcs.
WR14 3PS

+ Department of Metallurgy and Science of Materials, University of Oxford,
  Park Road, Oxford, OX1 3PH

* Present address: RSRE, St Andrews Road, Malvern, Worcs. WR14 3PS

Abstract   Double crystal X-ray topography has been used to examine the
grown-in dislocation distribution in an oblique <111> section cut from
a Si-doped LEC GaAs boule. Dislocations in the plane of the surface
have been characterised by imaging with different diffraction vectors
and the Burgers vector and line direction have been identified. The
same area has then been examined using SEM cathodoluminescence and
chemical defect etching using the AB etch. A good correlation has been
established between the three techniques and the images produced by
each technique are discussed.

1.   Introduction

The majority of the $10^5 cm^{-2}$ dislocations found in two inch diameter LEC
GaAs are introduced by the thermal stresses generated as the crystal cools
after growth. Since the preferred slip system for GaAs is <110> {111},
samples cut parallel to a {111} slip plane should reveal long segments of
dislocations in the plane of the surface. The density of dislocations is
far too low for characterisation by TEM so double crystal reflection X-ray
topography which has high strain sensitivity and good spatial resolution
(Jones et al, 1981) is more appropriate. Furthermore, by imaging the dislocations in the surface planes with three <440> reflections from the
three sets of <110> planes symmetrically disposed 35° from the surface
plane normal, the Burgers vector (and therefore dislocation line character)
can be deduced. Subsequent SEM cathodoluminescence (CL) and defect etching
studies should then reveal what effect, if any, dislocation character has
on the electrical behaviour of dislocations.

2.   Experimental Procedure

A 5 mm thick <111> slice was cut from the top cone of a Si-doped two inch
diameter LEC-grown crystal with a <100> growth axis (supplied by Cambridge
Instruments). A thick slice was necessary to minimise the effect of any
elastic strain introduced when mounting the slice for double crystal
topography. The measured carrier concentration (manufacturer's Hall
measurements) was $5 \times 10^{17} cm^{-3}$ which is significantly lower than the threshold for the precipitation of microdefects (Darby et al, 1979). The slice
was carefully orientated to within one minute of <111> and the {111} As

face was polished chemomechanically with a final polish in a solution of
bromine in methanol to remove any residual work damage. Details of the
double crystal technique have been described elsewhere (Jones et al, 1981).
The asymmetric (+ -) dispersive arrangement was used with high quality
silicon as the reference crystal. Cukα_1 radiation was used and the reflections used were <440> GaAs and <422> Si.

The CL studies were performed on a Stereoscan S4-10 at 30 kV (15 nA) and
10 kV (45 nA) at 88K and using a cooled S1 photomultiplier as detector.
All of the micrographs shown are 'total light' images. TEM samples were
prepared by etching the {111} As surface for 30 s.in AB and ion beam thinning from the {111} Ga surface and the microscopy was done on a JEOL 120C
at 120 kV accelerating voltage. Chemical defect etching with AB etch
(Abrahams and Buiocchi, 1965; Stirland, 1977) was done at room temperature.
The etch rate for {111} As at this Si concentration was determined as
4 µ/min.

3.   Results

3.1   X-ray Topography

The topographs shown in Fig. 1 for the 044 and 440 reflections clearly
demonstrate that the dislocations which are lying in the plane of the
surface exhibit weak contrast for the 440 reflection (Fig. 1b) but strong
contrast for 044 (Fig. 1a). The 404 reflection also gave strong dislocation contrast but is not shown here. If it is assumed that the dislocations
lying in the plane of the surface, (111), have moved in that plane by glide
then they must have a Burgers vector which also lies in that plane and
since they are virtually invisible for a diffraction vector of 440 then the
Burgers vector must therefore be a/2 [$\bar{1}$10] .

It is also apparent from Fig. 1a that the dislocations are roughly in the
shape of elongated open-ended loops with the open end towards the surface
of the crystal. The direction of elongation is [01$\bar{1}$] and therefore corresponds to the 60° orientation. The fact that the open ends of these loops
are invariably towards the edge of the slice implies that they were generated either at or close to the crystal surface. However, slip in the
planes parallel to the surface has clearly developed inhomogeneously since
a high density of loops is only found in discrete areas. The area shown
in Fig. 1 (and Fig. 2a) was the largest area of such localised activity,
and only a few such areas were observed. Localised activity on other slip
systems had also occurred (eg. the top right hand part of Fig. 1). Also
evident in Fig. 1 are large cusps which have presumably formed by the
dislocations bowing out between immobile jogs. This is further evidence
that glide has occurred. The dislocations are not straight and so are not
rigidly confined to Peierls valleys. They are also frequently broken which
suggests that they are meandering out of the slip plane in a vertical sense
over quite large distances (a few microns). This suggests that climb
processes have occurred as well as slip.

3.2   SEM Cathodoluminescence

Fig. 2 shows three micrographs of the same area comparing the three methods
of imaging. Fig. 2b is the SEM CL micrograph and the dislocations are
lines of dark contrast due to non-radiative carrier recombination. There
is a good correlation between the three techniques and the CL micrograph
is very similar to the optical micrograph of the chemically etched surface.

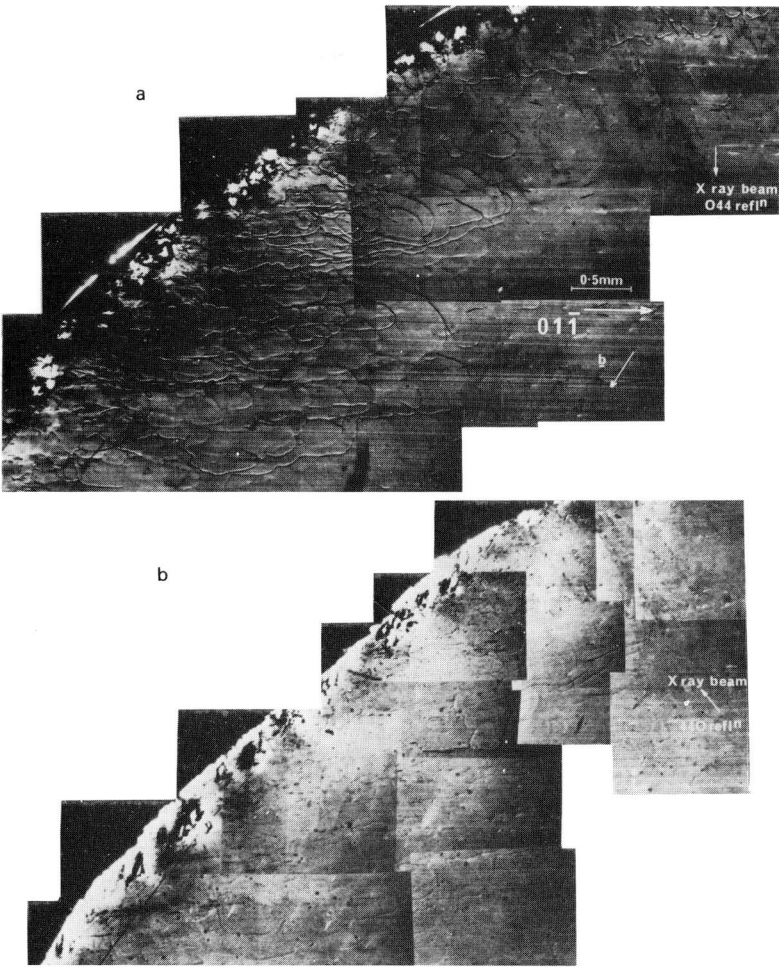

Fig. 1 Double crystal reflection topograph for 044 and 440 reflections

However, more dislocations are present on the X-ray topograph from which it can be inferred that the Cu$k\alpha_1$ X-rays penetrate further into the surface than the 3 μm generation depth of 30 kV electrons.

There are several interesting points which emerge on closer examination of Fig. 2b. It would appear that there is no marked effect on CL image as dislocation orientation changes. Some dislocations (eg. X on Fig. 2b) curve from 60° through to edge with little effect on the CL image width or intensity. The images are also sometimes weak and even disappear in some cases and a higher magnification example of this is shown in Fig. 3a. Successive etching in AB suggests that this is due to a depth effect. Removal of 0.5 μ (Fig. 3b) shows a broken etch feature which corresponds to a decrease in contrast and width of the CL image (Fig. 3a). Further etching to remove a total of 2.0 μ from the surface reveals a filling-in of the

etch feature suggesting that this
part of the dislocation lies
below the rest of the dislocation line. It was also possible
to confirm these observations by
varying the depth of the generating volume. Parts of dislocation lines which were not
apparent for an accelerating
voltage of 10 kV (which corresponds to a generation volume
extending 0.5 μ below the
surface) were present at 30 kV
which corresponds to a generation depth of 3 μm. These
observations confirm that dislocations are moving out of the
glide plane indicating that
climb has occurred.

Closer inspection of some dislocations reveals that at
points along a dislocation the
image can be modulated in width
and intensity (Fig. 4b). An
equivalent modulation is
observed in the etch feature
(Fig. 4a) but is barely detectable in the X-ray topograph.
The measured fluctuation in CL
image width is between 4 and
10 μm (this micrograph was
recorded for a 10 kV accelerating voltage). The width
of the non-radiative part
of the images of dislocations
lying in the plane of the

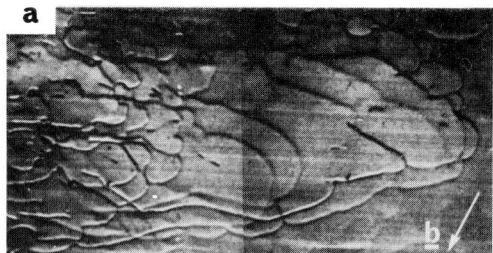

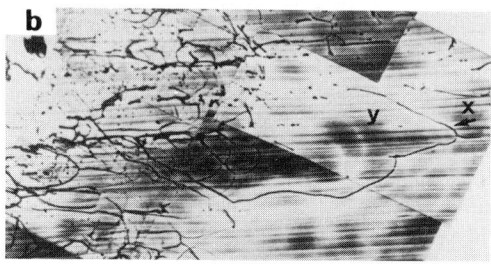

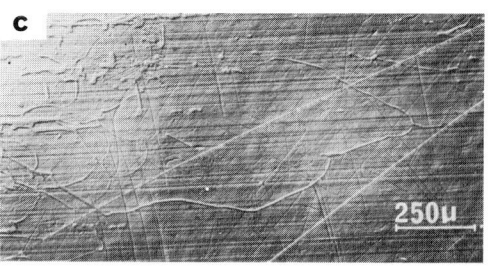

Fig. 2 A comparison of (a) double crystal
reflection X-ray topograph (b) SEM CL
micrograph at 30 kV (c) optical micrograph of AB etched surface (2 μm removed)

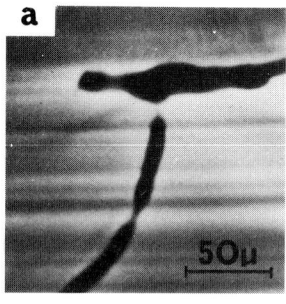
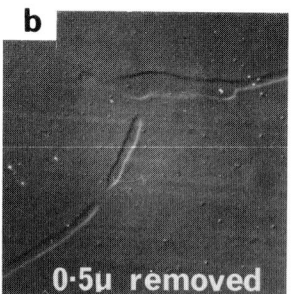
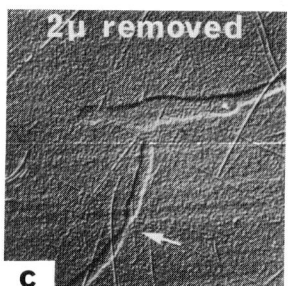

Fig. 3 (a) SEM CL micrograph at 30 kV (b) optical micrograph of
AB etched surface, 0.5 μm removed (c) 2 μm removed.

surface (eg. Fig. 2b), as measured by line traces, was up to 10 μ at 30 kV.
This is much larger than image widths estimated from previous CL studies

on Si-doped GaAs (grown by the
horizontal Bridgman technique)
which were of the order of 3 μm
at 30 kV (Darby et al, 1979;
Chu et al, 1981).

A further characteristic of the
CL images is the enhanced
luminescence which surrounds
the non-radiative part of the
image. This is particularly
apparent where the non-
radiative part of the image is
situated below the generation
depth whilst the luminescent
'atmosphere' falls within the

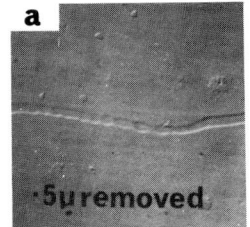 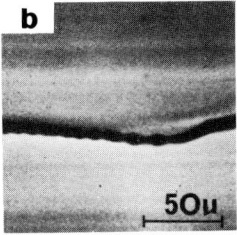

Fig. 4 (a) Optical micrograph of AB
etched surface, 0.5 μm removed (b) SEM
CL micrograph at 10 kV.

generation volume (eg. Y in Fig. 2b). The width of this luminescent part
of the dislocation image was measured from a line trace and found to be 35μ.
This effect has been noted before for n^+ GaAs (eg. Darby et al, 1980;
Casey, 1967; Balk et al, 1976) and has been called the 'dot and halo'
effect for dislocations viewed at acute angles to the surface. However it
has not previously been observed in Si-doped material although most Si-
doped material previously examined by SEM CL was grown by the Bridgman
technique. In a detailed examination of Te-doped GaAs, Chu et al, 1981,
showed that images similar to those described above were present in material with the dopant concentration below the level at which precipitation
occurs. This solubility threshold has been shown to be of the order of
$3.5 \times 10^{18}cm^{-3}$ for Si in Bridgman-grown GaAs (Darby et al, 1979). The
material used in this study should therefore contain none of the microdefects associated with exceeding this solubility limit. To confirm this
TEM samples were prepared and it can be seen in Fig. 5a that no microdefects were observed. The observations here are therefore consistent with

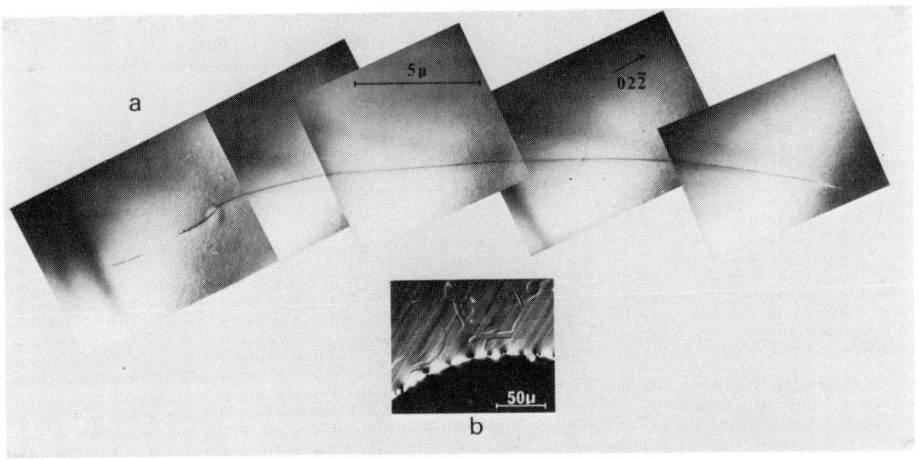

Fig. 5 (a) TEM micrograph of a dislocation (b) corresponding
optical micrograph of AB etched surface, 1 μm removed.

those for Te-doped GaAs when below the solubility limit. The images have been explained as being due to the gettering of impurities (eg heavy metal atoms) by the dislocation which decreases the number of non-radiative recombination centres in the zone around the dislocation thereby increasing the luminescence locally (Chu et al, 1981).

3.3 Chemical Defect Etching

The action of the AB etch on GaAs surfaces has been described by Abrahams and Buiocchi (1965) and Stirland (1977). Its etching behaviour on (111) is such that it attacks the matrix preferentially when dislocations are in the plane of the surface leaving a raised etch feature corresponding to the dislocation. This effect is shown in Fig. 2c which is an optical micrograph (using Nomarski interference contrast) of the same area imaged using X-ray topography and SEM CL after etching 2 μm from the surface. Many similarities exist between this etched micrograph and the CL micrograph (Fig. 2b). In particular the effect of dislocation depth (Fig. 3) and variations in image width (Fig. 4) which have both been discussed above.

The TEM sample (Fig. 5a) was prepared by thinning from the (111) Ga face with the (111) As face etched in AB. This enabled a correlation between the observed etch features and TEM. Fig. 5b shows the etch feature corresponding to the dislocation line in Fig. 5a. It is intended to extend this study to include TEM of the individual defects illustrated above and it is also intended to study semi-insulating GaAs.

4.   Conclusions

The main observations for dislocations in two inch Si-doped LEC GaAs are:-

  i) dislocations are generated within the growing crystal in an inhomogeneous manner and are possibly generated at the edges of the crystal,
 ii) loops are elongated along the 60° direction and are open-ended at the crystal surface,
iii) dislocation lines are not straight and thus not rigidly confined to Peierls valleys,
 iv) the SEM CL images were frequently broken or became weak along the length of a dislocation line and this has been attributed to a variation in the depth of the dislocation line below the surface,
  v) a modulation in width and intensity of the SEM CL image has been observed at some points along dislocation lines,
 vi) an 'atmosphere' of enhanced luminescence is observed around dislocations (~35μ wide) which has not previously been observed in Si-doped GaAs,
vii) TEM has shown that there are no microdefects present in this material.

References

Abrahams, M and Buiocchi, C J, 1965 J.Appl.Phys. 36 2855.
Balk, L J, Kubalek, E and Mezel, E. 1976 Proc. 9th SEM Symposium IITRI, Chicago, 257.
Casey, H C. 1967 J.Electrochem.Soc. 114 153.
Chu, Y M, Darby, D B and Booker, G R. 1981. Inst.Phys.Conf.Ser. 60 331.
Darby, D B, Augustus, P D, Booker, G R and Stirland, D J. 1979 J. of Microscopy, 118 343.
Jones, G R, Young, I M, Cockayne, B and Brown, G T. 1981 Inst.Phys.Conf. Ser. 60 265.
Stirland, D J. 1977 Inst.Phys.Conf.Ser. 33a 150.

Inst. Phys. Conf. Ser. No. 67: Section 8
Paper presented at Microsc. Semicond. Mater. Conf., Oxford, 21–23 March 1983

377

Acoustic and photoacoustic microscopy: application to device diagnostics

H K Wickramasinghe

Department of Electrical Engineering, University College London,
Torrington Place, London WC1E 7JE.

Abstract   In this paper, we review some recent advances in the field of
scanning acoustic microscopy such as quantitative measurements and
interior imaging with particular reference to microscopy of semiconducting devices. In addition, a new form of Photacoustic or Thermal
Wave microscope is described which is capable of detecting current
in a non-contacting fashion. The technique, which is based on an
optical interferometer, has been used to study electromigration and
detect faults in integrated circuits.

1. Introduction

The Scanning Acoustic Microscope (SAM) (Lemons and Quate 1974) has now
been developed to the point where its resolution operating in liquid
Helium at 4 GHz is better than 500 Å. (Quate 1982). Several review
articles have been published recently which clearly outline the progress
that has been achieved over the past decade. (Quate et al 1979,
Wickramasinghe 1983, Sinclair et al 1982). In this article, we review
some recent advances in the field such as quantitative measurements and
interior imaging with particular reference to microscopy of semiconducting
devices. In the following section, we present the basic operating principles of the instrument together with examples of images obtained in
reflection acoustic microscopy.

The problem of imaging the interior of solid samples is discussed in
Section 3 and some solutions are considered. Contrast observed in acoustic
micrographs can be related to density, elasticity and viscosity of the
sample being investigated. Techniques for obtaining quantitative
information will be discussed in Section 8.

Following a brief description of the Thermal Wave microscope, Section 5
will be devoted to a variety of experiments carried out on the detection
of current in thin film circuits, such as electromigration and defect
detection.

Finally, in Section 6, we present some brief concluding remarks.

2.   Principles of Operation

The heart of the acoustic microscope is a sapphire lens (Fig. 1).   A
short RF pulse (approxiamtely 30 ns in duration) is applied to a
piezoelectric transducer deposited on the back surface of this lens.   An
acoustic pulse propagates down the sapphire rod and is focused sharply
into the coupling liquid (usually water) by the spherical lens.   Whereas
in optics such a lens would result in severe spherical aberration, in
acoustics these aberrations are negligible due to the fact that there is
a sevenfold reduction in sound velocity in going from sapphire to water;
the focal spot is diffraction limited, and for a well designed lens its
diameter approaches the acoustic wavelength (1.5 µm at 1 GHz in water).
In the reflection acoustic microscope the object to be imaged is placed
at the focus of the lens.   Reflected acoustic waves return along the
incident path and are converted back into an electrical pulse by the
transducer.   The strength of this pulse is proportional to the acoustic
reflectivity of the object at the point being investigated.   An image is
formed by mechanically scanning the object in a raster fashion and using
the detected acoustic signal to modulate the brightness of a scan-
synchronised CRT display.   A transmission system can be constructed using
two such lenses arranged in a confocal geometry.

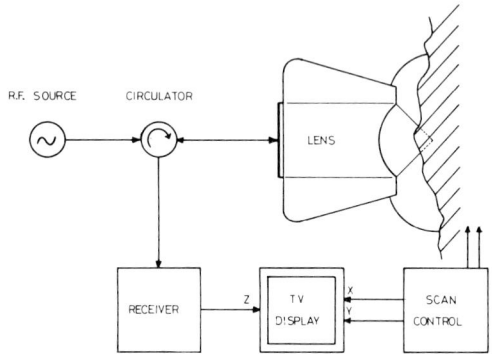

Fig. 1   Reflection scanning acoustic microscope.

2.1 Microscopy of semiconducting devices.

In this section, we shall present a series of examples on the application
of acoustic microscopy to semiconducting devices.   As the acoustic
microscope responds to changes in density, elasticity and viscosity, one
should expect it to provide us with new information about the samples
being investigated.   In what follows, we show a range of images taken
by M. Nikoonahad using a 1 GHz reflection instrument at University College
London.   Fig. 2 shows the acoustic micrograph of the control logic in a

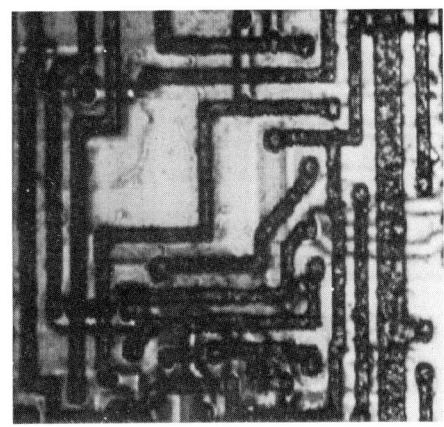

Fig. 2   Acoustic micrograph of the control logic in a 64 K
RAM chip; the field of view if 250μm x 250μm

64 K RAM chip.   Since the acoustic waves can penetrate the aluminium
metallisation, it is often possible to observe detail just beneath the
surface such as defects in the oxide layer and poor metal-film adhesion,
(Bray et al 1980).

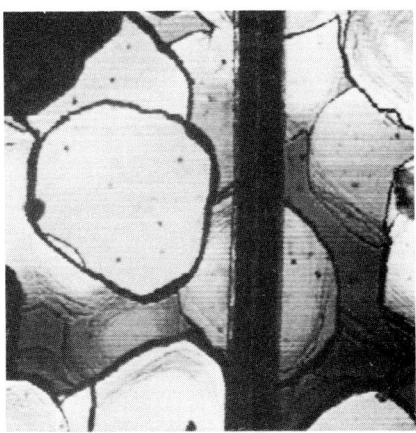

Fig. 3   Acoustic micrograph of the surface of a solar cell;
Field of view is 250μm x 250μm

Fig. 3 shows an acoustic micrograph of the surface of a solar cell.
Individual silicon grains are clearly visible in this image. Contrast in
this image could either be due to topographical variations on the surface
or variations in elastic properties from grain to grain. Further
investigation by other techniques indicated that the contrast was at least
partially caused by topographical variations. However, a unique property
of the acoustic microscope is that even on a perfectly flat sample,
contrast changes should be observed from grain to grain due to variations
in the 'acoustic reflectivity' via the V(z) effect. (Atalar et al 1977,
Wickramasinghe 1978, Atalar 1978, Parmon and Bertoni 1979). Fig 4 is
one such example; it is an acoustic micrograph of a polished steel surface
which looks uniform under the optical microscope. As we can see, the
grain structure is clearly visible. The grain structure can be revealed
optically after chemically etching the surface.

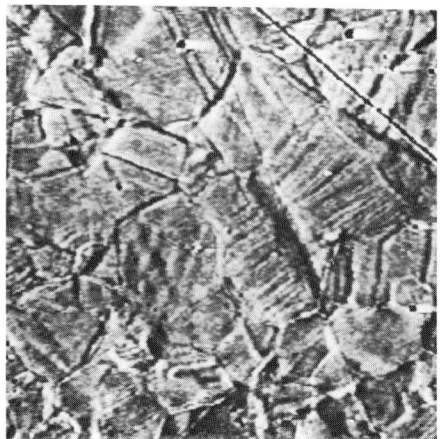

Fig. 4   Acoustic micrograph of a polished steel surface
showing grain structure; the field of view is
250 μm x 250 μm.

The micrographs presented so far were taken with the acoustic microscope
focused on or very close to the surface of the sample. Contrast in these
images therefore derived primarily from the elastic variations within a
wavelength or so from the surface. In the following section we shall
discuss the problems associated with imaging the interior of a sample.

3.   Interior Imaging

The basic problem that one faces when attempting to image the interior
of solids can be understood by referring to the ray diagram shown in
Fig. 5. Due to the large velocity ratio between solids and liquids,
converging spherical waves are seriously distorted at the solid-liquid
interface; paraxial rays come to a sharp focus at f, while rays incident

at large angles come to a focus closer to the surface. Due to this effect,
lenses that provide maximum refraction angles greater than 30° within
the solid, suffer from severe spherical aberration, and the resulting
resolution is not diffraction limited.

Several approaches have been suggested for overcoming this problem, the
simplest one being to reduce the lens numerical aperture to the point
where the maximum refraction angle within the solid is restricted to about
30°. (Pino et al 1981, Nikoonahad et al 1982, Nikoonahad 1983).
Calculations show that in this case the focal spot is diffraction limited
and its diameter is approximately two acoustic wavelengths within the
solid.

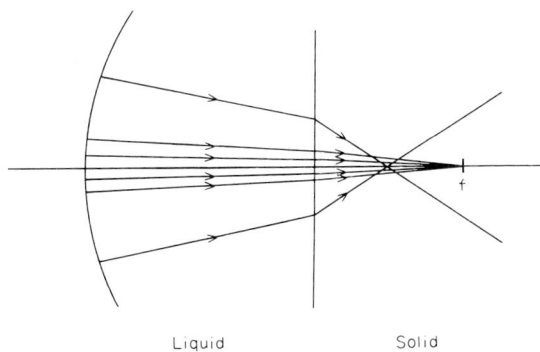

Liquid   Solid

Fig. 5   Ray diagram showing the aberrations introduced on to a
spherical wave focused inside a solid sample.

An example of an acoustic image taken using a reduced aperture lens is
shown in Fig. 6 and Fig. 7. The image was taken through a 1 mm thick
copper heat sink, and is of the bond region between it and the audio
power transistor chip TIP 33A; the acoustic frequency was 40 MHz. Fig. 6
shows a relatively uniform bond while Fig. 7 shows a poor bond.

Reduced aperture lenses work well for imaging through planar surfaces,
and provided that maximum resolution is restricted to about two solid
wavelengths. A mchanically scanned B-scan system has been used for
imaging through curved surfaces and achieving resolutions down to about
one wavelength within the solid. (Wickramasinghe 1981, Smith et al 1982).
The system utilises a second spherical surface in order to reduce
spherical aberrration. This second convex spherical surface is ground
onto a coupling element, preferably made of the same material as the
sample which is in acoustic contact with the sample to be imaged.

Fig. 6 Acoustic image (40 MHz) of power transistor header bond
        imaged through its 1mm thick copper heat sink; field
        of view is 6mm × 6mm (after Nikoonahad)

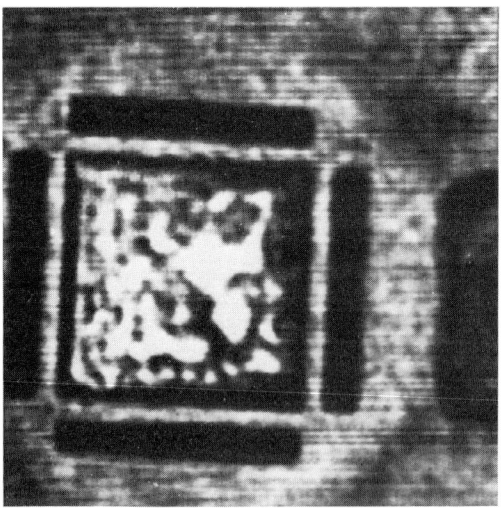

Fig. 7 Acoustic image of the same type of transistor as in
        Fig. 6 showing poor bonding. (after Nikoonahad)

4. Quantitative Measurements

Although acoustic micrographs in themselves provide us with a great
wealth of new information, sometimes it is useful to examine a particular
region of an object in greater detail using quantitative methods. One
method is to measure the V(z) response. (Atalar et al 1977). If one
stops the raster scan, and plots the detected acoustic signal as a function
of the defocus distance (as measured from the geometrical focus), we find
that each material surface produces a characteristic response. In
particular, nulls in the detected signal are observed spaced regularly
along the defocus direction. The null spacing can be directly related to
the Rayleigh wave velocity at the point being investigated. Such
measurements not only provide useful information about the elasticity of
the sample but also can be used to measure anisotropy. (Kushibiki et al
1981). More recently, another method has been proposed for measuring
elastic constants on a point by point basis. It is based on the fact
that the far field pattern of the focal distribution can be realted to
the spatial frequency transfer function of the object at the point being
investigated. (see Fig. 8). The pattern is recorded using an acoustic
probe which in itself could be a lens. Fig. 9 shows an example of a far
field pattern through a 2.2. wavelength plate of Aluminium using 4 MHz
sound waves; the recording plane was 40 mm away from the object. The
position of the rings in this image can be related to the shear and
longitudinal velocity in the plate at the point being investigated; the
accuracy achieved is typically around 6%. (Smith et al 1980, Smith et al
1981).

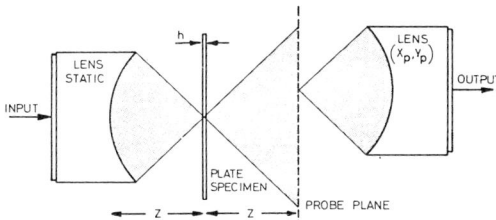

Fig. 8 Geometry for recording spatial frequency transfer function

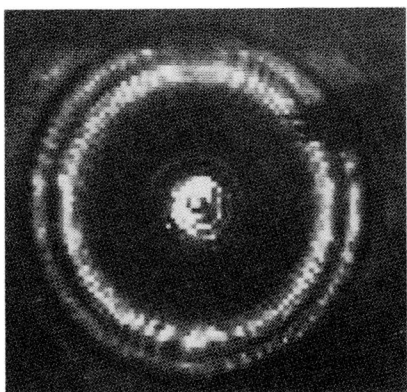

Fig. 9 Far field pattern for an Aluminium plate at 4 MHz

5. Thermal Wave Microscopy

Since the first photoacoustic micrograph was published (Wickramasinghe
et al 1978), several novel imaging systems have been reported. (Favro
et al 1980, Luukkala and Penttinen 1979, Ameri et al 1981). In this paper
we present recent results on a novel thermal wave imaging system for
measuring current in thin film circuits. (Wickramasinghe et al 1982).
The basic principle is illustrated in Fig. 10 Periodic current flowing
down the thin film track results in periodic expansion due to Joule
heating; this expansion can then be detected using an optical heterodyne
interferometer which is focused and scanned across the surface. Either
the phase or amplitude of the vibration can be recorded.

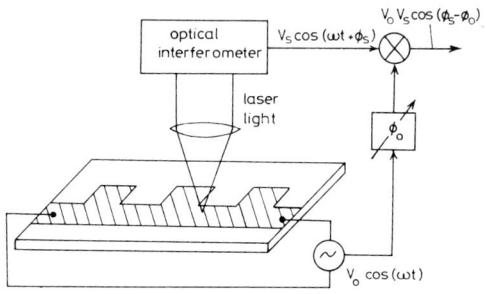

Fig. 10 Configuration for detecting current in thin film circuits

The form of the track used in our initial experiments is shown in Fig. 10.
The aluminium track was 0.4 um thick and its width varied from 20 μm to
40 μm with a periodicity of 20 μm; the peak current at   5 KHz was 60 mA
Fig. 11 shows the amplitude and phase plots across the track. It can be
easily shown that the amplitude signal is approximately proportional to
the square of the current density in the track. The asymmetrical pattern
is consistent with current crowding at the edges of the track.

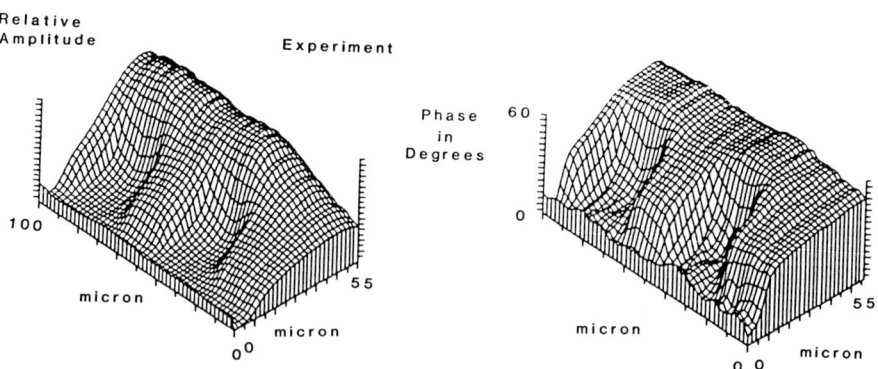

Fig 11   Vibration amplitude and phase plots across track shown in
          Fig. 10; detection frequency is **10 KHz.**

Theoretical calculations for the vibration amplitude and phase agree well
with the experiment. We have studied the effect of electromigration
using this technique. A 160 mA uni-directional current was passed through
a 100 μm wide aluminium track on glass for 70 hours. The vibration
amplitude was measured in the usual way using an AC current with a peak
amplitude of 60 mA. Fig. 12 shows amplitude scans along the track
before and after the uni-directional current was applied for 70 hours,
demonstrating clearly the effect of electromigration. (Wickramasinghe
et al 1983). Finally we show an example of defect detection in an
·integrated circuit. A fault was found between one of the input pins and
ground of a data conversion chip. A scan along the relevant current
carrying track clearly indicated a higher signal near the defect region
as shown in Fig. 13.

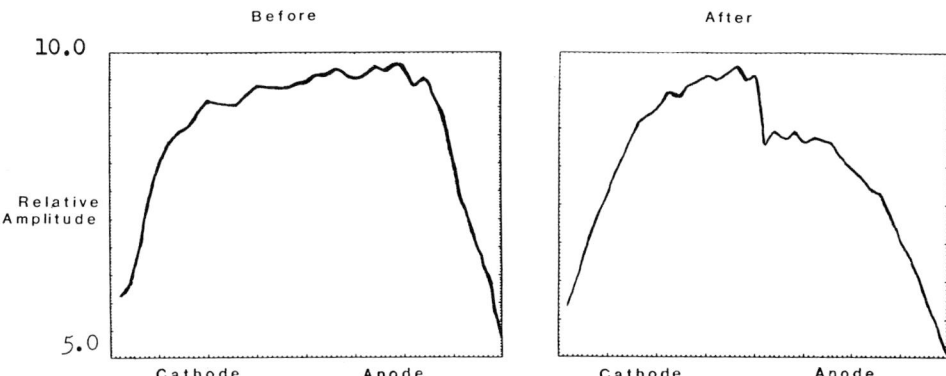

ELECTROMIGRATION

Fig. 12   Observation of electromigration in a 100μm Al track.

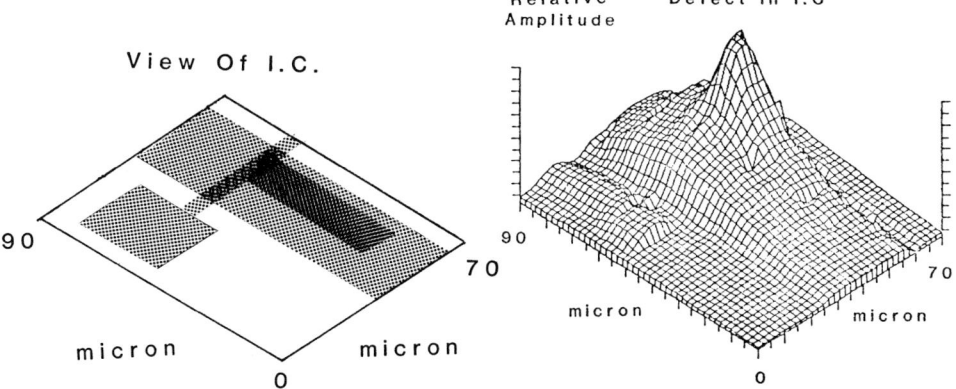

Fig. 13   Observation of a defect in an integrated circuit.

6. Conclusion

Recent advances in acoustic microscopy have been described with particular
reference to semiconductor devices. A new form of thermal wave
microscope has been used to study electromigration and defects in
integrated circuits. It is hoped that the technique will find
application in IC diagnostics.

References

Ameri S., Ash E.A., Neuman V & Petts C.R. (1981), Elect Lett, 17, 337.

Atalar A., Quate C.F. & Wickramasinghe H.K. (1977), App. Phys. Lett, 31, 791.

Atalar A. (1978) J. Appl. Phys., 49, 5130.

Bray R.C., Galhoun J., Koch R. and Quate C.F. (1980), Thin Solid Films, 74, 295.

Favro L.D., Kuo P.K., Pouch J.J., & Thomas R.L. (1980), Appl. Phys. Lett. 36, 953.

Kushibiki J., Ohkubo A. and Chubachi N., (1981), Proc IEEE Ultras. Symp., 552.

Lemons R.A. and Quate C.F. (1974), Appl. Phys. Lett., 24, 163.

Luukkala M. and Penttinen A. (1979), Elec. Lett., 15, 325.

Nikoonahad M., Yue C.Q. and Ash E.A. (1982), Proc DARPA/AFML Review of Progress in Quantitative NDE, (in press).

Nikoonahad M. (1983) in 'Research Techniques for Non-destructive Testing' (Ed. R.S. Sharpe), Vol 7 (in press).

Parmon W. & Bertoni H.L. (1979), Elect. Lett. 15, 684.

Pino F., Sinclair D.A. & Ash E.A. (1981), Proc Ultrasonics Intl., 193.

Quate C.F. (1982) Private Communication.

Quate C.F., Atalar A. & Wickramasinghe H.K. (1979), Proc IEEE 67, 1092.

Sinclair D.A., Smith I.R. & Wickramasinghe H.K. (1982), The Radio & Electronic Engineer, 52 (10), 479.

Smith I.R., Sinclair D.A. & Wickramasinghe H.K. (1982), in 'Acoustical Imaging', Vol. 12, Ed. E.A. Ash & C.R. Hill, 113.

Smith I.R., Sinclair D.A. & Wickramasinghe H.K. (1980), Proc Ultrason Symp. 677.

Smith I.R., Sinclair D.A. & Wickramasinghe H.K. (1981), Proc Ultrason Symp. 591.

Wickramasinghe H.K. (1983), J of Microscopy, Vol 129, Pt.1, 63.

Wickramasinghe H.K. (1978), Elect. Lett., 14, 305.

Wickramasinghe H.K. (1981), Appl Phys Lett., 39, 305.

Wickramasinghe H.K., Bray R.C., Jipson V., Quate C.F. & Salcedo J.R. (1978), Appl. Phys. Letts., 33, 923.

Wickramasinghe H.K., Martin Y., Ball S. & Ash E.A. (1982), Elect. Lett., 18 (16), 700.

Wickramasinghe H.K., Martin Y., Spear D.A.H. & Ash E.A. (1983), Invited Paper, 3rd International Topical meeting on Photoacoustic and Photothermal Spectroscopy. (in press).

Acknowledgements

This work was supported by the Wolfson Unit for Micro NDE at University College. The thin film circuits were fabricated by Mr F. Stride and the defective integrated circuit was supplied by British Telecom. Advice and encouragement from Professor E.A. Ash is gratefully acknowledged, on many aspects of the work described here.

Inst. Phys. Conf. Ser. No. 67: Section 8
Paper presented at Microsc. Semicond. Mater. Conf., Oxford, 21–23 March 1983

Techniques for scanning electron acoustic microscopy

L J Balk and N Kultscher

Universität Duisburg Fachgebiet Werkstoffe der Elektrotechnik
Leiter: Prof. Dr-Ing E Kubalek, Kommandantenstraße 60, 4100 Duisburg 1 FRG

Abstract Techniques for scanning electron acoustic microscopy are introduced, being applicable within a frequency range from less than 100kHz up
to the GHz regime. At lower frequencies lock-in amplifiers are applied for
signal processing, whereas at high frequencies boxcar averagers are used.

1. Introduction

Scanning acoustic microscopy has become most important for determining material properties associated with generation, attenuation, and propagation
of ultrasonic waves within the examined specimen region such as heat conductivity and capacity, thermal expansion coefficient, elastic constants,
piezoelectricity, and space charges. Inhomogeneities influencing these quantities can be imaged by acoustic microscopy. In the past, techniques have
been developed, mainly differing in the primary generation of the sound
wave. Most important techniques are: pure scanning acoustic microscopy
(Quate 1980) implying a focused sound wave. As an advantage the primary
sound "beam" can be focused into any depth of the specimen, enabling micrographs of arbitrary penetration depth. A second technique is scanning photoacoustic microscopy (Wong 1980) in which a laser is used for sound generation by thermal heating. The advantage of this method is that the laser can
be focused more precisely. Further not only the sound signal is used for
production of micrographs, but also photoluminescence and laser beam induced
currents. Shortcomings are that it can only be applied for materials not being transparent to the laser wavelength and that a modulation of laser
beams over a large frequency range is difficult. Here scanning electron
acoustic microscopy (SEAM) has a maximum of advantages (Cargill 1980). The
primary electron beam can be focused to diameters much smaller than with
other techniques, its intensity can be varied over many orders of magnitude allowing optimization of the experiment; it can be chopped from very
low frequencies up to the Gigahertz range without changing the experimental
conditions and with arbitrary duty cycle. Finally with SEAM many micrographs can be produced simultaneously, such as X-ray-microanalytic patterns,
Auger electron micrographs, secondary electron images, electron beam induced current, and cathodoluminescence. A shortcoming of this method is
that the beam cannot be focused as deeply as with acoustic microscopy, but
by altering the primary energy the mean sound generation depth can be varied
from the surface down to ~ 5-$10\,\mu m$. Spatial resolution is typically about
several micrometers, but it can be decreased to $\sim 0.2\,\mu m$. This depends on:
the primary energy dissipation volume, as a function of primary energy and
stopping power of the material, and the heat dissipation volume, which can
be considered the region of thermal waves creating the sound and which decreases with frequency (Rosencwaig 1980).

For adaptation of SEAM to scanning electron microscopes one has to apply an electron beam chopping system and to install a specimen stage allowing detection of sound waves emitted by the specimen. Electron beam chopping can be done by use of plate condensors (Menzel and Kubalek 1979). Use of square wave chopping is to be prefered instead of a sine wave, as it is easier to be established and no significant chopping degradation occurs. The transducer assembly must be constructed to avoid stray influences from other signals. Fig. 1 shows a completely electrically shielded arrangement allowing change of transducer slices. In this work only transducers have been used for detection of longitudinal modes. Signal levels for properly adjusted chopping conditions and for normal electron beam settings (100-200 μm final aperture, beam diameter < 0.1 μm) are in the mV regime. In SEAM techniques reported until now the electron beam has been chopped by a square wave, the ultrasonic signal has been amplified by a RF-(Cargill 1980) or a lock-in-(Rosencwaig 1980)-amplifier at the same frequency. In this manner only linear interactions between sound wave and specimen can be investigated. In this work techniques could be used giving information on nonlinear interactions, too. These methods imply lock-in amplification for different frequency ranges, not only tuned to the chopping frequency, but also to its harmonics. Boxcar averaging techniques allow both analysis of the acoustic waveform and production of micrographs with subnanosecond time resolution.

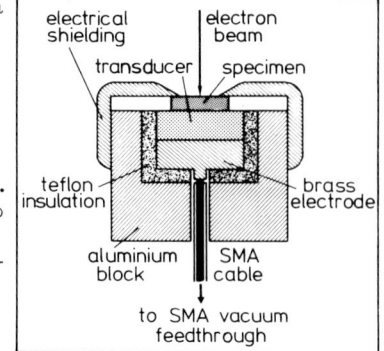

Fig.1.Transducer assembly

For the example of an InP substrate with Zn-doped regions of about 4 μm diffusion depth the efficiency of the different techniques is demonstrated.The surface has been coated with a 30nm gold layer to avoid charging. As doping concentrations are related to changes in heat conductivity (Rosencwaig and White 1981), the diffused regions can be imaged by SEAM. As InP is piezoelectric, a nonlinear coupling occurs caused by acoustoelectric interaction of sound wave, electric fields and space charges involved with sound propagation, leading to generation of harmonics; amplitudes and phases of which being determined by the nonlinear interaction.

2. Low Frequency Electron Acoustic Imaging

For chopping frequencies up to 100kHz two lock-in amplifiers (Ithaco Dynatrac 3) have been used as sketched in fig.2. The video signal can be either amplitude (A) or phase ($\emptyset$) at the ground wave or the second harmonic, additionally the phase difference can be monitored. The electron acoustic (EA) images are compared to secondary and reflected electron (SE+RE) micrographs under identical operation conditions. SE+RE contrast has been about 1% necessitating a high signal offset,whereas the EA signal has shown strong contrast. Comparing fig.3b and 3c shows that the second harmonic amplitude reveals more details than the ground wave, indicating inhomogeneities in the depth of diffusion, as this effect could not be observed at low primary

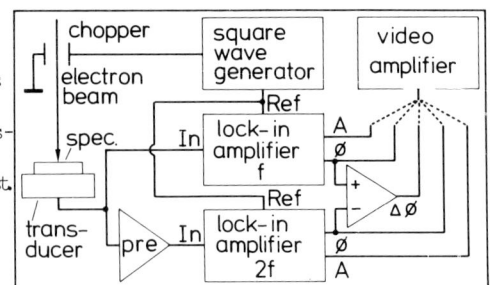

Fig. 2. Set-up for different modes of low frequency acoustic images

energies. The phase images only showed that the diffused regions are homogeneously oscillating with a constant phase shift compared to the substrate.

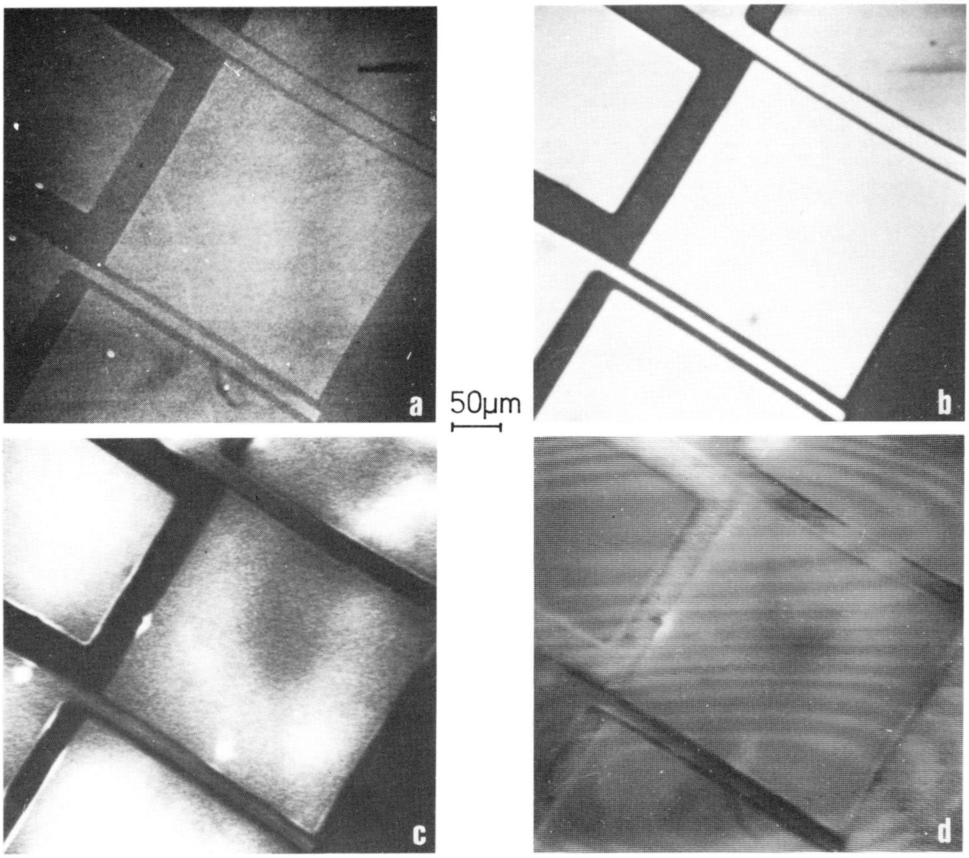

a) SE+RE   b) EA: A(f)   c) EA: A(2f)   d) EA:phase difference $\Delta\emptyset=\emptyset(f)-\emptyset(2f)$
Fig. 3. Electron acoustic images of Zn-diffused regions in InP for primary
electron energy of 30keV and a chopping frequency of 100kHz (square wave)

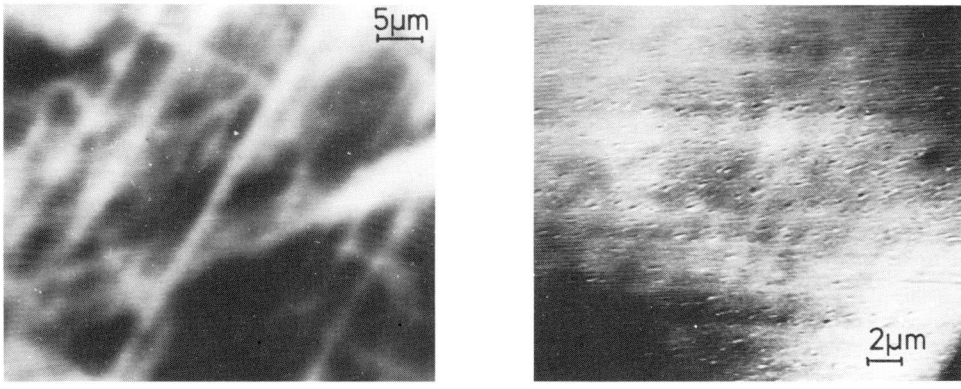

a) 30keV   b) 5keV
Fig. 4. EA images -A(2f)- of crystal imperfections within the diffused
regions at different primary electron energies

The phase difference (fig. 3d) reveals more inhomogeneities, especially in the narrow doping regions.Fig. 4a shows crystal imperfections imaged in the A(2f)-mode, which are assumed to be substrate dislocations, as they do not give comparable contrast in the corresponding 5keV micrograph (fig. 4b).Instead of this, pits of ∿0.3 μm size show up, possibly due to dislocations crossing the surface.

3. High Frequency Electron Acoustic Imaging Using a Lock-In Amplifier

For high frequency measurements a suitable lock-in amplifier has to be used (PAR 5202).To get micrographs of different harmonics, the lock-in amplifier has to be triggered with a corresponding multiple of the ground wave. This is done by a word generator (HP 8016A) which controls square wave generator and lock-in amplifier (fig. 5). By delay units of the word generator phase shifts between the harmonics in the reference line can be avoided. The lock-in amplifier output gives either AsinØ or AcosØ.

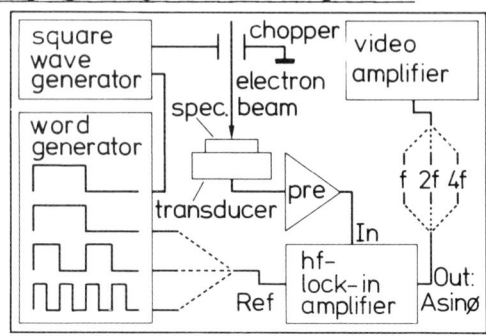

Fig. 5.Set-up for high frequency EA images with fixed phase settings

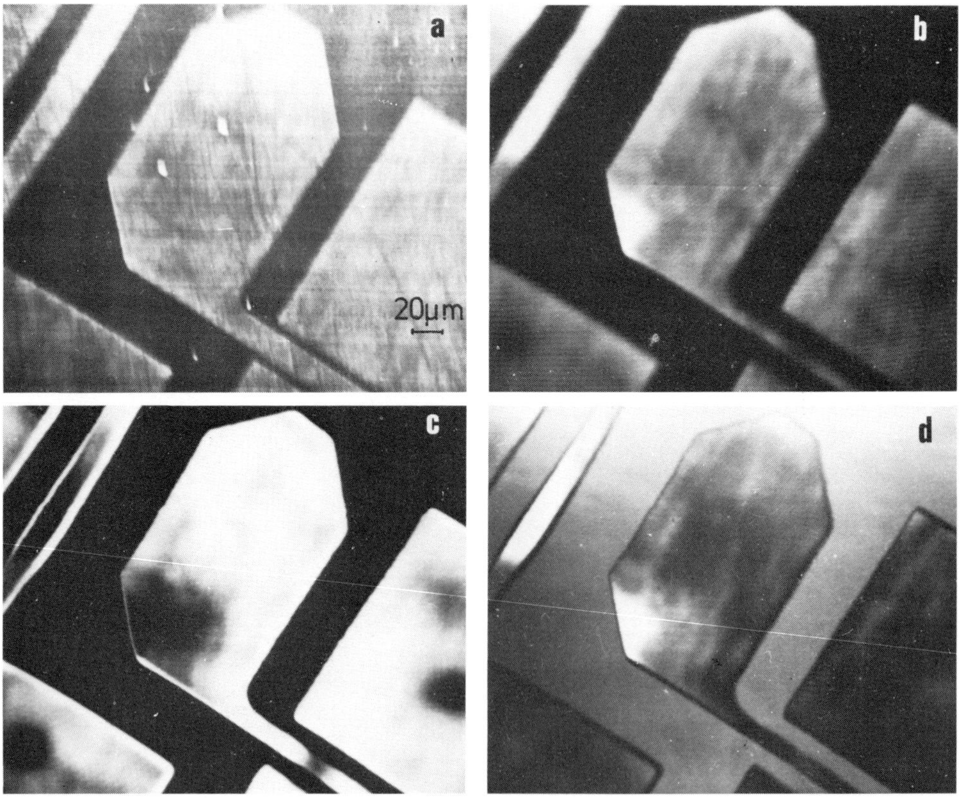

a) SE+RE b) EA: AsinØ at f c) EA: AsinØ at 2f d) EA: AsinØ at 4f
Fig. 6. High frequency EA images at different reference modes for a chopping frequency of f=5,6MHz and a primary electron energy of 5keV

Fig. 6 shows EA micrographs of ground wave, second and fourth harmonic at 5,6MHz chopping frequency, the typical signal relation is in the order of A(f):A(2f):A(4f)=1000:30:1. Comparing the EA micrographs one can notice that the spatial resolution increases with the number of the harmonic and more details are visible in the higher frequency modes. Comparison to the SE+RE image of fig. 6a shows that the borders of the diffused regions do not coincide with the structures of the EA images, but differ by $\sim 4\,\mu m$.

This effect, already to be noted when comparing fig. 3a and 3b, is due to the dependence of the amplitude of the acoustic wave on the effective charge concentrations. Thus the EA amplitude decreases in a similar way,as the electron beam induced current increases, when the beam is getting close to a pn-junction.

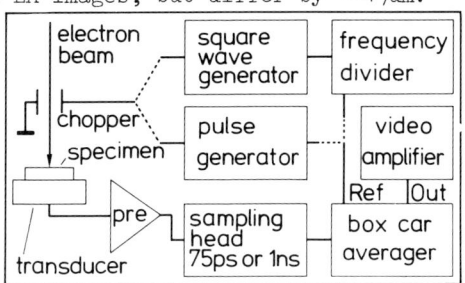

Fig. 7. Set-up for time resolved EA images

4. High Frequency Electron Acoustic Imaging Using a Boxcar Averager

As lock-in amplification is only applicable up to 50MHz, for higher frequencies or high time resolutions a boxcar averager based method has been developed (fig. 7). Triggering of the boxcar averager (PAR 162) has to be done via a frequency divider in the case of square wave chopping, as the boxcar averager only accepts trigger rates up to some MHz. For pulse mode chopping this is not necessary. For high temporal resolution a 75ps-sampling head is used,whereas in the case of lower signals a 1ns-head has been applied. Preamplification has been realized by cascading low noise preamplifiers with an overall gain of 86dB.

Fig. 8 shows an EA micrograph for square wave chopping at 8MHz.The image differs very strongly from EA images of the previous sections, as a nearly "topographic" structure seems to be revealed. This is due to the fact that for 8-10MHz the space charge within the pn-junction, i.e. the borders of the diffused regions, couples resonantly with the electric field associated with the sound wave.As this resonance starts as soon as free carriers are injected into the space charge region,the width of the region of resonance corresponds with the width of measurable electron beam induced current,in this example $\sim 8\,\mu m$.This effect is proven by images with lock-in amplification at 10MHz showing the same contrast.Furthermore a pulsed mode experiment using boxcar

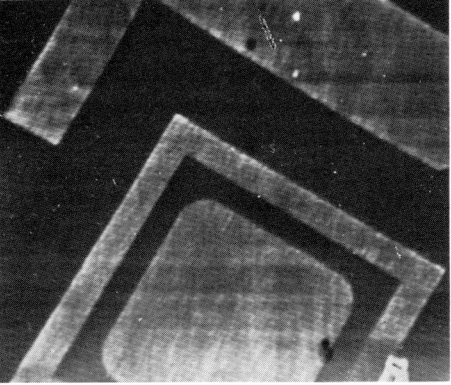

a) SE+RE   ⊢—20µm

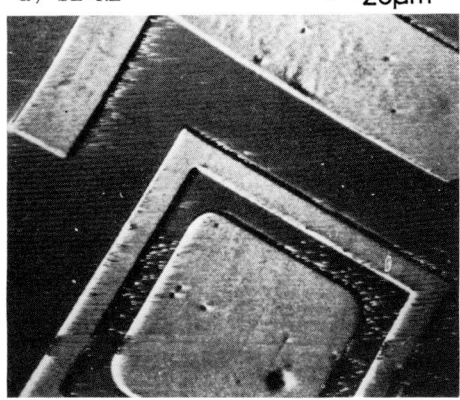

b) EA: 75ps-head, 8MHz square wave, time delay at signal maximum

Fig. 8. Time resolved EA image with periodic excitation at a primary electron energy of 5keV

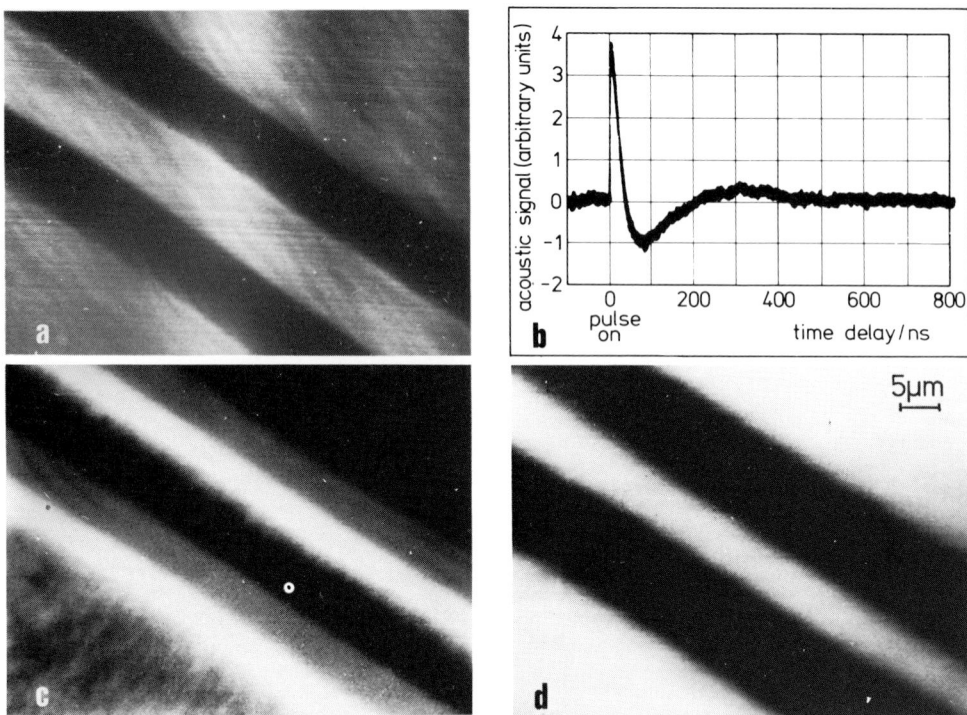

a) SE+RE   b) specimen response to a single primary beam pulse
c) EA with delay time smaller than 100ns   d) EA with delay time = 200ns
Fig. 9. EA images at different delay times in pulsed mode boxcar averager
measurements: pulse width=10ns, $t_{sampling}$=1ns, primary electron energy=5keV

averaging has shown for time delays below 100ns an identical contrast (fig.
9c), whereas for larger delay times, corresponding to smaller frequencies
than 10MHz, this effect cannot be observed (fig. 9d). Fig. 9b is an example
of the complete acoustic signal as a response to a single beam pulse. It
should be mentioned that the width of this acoustoelectric resonance correlates directly to the mismatch of the EA micrographs in fig. 6.

5. Acknowledgements

The authors like to thank Prof.Dr.E. Kubalek for helpful discussions, Prof.
Dr. K. Heime and R. Schmitt from the Fachgebiet Halbleitertechnik der Universität Duisburg assisted this work with InP samples, the transducer elements were supported by Mr. van der Linden from Siemens, Redwitz.

References

Cargill G S 1980 in Scanned Image Microscopy, ed. by E A Ash,
    Academic Press pp319-30
Menzel E and Kubalek E 1979 Scanning Electron Microscopy $\underline{I}$,
    by SEM Inc, AMF O'Hare MI   pp305-17
Quate C F 1980 in Scanned Image Microscopy pp23-55
Rosencwaig A 1980 in Scanned Image Microscopy pp291-317
Rosencwaig A and White R M 1981 Appl Phys Letts $\underline{38}$ pp165-7
Wong Y H in Scanned Image Microscopy pp247-271

Inst. Phys. Conf. Ser. No. 67: Section 8
Paper presented at Microsc. Semicond. Mater. Conf., Oxford, 21–23 March 1983

Aspects of SAM imaging of semiconductor devices

A J Miller

GEC Research Laboratories, Hirst Research Centre, Wembley, UK

Abstract   This   paper describes   progress in   the   assessment of
semiconductor devices and materials using a 900 MHz scanning acoustic
microscope (SAM). Images are presented for various types of device, and
visibility and distinguishability of materials are discussed in relation
to   V(z)   characteristics,   which   embody   information   about   elastic
properties.   Interference   fringe   patterns   parallel   to   surface
discontinuities   are discussed with reference to   characterisation, and
compromises   between   operating   frequency   and   lens   aperture   are
described.

1   Introduction

The   scanning   acoustic   microscope   has   been   under   continuous development
since the first demonstration by Lemons and Quate (1974), and commercial
versions are starting to appear.   Because of the unfamiliarity of the
images, and because of properties peculiar to acoustic microscopy such as
the limited depth of field, contrast due to acoustic properties and edge
reflections, work needs to be done so that object features of interest can
be recognised.   Some of this is described here, and it becomes obvious that
SAM images, while impressively detailed, include some false detail that
could cause confusion.

The present lateral resolution of the SAM is better than 1.5 µm, which is
as expected from an operating frequency of 900 MHz.   This resolution is
acceptable for surface and near-surface imaging, as is most   often required
for modern integrated circuits.   The microwave electronics is suitable for
frequencies of up to 2 GHz, and higher frequencies than are currently being
used will be implemented as required.   The details generally follow common
recent   practice, using a PIN modulator   or fast SPDT switch to   select the
echoes coming from the specimen, and a fast sample-and-hold circuit to form
the video signal which is stored in an analogue scan converter memory, and
subsequently displayed and recorded.   The sapphire lens used has a radius
of 120 µm and is heated to reduce attenuation in the coupling drop of
water, thereby improving the S/N ratio.   The time taken to acquire and
display an image approximately 250 µm on a side is twenty seconds.   The
focussing or Z direction is adjusted using a differential micrometer and
piezoelectric pusher.   There are facilities for obtaining V(z) data from a
chosen site on the specimen and images can also be displayed as Y-modulated
traces.

2   Image formation

The process of image formation in the acoustic microscope involves two
particular ray paths. One is axial and undergoes reflections at each
interface, dictated mainly by acoustic impedance mismatching. The other is
peripheral, and strikes the specimen at the appropriate angle for
conversion from longitudinal waves in water to leaky Rayleigh waves across
the specimen surface, which are focused (Smith et al 1983), and from which
energy leaks back to the lens. The result is a distortion in the
wavefronts returning to the transducer, which adds these varying
distortions across its face and produces a greater or lesser output.
Contrast in the image at a point is thus due to an interference process,
and significantly the peripheral rays have sampled the acoustic properties
of the specimen in a small area around the focus of the lens. At one
specimen point, defocusing the lens (bringing the lens and the specimen
closer together) produces the characteristic reversals of contrast, which
when plotted are referred to as a V(z) curve. The V(z) curve embodies two
pieces of information about the acoustic properties at a point, namely the
surface wave velocity and attenuation, both of these being averaged over
the XY plane. The garnering of accurate information from this approach in
the form of quantitative differences in acoustic properties in parts of an
image worthy of detailed examination is possible, but involves not only a
large amount of computing but also reliance on assumptions about the
perfection of the lens and the exactness of the theory.

3   Image contrast

Some earlier images from the present SAM have previously been published
(Miller 1982), since when image quality has been improved. A finished
integrated circuit is a formidably detailed object for SAM imaging (Figure
1) because of the layers of different materials of differing thickness,

Fig 1   General view of silicon-on-sapphire finished device

arranged in patterns with prominent edge detail. A layer of passivation on
the top surface will cause further difficulties but does at least protect
the circuit from the water coupling drop. Just as for electrical testing,
test structure patterns are useful for the analysis of possible image
features and the recognition and distinction of such features on real

specimens. Figures 2 and 3 show patterns on a SOS test insert, showing slight differences in contrast by virtue of having been imaged at alternative focal positions.

Fig 2   SOS test insert pattern

Fig 3   As Figure 2 but defocused slightly

The contrast between two points on a specimen arises from two causes: material and lens-object distance, and in the case of semiconductor devices, it is obvious that both are likely to occur frequently and that there is a great potential ambiguity. This is the central problem of practical acoustic microscopy. Two adjacent image points showing different brightness could be different points on the same, material-dependent $V(z)$ curve, indicating topographic changes, or at the same focusing distance on two different $V(z)$ curves, indicating material changes, or indeed a mixture of both. Figure 4 and 5 illustrate a change in contrast with defocusing due to a thickness change in a buried layer, implying a surface topography change. This explanation is not obvious from Figure 4 alone, which shows a generally consistent aluminium layer, or from Figure 5 alone, which might be construed as showing a difference in materials where none in fact exists.

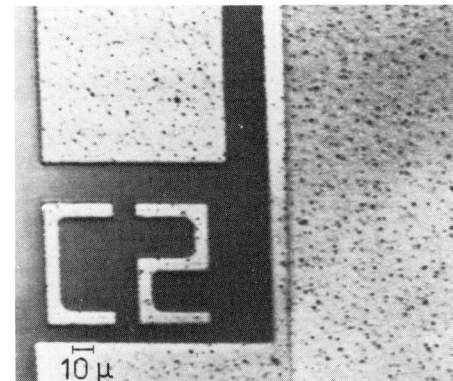

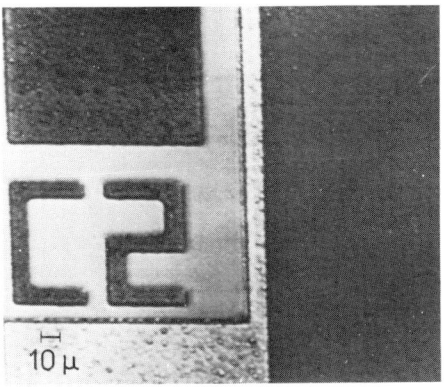

Fig 4   Test capacitor on bulk silicon with aluminium top surfaces evident

Fig 5   Change in contrast due to step in thickness of buried vapox layer

There is thus a problem of experience and experiment in the separation of this interlinked information for practical purposes. It is unproductive to try and calculate the relative effects on the acoustic signal of material and sample focus distance changes since the gradient of V(z) in general undergoes considerable and repeated changes. A recently published scheme to separate topographic information from surface wave velocity perturbation (i.e. material) information holds out some promise (Liang et al 1982) but is at present limited in spatial resolution. There is also the pitfall (Wilson and Weglein 1980) of particular relevance to integrated circuits, that a layer λ/4 thick for one SAM operating frequency may obscure V(z) detail entirely.

4 Further studies

What the high-resolution SAM should be particularly useful for is the visualisation of fine delaminations or cracks in devices. In the case of horizontal features very near the surface, acoustic impedance mismatching will ensure some reflection and contrast change in the image. This is relevant to the adhesion, or lack of adhesion, of thin films (Bray et al 1980). For vertical or angled surface-breaking cracks the leaky Rayleigh waves involved in image formation reflect from the side of the crack and cause additional interference, the net result in the image being fringes parallel to the crack, the fringe spacing being λ/2 for the surface waves on that material adjacent to the crack. This has been demonstrated for several non-semiconductor materials (Briggs et al 1982, Yamanaka and Enomoto 1982). Figures 6 and 7 show respectively optical and acoustic pictures of the same area at the sawn edge of an IC die. The acoustic picture clearly shows two features, which are optically invisible. While there are no fringes as such, the gradual change in contrast on one side of the probable position of the crack indicates an asymmetric feature, most likely a crack at a shallow angle, both by analogy with a subsurface horizontal interface, and because similar features have been observed by cathodoluminescence (after more specimen preparation).

Fig 6 Optical micrograph of
       sawn edge of silicon-
       -on-sapphire IC die

Fig 7 Acoustic micrograph of
       same area, revealing
       crack features

It is also possible for reflection at an edge or step to cause fringes.

Figure 8 shows fringes around and within features on a test IC. On SOS, they are more prominent around aluminium and silicon areas and within polysilicon areas, whereas on bulk silicon, they are more prominent within aluminium tracks. The reasons for these features are not simply related to acoustic impedance differences, as one might then expect fringes around all the surface markings since they all have a Rayleigh wave velocity less than that of sapphire. Other modes may be involved, but computer modelling is essentially intractable for all but the simplest cases when studying interactions of acoustic waves with objects whose dimensions are of the order of a wavelength, and therefore observation of these fringes on semiconductor devices will not be a simple aid to image interpretation.

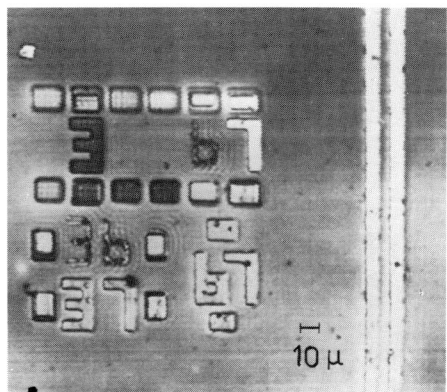

Fig 8   Acoustic fringes observed
         on SOS test insert

Fig 9   Optical micrograph of IMPATT
         device

Some subsurface imaging is possible with the high resolution instrument (Attal et al 1981), and in the present work, concerns IMPATT structures. Figures 9, 10 and 11 show respectively optical, acoustic surface-focussed, and acoustic subsurface-focused images. The central part has a gold contact layer. Figure 11, although defocused by 20 μm, which means that the background is in focus and therefore bright, illustrates defect detail only about 5 μm beneath the gallium arsenide surface because of refraction. Detail beneath the gold layer is difficult to see because surface features always dominate the image when a high-resolution, large-aperture SAM is defocused through a rough surface.

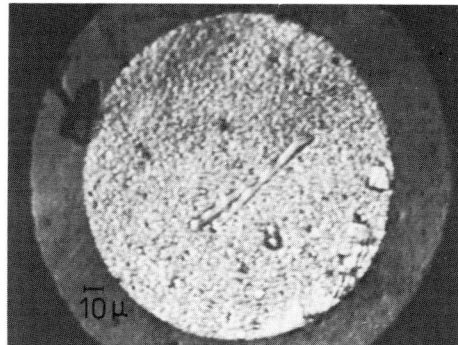

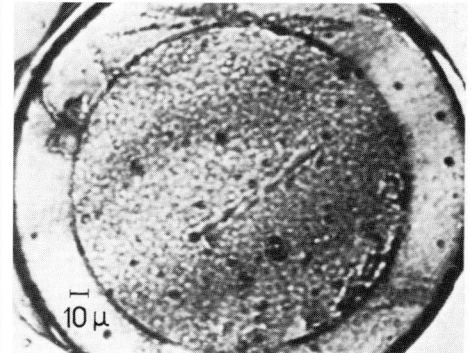

Fig 10 Acoustic micrograph of
       IMPATT surface-focused

Fig 11   As Figure 10 but defocused
         by 20 μ (~5 μm under GaAs
         surface)

The solution to this problem is to have the top surface absolutely flat, or go to a lower resolution instrument.   In this case, a narrow-aperture lens ensures that the leaky Rayleigh waves are not excited, and only detail from gross surface and subsurface features is seen, which may be separated by time-gating.   Such an instrument, operating at 40 MHz, has seen interface detail between a silicon power transistor and its heat sink (Nikoonahad et al 1982).   Further improvement may be possible using a pulse-compression system (Guangqi et al 1982), so that the quality of the contact between an IC die and its package might be examined non-destructively.

5   Conclusions

The development of the SAM technique for imaging would seem to be directed along two paths - the high resolution instrument for very detailed surface and near surface work, probably with sub-optical resolution, and the lower frequency narrow-aperture lens instrument for examining interface detail below a flat surface.   Other innovations such as on-line calculation of local elastic properties from V(z) curves and directional acoustic microscopy (Hildebrand and Lam 1983) will undoubtedly come, leading eventually to an 'elastic microprobe'.   In the short term, image analysis by computer and contrast enhancement would be a help in the case of, for example, laser-annealed polysilicon where image contrast is small, and phase, rather than amplitude imaging (Wickramasinghe and Hall 1976) would also be expected to be of assistance.

Acknowledgement

The author wishes to acknowledge the assistance of Dr H K Wickramasinghe, University College, London, in the design and construction of the scanning acoustic microscope.

References

Attal J, Truong Quang N, Cambon G and Saurel J M 1981 Microsc Semicond
    Mater Conf 1981 (Inst Phys Conf Ser No 60) pp441-446
Bray R C, Quate C F, Calhoun J and Koch R 1980 Thin Solid Films 74,
    295-302
Briggs G A D, Ilett C and Somekh M G 1982 Acoustical Imaging Vol 12
    ed E A Ash and C R Hill (New York: Plenum) pp89-99
Guangqi Y, Nikoonahad M and Ash E A 1982 Electronics Lett 18, 767-769
Hildebrand J A and Lam L K 1983 Appl Phys Lett 42, 413-415
Lemons R A and Quate C F 1974 App Phys Lett 25 251-253
Liang K, Bennett S D, Khuri-Yakub B T and Kino G S 1982 Appl Phys Lett
    41 1124-1126
Miller A J, Acoustical Imaging Vol 12 ed E A Ash and C R Hill (New York:
    Plenum) pp67-78
Nikoonahad M, Guangqi Y and Ash E A Proc DARPA/AF Review of Progress in
    Quantitative NDE, La Jolla, California August 1982 (to be published)
Smith I R, Wickramasinghe H K, Farnell G W and Jen C K 1983, Appl Phys
    Lett 42 411-413
Wickramasinghe H K and Hall M 1976 Electronics Lett 12 637-638
Wilson R G and Weglein R D 1980 Ultrasonic Materials Characterisation
    ed H Berger and M Linzer (NBS Publication 596: Gaithersburg) pp345-355
Yamanaka K, Enomoto Y 1982 J Appl Phys 53 846-850

Inst. Phys. Conf. Ser. No. 67: Section 8
Paper presented at Microsc. Semicond. Mater. Conf., Oxford, 21–23 March 1983

399

Optical stimulation of semiconducting devices in the SEM

E F Maher and P J Howard

ERA Technology Limited, Cleeve Road, Leatherhead, Surrey KT22 7SA

Abstract   ERA has recently developed* a new scanning optical microscope which is based on the SEM, and shares the same instrumentation. The 'SOMSEM' consists of a special specimen stage which incorporates a phosphor and optical system for the conversion of the scanning electron beam into a scanning optical beam.

The basic principles behind this new instrument will be discussed together with its application to the assessment of electronic materials and devices.

*MAHER, E F: British Patent Application No. 8224510
        Filed 26th August 1982

1.   Introduction

The use of scanning electron microscopes is now very widespread because of their versatility compared with other imaging systems. Apart from the vast range of magnification available, this versatility is a direct result of a) the depth of focus, b) the number of imaging modes within the same basic instrument and c) the control over scanning rates and video presentation. Although scanning optical microscopes should in principle have similar advantages, they have never achieved the same popularity. This is because relatively cumbersome means of scanning the optical beam have been developed e.g. rotating mirrors and prisms (Potter and Sawyer, 1967), or more recently specimen vibration (Sheppard, 1980). On the other hand, a beam of electrons, being charged, is easily deflected using magnetic or electrostatic lenses.

Whilst the ultimate resolution of any optical microscope can never approach that of an electron microscope, because the wavelength of kV electrons is very much smaller than that of light photons, the scanning optical microscope (SOM) does have important advantages in other respects. For example, the specimen in the SEM must be maintained in a vacuum, and must be resistant to electron beam damage. This restricts the range of samples considerably and biological specimens, in particular, are not well suited to examination without preparation. Furthermore, the penetration depth of kV electrons is very restricted, and, in general, scanning electron microscopy remains essentially a surface technique. In contrast, the scanning optical microscope does not require a vacuum and will not normally degrade the specimen. Consequently living organisms may be examined without the depth of focus restriction associated with conventional optical microscopes. Even more important is that with appropriate choice of wavelength of light, the investigation of structures

below the specimen surface becomes possible, provided those structures
have different optical characteristics compared with the remainder of the
specimen. For example, particles suspended in a liquid could be
examined.

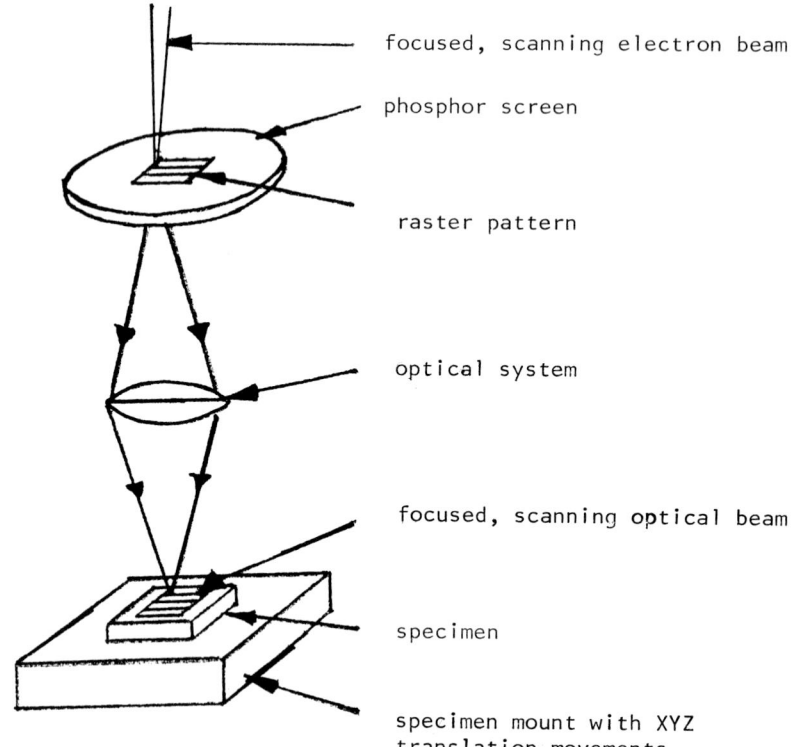

focused, scanning electron beam

phosphor screen

raster pattern

optical system

focused, scanning optical beam

specimen

specimen mount with XYZ
translation movements

Fig. 1 Schematic for the SOMSEM system.

The realisation that the instrumentation required for processing the
signal and ultimately displaying it on a video monitor is the same for
both SOMS and SEMS gave rise to the present invention whereby the
scanning optical microscope is proposed as an accessory compatible with
the SEM. In order to change the scanning electron beam into a scanning
optical beam, a phosphor screen is used together with a lens system
which projects the raster of light onto the specimen. This results in a
hybrid instrument - the 'SOMSEM'.

2   System Design

The basic principles of the SEM-based SOM are illustrated with
reference to Fig. 1. The scanning electron beam is focused on a phosphor
coated screen situated above the specimen within the evacuated column of
the SEM. The keV energy electrons are thus converted into photons
whose wavelength depends on the phosphor selected (typically 2000Å to
8000Å). A lens is placed between the screen and the specimen such that
the raster of emitted light is focused onto the specimen, which may be
contained in an environmental cell. Time-dependent processes in the
specimen can be investigated by varying the scanning rate or pulsing the

electron beam using a beam blanking unit; a range of phosphors with different wavelengths and persistence values is available. The information from the specimen, whether this be reflected light, transmitted light, luminescence or beam induced current, is displayed with the instrumentation of the SEM in the normal way, using a detection system appropriate to the signal.

The magnification obtained using the simple system of Fig. 1 is exactly analogous to the SEM case i.e. the ratio of the display screen size to the size of the raster scanned by the electron beam. The lens gives no additional magnification in this symmetrical case. However Fig. 2 shows asymmetrical arrangements where the magnification of the optical system is not unity. In Fig. 2a, the projected raster on the specimen is larger than the raster scanned by the electron beam, resulting in a reduction in overall system magnification compared with the symmetrical case. In Fig. 2b, the converse is true. The configuration of Fig. 2b is particularly desirable when the effects of phosphor grain size would otherwise marr the resolution of the system. Here a small raster of light is generated on the specimen compared with the larger raster scanned by the electron beam on the phosphor screen. The overall magnification will, of course, be higher than that indicated on the SEM micrograph. A further advantage of this configuration is that the reduced electron current density on the phosphor screen is less likely to cause degradation of the phosphor through burning.

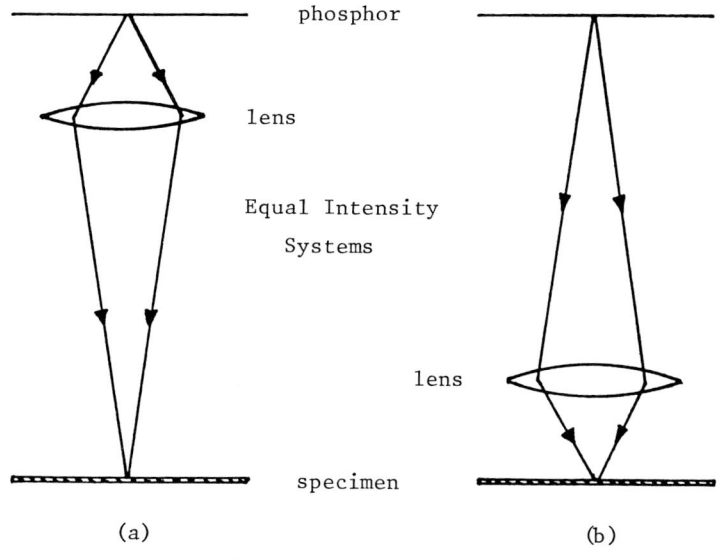

phosphor

lens

Equal Intensity
Systems

lens

specimen

(a)   (b)

greater depth of field   smaller depth of field
lower resolution   higher resolution
lower magnification   higher magnification

Fig. 2 Asymmetrical Configurations

In any of the above configurations, an iris may be introduced which limits the aperture of the optical system. Thus the trade-offs between light intensity, depth of field, resolution and lens aberration are under operator control. For the SOMSEM, the illumination from the phosphor screen is optimized by first focusing the electron beam on the phosphor using the normal secondary electron image. The accelerating voltage for a particular phosphor is then chosen to give the best OBIC image of a silicon PIN diode consistent with the phosphor burn characteristics. For example, very thin phosphor films give more light output when comparatively low accelerating voltages are used. It can be an advantage to defocus the electron beam slightly to obtain more uniform stimulation of the phosphor screen.

A range of phosphor screens mounted on a rotating carousel can be incorporated into the design, together with a selection of lenses. The plane to the phosphor screen is regarded as fixed, with positioning and focusing of the specimen achieved by XYZ translation movements. Movement of the lens in the Z-direction allows the necessary extra degree of freedom for the asymmetrical configurations and different lenses. In practice, a compound lens system is preferable to reduce working distances and allow the incorporation of an iris at the subsidiary focus. Straight-forward SEM operation, for comparison with optical images, is resumed by swinging the lens and phosphor out of the electron-optic axis.

3   Experimental

In order to demonstrate the feasibility of the SOMSEM design, a simple system has been constructed at ERA and used to image, via the OBIC mode, some electronic devices. The basis for the stage was the 100 Series specimen stage of a Steroscan 11A scanning electron microscope, and a stainless steel adapter plate and rails assembly allows the stage to be inserted into a Cambridge Stereoscan 250 Mk 2 microscope.

Fig. 3   Prototype
SOMSEM system.

Although modern SEMs tend to have very much more space in the specimen chamber than their previous counterparts, the use of an earlier design of stage meant that the prototype optical system had to be extremely compact (see Fig. 3). The total working distance from the phosphor screen to the base of the specimen stage was less than 10 cm and considerable modifications were necessary to realise a workable system in this space. The specimen tilt and rotation facilities were discarded, but the XYZ specimen movements were retained using an adjustable mounting bracket, suspended from the original specimen base plate, to position the specimen nearer the floor of the stage. Thus the specimen position could be externally controlled by the operator with the SOMSEM stage in the microscope.

The phosphor screen (P11) and lens (a X10 microscope objective) were mounted on a bridge-piece in a telescopic aluminium construction so that the phosphor-to-lens distance was adjustable. Phosphor screens and lenses were readily interchangeable.

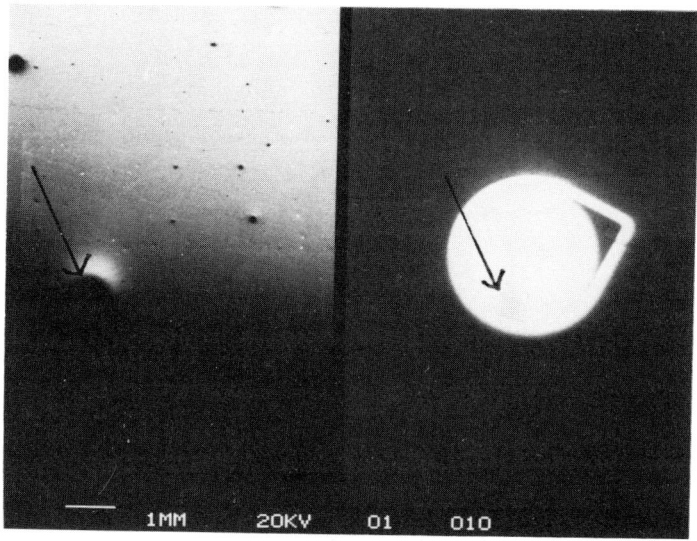

Fig. 4 Comparison between secondary electron image of phosphor and OBIC image of diode.

To date, all micrographs have been obtained using OBIC mode- that is by monitoring the current produced due to the optical stimulation as a function of position on the specimen. Fig. 4 shows the one-to-one correspondence relationship that exists between the phosphor and the specimen. Defects in the phosphor appear in both the secondary electron image of the phosphor and the OBIC image of the silicon photodiode (arrowed on micrographs). In the OBIC case there is an apparent discontinuity in the outer region of the diode, near the apex, which is in fact due to the shadowing effect of a bonding wire.

A comparison between a conventional reflection optical micrograph and an OBIC picture of a silicon MOS device is shown in Fig. 5. The device would be damaged if examined with an electron beam.

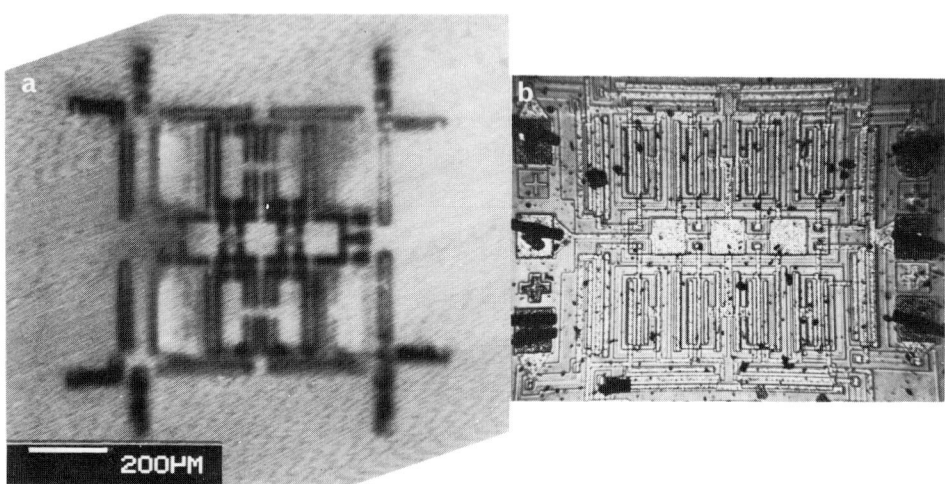

Fig. 5   Comparison between (b) conventional (reflection) optical micrograph and OBIC image (a).

4   Discussion

A flying spot scanner based on a miniature cathode ray tube has previously been developed at ERA by Singmin (1978) following an earlier design (Roberts and Young, 1952).   In this instrument, the raster of light generated on a high intensity CRT screen was projected onto electronic devices to image their photoelectronic properties by modulating the intensity of a video monitor scanned in synchronisation with the CRT. Again this results in a virtually inertialess scanning system, with an enormous range of scanning rates and recording times, in contrast to systems based on mechanical scanning.   However, with the SOMSEM, phosphors can easily be interchanged giving different wavelengths and persistence values, apart from the obvious convenience and cost-effective advantages of a hybrid instrument.

The principal problem with a scanning optical microscope based on phosphor excitation is the lack of light output from the screen compared with laser-based systems.   However, it should be noted that phosphors are available with conversion efficiences of 25%, and with careful optical design of the SOMSEM stage the specimen should receive a significant proportion of the energy density available.   In addition, comparatively large electron beam spot sizes are suitable for this purpose since the resolution of the optical system will in any case be limited to about one micron, which is also the typical resolution for the SEM EBIC mode.

Thus the optical stimulation of the specimen should be approximately as energetic as the electron beam stimulation would be if normal SEM operation were resumed. For an application requiring exceptionally high light intensity, it may be possible to substitute an electroluminescent panel for the phosphor screen so that field assisted cathodoluminescence could be used.

5 SOMSEM and its Application to Semiconducting Materials and Devices

With regard to the SEM, several considerations prevent the full power of its signal-processing capability being exploited in the point-by-point electrical and optoelectronic characterisation of materials and devices. Firstly, many of the more exotic semiconducting compounds for the latest generation of devices in optoelectronics are damaged by the electron beam. Secondly, a significant and increasing proportion of modern devices rely on insulating layers for device operation (e.g. MOS devices), and the examination of such technologies with a charged beam poses problems in interpretation and in situ device operation. Similarly, the presence of passivating surface layers poses a problem. Finally, most forms of electron microscopy are essentially constrained to surface examination, even when devices are specially sectioned. Although modern devices tend towards exploitation of planar technologies near the surface, substrate properties are vitally important, and in any case there is a parallel tendency for increased use of dielectric layers, which as mentioned above, are subject to charging effects.

In the past, much characterisation and fundamental research into electronic materials properties has relied on optical techniques, such as photoconductivity, absorption and more recently photo-DLTS (Deep Level Transient Spectroscopy). The wavelength of the light can be chosen more or less selectively to induce specific electronic transitions, whether band-to-band or impurity related. The control over wavelengths can be indirectly exploited in the sub-surface investigation of devices and materials, since the absorption coefficient is wavelength-dependent.

These optical techniques have generally employed 'flood' illumination and may give rise to erroneous results particularly in the case of closely compensated semiconductors (junctions are, of course, by definition closely compensated).

The SOMSEM provides a means of combining:

a) optical techniques for the investigation of electronic materials and devices

b) point-by-point application of (a) for homogeneity studies etc.

c) the signal processing and video presentation capability of the SEM

d) control over depth of focus for sub-surface studies.

A wide variety of phosphors (including some single crystal phosphors) is available with different wavelengths and persistences. Together with a beam-blanking accessory for the SEM, and temperature control for the stage, this flexibility results in a highly programmable light probe for the investigation of the spatial variation of band-gap properties or deep level parameters, or transmission and luminescence properties. Specific materials and device areas of current interest include:

- deep level studies in various devices: GaAs FETs, GaP LEDS, InP microwave devices

- band-gap variations in devices and materials: CMT infrared detectors, GaAlAs lasers

- inhomogeneities of thin film electroluminescent devices and magneto-optical films

- characterisation of semiconductor layers: epitaxial: ion implanted and laser annealed layers

- array assessment: detector arrays, solar cell arrays

- transmission and uniformity properties of optical components and coatings.

In most of the investigations referred to above, the SOMSEM would be used in the OBIC (optical beam induced current) mode, but luminescent, reflection and absorption modes are also of interest.

6   Conclusions

A simple design based on SOMSEM principles has demonstrated the feasibility of a hybrid microscope capable of both optical scanning and electron beam scanning. Conventional SEMS can be modified for optical scanning by incorporating a purpose-built specimen stage. Thus electronic materials and devices can be imaged using the optical beam induced current (OBIC) mode, which has several advantages when compared with EBIC. Other imaging modes possible using this system are the luminescent, reflection and absorption modes, each of which has its particular area of application. Further work must be undertaken to realize these additional imaging modes and to optimize phosphors and lens configuration. The result should be a valuable (and cost effective) investigative tool for the microscopy of semiconducting materials and devices.

Acknowledgements

The authors wish to thank Dr G Hill and Mr M Barnett of ERA for many helpful discussions, and the directors of ERA Technology Ltd for permission to publish.

References

Potter C N and Sawyer D E: 'Optical Scanning Techniques for Semiconductor Device Screening and Identification of Surface and Junction Phenomena' RADC Series in Reliability, Physics of Failure in Electronics, Vol. 5, Edited by Shilliday T S and Vaccaro J: June 1967, pp. 37-51.

Roberts F and Young J Z: 'Flying-spot microscope', Proceedings of the IEE, Vol. 99, PT.111A, 1952, pp.747-760.

Sheppard C J R: 'Scanning optical microscope', Electronics and Power, Vol. 26, No. 2, February 1980, pp. 166-172.

Singmin A: An Optical scanning system for semiconductor junction examination, Microelectron Reliability, Vol. 17, 1978, pp.399-402.

Inst. Phys. Conf. Ser. No. 67: Section 9
Paper presented at Microsc. Semicond. Mater. Conf., Oxford, 21–23 March 1983

Tracing of design weaknesses in VLSI circuits using the electron probe

E Wolfgang

Forschungslabcratorien der Siemens AG, Otto-Hahn-Ring 6, D-8000 München 83

Abstract   Tracing of design weaknesses is carried out during the design
phase of an IC. The elimination of weaknesses improves the quality of
the device which results in higher yield and reliability. A logic state
analyzer and a sampling oscillograph using an electron probe were
implemented. New techniques are described of comparing nominal data
obtained from computer simulations and actual data obtained from logic-
state mappings as well as from waveforms. The "delay mapping" allows
immediate detection of interconnections which cause additional delays.

1. Introduction

Very large scale integrated circuits (VLSI) are evaluated and designed by
means of computer-aided design tools. After the detection of all fatal
errors by functional testing with an IC tester and an electron beam tester
and their subsequent elimination in a redesign process, VLSI circuits still
tend to suffer from design weaknesses which limit their yield. Tracing and
elimination of these weaknesses is thus a prerequisite for their cost-effective manufacture.

Design weaknesses which occur frequently in practice and which can be located and determined quantitatively by means of an electron probe are described in section two. In section three we take a brief look at the electron beam tester and the test preparation procedure. The two most important
tracing techniques, logic-state mapping (a qualitative localization procedure), and waveform measurement, which provides a quantitative representation of the voltage-time relationship, are described in section four.
Articles presenting an overall review of the field of electron beam testing
have been published by Wolfgang et al (1979), Wolfgang (1980), Menzel and
Kubalek (1981), Ura and Fujioka (1982), Rehme (1982) and Feuerbaum (1983).

Feuerbaum and Hernaut (1978) were the first to show the significance of
comparing the actual waveforms - measured using the electron probe - and
nominal waveforms as a means of improving the quality of the simulation
models. These comparison techniques will be treated in detail in section 5,
for improvements in electron beam measurement techniques have increased
not only the accuracy but also the speed of measurement. The profusion of
results obtained must be processed in a way which provides the designer
with a clear picture and allows him to take rapid steps to improve the
quality of the circuit.

2. Design Weaknesses

Design weaknesses limit the yield of VLSI circuits whose logic functioning
is faultless. Three cases which occur in practice and can be detected with
the aid of an electron probe are shown schematically in Fig. 1. The continuous curve presents the calculated nominal waveform. The broken curve is
the measured actual waveform.

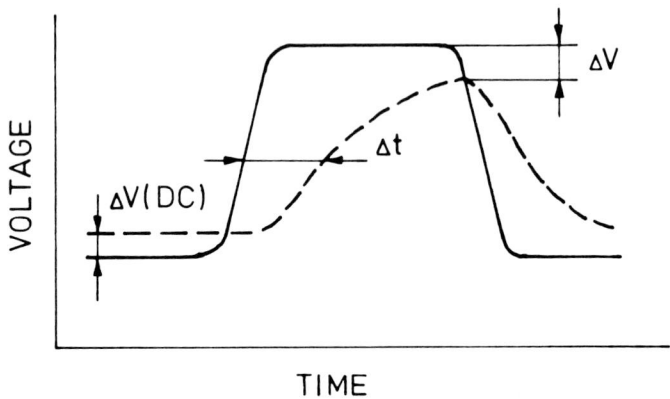

TIME

Fig. 1 Three possibilities of differences between calculated nominal (——)
and actual (——→) waveforms which lead to design weaknesses: Delay Δt,
voltage swing ΔV and offset ΔV(DC).

It can be frequently observed that a large number of ICs fail to reach
their maximum specified operating frequency. If we consider two extreme
cases, then the speed limitation can be due either to a generalized small
delay in all the transistors (technologically-conditioned) or to a relatively large delay in a single transistor (design-conditioned). Fig. 1
shows the delay Δt between the nominal and actual rising edge of the waveforms. If the rising edge of the actual waveform is flatter the nominal
voltage swing is not reached (voltage difference ΔV). The third case
applies mostly to analog circuits in which an incorrect adjustment of offset voltages impairs the quality of the circuits.

3. Electron Beam Testing

Various electron beam testers have been described in recent years (Fazekas
et al 1978, Menzel and Kubalek 1979, Ura and Fujioka 1982). We will therefore mention only the most important parts of such a system with reference
to Fig. 2. The basic instrument is a scanning electron microscope which
has been modified as follows: its electron optical column was turned
around (Otto et al 1978, Otto and Winterstein 1980), and a secondary electron spectrometer (Feuerbaum 1979) and a beam blanking unit were added
(Feuerbaum and Otto 1978). In the so-called "minichamber", the ceramic
package of the device under test (DUT) forms a part of the vacuum chamber:
the chip is thus exposed to the vacuum and the pins to air. This allows
the pin electronics (PEL), which generate the control signals, to be fitted
directly onto the pins. The test patterns which must be cyclic, are generated in the IC drive unit.

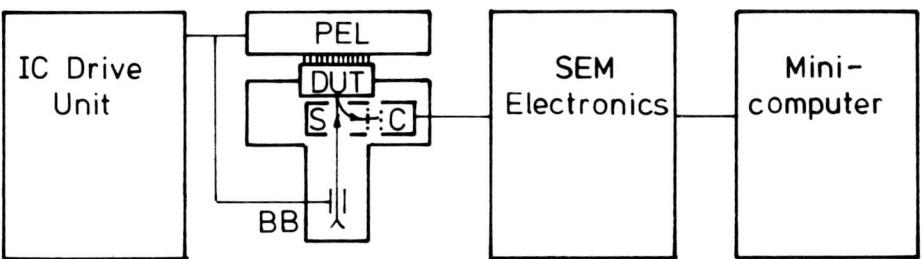

Fig. 2 Block diagram of an electron beam tester. PEL pin electronics, DUT device under test, S secondary electron spectrometer, C secondary electron collector, BB beam blanking unit.

The SEM electronics contains, in addition to the standard assembly, the signal chain for waveform measurements (Feuerbaum and Otto 1982), pulse generators for the beam blanking unit and the secondary electron spectrometer supply. A minicomputer controls the system and is able to display and compare actual and nominal waveforms.

As was mentioned above, the device under test (DUT) is actually a part of the minichamber. Consequently, the electron beam tester shown in Fig. 2 can be used to test only ICs which are mounted in ceramic packages. Before measurements can begin, however, a number of test preparations must be made (Fig. 3).

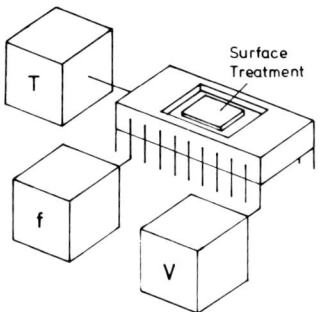

Fig. 3 Test preparation includes removing the passivation layer and cleaning the surface by means of plasma etching. The design weakness has to be reproduced by varying temperature T, frequency f and supply voltages V.

In the first instance, the chip surface must be treated (Fazekas et al 1983). It has proved of great advantage to remove the passivation layer by means of plasma etching. If the same IC is to be tested over a longer period, it is recommended that its contamination layers be removed by plasma cleaning (Beall 1977) and that the charged up parts of the chip be neutralized.

The final step in the preparation procedure consists of operating the DUT so that design weaknesses become clearly apparent. This operating procedure can be monitored by making electrical measurements at the pins. The temperature, frequency and supply voltages have to be varied.

4. Tracing Techniques

As a rule, two steps are required for tracing design weaknesses:
i) Localization with the aid of techniques which show up electrical functions within of larger areas (e. g. 0.5 x 0.5 mm^2) and over longer time
periods (e. g. 50 μs). Logic-state mapping has hitherto proved to be most
effective for this purpose (Crichton et al 1980, Fazekas et al 1981). When
run in TV mode, it is known as voltage coding (Lukianoff and Touw 1975).
ii) Quantification in the form of waveform measurements (Feuerbaum 1979,
Feuerbaum and Otto 1982).

4.1 Logic-state mapping

The results of voltage coding and logic-state mapping are in principle the
same, namely the projection of the logic states "0" and "1" as bright and
dark bars onto the interconnections (Fig. 4c). The contrast between "0"
and "1" results from the well known voltage contrast (Gopinath et al 1978).
In order to judge the effectiveness of both these techniques and to see
their difference more clearly, the principles of voltage coding (Fig. 4a)
and of logic-state mapping (Fig. 4b) are shown schematically and the characteristics of these two techniques are listed in the table in Fig. 4 below.

The voltage-coding can be observed directly on the screen in TV mode. The
maximum operating frequency of 500 kHz and the time resolution of 1 μs are,
however, insufficient for detecting design weaknesses which limit speed in
the case of most ICs. Logic-state mapping in stroboscopic mode allows temporal processes to be mapped in a very broad frequency range - from 0.1
to 10^5 kHz - with a time resolution of 1 ns. However, it takes between
50 to 1000 s to record such mappings.

4.2 Waveform measurement

The effectiveness of waveform measurement on 4 μm wide interconnections
is shown in Fig. 5. The minimum deviations in three different cases can
be obtained using different measuring times. These are shown in the table
in Fig. 5 for specified values of probe current i = 2.10^{-9} A, duty cycle
c = 10^{-3}, probe diameter d = 0.8 μm and primary electron energy E = 2.5 keV.
The long positioning and measuring time of 30 min. required for the offset
measurements V(DC) results from the need to carry out these measurements
at a minimum of three points - the measuring position and two supply
lines - under conditions which must be as similar as possible (Fazekas et
al 1983).

5. Comparison Techniques

The main problem currently encountered in electron beam testing is that of
evaluating the measured values. The great complexity of the circuits makes
it almost impossible for the designer to decide whether or not the circuit
operates correctly at an internal node. To do so he needs the calculated
nominal data, which are all stored in files and must be extracted in
suitable form. This is done by presenting the nominal and the actual data
on a plot or screen.

Fig. 6 shows one way of evaluating the logic-state mapping in TV mode
(voltage coding). It is done by superimposing one or several signals,
which are derived from the logic simulation, onto the TV image in the
form of bright pulses (only one of it is shown in Fig. 6).

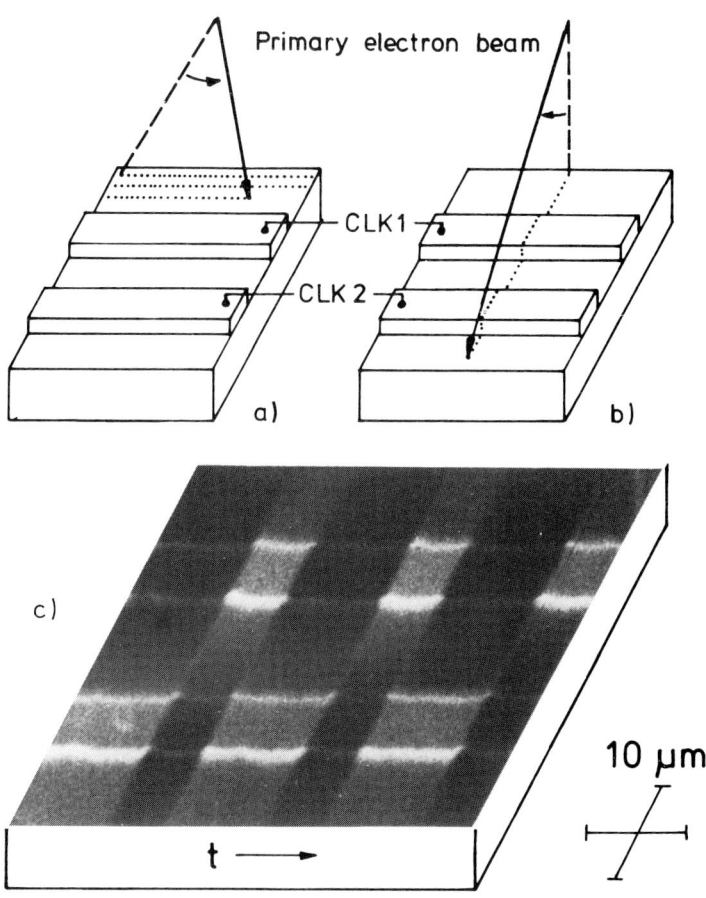

Fig. 4 Projection of logic states onto metal interconnections

	a) Voltage coding	b) Logic-state mapping
Primary electron beam	always on	pulsed
Pulse frequency	–	f_{clk}
Scan frequency f_{scan}	$f_{scan} = n \cdot f_{clk}$	
Scan type	square area	line scan
f_{clk}	10 – 500 kHz	$0.1 - 10^5$ kHz
Recording time	on line	50 – 100 s
Detectable voltage spikes	≥ 1 µs	≥ 1 ns
Voltage resolution	1 V	0.5 V

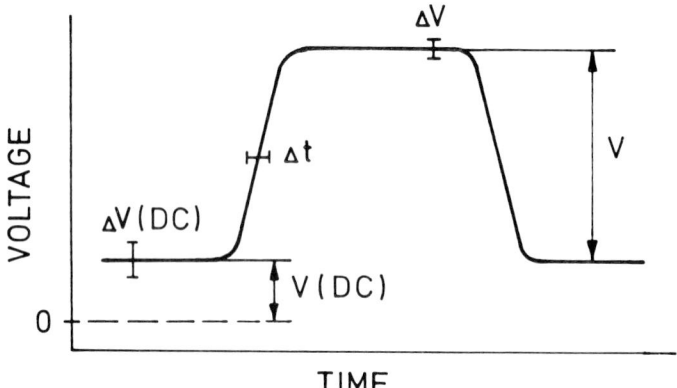

	Propagation delay Δt	Voltage swing ΔV	Offset $\Delta V(DC)$
Deviation	+/- 0.1 ns	+/- 10 mV	+/- 100 mV
Positioning time	10 s	10 s	15 min
Measuring time	30 s	5 min	15 min

Fig. 5 Accuracy of waveform measurements at a 4 µm aluminium interconnection. Probe current i = 2.10^{-9} A, duty cycle c = 10^{-3}, probe diameter d = 0.8 µm, primary electron energy E = 2.5 keV.

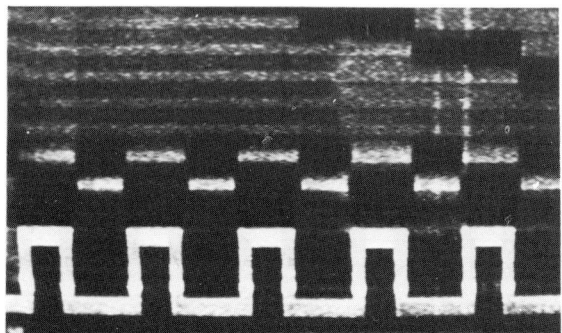

Fig. 6 Voltage coding and one nominal clock super-imposed

10 µm

Nominal clock (75 kHz)

The procedure used to verify a NAND-gate is demonstrated in Fig. 7. The results of the first simulation, which was carried out even before the manufacture of the IC, is shown in Fig. 7a. A falling edge of 1 V/ns with a 4 V swing was selected as the input signal I. The output signal O has a delay of 8.7 ns at a threshold voltage of 1.8 V. The measurements made using the electron probe are shown in Fig. 7b. It can be seen that the input signal is significantly flatter and that the delay time is 6.5 ns. The second simulation was now carried out with the actual input signal, the result being shown in Fig. 7c together with the measured output signal. As can be seen, there is good agreement between the nominal and actual output signals. This means that the NAND-gate has been correctly simulated, i. e. that the simulation models used is valid. In this case the comparison was made on the graphic display of the electron beam tester.

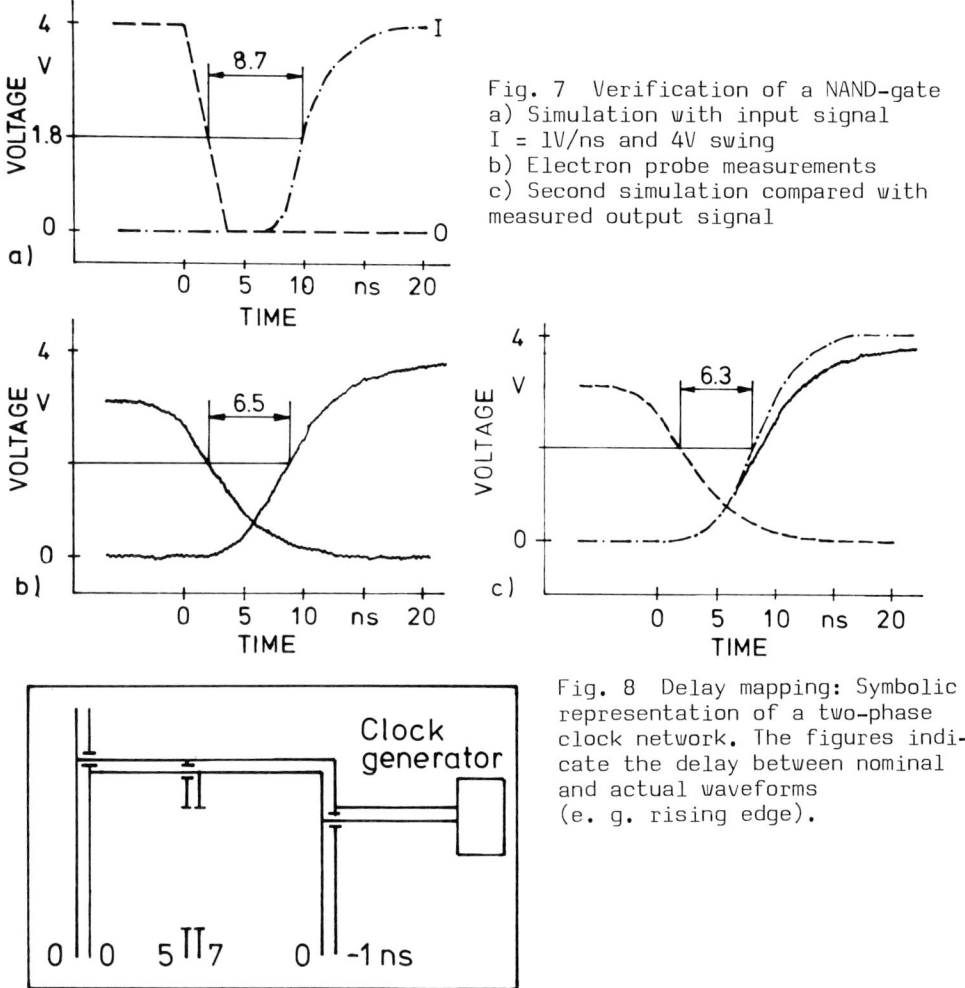

Fig. 7 Verification of a NAND-gate
a) Simulation with input signal
I = 1V/ns and 4V swing
b) Electron probe measurements
c) Second simulation compared with
measured output signal

Fig. 8 Delay mapping: Symbolic
representation of a two-phase
clock network. The figures indicate the delay between nominal
and actual waveforms
(e. g. rising edge).

It is obvious that a large number of waveforms appear in the procedure
demonstrated above. It is thus important for the designer to obtain the
results in a compressed form which allows him to detect design weaknesses
rapidly. One way of doing this is shown in Fig. 8, in which a clock network is represented in a symbolic form. Two clocks are generated by an
internal clock generator. The interconnections of the upper level metaliza-
tion are drawn as continous lines whereas those of the lower level are
indicated by bridges (⊣ ⊢). The deviations from nominal data are entered
in ns at different places. A zero means that actual data are in accordance
with nominal data. A positive figure (e. g. 5 ns) means that the clock is
slower and a negative figure (e. g. -1 ns) means that it is faster than
the nominal value. A glance at this representation - we call it "delay
mapping" - then allows immediate detection of the bridges which cause additional delays. The comparison is only allowed, however, if calculation
and measurement apply to the same temperature.

6. Conclusion

Tracing design weaknesses during the design phase results in an improvement of the quality of the device and of the computer simulation. This leads to higher yield and reliability, which are prerequisites for the cost-effective manufacture of VLSI circuits.

The tracing tools available today include a logic state analyzer and a sampling oscilloscope with a nondestructive and nonloading electron probe. Logic-state mapping and waveform measurement techniques are so effective that it is the evaluation of the acutal data which represents the bottleneck in the overall procedure. Since the nominal data are all stored in files, they can be compared with the actual data. It is important for the designer to detect the design weaknesses from these comparisons so that they may be eliminated by means of specific operations. Compressed representations, such as delay mapping, are valuable aids for this detecting procedure. In order to deal with the actual data evaluation, the electron beam tester will have to be connected up to the CAD system.

Acknowledgements

1 would like to thank my colleagues W. Argyo, P. Fazekas, Dr. H.P. Feuerbaum, F. Fox, Dr. N. Lieske, H. Mulatz, Dr. Müller-Glaser, J. Otto, Dr. A. Papp and F. Widulla for their helpful advice, Mrs. Reidt for her photographic work and Mrs. Kaczmarek for typing the manuscript. Special thanks go to Dr. E. Fuchs for reading the manuscript and his general support.

References

SEM = Scanning Electron Microscopy, Proceedings of the annual conference, (1967-1977) ITTRI Chicago, (1978), SEM Inc., AMF O'Hare, IL, U.S.A.
BEDO = Beitr. elektronenmikroskop. Direktabb. Oberfl.

Beall I R 1977 Proc. ATFA-77 pp 121-131
Crichton G, Fazekas P, Wolfgang E 1980 IEEE Test Conference pp 444-449
Fazekas P, Lindner H, Lindner R, Otto J, Wolfgang E 1978 SEM pp 801-806
Fazekas P, Feuerbaum H P, Wolfgang E 1981 Electronics $\underline{54}$ pp 105-112
Fazekas P, Fox F, Papp A, Widulla F, Wolfgang E 1983 SEM to be published
Feuerbaum H P, Otto J 1978 J. Phys. E.: Sci. Instr. $\underline{11}$ 529
Feuerbaum H P, Hernaut K 1978 SEM pp 795-799
Feuerbaum H P 1979 SEM pp 285-296
Feuerbaum H P, Otto J 1982 SEM to be published
Feuerbaum H P 1983 Scanning to be published
Gopinath A, Gopinathan K G, Thomas P R 1978 SEM pp 375-380
Lukianoff G V, Touw T R 1975 SEM pp 465-571
Menzel E, Kubalek E 1979 SEM pp 297-304
Menzel E, Kubalek E 1981 SEM pp 305-322
Otto J, Lindner H, Feuerbaum H P 1978 BEDO $\underline{11}$ 73
Otto J, Winterstein A 1980 BEDO $\underline{13}$ 127
Rehme H 1982 Phys. Bl. $\underline{38}$ 253
Ura K, Fujioka H 1982 Japan Annual Reviews in Electronics, Computers and
    Telecomunications: Semiconductor Technologies (Amsterdam: North Holland)
    pp 308-327
Wolfgang E, Lindner R, Fazekas P, Feuerbaum H P 1979 IEEE J. Solid-State
    Circ. $\underline{Sc-14}$ 471
Wolfgang E 1980 Microcircuit Engineering (Cambridge University Press)
    pp 409-438

Inst. Phys. Conf. Ser. No. 67: Section 9
Paper presented at Microsc. Semicond. Mater. Conf., Oxford, 21–23 March 1983

SEM voltage contrast techniques for CMOS device assessment

S M Davidson

GEC Research Laboratories, Hirst Research Centre, Wembley, UK

Abstract   SEM voltage contrast techniques have been developed for CMOS
device assessment.   Facilities for voltage contrast mapping, synchronous
and stroboscopic imaging and waveform recording are being applied to a
range of problems in device diagnostics.   The SEM operating conditions
and equipment requirements are discussed for each type of measurement.

CMOS devices are being increasingly used in digital and analogue electronic
systems.   This decreasing feature size –5 μm at present, 2–3 μm next year,
and 1–1.5 μm by the end of the decade – poses serious problems for the
engineer concerned with CMOS design, development, testing and failure
analysis.   The emergence of the CMOS custom device business, mostly using
ULA or standard cell designs also brings the component user closer to the
problems associated with device specifications, assessment and diagnostics.
A requirement therefore exists for techniques which will perform CMOS device
verification and failure analysis, and which can be used by both
semiconductor manufacturers and component users.

The most usual method of internal testing is microprobing, but this has a
number of drawbacks for small geometry CMOS devices.   The probe will obscure
part of the device, it may cause damage to the metal or oxide, passivation
often has to be removed and most important, the probe will capacitatively
load the internal device nodes, seriously disturbing internal timing and
sometimes completely disrupting device operations.

The alternative to probe–testing is electron beam testing, using the voltage
contrast mode of operation of the scanning electron microscope.   These
techniques have been used within the semiconductor industry for many years
for device and wafer inspection and for simple voltage contrast work.
Recently, however commercial equipment manufactured by Lintech Instruments,
has become available which greatly extends the versatility of the voltage
contrast method.   Voltages can be measured, stroboscopic images recorded and
waveforms at internal nodes monitored.   Electron beam–induced current (EBIC)
techniques also have value for CMOS assessment, although these are perhaps
more appropriate to device process development than failure analysis.   In
this paper we describe the philosophy behind the use of SEM voltage contrast
techniques for CMOS characterisation, and the points to be borne in mind
during their application.

CMOS devices are particularly amenable to electron beam testing because of
their large internal voltage swings.   In normal operation, either the
p–channel or the n–channel transistor of a complementary pair will be off.
The internal nodes will thus always sit either at V_{SS} (zero volts) or V_{DD}

(positive rail).   In a static voltage contrast map of a CMOS device all the tracks (nodes) will either be dark (logic high) or bright (logic low).   No intermediate levels of contrast should be observed.

The voltage contrast can be observed over a wide range of beam voltages (kV), but not all can be used for CMOS assessment.   The main factors to be taken into account are (a) the possibility of damaging the device by electron irradiation, (b) the build-up of charge in the device field or isolation oxide, (c) the presence or otherwise of passivation, (d) the performance of the SEM, and (e) the type of measurement to be performed.

Damage to CMOS devices can occur if the electron-beam penetrates the gate. Excess charge will be generated in the gate oxide, and additional surface states are created at the oxide-silicon interface. The effect in both cases is to shift the n-channel and p-channel transistor thresholds negatively. The former will be turned on all the time, while the latter will require a higher drive voltage.   The first effect observed is a dramatic increase in the device quiescent current, because n-and p-channel transistors in the same gate will be on at the same time; soon after, the device will fail to function correctly.   It is sometimes possible to anneal out the charge in the gate oxide, but the excess surface states are more difficult to eradicate.   The only way to avoid electron beam damage is to ensure that the electron energy is such that high energy electrons never see the gate oxide, typically below 10 kV.   The second effect, build-up of charge in the field oxide, occurs when the electron beam comes to rest within the field or isolation oxide, and when the incident energy is such that the secondary electron emission coefficient is less than unity. The oxide charges up to a high negative potential, eliminating any voltage contrast effects from the adjacent metal tracks.   The secondary emission coefficient is only greater than unity for incident energies between ~600V and 3 kV.   All voltage contrast work therefore needs to be performed in this kV 'window'.   Charge build-up can also be avoided by operating at high kV, where the electron beam penetrates the oxide.   However under these conditions the MOS transistors are damaged by the electron beam.

The third consideration is whether or not the device is passivated.   Most commercial devices are protected by means of a passivating layer of silicon dioxide, silicon nitride, or polyimide.   Only the bonding pads are left exposed.   If the device is not passivated, good voltage contrast can be obtained from the metal tracks at operating voltages up to 2.5 kV.   Since SEMs generally operate better at higher kV (high beam current or smaller spot size) it is advantageous to operate close to the second crossover, between 2 kV and 3 kV. The fourth consideration, the performance of the SEM at low beam voltage is also relevant here.   The latest SEMs perform significantly better than their predecessors in this respect, so operation at as high a voltage as the device will allow may not be so necessary.

If the device is passivated, signals on the metal tracks underneath the passivation will be capacitatively coupled to the passivation surface. Electron beam irradiation will discharge the surface potential, and thus 'erase' the voltage map. The time for this to occur will depend on the beam voltage, beam current and SEM magnification, but is typically milliseconds to seconds.   Operating the SEM under conditions where the passivation surface charge is disturbed least thus implies the use of even lower beam voltages, ideally between 500V and 1 kV.

The fifth consideration, the type of measurement to be performed, will also

influence the SEM operating conditions. Voltage contrast mapping of
unpassivated devices can be carried out at 10^{-9}A. However, to avoid
disturbing the surface charge, passivated devices are often best examined at
lower currents, say 10^{-10}-10^{-11}A. For waveform measurements, where the beam
is positioned at a point on the device, this consideration is even more
important. It is sometimes necessary to record waveform measurements from
passivated devices at different beam currents to check that artefacts are
not being introduced into the waveform by electron beam discharge effects.

On the JSM-35C in use at the GEC Hirst Research Centre, we mostly examine
unpassivated devices at 2 kV, and passivated devices at 1 kV. Good TV-rate
images can be achieved at beam currents $\sim 10^{-10}$A with a 600 μm final
aperture, and the device tilted at $\sim 20°$ from horizontal.

The equipment requirements for electron beam CMOS characterisation are listed
in Table I according to the type of measurement performed. Simplest is
static voltage contrast mapping, which can be used for detecting logic
faults and breaks in unpassivated devices. All that is required is a normal
SEM which can operate at low beam voltages (1-2 kV), and some means of
monitoring the device in the specimen chamber and connecting it to voltages
and signals. A dual-in-line socket connected by ribbon cable to a multi-way
vacuum feedthrough is the most common device mount, although coaxial cable
may be used for high frequency signals. Figure 1 shows the use of static
voltage constrast for logic fault detections. The device is a CMOS ULA,
where the logic unit is a AND-NOR cell. In Boolean representation $D=\overline{A.B+C}$,
when output D is low if either A and B, or C are high. Functional testing
showed that the device was non-operational. SEM voltage contrast traced the
fault to the cell illustrated. It can be seen that A and B are both high
(dark), therefore the output (D) should be low. In fact, D is high. Altering
the logic inputs to force the output high sometimes caused the output to go
low. Further examination showed that the fault was caused by a short to an
adjacent track.

Voltage coding is the term given to the display of signals as bar patterns
along metal tracks. Again, only a SEM plus device holder and drive capacity
are required. Normally, the device signals would not by synchronised with
the SEM, giving a tilted or rolling pattern. Although some information
regarding frequency and phase may be derived from such images, their main
applications to CMOS diagnostics is for continuity checking. The technique
is applicable to passivated devices because AC signals will continuously
refresh the voltage on the passivation, although steady voltages, on, say,
the power supply lines will be erased. Figure 2 shows an frequency divider.
Correct operation is verified by observation of the bar spacing, which
doubles at each cell.

Another method of deriving voltage contrast information from passivated
devices is to synchronise the device voltage with the SEM scan, by
triggering a pulse generator from the line or frame scan ramp. Figure 3
shows part of a passivated ULA. The band in the centre of the picture
corresponds to the device in the on-state, while the regions on either side
are in the off-state. Voltage contrast on the metal tracks can be seen as
if the device was unpassivated. Contrast from the diffused region in the
silicon can also be detected although this is rather diffuse. The spatial
resolution in general is worse than for an unpassivated device, although the
extent of the degradation will depend on the passivation thickness. This
technique has also been used for detecting shorts between polysilicon
electrodes on a charge coupled device (CCD), and investigating latch-up at

Table 1

Mode	Information	Equipment
Static VC	Voltage distributions logic levels	SEM + device holder + power supplies
VC voltage coding	Logic signals frequency/phase	+ pulse/pattern generator
VC stroboscopic imaging	Dyanmic voltage distributions	+ beam pulsing
VC waveform observation	Frequency/phase Time relation-ships Rise/fall time Propagation delays	+ delay genertaor + waveform recording/ averaging system
VC measurement	Voltage levels Waveform amplitudes (passivated devices)	+ electron energy filter/control
Multi-point VC waveform recording	Waveforms/voltages at sequence of points	+ vector generator

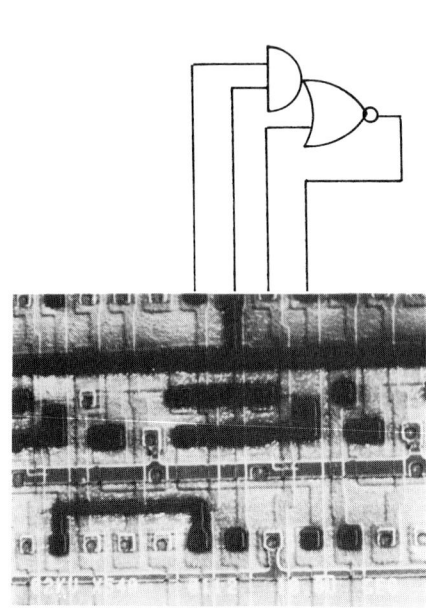
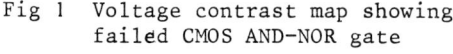

Fig 1   Voltage contrast map showing failed CMOS AND-NOR gate

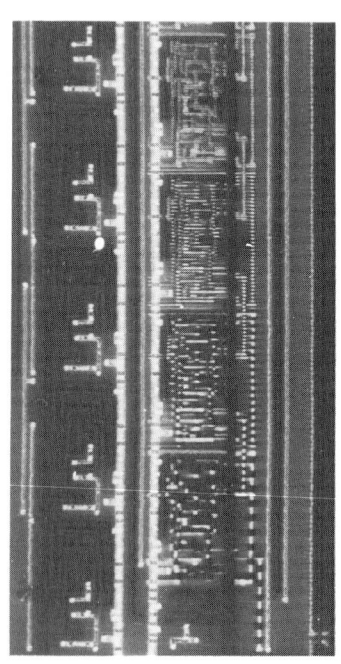

Fig 2   Custom CMOS frequency divider

CMOS input and output structures. These are reported elsewhere.

The next mode of electron beam testing to consider is stroboscopic voltage contrast imaging. The electron beam, is now switched on in synchronism with the device operation. The equipment requirement is a beam blanking or pulsing system, which can be triggered by the device. A delay generator, built into the system, is also necessary to vary the phase of the electron beam with respect to the trigger. A stroboscopic voltage constrast map can then be recorded at any instant in the device operating cycle. Stroboscopic images of unpassivated or passivated devices can be recorded. Long recording times are often necessary if the electron beam pulse length is much shorter than device cycle time, i.e. at low beam duty cycles, because the images are rather noisy. The average electron beam current is reduced in proportion to the duty cycle. Figure 4 shows strobroscopic images of a CMOS-SOS inverter chain, recorded at intervals of 5 ns. The propagation of the signal is clear.

Stroboscopic images are of limited value for fault diagnosis and design validation. Much more useful is the ability to record waveforms at internal device nodes. In CMOS devices these nearly always correspond to metal tracks. The electron beam is now positioned at one point on the device. The delay between device trigger and electron beam pulse is swept through the appropriate part of the signal and the waveforms recorded by plotting the SEM secondary electron detector output against delay time.

Figure 5 shows waveforms recorded from a custom CMOS cell structure device. The problem was one of limited operating frequency, the device failing some 2 MHz below specification. Waveforms recorded from a nominally synchronous frequency divider are shown in Figure 5 (10 ns time resolution). It can be seen that the clocks are not synchronous. This was traced to delay in one of the clock lines along a polysilicon interconnect. Further measurements at the D-inputs of the next stage of the divider are also shown. These show the terminal count signal (TC) generated when the count drops to zero rippling through a chain of AND gates. By the end of the chain the delay from the clock is 130 ns – coinciding with the frequency at which the device failed. Both the circuit design and the layout were changed following this investigation. Waveform recording techniques have also been applied to propagation delay problems in CMOS ULAs and access time assessment in CMOS memories. In all cases the use of electron beam testing was essential to avoid loading the internal nodes. Microprobing would have been virtually useless.

To conclude, the electron beam testing system devised at GEC for CMOS testing and diagnostics comprises JEOL JSM-35C SEM, Lintech stroboscopic voltage contrast system, Apple microcomputer and a special device holder/ matrix board connector. Desirable future developments include a vector generator to facilitate multiple waveform acquisition, scan converter/frame store for image processing and storage, wafer prober to test devices on the slice and micromanipulator probe to perform accurate voltage measurements. This extended equipment should be capable of diagnosing most problems in CMOS development and operation.

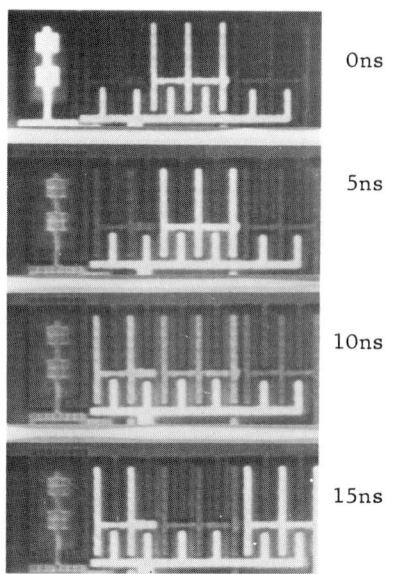

0ns

5ns

10ns

15ns

Fig 3   Synchronous voltage contrast
map of passivated CMOS ULA

Fig 4   Stroboscopic voltage
contrast images of CMOS-SOS
clock driver

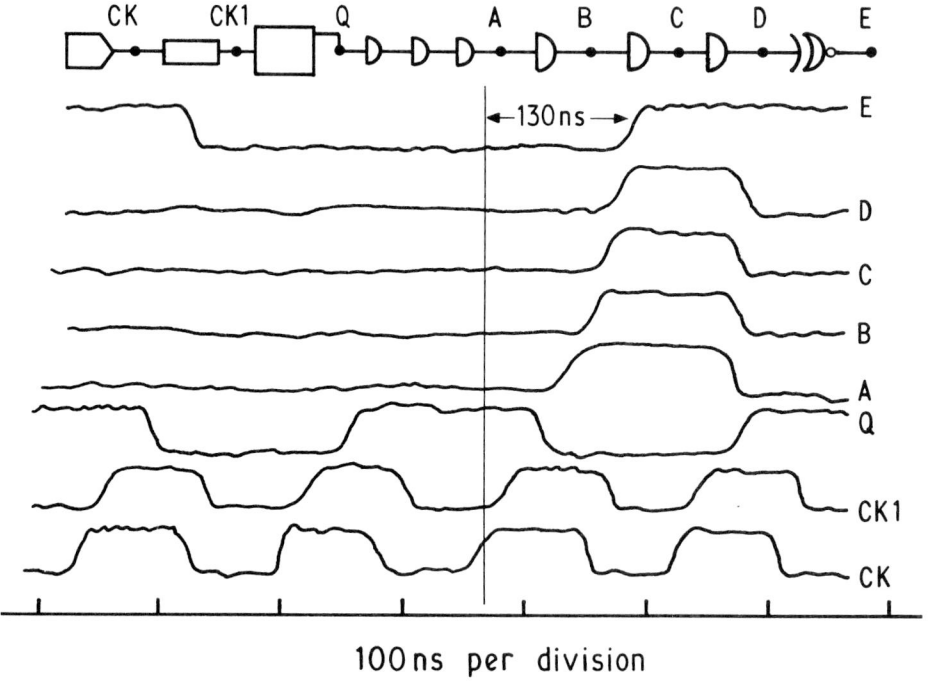

100 ns per division

Fig 5   Voltage contrast waveforms recorded from a passivated CMOS-SOS
divider

Inst. Phys. Conf. Ser. No. 67: Section 9
Paper presented at Microsc. Semicond. Mater. Conf., Oxford, 21–23 March 1983

Multichannel logic state analysis of IC-internal signals by an electron-beam probe

M Ostrow, E Postulka, E Menzel, E Kubalek
Universität Duisburg
Fachgebiet Werkstoffe der Elektrotechnik
Leiter: Prof. Dr.-Ing. E. Kubalek
Kommandantenstr. 60
4100 Duisburg 1, F.R.G.

Abstract Electron beam logic state analysis techniques like Voltage
Coding, Logic State Mapping and Logic State Timing Diagram allow the
measurement of IC-internal logic states of large and very large scale
integration circuits. These techniques, however, suffer from the fact
that signals with low repetition rates or single events may not be detected, and that long data aquisition times are required. A Multichannel Logic State Analysis method which overcomes these drawbacks was developed. This paper reports the principles of this method, describes
the experimental setups and shows the achieved results.

1. Introduction

Generally logic-state analysis of ICs is carried out as a real time measurement at the IC-external leads by use of logic state analyzers. The demand for additional testing of IC-internal logic states has led to the
development of different methods based on the so called"Voltage Contrast"
in a scanning electron microscope using the electron beam (e-beam) as a
nondestructive, nonloading fixed or scanned microprobe.
These methods like the Voltage Coding (Lukianoff et al.1975), Logic State
Mapping and Logic State Timing Diagram (Fazekas et al.1981) require repetitive signals, e.g. short program loops and do not allow real time logic state analysis. This is because the e-beam is scanned across the IC
surface and/or is chopped, so that interference pulses, missing or additional bits etc. are only detected by chance. Signals with low repetition
rates or nonrecurring signals can not be detected.
These drawbacks has led to the development of the Real Time Logic State
Analysis (RTLSA) (Ostrow et al.1981) technique in which the e-beam acts
as an additional channel for a logic state analyzer. The RTLSA technique
uses a continous e-beam probe that is positioned to the node of interest
where it detects the time dependent IC-signal in a real time measurement.
In addition and on the basis of the RTLSA single channel operation a Multichannel Logic State Analysis method (MLSA) was developed, allowing the
extremely fast detection of logic states at a number of freely programmable IC internal nodes (Ostrow et al.1982), and the applicability of this
method was shown in practice.
Two techniques can be used. In technique 1 the bit sequence of each IC-
node is detected in real time and is performed sequentially at the different IC-nodes.
In technique 2 the bit sequence is detected by positioning the e-beam on
the different IC-nodes during a clock period of the IC-clock.

This paper reports the principles of these two MLSA-techniques in detail and describes the corresponding experimental setups. The achieved experimental results are demonstrated and the limitations, and further improvements of technique 1 are discussed.

2. Principles of the Multichannel Logic State Analysis (MLSA) Method

The RTLSA-technique only allows the analysis of logic signals at one internal point of an IC but in many ICs, e.g. microprocessors, it is necessary to be able to compare logic states at different IC nodes in the form of a timing diagram or a trace list. Therefore it was necessary to develop an IC-internal MLSA-method.
There are two principle techniques, which complete one another, so that for a practical MLSA, there must be the possibility to perform both techniques.

2.1 Technique 1 (MLSA-1)

This technique is based on the RTLSA-technique and uses a continuous e-beam. The bit sequence of each IC-node, e.g. a data bus, is detected in real time and performed sequentially at the different IC-nodes, so for an analysis of eight nodes, a program must be repeated 8 times. The serially measured data are read into a buffer. At the end of the measurement the contents of the buffer is read out parallel into a logic state analyzer or a computer. However, the actual timing of the signals at the IC-nodes is lost by this technique.

2.2 Technique 2 (MSLA-2)

In contrast to the MSLA-1 technique the measurement of logic states is done here during one clock period of the IC-clock by positioning the e-beam from one conductor line to another. The serial measured signals are transformed by a demultiplexer into a parallel signal and then fed to a logic state analyzer or a computer. For an analysis of data at eight nodes a program must only start one time. The actual timing of the measured signals is preserved but pulses (glitches) which are shorter as the clock rate are not detected.

3. Experimental setup of MLSA-1 technique

The experimental setup is shown schematically in figure 1. The IC is operated from the IC-drive unit. The secondary electrons (SEs) are analyzed in a retarding field spectrometer (Feuerbaum 1979), which is operated in an open loop configuration with no linearization. The retarding field voltage is adjusted to get a maximum contrast between a logical "0" and "1". The signal is detected with a conventional Everhart-Thornley detector and then fed to the signal processing unit. The e-beam is posi-

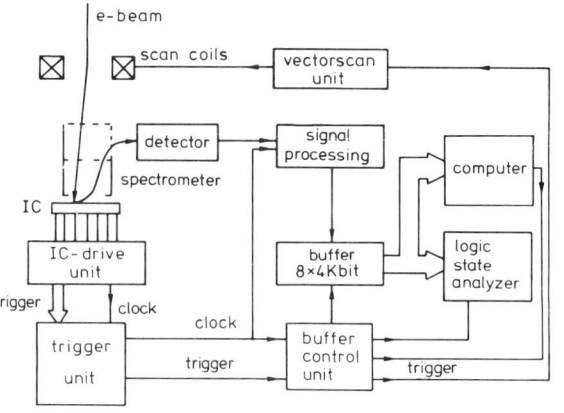

Fig.1 Experimental setup for MLSA-1 technique

tioned to a predetermined IC-node by the vector scan drive unit controlled
by the buffer control unit. At a freely selectable trigger condition (word
or single pulses), obtained at any desired location within the IC drive
unit, the trigger unit is initiated and causes the buffer control unit to
read the data from the signal processing unit into the buffer (first channel). If this channel is filled up to (4096 bits) the e-beam is positioned
to the next node and at the next trigger condition within the restarted
program or program loop the sequence is repeated until all 8 channels are
filled. Hereafter, the contents of the buffer memory are read again into
a logic state analyzer or into a computer by a trigger pulse.
The signal processing for the MLSA-1 technique to detect logical signals
in a real time measurement is shown in fig. 2. The signal coming from the
fast photomultiplier is am-
plified by a preamplifier and
then passed through an tunable
lowpass filter. A comparator
decides whether the signal is
a logic "0" or "1". A sample
and hold unit samples the
signal on the positive or negative slope of the IC-clock.
After sampling, the signal is
transferred to a buffer.
A tunable delay compensates
time delays between the IC-
clock and the detected signal. These time delays are

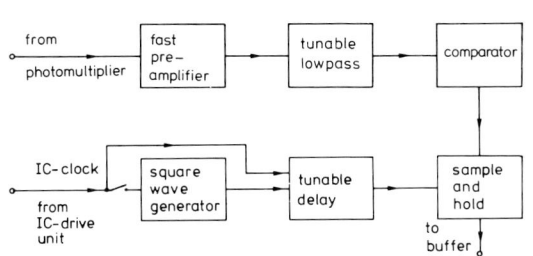

Fig.2   Experimental setup for signal
        processing

caused due to the photomultiplier transit time, the SE-flight time and the
lowpass filtering, etc. The sampling of the detected signal can be done by
using the original IC-clock from the IC-drive unit or a multiple of the
IC-clock (synchronous sampling) or a free running square-wave from a frequency generator (asynchronous sampling). Fig. 3 shows the principle of
the signal sampling.

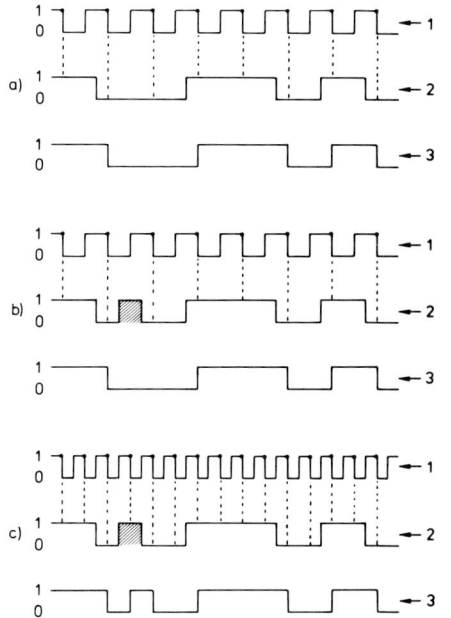

1 → clock signal

2 → signal on IC-data line

3 → signal after sample and hold

⌐ → sample slope

Fig.3   Principle of signal
        sampling

        a) without additional pulse

        b) additional pulse is not
           detected

        c) additional pulse is de-
           tected

The sampling of the signal occurs on the negative slope of the IC-clock (figure 3a). At this slope the logic state ("0" or "1") of the signal is held in the sample and hold unit (figure 2) and is only changed at the next sample point when the logical state changes.
In figure 3b an additional pulse (e.g. a glitch) appears on the IC-line. It is not detected, because the appearance of the pulse is not in coincidence with the sample slope. This pulse only is detected by sampling the logical state with a multiple of the IC-clock frequency.

Although this technique measures the bit sequence at each node in a real time manner, it may not be regarded as a real time technique, because the data of different nodes were obtained one after the other.
The main advantage of this technique is its high data acquisition speed. If for example a part of a program (which has not necessarily to be a loop) of 4096 x 8 bit words has to be analyzed at any 8 nodes on the IC, the clock frequency being 2.5 MHz = 0.4/us and the settling time of the scan coils being about 10/us, then the total data acquisition time for MLSA-1 is

$$t = (4096 \times 0.4/us) \times 8 + 7 \times 10/us = 13.2 \text{ ms}.$$

This is a very short time in comparison with the other techniques which e.g. for a 50 bit long bit sequence ranges from typically 100 s in the Voltage Coding and Logic State Mapping technique to about 1 s in Logic State Timing Diagram technique.
A further advantage is the easy data evaluation in the logic state analyzer or in a computer, which allows automatic measurements and tests of the measured data.

4. Experimental setup of MLSA-2 technique

Figure 4 shows the experimental setup for the MLSA-2 technique. The IC is operated from an IC-drive unit.

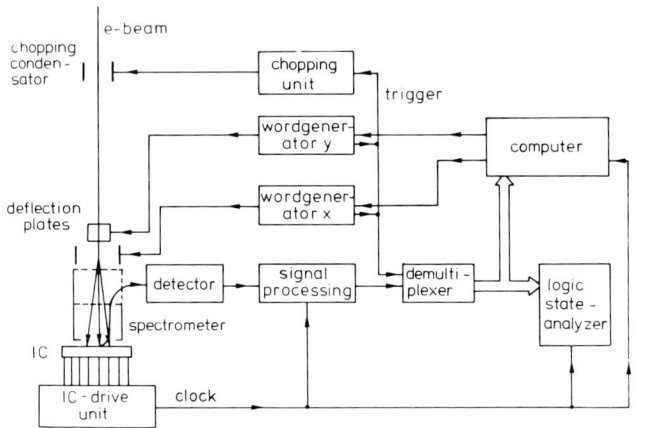

The e-beam is positioned to predetermined IC-nodes by a fast electrostatic deflection unit. The SEs are analyzed in a spectrometer, detected with a conventional Everhard-Thornley-Detector and fed to the signal processing. The deflection voltage for the deflection of the e-beam is generated by fast wordgenerators, which are programmed by the computer. The demultiplexer syn-

Fig.4 Experimental setup for the MLSA-2 technique

chronously with the positioning of the e-beam feeds the data to the channels of the logic state analyzer or directly to the computer. During the positioning of the e-beam, the beam is chopped to get an information only when the beam impinges on a conductor line. First experimental results of this technique are presently being obtained and will be published elsewhere.

5. Experimental results of the MLSA-1 technique

Initial measurements were performed at 8 nonpassivated 8 /um wide parallel IC-lines of a test-IC, which was directly connected to the external data lines of a Z - 80 microprocessor (/uP) unit. This test procedure was chosen to be able to compare the e-beam measurements, which were done by a computer and a logic state analyzer (LSA), with electrical measurements and thereby test the reliability of this technique.

In principle the measured data can be performed variously, e.g. as a block of data in hexadecimal form or in a timing diagram or in a trace list. The comparison of known and unknown data is done by a computer. Figure 5 shows the performance of the first 35 bytes of a program which was 4096 bits long in a trace list and in a timing diagram.

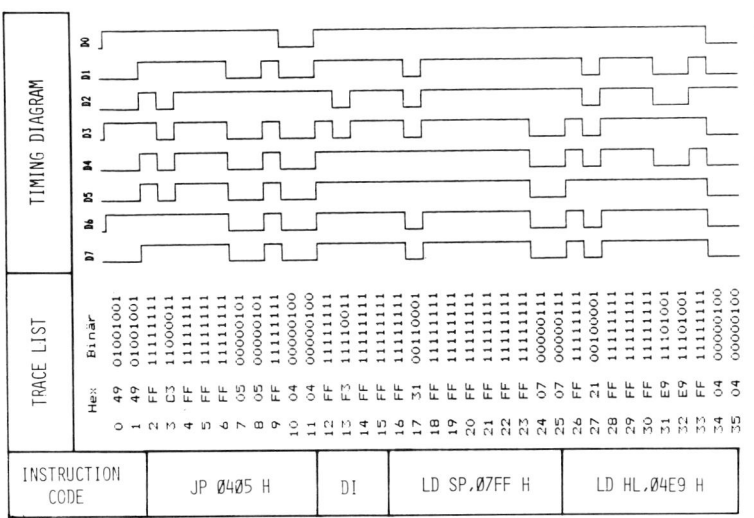

Fig.5
Performance of the first 35 bytes of a 4096 bits long program

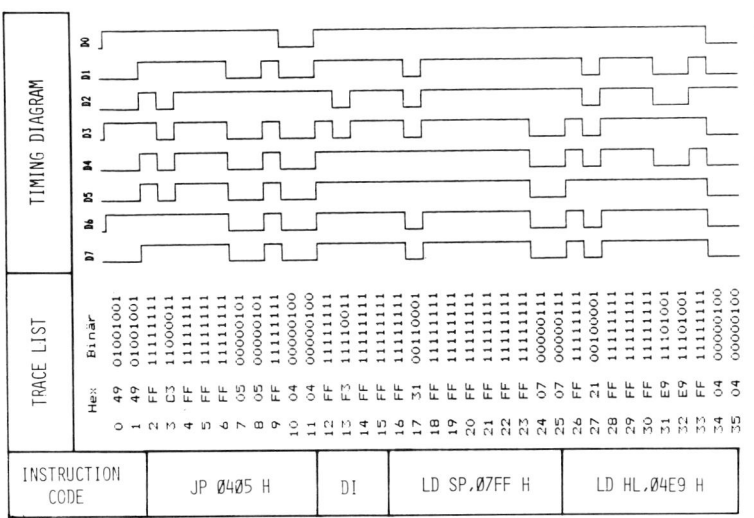

Fig.6 A LSA-display of the same program as in fig.5, sampled with the M1-cycle signal

Measuring specified events in a program is possible by a suitable choice of the clock and the sample point.
To analyse only the instructions of a program the sampling of logical states has to be done with the M1-cycle of the /uP. With this signal the /uP orders new instructions from the memory.
Figure 6 shows a LSA-display of the same program as in figure 5.

The correct bit pattern in figures 5 and 6 were obtained at a clock frequency of 2.5 MHz using a nonloading and noncharging e-beam probe at the point of the $\delta + \eta$ yield (δ = SE yield, η = backscattered electron yield) where $\delta + \eta$ = 1. (The measurements were done at the second crossover). The used e-beam parameters were primary beam energy E_{PE} = 750 eV, primary beam current I_{PE} = $1.5 \cdot 10^{-8}$A.

The maximum time resolution, which was achieved with this technique was about 50 ns by a voltage resolution of 2.5 V with a necessary e-beam current of $I_{PE} \sim 10^{-7}$A. To obtain a nonloading beam by this e-beam current is to operate at the second crossover of $\delta + \eta$ (Ostrow et al. 1982). The present achievable e-beam diameter by a beam current of $I_{PE} \sim 10^{-7}$A is about 3/um with a LaB$_6$ electron gun.

6. Future improvements of the MLSA-1 technique

In view of Very Large Scale Integration - structures our further work is to overcome the present limitations of this technique. It is required to decrease the spot size of the electron beam and to increase the time resolution which can be improved by further noise reductions, e.g. by making the S-curve of the spectrometer steeper, optimization of the SE-detection and signal filtering systems.

Acknowledgements

Thanks are due to the federal ministry of research and technology (BMFT) Germany and to the ministry of science and research (MWF) of North-Rhine Westfalia for financial support of some parts of this research program.

References

Fazekas P, Feuerbaum H P, Wolfgang E Electronics 14 1981 pp 105-112
Feuerbaum H P Scanning Electron Microscopy 1979 pp 285-296
Lukianoff G V, Touw T R Scanning Electron Microscopy 1975 pp 465-471
Ostrow M, Menzel E, Kubalek E Microcircuit Engineering Lausanne 1981
    pp 514-521
Ostrow M, Menzel E, Postulka E, Görlich S, Kubalek E Scanning Electron
    Microscopy 1982 pp 563-572

Inst. Phys. Conf. Ser. No. 67: Section 9
Paper presented at Microsc. Semicond. Mater. Conf., Oxford, 21–23 March 1983

Quantitative voltage contrast: instrumentation and signal processing

B Gilhooley and A R Dinnis

Department of Electrical Engineering, University of Edinburgh, The King's
Buildings, Mayfield Road, Edinburgh. EH9 3JL., Scotland.

Abstract   The design and operating principles of an instrumentation system for real-time testing of L.S.I. integrated circuits, using the electron beam probe in the voltage contrast mode, are described. The system
utilises coherent detection and second harmonic suppression to realise a
linear relationship between the surface voltage and the error signal   in
the feedback loop. The preliminary evaluation results are presented.

1. Introduction

The electron microprobe utilised in the voltage contrast mode has established itself in semiconductor research laboratories as a practical and
precise instrument for the detection and characterisation of timing faults
in the early design stages of integrated circuits (Wolfgang, 1982, Lukianoff, 1981, Feuerbaum, 1979).   With the advent of high density, high
impedance V.L.S.I. circuits and the resultant economic penalty of high pin
count packages, it will become necessary to implement internal nodal testing at an early stage of production to verify functional devices. If functional testing is to be implemented by the electron beam probe, a fully
automated test system, with a real-time 3dB bandwidth of approaching five
times the minimum clock frequency of the process will have to be realised.
High voltage and temporal resolution will not constitute major design constraints, as in the parametric testers, but rather the ability to discriminate between "1"s, "0"s and bad levels. High speed open loop systems have
been developed (Ostrow, 1981, Macari, 1982), but suffer from being both
qualitative and requiring user set-up time.

To this end an electron beam test system is being developed to test dynamic
nmos integrated circuits, fabricated using a 5μm polysilicon gate process
at Edinburgh. The system is being designed to complement rather than
replace conventional test systems and will be able to implement both functional and parametric testing. The system consists of IEEE 488 controlled,
2 phase clock generator, plus and minus power supplies, test vector generator and a data acquisition system. The stimuli are interfaced to the
device-under-test via a patch board and are controlled by a microcomputer
using the Pascal language. This paper addresses the problem of deriving
quantitative voltage measurements from the surface of integrated circuits,
while maintaining a system bandwidth capable of resolving minimally clocked
circuits.

2. Operating Principles

The system is based on the surface voltage modulation of the secondary

electron distribution and incorporates an energy/velocity filter
(Ranasinghe, 1983). For signal processing purposes, the output of the
video head amplifier can be considered as a function of the integral of the
product of the energy filter´s impulse response (H(E)) and the secondary
electron distribution (N(E)):

$$Vo = Zm.T. \int_0^{50eV} \frac{H(E-e)N(E-f(Vs))dE.Isec}{N(E)dE}$$

where Zm is the transimpedance of the video head amplifier, T the transport
efficiency of the filter, e the filter cut-off energy, Vs the surface voltage and Isec the secondary electron current. Although bandpass energy
filters have been used (Hannah, 1974), the signal-to-noise ratio and convolution errors involved have been such that they have been superceded by
highpass, retarding grid filters. This type of filter was used throughout
these experiments. If the energy filter´s cut-off is made time dependent,
the characteristic "S"-curve can be generated by convolving the impulse
response of the filter with the secondary electron distribution. Ideally,
the energy filter´s impulse response should be a step function and independent of the secondary electron angle of emission.

In order to obtain a linear relationship between the indicated value and
the actual voltage on the specimen, Flemming(1970) stated the now common
practice of applying a feedback voltage to the detector. The effectiveness
of this depends on the surface conditions and the electron optical performance of the detector. Since feedback to the retarding grid was utilised,
an operating point had to be selected and to this end the characteristics
of the secondary electron distribution were assessed. The "S"-curve is a
monotonic function of energy and is the result of a stationary statistical
process: i.e. secondary electron distribution. Owing to these facts, the
peak of the distribution was selected, as this should remain fixed independent of beam current and time. Inherently, the effects of surface fields
and the energy filter itself are assumed to be constant for each measurement point, these being a function of energy filter design and not signal
processing.

Employing a bandpass energy filter, Hannah(1974) used differentiation followed by zero-crossing detection to estimate the point of inflection on
the "S"-curve. However, this method degrades the signal-to-noise ratio and
is not suited to a feedback system because of the required scanning. The
method employed was spectral differentiation. That is, if a Taylor expansion of an estimating polynomial of the"S"-curve is taken, it can be shown
(Taylor, 1969) that under the correct conditions, the second term in the
expansion can be made proportional to the first derivative of the "S"-curve
and the third component to the second derivative (figure 1). Since the system had to be compatible with the sampling parametric tester, coherent
detection was employed to eliminate the detection system´s 1/f noise and
improve the signal-to-noise ratio, at low system bandwidths. By applying
the modulation function to the retarding grid and detecting the first and
second harmonics the first and second derivatives could be estimated and
thus, the point of inflection in the "S"-curve. The feedback loop was constructed to null out the second harmonic by applying a correction voltage
to the analyser´s retarding grid and therefore, to track the zero point of
the second derivative. By employing a modulating function, the system
bandwidth is subject to the limitations set by the Sampling Theorem and the

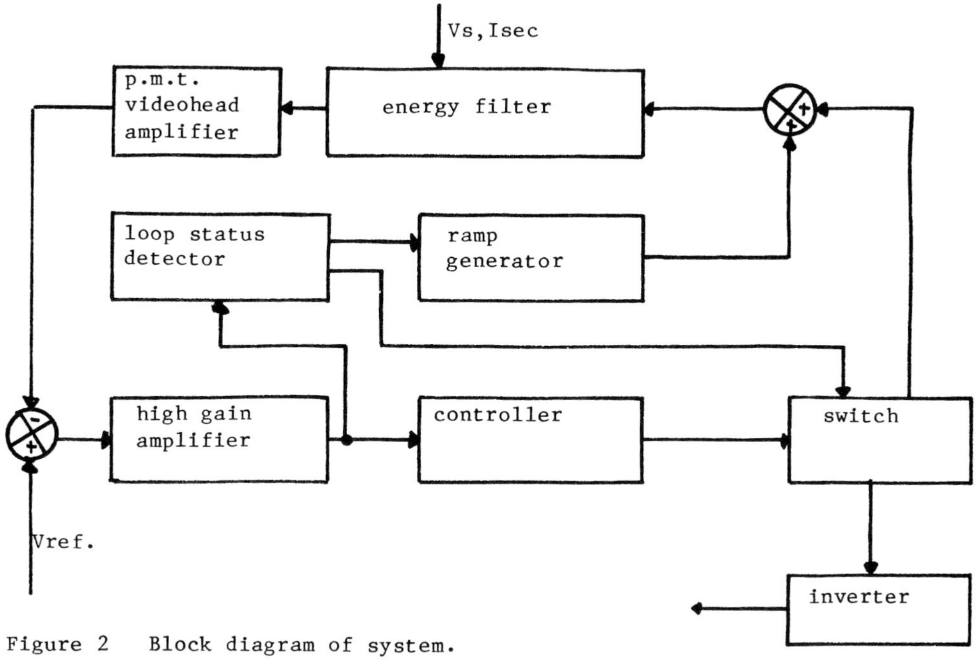

Vs,Isec

| p.m.t. videohead amplifier | ← | energy filter | ← | ⊗ |

| loop status detector | → | ramp generator |

| high gain amplifier | → | controller | → | switch |

Vref.

| inverter |

Figure 2   Block diagram of system.

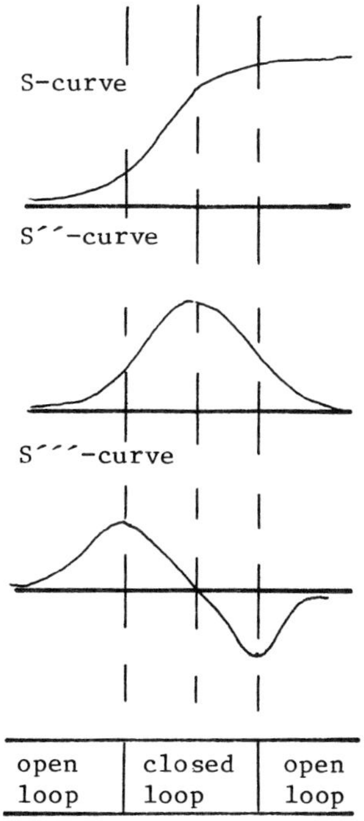

S-curve

S´´-curve

S´´´-curve

| open loop | closed loop | open loop |

Figure 1   System operating characteristics.

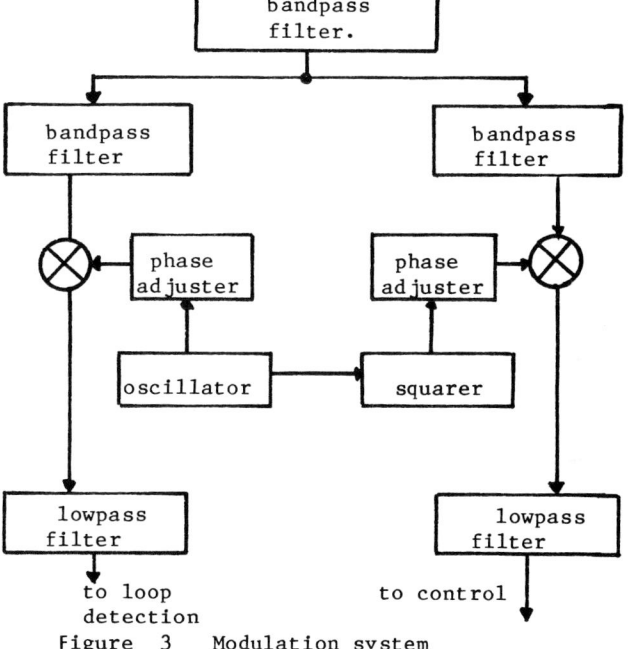

from video head amplifier.

| bandpass filter. |

| bandpass filter | | bandpass filter |

| ⊗ | ← | phase adjuster | | phase adjuster | → | ⊗ |

| oscillator | | squarer |

| lowpass filter | | lowpass filter |

to loop detection

to control

Figure 3   Modulation system block diagram.

signal-to-noise ratio is effectively halved for large bandwidths.

3. System

A block diagram of the system is presented in figure 2. Initially, the system operates in open loop and a 200kHz sine wave superimposed on a slow slewing ramp(10Hz) is applied to the retarding grid of the energy filter. The first and second harmonics are detected and when the peak of the first derivative is encountered, a level detector circuit samples the ramp voltage and closes the loop. The feedback loop pulls the error signal into the null point and tracks the surface voltage. The system is a.c. coupled to the video head amplifier with a bandlimiting predetection filter (figure 3), to prevent saturation in the detection circuits. Further filtering is employed to prevent cross-talk between the slow first and fast second harmonic channels. To maximise the output signal-to-noise ratio these filters have to be matched closely with the signal band. The phase response of these filters is critical, since drift in the modulating frequency can cause loss in coherence. Since the feedback loop gives no improvement in signal-to-noise ratio and the video head amplifier has a bandwidth of 5MHz, the resolution of the system is determined by the amount of bandlimiting employed. This is controlled by the bandwidth of the lowpass filters employed at the output of the modulation system, which are second order Butterworths, tunable from 1-100Hz in decades. With a signal band of 20kHz, 5MHz video bandwidth and white noise, the process gain is about 30dB. The accuracy with which the system can track a surface voltage is dependent on the Loop Gain of the system, which is predominantly controlled by the photomultiplier tube gain.

4. Results

Preliminary experiments to evaluate the system's performance were carried out using a copper stub and a beam accelerating voltage of 2kV on a Cambridge Instrument's S150 scanning electron microscope. With the system operating in the open loop mode, the "S"-curve, first and second derivatives were obtained as shown by figure 4.a. Filter cut-off frequencies of 100Hz and 1kHz were used to achieve the first and second derivatives, respectively, while the "S"-curve itself was obtained directly from the video-head amplifier output. When +1.5V level was applied to the specimen, a corresponding shift in the energy domain occurred. This is demonstrated by the right lateral shift of the inflection point, shown by the curves in figure 4.b. as compared with its position in figure 4.a. Since a positive slewing 15V peak-to-peak ramp waveform, with a period of 35mS was used to scan the energy domain, a 1:1 correspondence between the applied surface voltage and the correction voltage can be inferred. While in the closed loop, tracking mode a 8V peak-to-peak triangular waveform was applied to the specimen to demonstrate the linearity of the system. Although a 10Hz cut-off output filter was used to detect the correction voltage, the output waveform was distorted by low frequency noise, as shown by figure 4.c However, a linear system can be extrapolated from the results. Due to the beam current noise and the gain of the system, the maximum closed loop tracking bandwidth achieved at this stage was about 1-2kHz, as shown by the sine wave response given in figure 4.d. Although there is only slight or no phase distortion of the output sinewave, above a cut-off frequency of 1kHz the system would pull out and in of lock unpredictably, before the system began to oscillate. These are preliminary results, without either the bandlimiting filters or the modulation depth having been optimised: i.e. the predetection filter had a 400kHz bandwidth, while the signal band

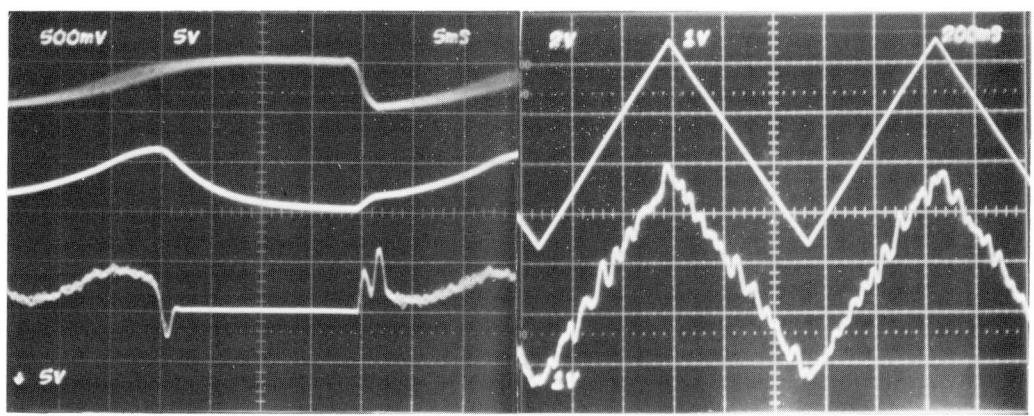

Figure 4 a   from top to
bottom: "S"-curve, 1st
derivative, 2nd deriva-
tive. Surface voltage =
0v.

Figure 4 c   Upper 8Vpp triangle
surface voltage. Lower output
votage / two.

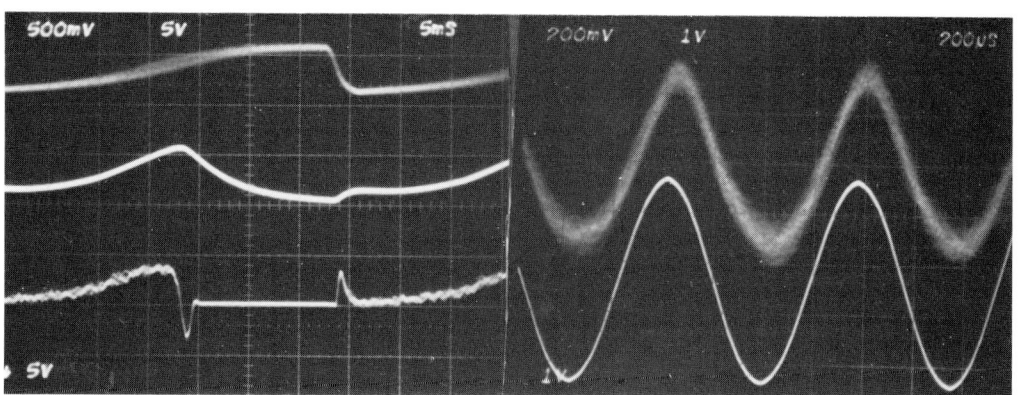

Figure 4 b from top to
bottom: "S"-curve, 1st
derivative, 2nd deriva-
tive, Surface voltage =
+1.5V.

Figure 4 d Upper sinewave
output voltage. Lower sinewave
surface voltage.

was 2kHz and the modulation is barely visible on the "S"-curves of figures 4.a and b. Therefore, it is believed that by accurately aligning the filters and optimising the modulation depth, the system will eventually be able to track the minimum process clock frequency of 5kHz.

5. Conclusions

A voltage contrast instrumentation system has been developed which can track surface voltages using second harmonic suppression. The preliminary results are encouraging with a system bandwidth of 1-2kHz which is adequate for a parametric sampling system. With accurate alignment of the system filters and optimisation of modulation depth an improvement in signal-to-noise ratio and system bandwidth should be attained.

References

Feuerbaum H P 1979 Proc. 12th SEM Symposium IITRI, Chicago 1 285
Flemming J P 1970 Proc. 3rd SEM Symposium IITRI, Chicago 467
Hannah J M 1974 Ph.D. Thesis University Of Edinburgh
Lukianoff G W, Wolcott S J and Morrissey J M 1981 Digest of Papers IEEE
    Int. Test Conf. 68
Macari M, Thangamuthu K and Cohens S 1982 Proc. Int. Phys. and Rel. Symp.
    163
Ostrow M, Menzel E and Kubalek 1981 Microcircuit Engineering 514
Ranasinghe D and Khursheed A 1983 Proc. Microscopy of Semiconducting
    Materials, to be published
Taylor N J 1969 Rev. Sci. Instrum. 40 6 792
Wolfgang E, Fazekas P, Otto J and Crichton G 1982 Hardware and Software
    Concepts in VLSI (Amsterdam: Van Nostrand)

Inst. Phys. Conf. Ser. No. 67: Section 9
Paper presented at Microsc. Semicond. Mater. Conf., Oxford, 21–23 March 1983

Design of a high efficiency secondary electron collector

D W Ranasinghe and A Khursheed*

British Telecom Research Laboratories, Martlesham Heath, Ipswich IP5 7RE,
Suffolk

*University of Edinburgh, Department of Electrical Engineering,
King's Buildings, Mayfield Road, Edinburgh EH9 3JL

Abstract   A computer simulation technique has been developed which by
modelling the electrostatic field distribution and plotting the electron
trajectories, enables the performance of electron collectors to be
theoretically predicted. In this paper the accuracy of simulation is
explored by comparing measured results from a known collector with the
theoretical values obtained from the program. The technique has been
applied to optimise electron collection. A new structure is described
and early results of its performance discussed.

1. Introduction

The scanning electron microscope in the voltage contrast mode is now widely
recognised as a powerful tool for design validation and fault diagnosis of
complex integrated circuits. Either the voltage contrast image (Ranasinghe
et al 1978), or more recently by using a modified electron collector,
(Ranasinghe et al 1981, Wolfgang et al 1979), waveforms of signals within
the circuit are displayed to enable faults to be located. Fundamental
to the performance of these diagnostic systems is the efficiency of the
electron collector used since the emission of secondaries from the
specimen is often extremely low, particularly where beam blanking has been
used to observe high frequency operation. Although a number of practical
solutions to the problem of electron collection have been suggested with
varying degrees of success (Feuerbaum 1979, Fujioka et al 1981, Gopinath
et al 1977, Lukianoff et al 1976 and Ura et al 1982), little theoretical
justification has been presented for the configurations described. This
paper attempts to rectify this omission by applying computer simulation
techniques to predict the performance of such structures. By solving
the electrostatic field distribution of the electrode assembly, the
electron trajectories through the field can be mapped. Hence, by
considering a range of typical secondary electron energies and launch
angles the fraction of secondaries transmitted which ultimately reach the
scintillator can be calculated, and the S-curves predicted. In this way,
various electrode spacings and layouts can be explored to derive an
optimum design for voltage contrast applications.

In this paper, the effectiveness of the simulation method is first
demonstrated by comparing theoretical with practical results taken on a

prototype Lintech SE-collector, (Plows et al 1981), incorporated in the BTRL SEM system. Application of the technique to the design of a new collector is then detailed and early measurements of its performance discussed. The transport efficiency calculated by simulation is the fraction of electrons which reach the collector from the total emitted by the specimen. In the experimental results measured, collected current has been displayed. Clearly there will be a direct correlation between transport efficiency and collected current, although how absolute values relate is not known since a precise relationship between measured current and collected secondaries is not known.

2. The Existing Collector

The prototype electron collector is shown in Fig. 1 and consists of 3 electrodes. The first electrode is used to extract the electrons from the specimen surface, the second is used to filter the electrons according to their energy. A suppressor grid is also incorporated to prevent backscatter electrons. The minimum working distance achieved using this structure is 25 mm from the final lens of the SEM. Thus the minimum probe diameter obtained is of the order of 2 μm. In this study the effect on transport efficiency for changes in the voltage on the extraction grid, retarding grid (or filter grid) and on the specimen are considered. Fig. 2 shows the dependence on retarding grid voltage with the specimen voltage changing from -5V to +10V and the extraction grid held at 1kV. The first feature to note is that the maximum transport efficiency is only 70%. Secondly the integral of the secondary electron spectrum or 'S-curve' expected for a retarding grid analyser has not been obtained, and the collection efficiency falls off for positive retarding grid voltages

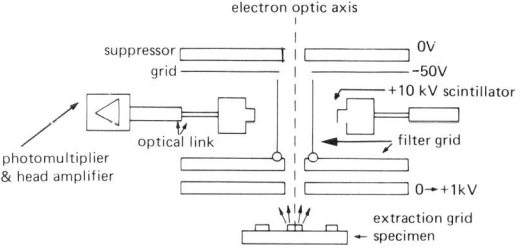

Fig. 1   Prototype Electron Collector

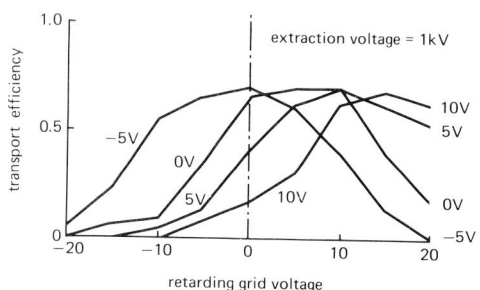

Fig. 2   Transport Efficiency Vs. Retarding Grid (Filter Grid) Voltage for Change in Specimen Voltage

rather than asymptoting to a maximum value. The experimental results obtained from measurements on the collector are shown in Fig. 3. Good agreement has been achieved between the predicted trends and those observed in practice. In Fig. 4, experimental and theoretical plots of measured current and transport efficiency have been superimposed for the different extraction voltages with the retarding grid and specimen voltage held at 0 volts. Again the observed behaviour of the collector has been reproduced by the simulation program. The maximum efficiency achieved under these conditions was 60%, with a marked deterioration as the extraction voltage is reduced below 600 volts.

To establish the reasons for the low collection efficiency it is necessary to study the electron trajectory plots for the complete range of SE-energies. However, the 3 eV energy level case illustrate the problem. The first trajectory plot, shown in Fig. 5, shows the electron paths for 1 kV extraction. A good collection efficiency is achieved. When the extraction voltage is reduced the secondary electrons are internally scattered, as shown in Fig. 6, and the number which ultimately reach the collector is significantly reduced. Similar results are obtained when other typical electron energies are analysed. For completeness, the transport efficiency plot for different retarding grid voltages with 300 volts on the extraction grid is shown in Fig. 7. Although some improvement can be obtained in efficiency with high retarding grid voltages, the overall performance is poor over the complete range of voltages considered. The final trajectory plot shown in Fig. 8 is for 50 eV electrons. It is seen that all the secondary electrons are deflected away from the scintillator. This is to be expected for high energy backscattered electrons.

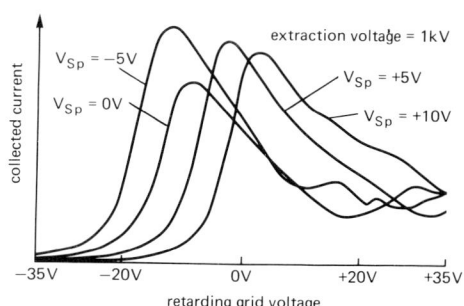

Fig. 3 Collected Current Vs. Retarding Grid (Filter Grid) Voltage for Change in Specimen Voltage

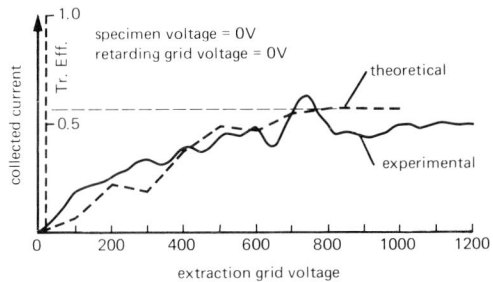

Fig. 4 Efficiency Curves at 1 kV Extraction

3. The New Collector Design

Having established that the trends were satisfactorily predicted for the prototype collector, an alternative assembly offering improved performance according to the simulation studies was constructed as shown in Fig. 9. The structure consists of 4 electrodes, an extraction grid, a retarding grid and 2 cylindrical focusing grids. The mimimum working distance achieved in the prototype is 16 mm from the final lens. As before, Fig. 10 shows

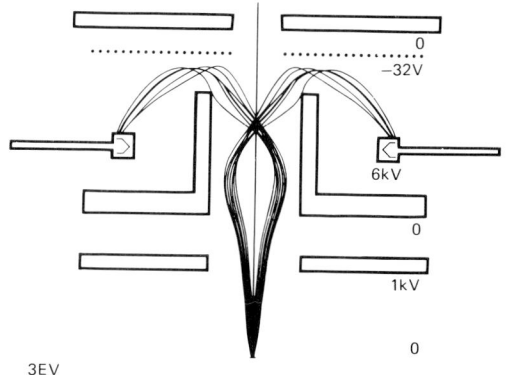

Fig. 5 3 eV Electron Trajectories

the relationship between transport efficiency and retarding grid voltage for different voltages on the specimen with 600 volts on extraction grid.

In this case the integral of the secondary electron spectrum or S-curves has been obtained, with no fall-off in efficiency with increasing retarding grid voltages. Furthermore, the shift is linear, with change in specimen voltage, and the maximum transport efficiency is now 99%. The corresponding experimental results are shown in Fig. 11 and confirms the improved behaviour indicated in the predictions. Even with 600 volts extraction voltage, excellent S-curves have been produced with linear shifts observed over the complete range of specimen voltages considered.

A typical electron trajectory plot for low energy electrons is shown in Fig. 12. The 3 eV electron trajectory plot and the overall transport efficiency has been shown to illustrate the collector performance; it is seen that virtually all the electrons emitted ultimately reach the scintillator. Also shown in the Figure is the typical secondary electron spectrum for 0 to 12 eV electrons. Superimposed on this curve is the fraction of those electrons which are collected. The 2 curves are virtually indistinguishable over the full range of energies indicating that the efficiency is maintained over the whole spectrum. The final Figure shows electron trajectories at 50 eV. As would be expected for the spectrum of Fig. 13, few of these electrons constitute the final collected current. Hence, a high rejection of backscattered electrons has been achieved.

4. Conclusions

To assess the validity of a computer simulation program, the performance of the prototype Lintech electron collector in the BTRL-SEM system has been analysed. Good agreement between the predicted and measured results was obtained. A number of shortcomings in the behaviour of the

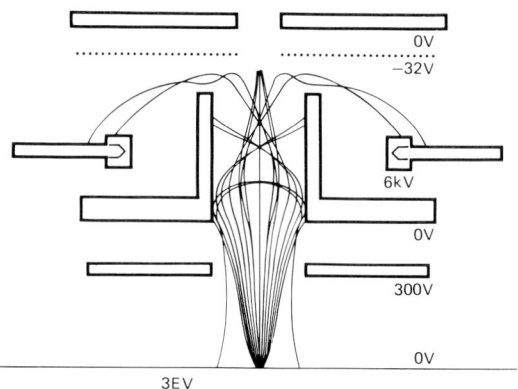

Fig. 6   3 eV Electron Trajectories. At 300 V Extraction

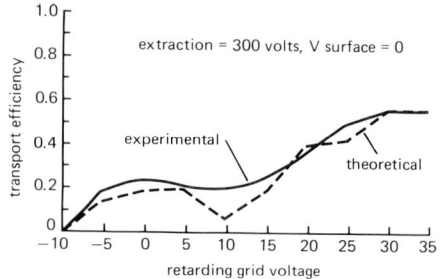

Fig. 7   Efficiency Curves. At 300 V Extraction

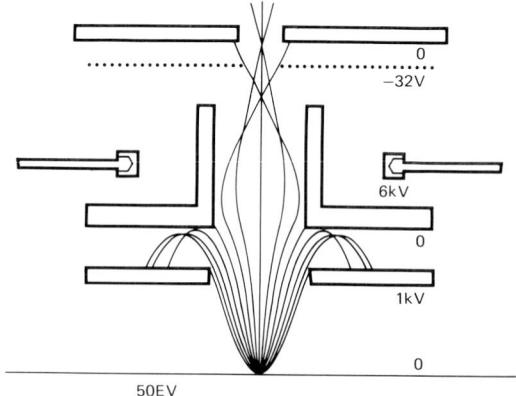

Fig. 8   50 eV Electron Trajectories

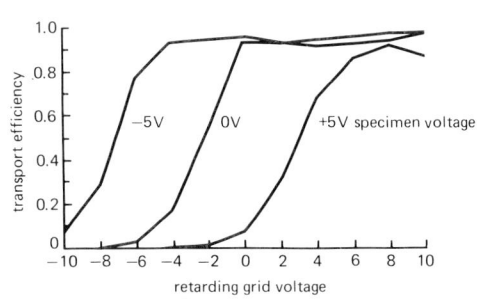

Fig. 9   The New Collector

Fig. 10   S-Curves for the New Collector

prototype was highlighted. In particular, poor collection efficiency at low extraction fields was described, and the voltage measurement relied on the displacement of a non-ideal S-curve and not by the displacement of the integral of the secondary electron spectrum. A modified collector configuration designed using the CAD techniques has also been assessed. Again the theoretical results were confirmed by measurements on the new design. Transport efficiencies approaching 100% over the secondary electron energy range of 0 to 12 eV was observed with good rejection of high energy backscattered electrons. Further uniform S-curves were produced which were displaced linearly as the specimen voltage was changed. An additional advantage of the new design is the reduced size, enabling shorter working distances, thus enabling reduced electron probe diameters to be used.

extraction = 600V

specimen voltage 0~7V in 1 volt steps

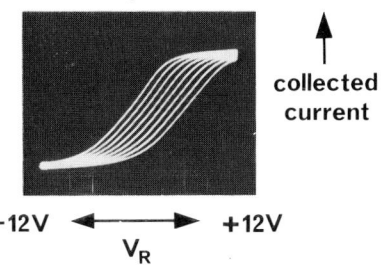

Fig. 11   Experimental S-Curves for the New Collector

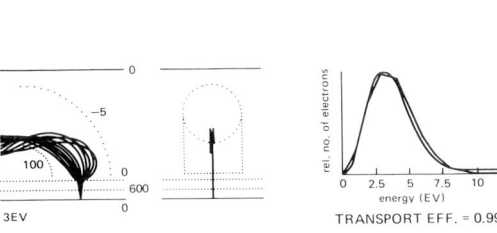

Fig. 12   3 eV Electron Trajectories and and Overall Transport Efficiency of the New Collector

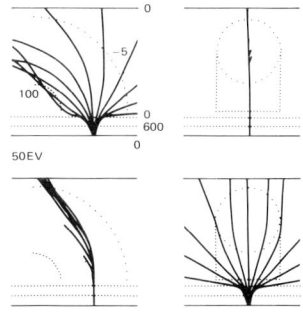

Fig. 13   50 eV Electron Trajectories for the New Collector

References

Feuerbaum H P 1979 Scanning Electron Microscopy Vol 1 (IL, USA: SEM Inc)
    pp 285-296
Fujioka H, Nakamae K and Ura K 1981 Scanning Electron Microscopy Vol 1
    (IL, USA: SEM Inc) pp 323-332
Gopinath A and Tee W J 1977 Journal of Physics E 10 6 pp 660-663
Lukianoff G V and Touw T R 1976 Voltage Contrast Detector for Scanning
    Electron Beam Instrument US Patent 3 961 190
Plows G W 1981 Lintech Instruments Ltd., Cambridge Science Park,
    Milton Road, Cambridge CB4 4BN, Lintech Sampling Electron Beams,
    Technical Brochure
Ranasinghe D W, Proctor G and Speight J D 1978 Digest of Technical Papers,
    ESSCIRC 78, Amsterdam pp 169-170
Ranasinghe D W, Proctor G and Richardson N 1981 Proceedings of Microcircuit
    Engineering 81, Lausanne, Switzerland (Swiss Federal Institute of
    Technology, Lausanne, Switzerland) pp 522-536
Ura K and Fujioka H 1982 Japan Annual Reviews in Electronics, Computers and
    Telecomms Vol 1 Semiconductor Technologies pp 308-327
Wolfgang E, Lindner R, Fazekas P and Feuerbaum H P 1979 IEE Trans Electron
    Devices ED-26 4 pp 549-559

Inst. Phys. Conf. Ser. No. 67: Section 9
Paper presented at Microsc. Semicond. Mater. Conf., Oxford, 21–23 March 1983

439

Electrically active defects in Si photodetector devices

M Lesniak and D B Holt

Department of Metallurgy and Materials Science, Imperial College of Science and Technology, London SW7 2BP.

Abstract   A significantly improved conductive mode detection system was built, calibrated and used in studies of Si photodetectors.   Micro-plasmas were seen in noisy reach-through avalanche photodiodes (APDs) under large reverse biases.   Misfit dislocation networks, precipitates and diffusion-retardation effects were observed and related to the device characteristics of the APDs.   Defects in small photodiode arrays and in large CCD arrays were also studied.

1. Introduction

It was argued previously (Holt, 1981) that purpose-built conductive mode detection systems (CMDSs) were needed for quantitative charge collection (CC) studies in SEMs.   Six types of CC signal can be interpreted to obtain values of semiconductor materials and device parameters.   The bulk and the barrier electron voltaic effect (beve) can both be detected as interpretable short-circuit currents, I_{sc} or open-circuit voltages V_{oc}.   The β-conductivity effect can be detected either as a varying voltage, V_β, with constant-current biasing or as a varying current, I_β, with constant-voltage biasing. Unambiguous, calibrated detection of one specific signal from this set is the essential requirement of the design of a CMDS.   A first design of such a detection system was reported by Unvala et al (1981).

2. The New Conductive Mode Detection System

As a result of experience with the first CMDS and the advances in the integrated circuits commercially available, a "mark II" modular CMDS, shown schematically in Figure 1, was built and calibrated.

There are two head amplifiers in a copper box inside a port of the SEM specimen chamber.   One detects beve V_{oc} signals.   The other detects beve I_{sc} (EBIC: "electron beam induced current') signals.   Interchanging the specimen and the feedback resistor across the first operational amplifier in this circuit enables it to detect constant-current β-conductivity, V_β, signals.

System noise was minimized by the selection of low-noise components, by firm mechanical mounting of components and printed circuit boards and by shielding each module in a separate metal box.   The provision of numerous controls on each module ensured convenient and versatile operation.   The power supplies and specimen biasing module uses rechargeable batteries.   This avoided a.c. ripple and allowed floating, high-voltage (up to 700V) biasing of specimens.   The rack-mounted module provides additional amplification, simple signal processing, digital multimeter monitoring of signals and operating parameters and a computer interface.   Measurements of all forms of signal in all sensitivity ranges were calibrated to give accuracies. of 1% or better.

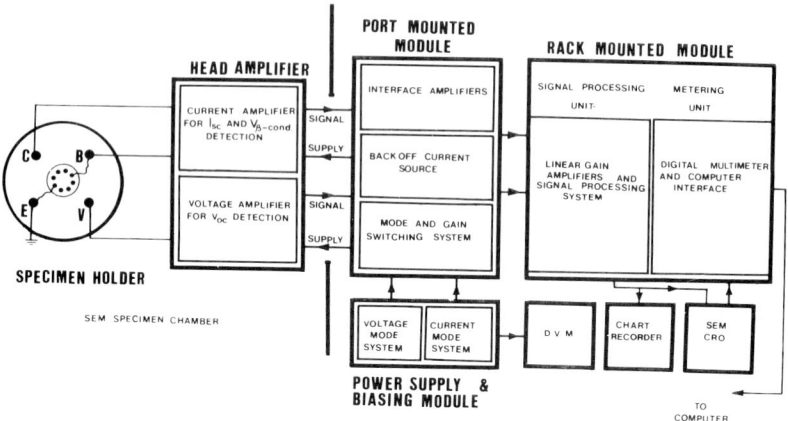

Figure 1.   Schematic diagram of the second modular conductive mode detection system.

3.   Defects in Reach-Through Avalanche Photodiodes

Reach-through avalanche photodiodes (APDs) are large discrete Si devices. Photons are detected as current pulses produced by the barrier photovoltaic effect.   SEM beve I_{sc} ("EBIC") observations are therefore directly relevant to device operation.   The structure of the APDs is shown in Figure 2(a).

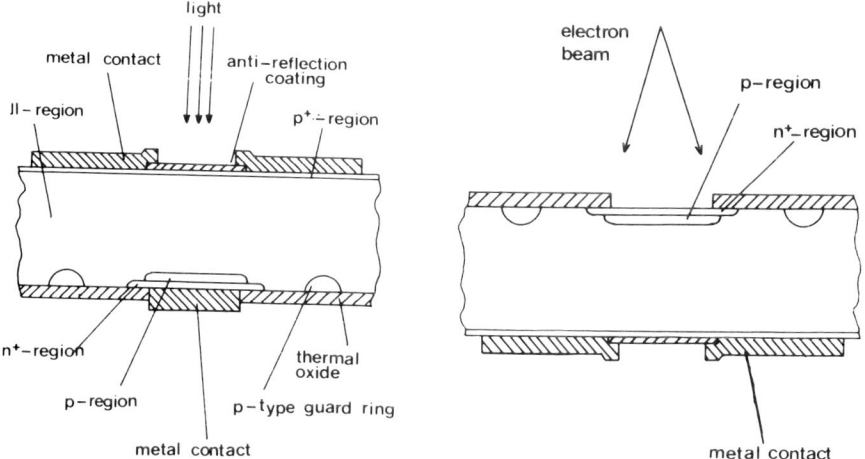

Figure 2.   Reach-through avalanche photodiode (a) as used to detect near infra-red light and (b) inverted for SEM studies of the active p-n+ junction region.

Under reverse biases of hundreds of volts the depletion region at the p-n$^+$ junction widens until it "reaches through" the whole thickness of the slice i.e. through the π region. The large field then produces hole-electron pair "avalanches" from each initial pair produced by the absorption of à photon. Gains of up to about 100X can be obtained. This requires fields which can only be sustained if the device is defect-free.

A number of APDs that exhibited premature, noisy breakdown were studied. Nothing was seen when the top surfaces of the devices, oriented as in Figure 2(a), were scanned. Some devices were supplied inverted as in Fig. 2(b), so the active p-n$^+$ junctions could be examined. Figures 3(a) and (b) are beve I$_{sc}$ micrographs of such an APD taken under conditions of low and high reverse bias. A dense network of slip lines can be seen in (a) and a number of microplasmas appear in bright contrast in (b). When lower densities of dislocations were present, more numerous microplasmas were found as in Figure 4(a).

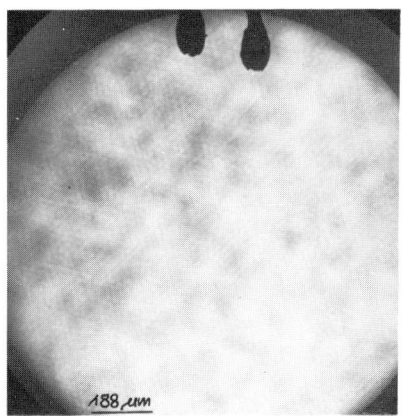
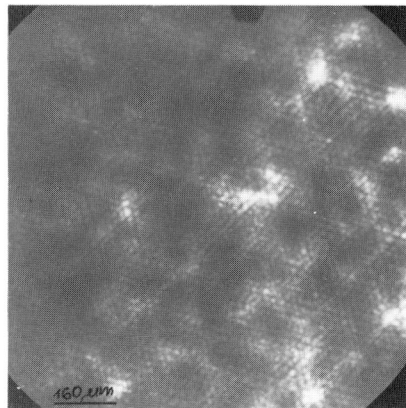

Figure 3. SEM CC micrographs of the p-n$^+$ junction of an inverted APD at biases of (a) 0V and (b) 312V.

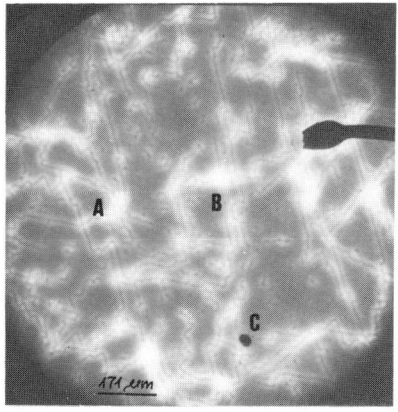
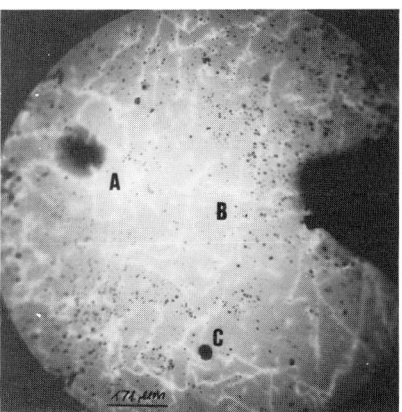

Figure 4. (a) An SEM CC micrograph of another inverted APD at a reverse bias of 300V. (b) An SEM CC micrograph of the same device after thinning to remove the misfit dislocations and precipitates as shown in Figure 6.

Varying the SEM beam voltage and hence the electron penetration range, the depth of the diffusion-induced misfit dislocations introduced by slip and seen in Figures 3(a) and 4(a) was found to be 2.4μm. Electropolishing from below followed by ion beam thinning from above produced specimens transparent in a 1 MeV TEM and containing the dislocation network. Figures 5(a) and (b) show typical areas. It was confirmed that the premature breakdown micro-plasmas occurred at precipitates and that the dark line defects in Figures 3(a) and 4(a) corresponded to misfit dislocation networks.

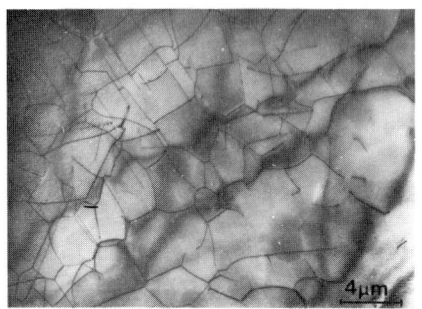 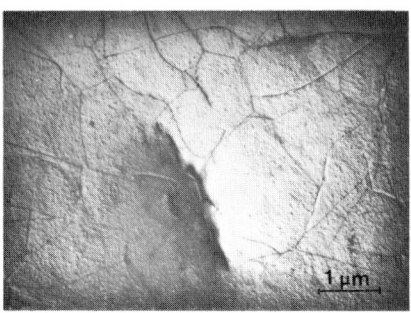

Figure 5. TEM micrographs of the misfit dislocation networks in (a) an area of high dislocation density and (b) an area with a precipitate near the centre.

By thinning down more than 2.4μm from the top, SEM specimens of the form shown in Figure 6 were produced. In the absence of the misfit dislocation networks and precipitates, the SEM CC micrographs appeared as shown in Fig. 4(b). Comparison with Figure 4(a) shows that the line patterns have the same geometry. However, in Figure 4(b) the central dark line contrast is gone leaving only the bright lines. It is concluded that defect-retardation of the diffusion of the n^+ dopant occurred. This produced raised lines in the $p-n^+$ junction as indicated in Figure 6. These in turn increased the effective charge collection by the junction for certain ranges of beam volt-age. Defect-retarded diffusion was first reported in Si bipolar transistors by Ashburn and Bull (1979).

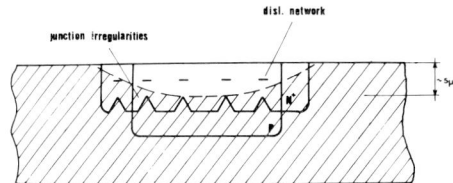

Figure 6. Schematic diagram of an APD thinned from the top only, to remove the layer containing the dislocation network and reveal the dislocation-retarded diffusion irregularities in the $p-n^+$ junction.

Thus the contrast due to the dislocation networks corresponds to locally en-hanced recombination as generally observed (e.g. Donalato 1981) and the lin-ear white-black-white contrast in for example Figure 3(b) and 4(a) is due to two separate mechanisms.

The relation of the microplasma CC current peaks to device noise was directly observed in I_{sc} line scan traces like those of Figure 7. The CC current varied with beam position but had well-defined, noise-free values for 360V reverse bias. At 365V bias, however, except at the highest microplasma peak, the signal became very noisy. 365V was the value at which the first of the premature, noisy breakdown steps occurred in the reverse current-voltage characteristic of this device in the dark. Corresponding phenomena were observed at voltages corresponding to the other breakdown voltages of this device.

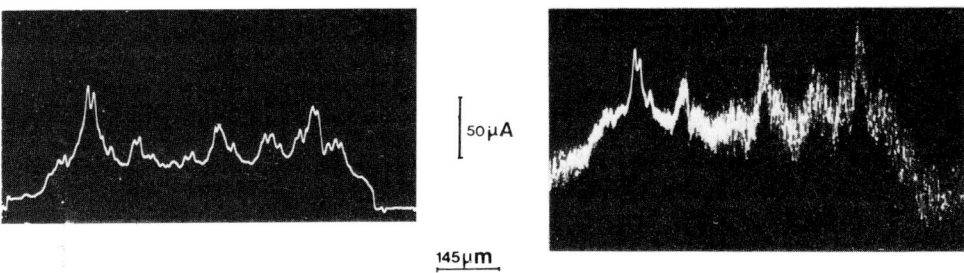

Figure 7. Barrier electron voltaic effect line scan traces recorded across a number of lines and microplasmas, as in Figures 3(b) and 4(a), for reverse biases of (a) 360V and (b) 365V.

Such direct correlations as these could not be made without the facilities of the mark II CMDS, although observations like Figures 3(a) and (b) have been made (e.g. Varker 1971 on Zener diodes). For a review of the literature on defects and p-n junction leakage see Holt (1979).

4. Defects in Small Photodiode Arrays

Several arrays of small numbers of photodiodes produced by several different processing procedures were examined. SEM CC micrographs (beve I_{sc} signals) showed up two different types of electrical inhomogeneities. Bright spots around the peripheries of the rectangular photodiodes occurred for all processing procedures. These were identified as due to small p doped pipes due to pin holes in the oxide during diffusion or implantation. A second type is shown in Figure 8. At low reverse biases these spots appeared dark, at higher biases they exhibited radial variation from bright to dark contrast. They were found to be due to larger holes in the oxide during an earlier diffusion producing phosphorus pipes extending below the junction. Such processing inhomogeneities are commonly found in rejected Si devices, rather than any subtle crystallographic defects.

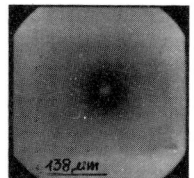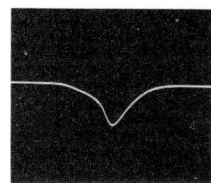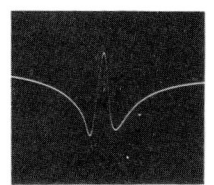

Figure 8. A round electrical inhomogeneity in a diffused p-n photodiode. (a) beve I_{sc} micrograph. The variation of I_{sc} across such a defect for reverse biases of (b) OV and (c) 30V.

5. Defects in Large CCD Si TV Sensors

These devices contain about half a million photodiodes in two square arrays in an area of 0.5 cm x 1 cm. Each diode is crossed by three polysilicon electrodes. These are used to transfer the detected charge in CCD (charge-coupled device) fashion from the quarter-million diode picture-sensing array in one 0.5×0.5 cm^2 area to another on which light is not allowed to fall. Here the video signal is read out while the next "frame" is being generated by the light falling on the exposed half of the total array. (Fig. 9).

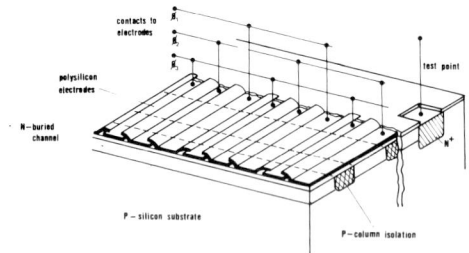

Figure 9. Schematic diagram of the structure of the CCD arrays. Each unit is crossed by 3 polysilicon electrodes at right angles to the inter-channel-isolation p-regions. CC current was taken via the test point and the bottom of the device. All the polysilicon electrodes were shorted together.

Exposure to a TV test card pattern and examination of the picture on a TV monitor is carried out at the GEC Hirst Research Centre.   A number of arrays with tabulated lists of TV defects such as black or white lines or black/white dots on the picture, were supplied.   Again most (but not all) could be accounted for by obvious ("cosmetic") processing defects.   These include the physical breaks in the long polysilicon electrodes in Figure 10.

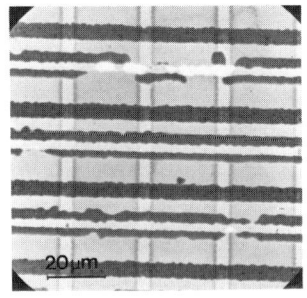 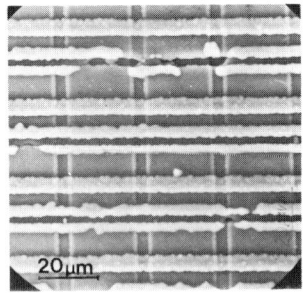

Figure 10. (a) CC and (b) secondary electron SEM micrographs of a small area of a Si CCD array. The channels run vertically and the polysilicon electrodes run in trios, horizontally.

The challenge in the case of these devices is presented by their size and complexity.   To speed analysis image processing techniques such as spatial filtering are being used (Holt et al 1983).

Acknowledgements

Thanks are due to Mr. A. Lucas of Centronic Ltd. for the supply of the APDs and small photodiode arrays.   We are also grateful to Mr. D. Burt of the GEC Hirst Research Centre for giving us the large CCD photosensors.

References

Ashburn, P. and Bull, C.J., 1979, Sol. State Electron. 22, 105.

Donalato, C., 1981, Microscopy of Semiconducting Materials, 1981 ed. A.G. Cullis and D.C. Joy, Conf. Series No. 60 (Bristol: Inst. Phys.)

Holt, D.B., 1979, J. de Physique Colloque C6, Vol. 40, pp.C6-189 to C6-199.

Holt, D.B., 1981, Microscopy of Semiconducting Materials, 1981, ed. A.G. Cullis and D.C. Joy. Conf. Series No. 60 (Bristol: Inst. Phys.)pp.165-178.

Unvala, B.A., Lesniak, M., Regan, B. and Holt, D.B., 1981, Microscopy of Semiconducting Materials, 1981, ed. A.G. Cullis and D.C. Joy, Conf. Series No. 60, (Bristol: Inst. Phys.)pp.223-228.

Varker, C.J., 1971, Proc. 9th Annual Conf. Reliability Physics, (New York: IEEE) pp.155-162.

Inst. Phys. Conf. Ser. No. 67: Section 9
Paper presented at Microsc. Semicond. Mater. Conf., Oxford, 21–23 March 1983

High resolution lithography and the role of secondary electrons

David C Joy

Bell Laboratories, Murray Hill, NJ 07974

Abstract   A Monte Carlo simulation including fast secondary electron
production has been used to determine the spatial resolution limits of,
and optimum exposure conditions for, electron lithography in polymer
resists.

1. Introduction

As the size of microcircuit elements continues to decrease it is important
to determine what factors limit the width of the finest line that could be
fabricated using electron lithography and conventional polymer resists,
and to attempt to find the optimum exposure conditions to realise this
resolution. Since the line profiles in the exposed and developed resist
follow the contours of equal absorbed energy density the necessary first
step is to model the interaction between the incident electrons and the
resist. This paper describes a new Monte Carlo simulation of the beam
interaction process which includes the effects of fast secondary electrons
(FSE). It is demonstrated that resolution and dose figures predicted in
this way are in excellent agreement with recent experimental results. The
model is then used to determine optimum exposure conditions for the
production of fine line (less than 1000A in width) conductors on bulk
substrates of Si and GaAs.

2. Theory

Conventional analytical, or Monte Carlo, models of electron beam solid
interactions assume that the incident electrons are elastically scattered
but continuously transfer energy to the solid through a variety of
processes represented by some mean ionization potential J in the Bethe
cross-section. However, it was recently pointed out by Murata et al (1981)
that the incident electrons can also scatter inelastically and lose energy
in knock-on collisions which produce fast secondary electrons (FSE) with
energies up to half that of the beam. These FSE are significant in the
study of high resolution lithography for three reasons. Firstly, PMMA and
other polymer resists can be exposed by electrons with as little as 5eV
energy so secondaries cannot be ignored. Secondly, the rate of energy
deposition (eV/cm) from an FSE is typically 10 to 50 times greater than
that from the incident electron because of the lower energy of the
secondary. So although the fractional yield of FSEs is only a few percent,
their contribution to the energy deposition profiles can be very
substantial. Thirdly, as a result of the mechanics of the knock-on
collision the FSEs leave with trajectories which are nearly normal to the
incident beam direction. Since the mean free path of an electron with a
few keV of energy is several hundred A it is clear that this could lead to

a resolution limit larger than that predicted from a consideration of the trajectories of the primary electrons alone.

A Monte Carlo simulation which follows the trajectories and energy deposition of both the incident electrons, and the FSE they produce, has been constructed and used to model the lithographic process. The cross-section for FSE productions used here was the classical one given by Evans (1955)

$$\frac{d\sigma}{d\Omega} = \frac{\pi e^4}{E^2}\left[\frac{1}{\Omega^2} + \frac{1}{(1-\Omega)^2}\right] \tag{1}$$

where e is the electron charge, E is the incident electron energy, and $\Omega.E$ is the energy of the FSE. The energy loss dE along a segment dS of the primary electron trajectory is

$$-\frac{dE}{dS} = \Sigma_i N_i Z_i \int_{\Omega_c}^{0.5} E\Omega.\frac{d\sigma}{d\Omega}.d\Omega \tag{2}$$

where N_i is the number per unit volume of atoms Z_i. The lower integration limit Ω_c avoids an infinite cross-section at zero energy and also removes from consideration very low energy secondaries which contribute little to the energy deposition but consume large amounts of computer time. The calculated energy deposition is not sensitive to the choice of Ω_c, so here a value of 0.01 was used. The FSE were assumed to lose energy by the usual Bethe process

$$-\frac{dE}{dS} = \frac{2\pi e^4}{E}.\Sigma_i N_i Z_i . \ln\left(\frac{1.166\ E}{J}\right) \tag{3}$$

where J is the mean ionisation potential and E is the instantaneous FSE energy. Tertiary production of secondary electrons by the FSE was not included although this would not be difficult to do. The inelastic scattering at knock-on causes a deflection α of the incident electron

$$\sin^2\alpha = 2\Omega/(2 + t - t\Omega) \tag{4}$$

where t is the kinetic energy of the incident electron in units of its rest mass (511 keV). For a 100keV incident electron producing a 2keV FSE α is about one degree. The FSE, however, leave the impact at an angle β given as

$$\sin^2\beta = 2(1 - \Omega)/(2 + t\Omega) \tag{5}$$

so the 2keV secondary is travelling at about 80° to the incident direction.

The fractional yield of FSE is determined by the ratio of the total scattering cross-section (i.e elastic and inelastic) to the elastic cross-section, and is typically between 1 and 5 percent. The operation of the program is such that when a primary electron is determined, by a random number call, to have produced an FSE the tracking of the primary trajectory is suspended and the FSE is followed until it escapes or falls below 100eV in energy, at which point the primary trajectory is restarted. The procedures as described here were incorporated into a single scattering Monte Carlo program (Joy 1982), written in BASIC, and run on an APPLE microcomputer equipped with a hardware arithmetic processor.

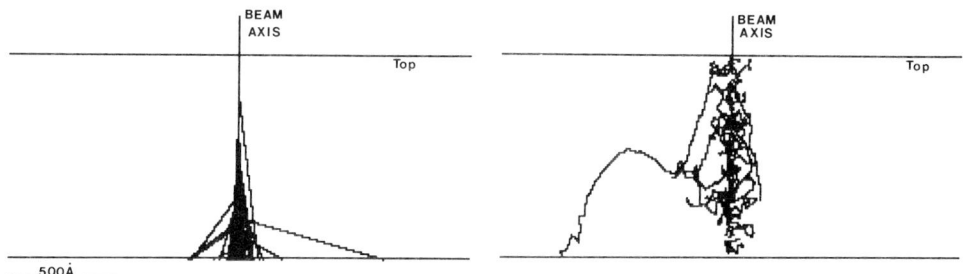

Figs.1 and 2 Typical trajectories for (1) 100keV primary and (2) fast secondary electrons in 1000A PMMA

Trajectories, computed at a typical rate of 6000 per hour, were displayed in real time on the monitor screen. Figures (1) and (2) compare plotted trajectories for 100keV primary electrons, and for the FSE they produce, in 1000A of PMMA. The contrast between the conical spread of the primary beam, and the roughly cylindrical spread of the FSE is clearly evident.

3. Spatial Resolution Limits in Lithography

The application of this model to the study of the spatial resolution limits of electon lithography can be illustrated by considering the idealised situation of a thin resist layer, either free standing or on a substrate of negligible scattering power, irradiated by a fine beam of high energy electrons. Figure (3) shows contours of absorbed energy density (eV/cm^3/electron) in a 1000A layer of PMMA exposed by a 100keV beam of electrons with a gaussian probe of diameter 20A and computed with this model using 250,000 trajectories. For comparison the equivalent profiles calculated assuming only elastic scattering and Bethe energy loss are also shown. It can be seen that the FSE Monte Carlo model predicts both an increase in the energy deposition, and contours which differ radically in form from those due only to the primary electrons. The contours for the FSE model are almost cylindrical about the beam axis, suggesting that reducing the thickness of the resist layer will not lead to any significant change in the line width produced after development under these conditions. This is confirmed by the results of other calculations which show that the energy absorbed within a few hundred angstroms of the beam axis varies little with either beam energy (in the range 30 to 100keV) or resist thickness (in the range 250 to 1500A).

The energy density at the exit surface of the resist resulting from a line exposure under these conditions is shown in figure (4). The absorbed energy is seen to be essentially constant (at about 800×5.10^8 eV/cm^2/el) from the beam axis out to a distance of about 60 angstroms. Simulations assuming a line source of zero width show an identical result, implying that this plateau is the result of the lateral FSE motion and that the near-axis energy deposition is dominated by these secondary processes. The line width actually achievable after exposure and development can be calculated using the threshold energy density model of Greeneich and Van Duzer (1971) which assumes the lower limit of solubility for PMMA to be 10^{22} eV/cm^3. On this basis the line width would fall rapidly as the exposure dose is reduced, and the narrowest line would

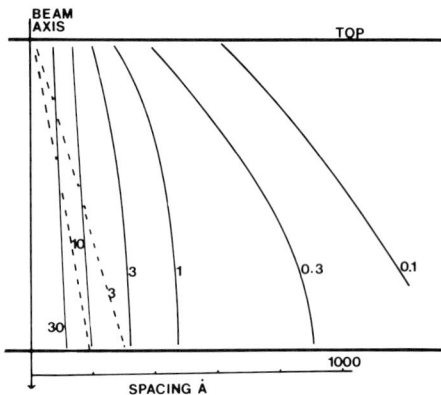

Fig.3 Energy contours in 1000A PMMA exposed by 100keV beam and 20A probe computed by Monte Carlo program with FSE production. The dotted lines are the corresponding contours for elastic scattering only. Energy plotted in units of 10^{15} eV/cm^3/el

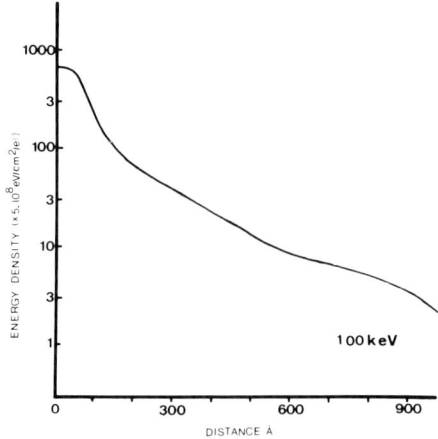

Fig.4 Absorbed energy contours at exit surface of 1000A PMMA layer exposed by 20A line source at 100 keV in energy units 5.10^8 eV/cm^2/el. Data computed from 250,000 trajectories. Note the plateau around the axis due to FSE lateral motion.

be obtained at an optimum exposure which would be that just sufficient to expose the central plateau to the threshold density. Using the data of figure (4) this is seen to require a dose of about 2.10^{-9} C/cm, at 100keV, which would result in a minimum line width of about 100 to 120A. These figures are in excellent agreement with those reported by Broers (1981) indicating that the theory is capable of accurately representing this experimental regime. At 50 keV the optimum dose is about thirty percent lower, but the minimum line width remains unchanged.

4. Fine Line Lithography on Solid Substrates

When a bulk substrate is placed beneath a thin resist layer then backscattered electrons will also contribute to the exposure dose. The effect of backscattering on high resolution lithography is readily calculated from the data obtained for a free standing resist by applying the reciprocity principle of Chang (1975). The radial distribution of absorbed energy in the resist supported on a substrate, using this principle, will be the convolution of the corresponding energy distribution in the unsupported resist and a function f(r) of the form:

$$f(r) = \left\{ \Delta_{(r=0)} + n \frac{\beta_i^2}{\beta_s^2} \exp\left[\frac{-r^2}{\beta_s^2}\right] \right\} \qquad (6)$$

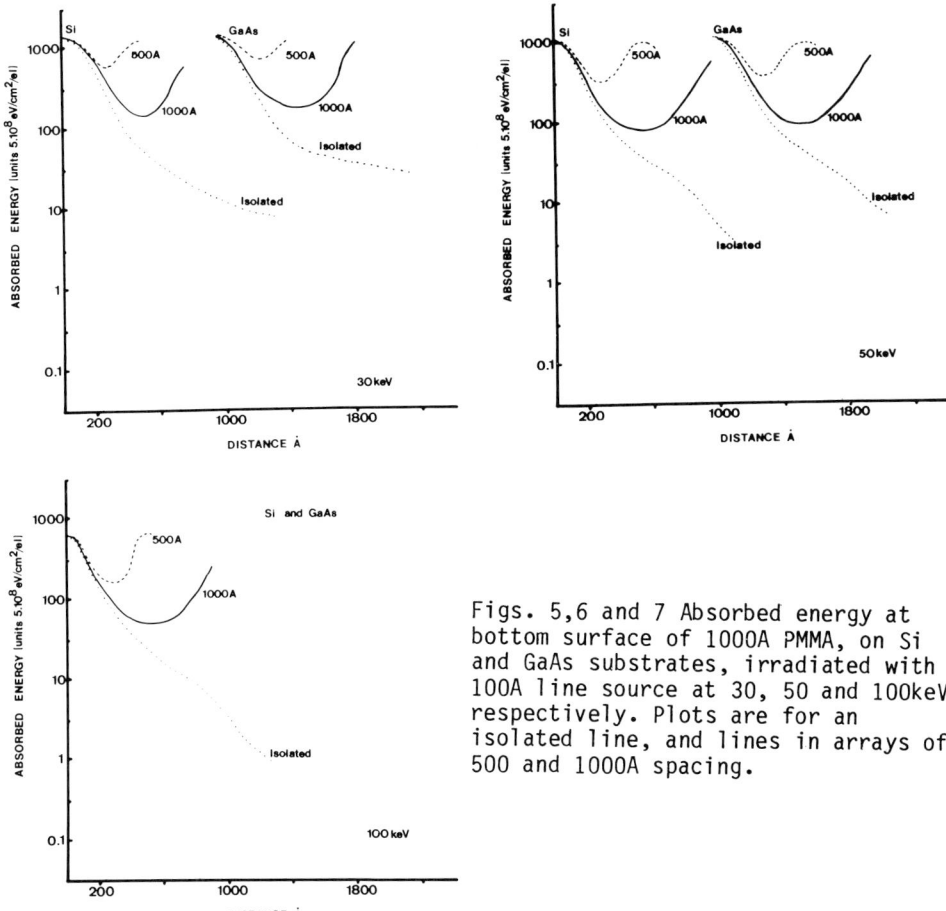

Figs. 5,6 and 7 Absorbed energy at bottom surface of 1000A PMMA, on Si and GaAs substrates, irradiated with 100A line source at 30, 50 and 100keV respectively. Plots are for an isolated line, and lines in arrays of 500 and 1000A spacing.

where Δ is a delta function at the axis, and β_i and β_s are characteristic widths for the scattering (approximated as gaussian) of the incident primary electrons in the resist and the backscattered electrons in the substrate respectively. n is an effective backscattering coefficient whose value reflects the fact that the backscattered electrons have lost energy and therefore have a higher rate of energy deposition. β_i was calculated directly from the Monte Carlo simulation, and β_s was taken to be 0.4 of the Kanaya - Okayama (1972) range. The values for n are not well established, and here estimates of 0.8 for Si, and 1.6 for GaAs, were assumed.

Figures 5,6 and 7 plot the computed absorbed energy, in a 1000A film of PMMA on Si or GaAs substrates, for either isolated lines or arrays of lines spaced 500 or 1000A and written by a 100A line source at 30, 50 or 100 keV. In all cases the periodicity of the array spacing is visible in the energy deposition, and so by choosing the exposure dose properly structures with the desired spacing could be developed although the allowable range of exposure to achieve this is seen to decrease at the lower energies and especially for the 500A spacing. The slope of the curves also decreases at low energies so increasing the exposure will lead to a more rapid increase in line width than at higher energies. Applying the threshold dose criterion the optimum dose is close to that found for the unsupported resist, varying from about 0.5 to 2.10^{-9} C/cm. The minimum line width produced for the optimum exposure dose is still of the order of 100A and, as before, reducing the probe size below 100A would affect no improvement. It would, however, also be undesirable to use a larger probe even if wider lines and greater array spacings were required. This is because the energy deposited by the FSE is localised around the incident axis and so will be spread over a larger volume if the probe is enlarged, leading to a fall in the peak energy absorbed in the written areas. Since the energy absorbed outside the patterned areas remains constant the maximum to minimum variation in the energy distribution will be reduced and the choice of exposure dose will become more critical. These predictions, including the optimum exposure dose values, are in close agreement with recent experimental results obtained by Craighead et al (1983).

5. Conclusions

Fast Secondary Electrons play a significant role in high resolution lithography. The lateral motion of the FSE, and their high rate of energy deposition, set a limit of about 100A to the attainable spatial resolution even in the most idealised conditions. However, this same effect is of value in the fabrication of structures on bulk substrates since it provides sufficient differential energy deposition between the written and unwritten areas to permit an exposure to be chosen which will only allow the written area to develop. The magnitude of the differential can be enhanced by reducing the current density of the backscattered component, i.e by increasing the energy of the incident beam and hence raising β_s, and by minimizing the size of the probe so as to concentrate the FSE energy in the patterned area. The excellent agreement between the predictions of the theory and experiment validate the assumptions of the model and should permit this approach to be used with confidence in the design of high resolution structures. Since the ideal instrument for high resolution lithography is seen to be one capable of generating a probe of the order of 100A and operating at an energy of 100 keV, a specification which is very different to that currently found in commercial beam writers, it can also be suggested that a new generation of instruments is necessary.

References

Broers A N 1981 J.Electrochem. Soc. 128 166
Chang T H P 1975 J. Vac. Sci. Tech. 12 1271
Craighead H G, Howard R E, Jackel L D and Mankiewich P M 1983 App. Phys. Lett. 42 38
Evans R D 1955 The Atomic Nucleus (New York:McGraw Hill) 576
Greeneich J S and van Duzer T 1971 IEEE Trans. Elect. Devices ED-21 286
Joy D C 1982 Proc. 40th Meeting EMSA ed G Bailey (Baton Rouge:Claitors Press) 746
Kanaya K and Okayama S 1972 J. Phys.D. Appl. Phys 5 43
Murata K, Kyser D F and Ting C H 1981 J. Appl. Phys. 52 4396

Inst. Phys. Conf. Ser. No. 67: Section 10
Paper presented at Microsc. Semicond. Mater. Conf., Oxford, 21–23 March 1983

451

Structural defects and their electronic effect in devices and unprocessed silicon

H Strunk

Max-Planck-Institut für Metallforschung, Institut für Physik,
7000 Stuttgart 80, FRG

Abstract   The paper presents examples of observations which reveal the correlation of electrical effects with crystal defects. The high pene-tration power of a high voltage electron microscope is essentially used. In present-day single crystal devices two electrical failure types are generally encountered: pipe diffusion along dislocation lines, shunting p-n junctions, and impurity decoration of dislocations, representing efficient carrier generation centres. Various mechanisms for introduc-tion of dislocations during processing are considered, some of which are very efficient in narrow devices. For the case of polycrystalline sili-con, minority carrier recombination, as measured by EBIC, and correlation with the boundary structure is briefly considered. An evaluation of ma-jority carrier transport through a boundary is shown to be possible in terms of a recent 'transistor-type' model which yields parameters on the charge distribution. Thus a new type of correlation experiment becomes feasible which is exemplified by measurements on a small-angle boundary of a bicrystal. The result suggests that the charges at the boundary are located at the observed secondary dislocations.

1. Introduction

Most of the silicon material that is processed in semiconductor industry consists of single crystalline wafers and contains, apart from desired p-or n-dopants impurities on a very low concentration level. These crystals are generally free of linear and planar lattice defects. The quality of this material can, however, frequently not be maintained during manufacturing of devices, such as integrated circuits or transistors. The various processing steps, like mask formation or dopant diffusion, and even wafer handling only can introduce crystal defects, especially dislocations, stacking faults, and precipitates, that eventually degrade the electrical characteristics of the devices. Improvement of the yield means actually to avoid the formation of such defects, at least in the sensitive device areas i.e. keeping the ini-tial defect-free state of the material.

The situation is somewhat different in polycrystalline silicon. Such mate-rials, being produced by various techniques (e.g. Dietl et al. 1981), re-ceived increasing interest in the last decade, mainly because they are con-sidered as potential material for inexpensive solar cell production. As thin films, they are also investigated for device technology application (e.g. Gibbons 1981, Keller and Timble 1982). The electronic properties of these materials are largely controlled by the electronic effects at the inherent grain boundaries. In addition, during processing, the grain boundary struc-ture and the electronic behaviour can change (Cunningham 1981, Redfield 1981, Kazmerski and Russell 1982). The basis for improving the electrical

performance of these materials is to understand minority and/or majority
carrier behaviour at boundaries. Such an understanding can be gained by
investigating individual grain boundaries and directly correlating electrical behaviour with structure. Minority carrier experiments of this type,
using the scanning microscope (SEM), have been performed for several years
starting with work on dislocations (Ourmazd et al. 1977) and, later, on
grain boundaries (Strunk and Ast 1980), whereas correlations in majority
carrier experiments were obtained only recently (Werner and Strunk 1982).

In the following, after a few remarks on the advantage of using high voltage
electron microscopes, the typical failure mechanisms of bipolar and MOS devices due to lattice defects will be considered. Emphasis will be placed
on dislocation multiplication processes that are induced or promoted by the
geometry of the devices. These are thought to become increasingly important
with the ever decreasing dimensions of the devices. The next sections consider briefly the experimental correlation between electrical (minority or
majority carriers) and structural properties of selected grain boundaries.

2. Experimental techniques
In high voltage electron microscopes, silicon specimens can be imaged
having a thickness of several μm. This thickness compares to that of the
electrically active region of devices and to that sampled by EBIC, so that
the whole volume of interest can be structurally analysed after suitable
thinning from the 'back side'. For devices a large-area thinning technique
was developed by Kolbesen et al. (1975). It permits, generally after strip-
ping of metallization and oxide-and nitride-layers, the preparation of specimens which are homogeneously thick over their whole areas (see Fig. 1).
Thus the analysis of electrically interesting features preselected on the
wafer is greatly facilitated. Under suitable conditions of surface orientations (texture), polycrystalline material can also be large-area thinned.
The availibility of such specimens might suggest to correlate simultaneously
electrical activity, as measured with EBIC, and structure by using such a
specimen in a high voltage scanning electron microscope. However, as shown
by Fathy et al (1979), radiation damage (generally not visible in silicon
by diffraction contrast) shows up in EBIC and prevents the analysis of the
electrical behaviour of original material.

3. Failures in devices
The examples are taken from investigations of bipolar devices with oxide-
well insulation (Fig. 2) (e.g. Murrmann 1976) and of MOS capacitors. Using
Si_3N_4 - masks, wells are etched into the silicon in which then SiO_2 is thermally grown to insulate the transistors from each other. Two side effects
are associated with this technique (Fig. 2):
i. During oxidation Si interstitials are injected into the silicon and can,
in presence of nuclei at the surface or in the surface-near volume, condense
into stacking faults (Hu 1974) (Oxide-induced stacking faults).
ii. The SiO_2 is formed under volume expansion by a factor of two. The resulting forces onto side walls and Si_3N_4 overhang cause eventually the activation of potential dislocation sources and the plastic relaxation of
these stresses. In MOS devices, generally no oxide wells are employed.

3.1 Typical defect mechanisms of electrical failures
The electrically malfunctioning bipolar devices can be recognized by measuring the reverse characteristics of the p-n junctions or more qualitatively by anodic decoration, which marks areas of unusually high current flow
(Hu 1977). In all cases of emitter, vase and collector. Fig. 3 shows an
extremely decorated, leaky transistor containing many such dislocations in

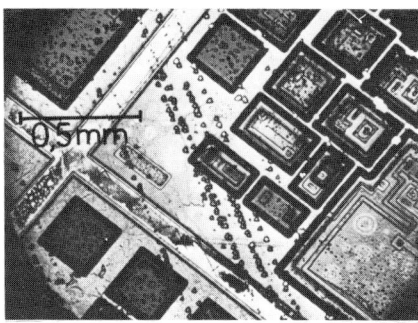

Fig. 1 Large-area specimen of a test pattern. Optical microscope in transmission. Etch pits delineate glide bands.

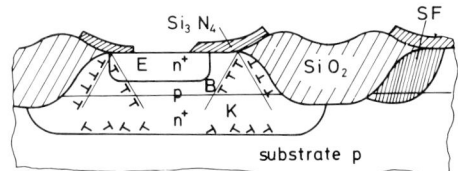

Fig. 2 Principle of bipolar technology with oxide isolation. SF: stacking fault, E: emitter, B: basis, C: collector. Dislocation symbols denote oxide-nitride-edge dislocations and misfit dislocations near the substrate/epitaxial-collector interface.

Fig. 3 Bipolar transistor, with oxide-nitride-edge dislocations, exhibiting collector-emitter-short.

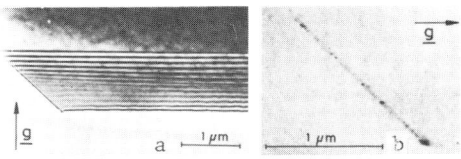

Fig. 4 Stacking fault in MOS condensor with low storage time (a), caused by microprecipitates of heavy metals at the partial dislocation (b).

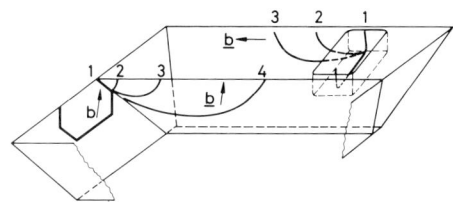

Fig. 5 Schematic of two multiplication mechanisms for oxide-nitride-edge dislocations. The planes are of $\{111\}$ type.

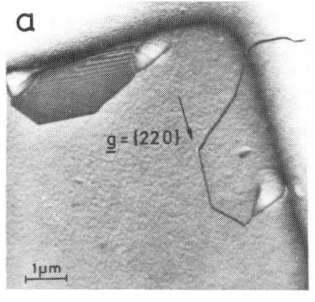

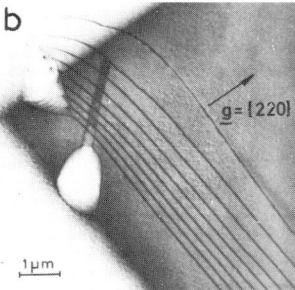

Fig. 6a Transistor with collector-emitter-short due to dislocations from stacking faults. One is unfaulted and forms a dislocation source (e.g. Fig. 5).
b:planar dislocation arrangement as expected from stacking fault induced dislocation source.

its two emitters. In many cases also oxide-induced stacking faults were observed to cause leakage (Fig. 6). Analysis showed that the emitter-base and
base-collector diodes of these transistors were generally not influenced by
these dislocations (Kolbesen and Strunk 1981). Such a behaviour can be ascribed to enhanced (pipe) diffusion of the emitter dopant along the dislocation lines into the collector producing an ohmic current path. This explanation is supported by the fact that no decoration of the dislocations
was observed which would, due to carrier generation, affect the reverse characteristics of these diodes. Pipe diffusion, of course, has also been observed to be caused by other defects joining emitter and collector such as
microtwins (Kolbesen and Strunk 1981) and needle-like precipitates (e.g.
Fig. 6b) (Augustus et al 1980). Microcracks that are sometimes formed at the
oxide-nitride edges, are always deleterious (Kranzer et al 1978).

MOS devices exhibit a different failure mechanism. MOS capacitors are used
to store information as charge. Since leakage mechanisms reduce this charge,
it must be refreshed cyclically. Capacitors that 'forget' during the refresh
time (i.e. have a too low storage time) have to be rejected. Very frequently
refresh failures were correlated with the presence of stacking faults in the
capacitors (Fig. 4). The analysis showed that in these cases at least one
partial was within the capacitor and was always visibly decorated by precipitates. Capacitors that contained only the planar fault were generally unaffected. Evidently the precipitates, containing very probably heavy metals
like Cu or Fe, act as generation centres for carriers that decharge the capacitor. This conclusion correlates with the qualitative findings that the
storage time is increased with lesser decoration (Strack et al 1979). The
harmful decoration can drastically be reduced by appropriate gettering techniques, whereby also the nuclei are reduced for stacking fault formation
(e.g. Murarka et al 1980). Thus the examples indicate these dislocations
and stacking faults cause device failures only indirectly by 'steering' dopant and impurity diffusion. Intrinsic electric effects arising for example from dangling bonds in the dislocation cores might become important in
MOS capacitors only with further decreased active volumes (and charges).

3.2 Dislocation multiplication and device geometry
Because of the ability of dislocations to multiply profusely, already the
formation of the first dislocations should be avoided in devices, in order
to increase the yield. This requires information about the dislocation
source mechanisms. Apart from 'general' dislocation sources, like surface
damage by undue handling, dislocations due to wafer deformation (e.g. Fig.1)
or punched-out by precipitates, there exist specific ones that are typical
of features in device processing steps and device geometry. Some of these
shall be considered briefly. Stacking faults, cannot , in general, multiply.
However, upon unfaulting, a perfect dislocation results that can act as a
dislocation source. Fig. 5 shows a sketch of a geometrically shaped unfaulted stacking fault. The dislocation segment at the surface that happens to
lie on its {111} glide plane may glide and multiply as a single-ended
source (Sequence 1 to 4). Fig. 6a shows a leaky transistor that contains
such an unfaulted oxide-induced stacking fault which already started to
emit a dislocation. Fig. 6b shows a planar arrangement of oxide-nitride
edge dislocations which is supposed to be created by this process. With
increasing packing density, i.e. decreasing dimensions and more complicated
structures, stress concentrations at corners and rims may become important
and cause further source mechanisms. Fig. 7 shows part of a transistor with
emitter-collector short. A contrast analysis established that the dislocations in the emitter and the misfit dislocations climbed into the collector contact had the same Burgers vectors. These misfit dislocations, al-

though generally not glissile, apparently emitted the glide dislocations.
This is possible, as sketched in Fig. 5, because of the relatively long
dislocation segments emerging steeply at the surface. These may lie in a
potential glide plane and can be activated as single-ended sources by the
stress fields around the contact volume. Another source is formed by the
misfit dislocation array near the substrate-collector interface, which is
in itself beneficial as gettering site, (e.g. Fig. 2). During emitter diffusion, the dopant concentration and thus the lattice parameter gradient
can become so high that it becomes energetically more favourable for the
misfit dislocations to move up into the emitter, which then, of course, immediately leads to a pipe. Fig. 8 shows dislocations that caused electrical
failure of this type.

4. Polycrystalline silicon, minority carrier behaviour

EBIC experiments performed on polycrystalline silicon specimens readily
suggest a qualitative classification of grain boundaries into groups of
different electrical activity, (see Fig. 9):
i. Grain boundaries with strong recombination. These are generally high-energy boundaries and their activity is thought to be mainly due to impurity
segregation, especially after heat treatment (Kazmerski and Russell 1982).
ii. Boundaries essentially inactive along their whole length. These are generally coherent first-order twin boundaries ($\Sigma=3$). An observed local activity could be correlated with curved partial dislocations in such boundaries
(Strunk and Ast, 1980), whereas straight ones, lying along $<110>$, gave no
contrast effect (Strunk et al 1981). These observations compare to Hirsch's
(1980) suggestion that kinks in otherwise reconstructed partial cores are
sites of electrical activity. It is interesting to note that straight grain
boundary dislocations in a second order twin boundary ($\Sigma=9$) were also inactive (Dianteill and Rocher 1982).
iii. Grain boundaries that change activity on changing their direction
(not Σ). The boundary A in Fig. 9b changes its contrast from zero (at a
and c) to a certain value at b (compare with trace of this boundary in Fig.
9a) Such an extreme contrast change permits, without the need for a detailed
EBIC contrast analysis, a direct correlation with microscopic structure
changes. A $\Sigma=9$ boundary was investigated this way by Dianteill and Rocher
(1982). However, as will be seen in this preliminary case, grain boundaries
selected only on grounds of their electrical properties are not always optimal for an analysis in the microscope. Fig. 10 shows TEM micrographs of
the electrically active segment (b), which is facetted on a μm scale below
EBIC resolution and the inactive (a) one, which intersects the foil surface along a straight line. Both segments contain a dislocation network of
curved dislocations. One might be inclined to attribute the electrical activity to the facetting. However, close inspection of the varying fringe
spacings in Fig. 10 a indicates that this segments is also facetted, but almost parallel to the specimen surface. Unfortunately the three-dimensional
facetting of the boundary prevented a crystallographic analysis thus far.
It is likely, however that the two types of facetting do not behave very
different electrically, so that the observed alternating behaviour might
have another reason. So far, it cannot be ruled out that the depth of the
p-n junction varies strongly due to differently enhanced dopant diffusion
along the two segments of the boudary, which would affect the EBIC signal.
This experimental uncertainty and the problem of possible impurity segregation due to the high temperature diffusion (which would require a chemical analysis) can be avoided by using Schottky diodes.

5. Grain boundary and majority carrier experiment

In this section it will be demonstrated how structural analysis may prove

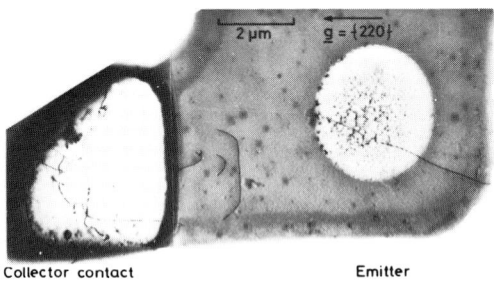

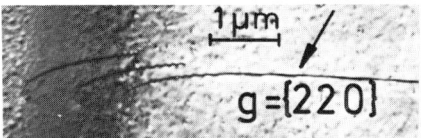

Fig.8 Dislocations that glided into the emitter from the misfit network in the epitaxial collector (e.g. Fig.2) and caused emitter-collector short.

Fig.7 Part of a transistor with collector-emitter short. The misfit dislocations in the collector contact are the source for the deleterious dislocations in the emitter area.

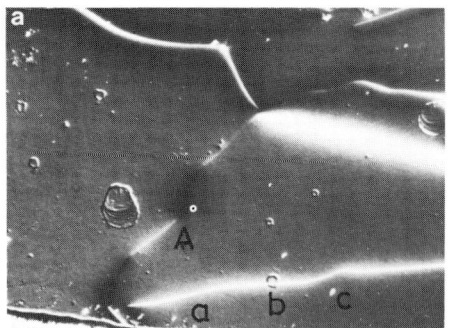

Fig. 9a Optical micrograph of a Silso-Wacker solar cell

Fig. 9b EBIC image of same area as in Fig. 9a.

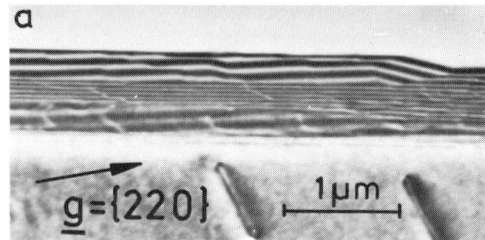

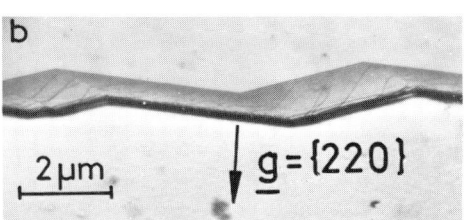

Fig.10a Electron micrograph of boundary A, segment a in Fig. 9

Fig.10b Electron micrograph of boundary A, segment b in Fig. 9

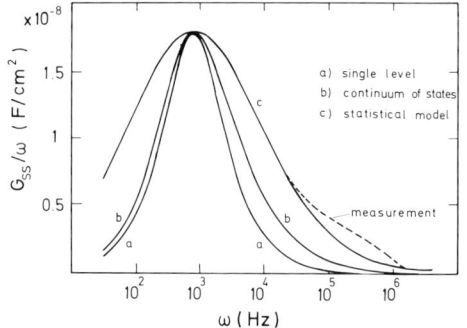

Fig.11 Measured conductance-over-ω behaviour (dashed line) in comparison to three different models:
a. single energy level
b. continuum of energy levels
c. continuum of energy levels and statistical fluctuation of potentials in space. This model coincides practically with experiment.

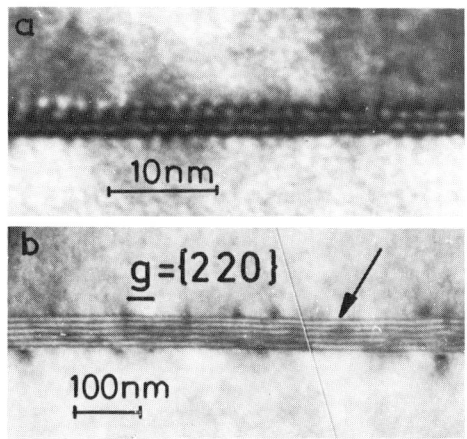

Fig. 12 High resolution micrograph (multibeam along <110> in upper grain) of grain boundary showing basic regular dislocation structure with periodicity of 1,8 nm.

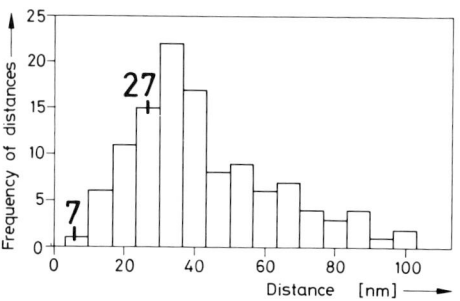

Fig. 13 Histogram of distances between secondary dislocations. The values of minimum (λ_{min}) and average distance (λ) as determined from the measured conductance behaviour (Fig. 11) are introduced.

and support models for the transport of free carriers through grain boundaries. Electrical measurements are usually described in terms of the model of a double-layer barrier, created by charge trapped in deep states in the forbidden gap. As shown by Werner and Strunk (1982) and Werner (1983), the energetical and spatial distribution of the charge at the grain boundary can be derived from an analysis of the frequency dependence of the ac-admittance. This is accomplished using a transistor-type model in which the measured current due to free carriers thermionically emitted over the potential barrier is controlled by the trapping and emission current at the boundary. Within this 'transistor'-model the conductance G_{ss} of the trapping/emission-current can be evaluated and then be analyzed using the methods of admittance spectroscopy (Nicollian and Goetzberger 1967). Measurements of a small-angle (7.8°, predominantly tilt) boundary of a CZ-grown silicon bicrystal (boron doped 10^{15}cm $^{-3}$) were evaluated. Fig. 11 shows that the measured G_{ss}/ω-behaviour can only be explained with a model using a statistical fluctuation of potentials at the boundary. From the standard deviation of the potentials (which is obtained from the broadness of the G_{ss}/ω-curve) a minimum distance of the statistically distributed charges λ_{min}= (7±2)nm can be estimated. In addition,from the total charge at the boundary an average distance λ=(27±1)nm is calculated (Werner and Strunk 1982). Fig. 12 summarizes the electron microscope observations. The boundary consists of a regular basic dislocation network (Fig. 12a) with a repetition distance of ~ 1,8 nm. As shown in Fig. 12b, this boundary also contains secondary dislocations that are irregularly distributed. A histogram of the distances between these dislocations (Fig. 13) indicates a statistical distribution. It can be seen that the minimum and average distances determined above fit well into this distribution. Thus, the secondary dislocations are compatible with the electrical characterization and have to be assumed to be the cause of the charges in the boundary.

According to this analysis, the regular basic dislocation network is electrically inactive. This might indicate a reconstruction of the dislocation
cores, which is disturbed at the secondary dislocation. Another possibility is indicated by the observation of Bourret and d'Anterroches (1982)
that dislocation cores in small-angle boundaries (<10°) are always oxygen-
decorated. This cannot be ruled out in our CZ-grown material and would indicate a passivating effect of oxygen. Such an effect was already considered
for a Σ=25 boundary by Dianteill and Rocher (1982). A chemical analysis
of the investigated boundaries is necessary for an answer to these problems. In any case, the correlation of structural and electrical properties of a boundary, as measured from majority-carrier behaviour, is a
promising technique, which can give information on a smaller scale than
EBIC.

6. Concluding remarks
In mono-crystalline silicon devices extended defects introduced during processing constitute a prominent cause for electrical malfunction. Dislocations and stacking faults act indirectly by causing pipe diffusion or by
gettering metal impurities that precipitate and form efficient carrier
recombination/generation centres. Intrinsic electrical properties do not
play a significant role. In view of the properties of polycrystalline silicon, the correlation of electrical and structural properties of the grain
boundaries becomes important. Some work performed to-date indicates the
difficulty of discriminating expected effects due to intrinsic properties
('dangling bonds') and impurity segregation. Thus the simultaneous application of spatially resolving chemical analysis techniques seems inevitable
in the future for a safe interpretation of minority and majority-carrier
behaviour at boundaries.

7. Acknowledgment
The author is indebted to D. G. Ast, B. Cunningham (Cornell University,
Ithaca), O. Hildebrand (University Stuttgart), B. Kolbesen (Siemens, Mu-
nich), J. Marek and J. Werner (Max-Planck-Insitut für Festkörperforschung
Stuttgart) for their contribution by collaboration and discussions.

References

Augustus P D, Knights J and Kennedy L W 1980 J. Microscopy 118 315
Bourret A and D'Anterroches C 1982 J. Physique Colloque 10 C1-1
Cunningham B 1982 40 Ann. Proc. Electron Microscopy Soc. Amer. ed
    G W Bailey Washington D. C. p. 434
Dianteill C and Rocher A 1982 J. Physique Colloque 10 C1-75
Dietl J, Helmreich D and Sirtl E 1981 Crystals Growth and Applications
    volume 5 ed Grabmeier J (Berlin Heidelberg: Springer) p. 43
Fathy D, Sparrow T G and Valdrè U 1979 J. Microscopy 118 263
Gibbons J F 1981 Inst. Phys. Conf. Ser. 60 431
Hirsch P B 1980 J. Microscopy 118 3
Hu S M 1974 J. Appl. Phys. 45 1567
Hu S M 1977 J. Electrochem. Soc. 124 578
Kazmerski L L and Russell P E 1982 J. Physique Colloque 10 C1-171
Keller G K and Trimble L E 1982 J. Physique Colloque 10 C1-353
Kolbesen B O, Mayer K R and Schuh G E 1975 J. Phys. E 8 197
Kolbesen B O and Strunk H 1981 Inst. Phys. Conf. Ser. 57 21
Kranzer D, Lemme R, Kolbesen B O, Mayer K R and Strunk H 1978 Rev. Phys.
    Appl. 13 803
Leamy H J, Kimerling L C and Ferris S D 1978 Scanning Electron Microscopy 1
    717

Murarka S P, Seidel E T, Dalton J V, Dishman J M and Read M H 1980
   J. Electrochem. Soc. 127 716
Murrmann H 1976 Siemens Forsch. u. Entwickl. Ber. 5 353
Nicollian E H and Goetzberger A 1967 Bell Syst. Techn. J. 46 1055
Ourmazd A, Darby D B and Booker G R 1977 Inst. Phys. Conf. Ser. 36 251
Redfield D 1981 Appl. Phys. Lett. 38 174
Strunk H and Ast D G 1980 38. Ann. Proc. Electron Microscopy Soc. Amer.
   ed E W Bailey (San Francisco)p. 322
Strunk H, Cunningham B and Ast D 1981 Defects in Semiconductors
   ed J. Narayan and T. Y. Tan (New York: North Holland)p. 297
Werner J and Strunk H 1982 J. Physique Colloque 10 C1-89
Werner J 1983 Dr.-thesis, University Stuttgart and to be publ.

Inst. Phys. Conf. Ser. No. 67: Section 10
Paper presented at Microsc. Semicond. Mater. Conf., Oxford, 21–23 March 1983

Details of microstructure and geometrical configuration of integrated circuits studied by transmission electron microscopy

H Oppolzer and V Huber

Siemens AG, Research Laboratories, Munich, FRG

Abstract TEM of thin cross sections represents a powerful tool for investigating technologic problems in IC fabrication. Examples for material problems with interconnections as well as edge configurations of
IC structures are presented. Phosphorus-doped poly-Si in contact windows to Si, the appearance of Si inclusions and protuberances in thermal oxides on poly-Si, and thermal oxidation of $TaSi_2$/poly-Si double
layers were studied. Regarding edge configurations of IC structures,
reoxidation of floating-gate transistors and technological details of
a 256K memory cell were investigated.

1. Introduction

Besides doping of the single crystal silicon in the electrically active
areas, the process steps for fabrication of integrated circuits (ICs) include the growth of thermal oxides and the deposition of conducting layers
for interconnections and also of additional insulating layers. With progressive integration the structural dimensions of ICs shrink both laterally
and vertically. The material properties and the edge configurations of the
various structured films have therefore an increasing influence on the
electrical parameters. For investigating problems of both types, i.e. material properties like microstructure or structure of interfaces on the
one hand and details in the geometrical configuration on the other, transmission electron microscopy (TEM) of thin cross sections has proven to be
a powerful tool (Sheng and Markus (1980), Henghuber et al. (1980)) since
it allows to visualize details as a function of sample depth directly.
Examples for both types of problems will be presented.

2. Experimental

In order to obtain the correct section plane when preparing a thin cross
section through an IC, the ion milling process has to be checked by observing the specimen in a light microscope (Oppolzer (1980)). The periodic
arrangement of the IC structures, e.g. memory cells, facilitates this procedure. The TEM investigations were performed at 100 kV and 200 kV beam
voltage.

3. Material problems with interconnections

3.1 Polysilicon-silicon contacts

The change in microstructure of polycrystalline silicon (poly-Si) films
with deposition temperature (Oppolzer et al. (1980)) and with phosphorus-
doping (Falkenberg et al. (1979), Oppolzer et al. (1981)) has been studied

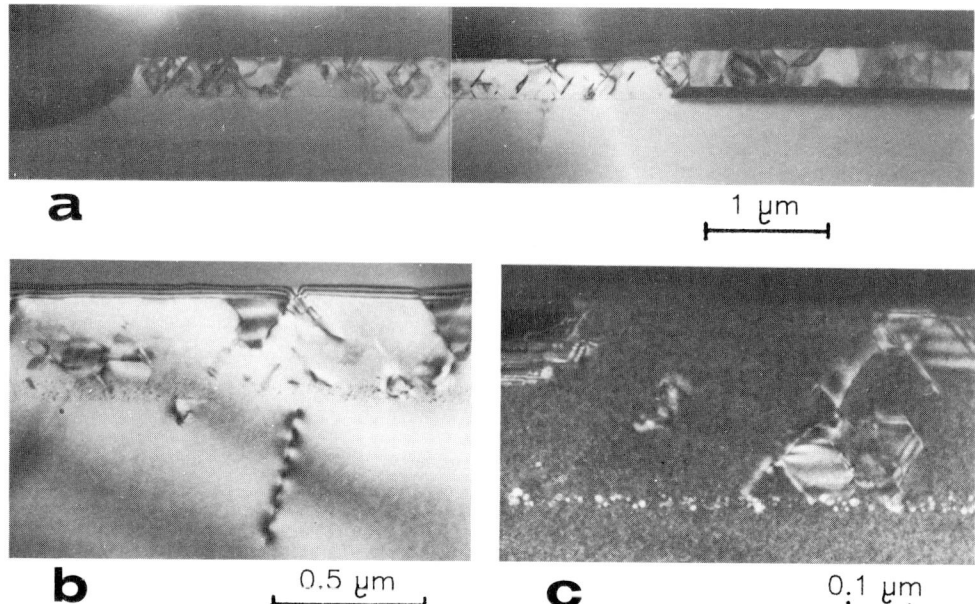

a

1 μm

b

0.5 μm

c

0.1 μm

Fig. 1: Cross section through contact window of phosphorus-doped poly-silicon to Si substrate (a). Interface tilted in bright field image (b) and weak beam image (c).

in detail. In contact areas of poly-Si on the single crystal silicon the poly-Si film is deposited on the native oxide of the Si substrate. At low annealing temperatures after phosporus diffusion the about 2 nm thick native oxide remains intact (e.g. 900°C in Fig. 5d). At higher annealing temperatures (e.g. 1000°C) epitaxial regrowth occurs within the contact area (left in Fig. 1a) compared to the polycrystalline Si on top of the oxide (right in Fig. 1a). The epitaxial Si layer contains numerous micro-twins. Misfit dislocations due to high phosphorus doping are observed in the Si substrate. Tilting of the specimen reveals dark resp. white dots at the interface (Fig. 1b,c). These particles which have a size of 3 to 10 nm form the residue of the native oxide and therefore should consist of SiO_2.

3.2 Thermal oxide on polysilicon

The lower dielectric breakdown strength of thermal oxides grown on poly-Si compared to oxides on single crystalline Si can be explained by its ir-regular interface morphology. Protuberances and Si inclusions were found by Markus et al. (1982) after wet oxidation in the temperature range of 950 - 1100°C. For wet oxidation at lower temperatures (870°C) grain bound-ary grooving is observed (e.g. at A in Fig. 2a) which is probably due to enhanced oxidation because of dopant segregation at the grain boundaries. By preferential oxidation at grain boundaries Si inclusions in the oxide can form (e.g. at B in Fig. 2a). After dry oxidation at 900°C protuberan-ces are observed (Fig. 2b). The difference in oxide thickness above and besides the protuberant Si grains cannot be fully explained by differ-ences in oxidation rate due to different crystallographic orientation. Dry oxidation at higher temperatures (1000°C) shows nearly uniform oxide thickness despite bumps in the poly-Si surface (Fig. 2c). No protuberances

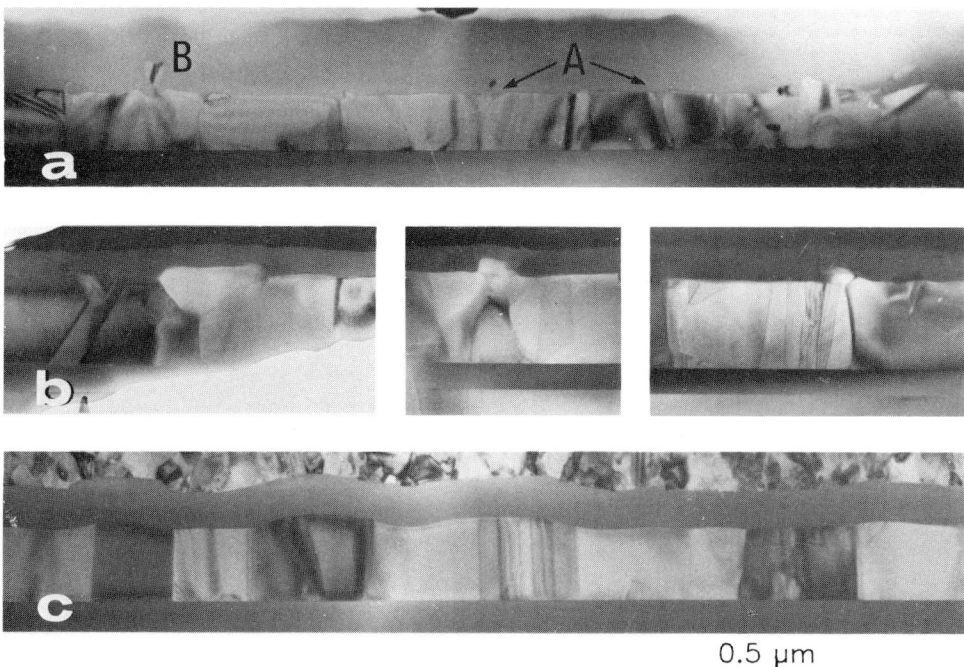

0.5 μm

Fig. 2: Thermal oxides on phosphorus-doped polysilicon. (a) wet oxidation at 870°C, (b) dry oxidation at 900°C, (c) dry oxidation at 1000°C.

or inclusions were observed in this sample. Preferential oxidation at grain boundaries which was found for wet oxidation at lower temperatures represents an additional mechanism for the formation of Si inclusions, compared to the model proposed by Markus et al. (1982) assuming a reduction of oxidation rate at protuberances due to intrinsic stress in the oxide.

3.3 Thermal oxidation of $TaSi_2$/polysilicon double layers

To achieve a lower sheet resistance at the gate level polycide layers, i.e. double layers of poly-Si and $TaSi_2$ are recently used for interconnections in VLSI circuits. The oxidation behaviour of polycide layers was studied, e.g., by Murarka et al. (1980) and Pawlik et al. (1982). The effect of the native oxide between the poly-Si and $TaSi_2$ films on oxide growth was investigated by TEM and Auger depth profiles by Pawlik et al. (1983). Fig. 3a shows the double layer after sputter-deposition of amorphous $TaSi_2$. If no etching with hydrofluoric acid (HF-dip) was performed before $TaSi_2$ deposition the native oxide on the poly-Si has a thickness of about 2 nm (Fig. 3b). After sintering at 900°C the $TaSi_2$ crystallizes but the $TaSi_2$/poly-Si interface does not change (Fig. 3c). If an HF-dip was applied both layers react and the individual $TaSi_2$ grains at the interface develop curved boundaries with the poly-Si grains (Fig. 3d). This reaction achieved after an HF dip is crucial for oxide growth since it allows Si from the poly-Si to diffuse through the $TaSi_2$ film and to form an SiO_2 layer on top of the

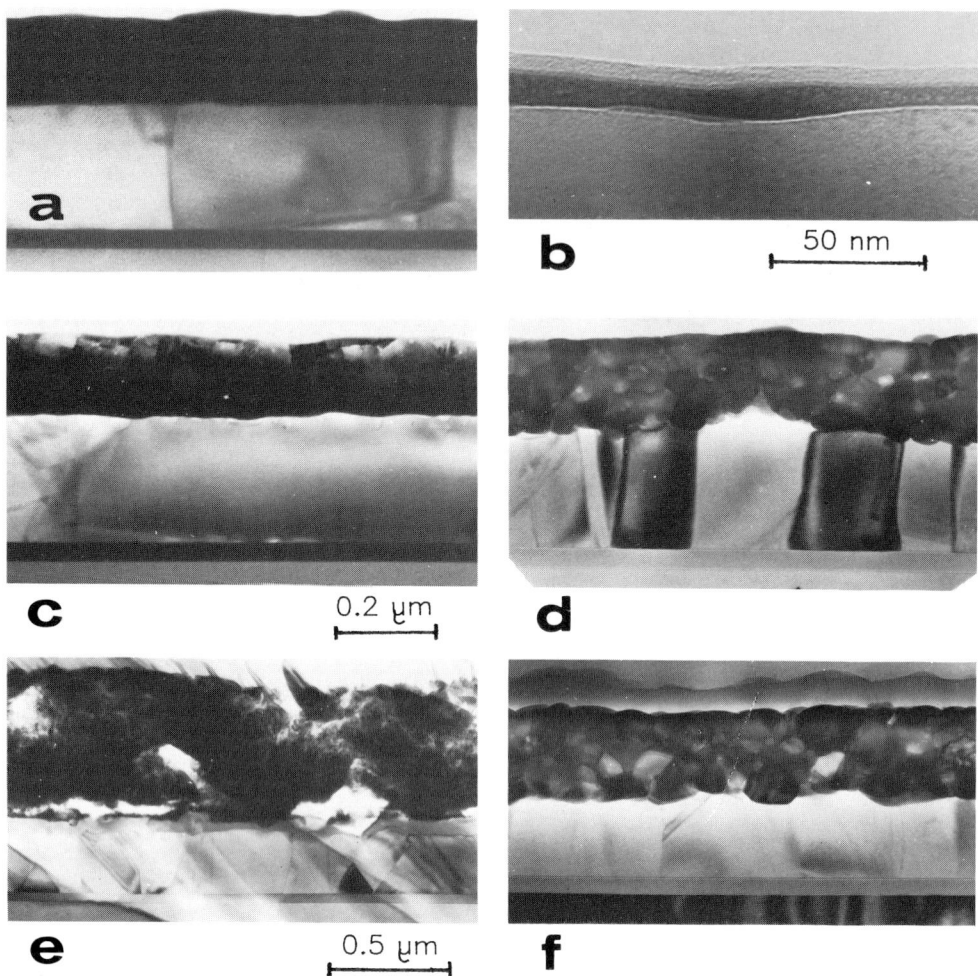

<u>Fig. 3:</u> TaSi$_2$/polysilicon double layers after various process steps.
(a) after TaSi$_2$ sputtering, (b) poly-Si/TaSi$_2$ interface of sample without
HF-dip before TaSi$_2$ sputtering, after sintering at 900°C: (c) no HF-dip,
(d) with HF-dip, after wet oxidation at 900°C: (e) no HF-dip, (f) with HF-
dip.

TaSi$_2$ film (Fig. 3f). If a sample without HF-dip is oxidized this Si diffusion is inhibited and the TaSi$_2$ film is oxidized forming a Ta-silicate
(Fig. 3e).

4. Geometrical configuration of IC structures

4.1 Reoxidation of double polysilicon layers

The floating-gate transistors in electrically programmable memories consist
of two poly-Si lines isolated from each other by the inter poly-Si oxide
and from the substrate by the gate oxide. After chemical etching of the

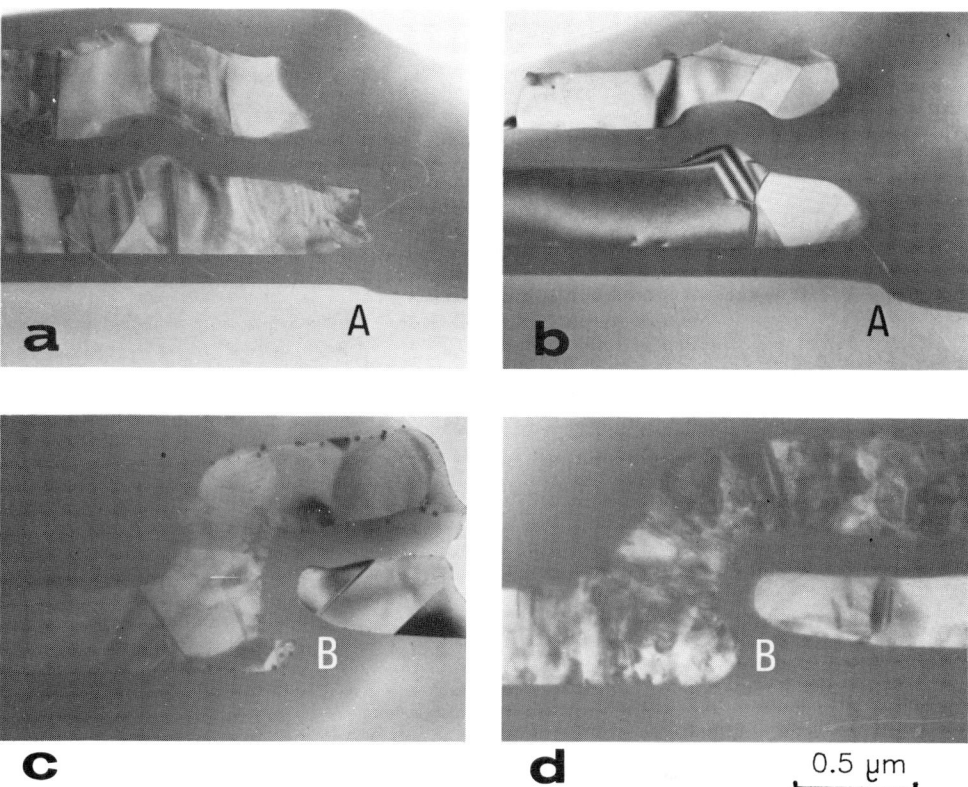

Fig. 4: Edge configuration of floating gate transistors in EPROM chip
after slight (a) and stronger (b) reoxidation. Inter poly-Si oxide at
edge of lower poly-Si film showing edged (c) and rounded (d) configuration.

poly-Si and oxide films a reoxidation process is performed to close the
undercuts of the oxides. Figs. 4a, b show the edge configuration of the
double poly-Si layers for slight (Fig. 4a) and stronger (Fig. 4b) reoxidation. The poly-Si thickness at the edges decreases and a step in the
Si/SiO_2 interface forms (at A in Figs. 4a,b). In Figs. 4c, d the configuration of the inter poly-Si oxide at the edge of the lower poly-Si film is
depicted for two different oxidation conditions. The underetching at the
poly-Si edge is basically preserved during oxidation and is filled by
poly-Si from the upper film, such forming a poly-Si "nose". For both oxidation conditions the oxide thickness at point B is reduced by about 20 %,
but in Fig. 4c the poly-Si "nose" shows sharp edges whereas the "nose"
is well rounded in Fig. 4d. At sharp edges the dielectric break-down
strength is reduced due to an increase of local electric fields.

4.2 256K memory cell

In the cross section of Fig. 5a various elements of the memory cell are
indicated. Several details are shown in Figs. 5b, c, d. The gate oxide
thinning at the tip of the bird's beak of the field oxide which is grown
by selective oxidation, is only approx. 10 % (at A in Fig. 5b). The bend-

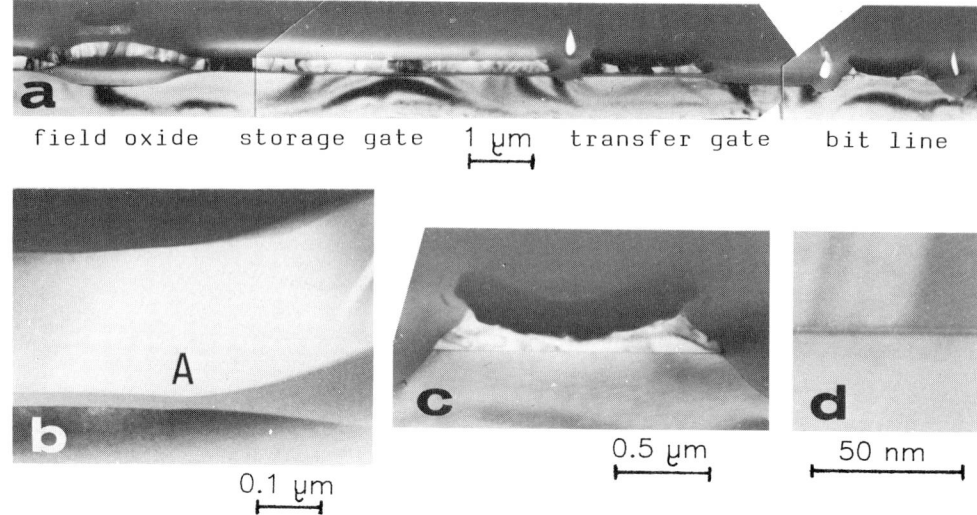

field oxide   storage gate   1 μm   transfer gate   bit line

A

b   0.1 μm

c   0.5 μm

d   50 nm

Fig. 5: Cross section through 256K memory cell, (a) overview, (b) tip of
bird's beak of field oxide, (c) bit line, (d) native oxide at poly-Si/Si
interface.

ing of the TaSi$_2$ film (bit line and transfer gate) is not typical for
polycide interconnections. During etching of the polycide layer by reactive ion etching (RIE) the poly-Si was slightly underetched (Fig. 5c).
Because there is no selectivity for RIE between poly-Si and Si, over
etching produced a groove in the Si substrate on both sides of the bit
line (Fig. 5a, c). The native oxide between the poly-Si of the bit line
and the Si substrate was not destroyed during phosphorus diffusion due
to process temperatures as low as 900°C (Fig. 5d).

Acknowledgements: The authors wish to thank H.D. Hecht, W. Müller, F. Neppl,
D. Pawlik and A. Scheibe for valuable discussions and providing samples as
well as L. Reidt for the photographic work.

References

Falckenberg R, Doering E and Oppolzer H 1979 Ext. Abstr. Fall 79 Meeting
    Electrochem. Soc. (Los Angeles) 79-2 1429
Henghuber G, Oppolzer H and Schild S 1980 Siemens Forsch.- und Entwickl.-
    Ber. 9 363
Markus R B, Sheng T T and Lin P 1982 J. Electrochem. Soc 129 1282
Murarka S P, Fraser D B, Lindenberger W S and Shinha A K 1980 J. Appl.
    Phys. 51 3241
Oppolzer H 1980 Proc. 7th Europ. Congr. Electron Microscopy (Den Haag) 326
Oppolzer H, Falckenberg R and Doering E 1980 J. Microscopy 118 97
Oppolzer H, Falckenberg R and Doering E 1981 Proc. 2. Microsc. Semicond.
    Mater. (Oxford) Inst. Phys. Conf. Ser. No 60, 283
Pawlik D, Doering E and Oppolzer H 1982 Ext. Abstr. Spring 82 Meeting
    Electrochem. Soc. (Montreal) 82-1 310
Pawlik D, Oppolzer H and Hillmer T 1983 to be published
Sheng T T and Markus R B 1980 J. Electrochem. Soc. 127 737

Inst. Phys. Conf. Ser. No. 67: Section 10
Paper presented at Microsc. Semicond. Mater. Conf., Oxford, 21–23 March 1983

TEM studies of small geometry silicon MOSFETs

M C Roberts, G R Booker, S M Davidson* and K J Yallup*

Dept of Metallurgy and Science of Materials, University of Oxford
* GEC Research Laboratories, Hirst Research Centre, Wembley, UK

Abstract TEM and SEM/EBIC examinations have been performed on NMOS polysilicon gate transistors to determine both the geometry of the devices and the location and nature of any defects. TEM etched cross sections have been used to delineate arsenic doped n^+-p source/drain junctions for a range of heat treatment times at 950°C to an accuracy of ±10 nm. The experimental values of junction depth and lateral spreading obtained have been compared with a two-dimensional dopant diffusion model. The level of agreement achieved is encouraging.

Introduction

Much of the current research in silicon integrated circuits is aimed at producing smaller feature sizes to increase device speed and packing density. This present study is part of a programme aimed at developing a fine dimension (~1 μm) complementary MOS process using <100> bulk silicon substrates with semi-recessed field oxide and channel-stop isolation, self-aligned polysilicon gate electrodes, thermally grown gate oxide, and implanted arsenic or boron source/drain regions. A schematic cross-section of the n-channel devices is shown in Figure 1. It is clear that accurate knowledge of the physical structure produced in device fabrication is necessary to enable a complete characterisation of the CMOS process. For example, it is important to know the precise geometry of the edge of the polysilicon gate, and the distance (Y_s see Figure 1) to which the source and drain n^+-p junctions extend laterally beneath the polysilicon gate. This information becomes especially important as the channel length is reduced to 1 μm or less. Other important physical device parameters include the source/drain junction depth (X_j), the gate oxide thickness (t) and the geometry of the field oxide. It should also be noted that the occurrence, nature and distribution of any fabrication-induced defects need to be determined because these could adversely affect the final device operation.

In the present study, TEM and SEM methods were used to obtain this geometrical and defect data. For the TEM examinations, both plan-view and cross-section specimens were prepared using ion-beam thinning (Ar^+, 5 keV, 35 μA); a polishing and ion-beam thinning procedure was developed so that cross-section specimens could be obtained corresponding to specific individual MOSFETs which had previously been electrically characterised. The plane of the cross-section specimens corresponded to (110). In order that accurate measurements of the device geometry could be made, it was

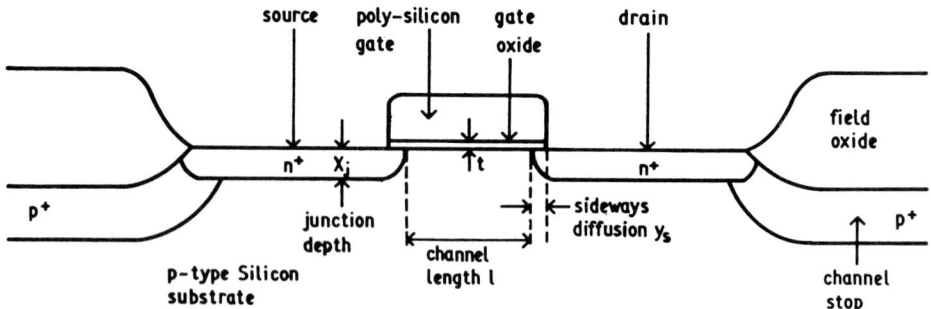

Fig 1   Schematic cross-section of small geometry n-channel MOS transistor

often advantageous to tilt the specimen in the TEM to the precise [110] pole (symmetrical many-beam condition) and record bright-field micrographs. This reduced any strong dynamical contrast effects, especially of interfaces, and also ensured that the interfaces were being viewed 'edge-on'. The TEM examinations were peformed using a Philips EM300 microscope at 100 kV. In order to reveal the n^+-p junctions between the source/drain regions and the underlying substrate, TEM cross-section specimens were lightly etched (5 to 10 seconds in HF 0.5% HNO$_3$ 99.5%) using the technique described by Sheng and Marcus (1981). The subsequently obtained TEM results were compared with theoretical calculations of the junction depth (X_j) and the lateral spread (Y_s) using the approach adopted by Tielot (1982). For the SEM studies, bulk cross-sections were prepared by diamond polishing, finishing with Syton polishing. The examination was performed in a JEOL 35X at 4 kV using the EBIC method.

Results

All of the results presented in the present paper correspond to structures fabricated in p-type 5 to 10 ohm cm Si slices. A TEM cross-section of an n-channel MOS transistor showing the polysilicon gate and the gate oxide is given in Figure 2. For ease of sample preparation, the device used in this instance had a channel region 20 μm

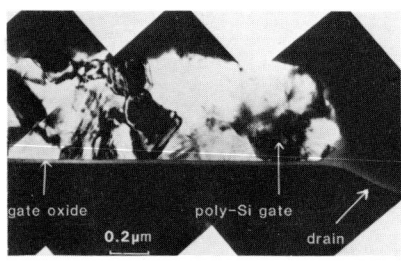

Fig 2   TEM micrograph of the edge of the polysilicon gate MOSFET taken with the incident beam direction parallel to the <110> direction of the cross-section

wide by 20 μm long. The source/drain areas were formed by ion implantation of arsenic (40 keV, 1 x 10^{16} cm^{-2}) followed by an oxidation/inert ambient drive-in (950°C wet oxidation 15 mins + 950°C dry nitrogen 15 minutes); the dip in the silicon surface between the channel and the drain region in Figure 2 results from the oxidation. Figure 3 shows the same cross-section at higher magnification from which a gate oxide thickness of 13.5 ± 1 nm is measured.   This result is in good agreement with the values obtained from capacitance   measurement   (12.5 ± 0.5 nm)   and   from   ellipsometry (13.0 ± 1.0 nm).

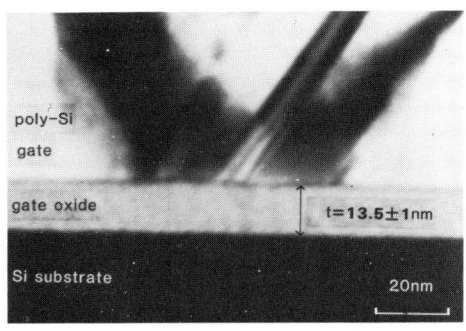

Fig 3   Higher magnification TEM micrograph of Fig 2 showing gate oxide of MOSFET

A TEM cross-section showing the   source/drain region is given in Figure 4. This is a composite micrograph, the left portion being taken from a specimen which has received the oxidation/inert ambient drive-in, and the right portion from a specimen which has received only an inert drive-in (950°C dry nitrogen 30 minutes).   In the right portion, the n^+-p junction has been delineated using the etching technique.   The unetched specimens allow any crystalline defects to be seen more readily.   Dislocation loops of diameter ∼50 nm occur in the region close to the surface, i.e. away from the n^+-p junction which is at a depth of ∼0.2 μm.   Previous work by Mader and Michel (1976) on analogous arsenic implantation (40 keV, 1 x 10^{16} cm^{-2}) into silicon suggests that these dislocations are formed at or just below the interface between the amorphous silicon arising from the implantation and the underlying substrate, and grow up to the surface.   Ashburn et al (1977) and Bull et al (1979) have shown that provided such dislocations remain within the n-type region and do not penetrate the junction, the electrical characteristics of the device are not significantly affected.

Fig 4   TEM composite micrograph shows the source/drain region of MOSFET. The micrograph on the left shows dislocations in the n^+ region, while the micrograph on the right shows   the delineated n^+ region

The etched specimen in Figure 4 reveals the source/drain to substrate n^+-p junction and in particular the profile beneath the edge of the polysilicon gate.   As an inert ambient was used for this heat treatment, the surface of the silicon does not show the dip present in Figure 2.   The boundary of the etched source/drain region can be measured to an accuracy of ∼10 nm.   Sheng and Marcus (1981) have shown that this delineation corresponds to a doping concentration of ∼10^{19} cm^{-2}.   However, as will be illustrated later, the concentration   dependence   of   the   arsenic   diffusivity   causes   its concentration to fall from 10^{19} cm^{-3} to 10^{15} cm^{-3} within 40 nm.   Hence, the position of the boundary in the TEM cross-section should be ∼40 nm above the n^+-p junction because the substrate doping is ∼10^{15} cm^{-3}.

An SEM/EBIC tilted cross-section bulk specimen showing the polysilicon gate and the drain region of a 2 μm nominal gate length n-channel MOS transistor is given in Figure 5.   The bright region beneath the drain corresponds to the depletion region of the n^+-p junction.   The upper boundary of this region is sharp because the depletion region does not extend far into the heavily doped drain.   Conversely, the lower boundary is diffuse because the depletion region extends a significant distance into the underlying more lightly doped silicon substrate.   The junction depth can been determined by measuring from the specimen surface to the upper boundary of the bright region, the accuracy being ~0.1 μm.   It can be seen that at the present level of development, this SEM/EBIC technique does not have sufficient resolution to allow the geometries of the source/drain junctions in fine dimension MOS transistors to be studied.   Further work to adapt the approach that was used in the TEM cross-section etching method for SEM analysis is at present underway.

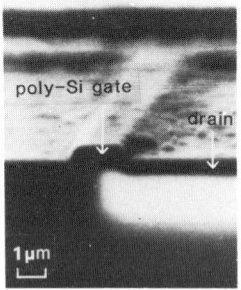

Fig 5   SEM EBIC micrograph of a MOSFET showing the depletion region of the drain to substrate junction

It has been shown that the etched cross-section TEM technique is able to provide delineation of n^+-p junctions in two dimensions.   A series of silicon samples was fabricated to compare the geometry of n^+-p junctions determined in this manner with one-dimensional and two-dimensional computer models of the arsenic diffusion.   The specimens were fabricated with 8 μm wide and 0.6 μm thick oxide strips spaced at 14 μm intervals on the substrate surface, the oxide patterning being performed using standard photolithography and wet etching.   The samples were implanted with arsenic (40 keV, 1×10^{16} cm^{-2}) and were annealed at 950°C for times in the range 30 minutes to 480 minutes.   Figure 6 shows the TEM etched cross-section

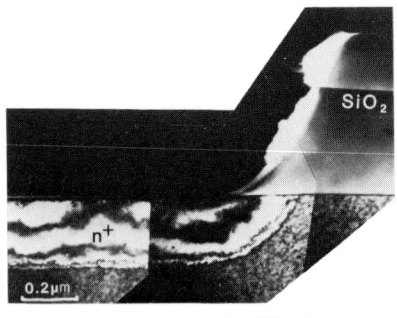

Fig 6   TEM etched cross-section of a 40 keV, 1×10^{16} cm^{-2} arsenic implant annealed 60 minutes at 950°C

for a heat treatment of 60 minutes.   From the micrograph it is possible to determine the junction depth (X_j) and the lateral spreading (Y_s).   The lateral spread is measured from the oxide edge.   Due to the tapered oxide profile obtained by the wet chemical etch, the arsenic implant will penetrate the oxide mask at its edge resulting in a lateral spread immediately after implantation. The consequent two-dimensional arsenic distribution before heat treatment has been modelled using the approach adopted by Runge (1977). The calculation shows that for the oxide mask edges used in this work, an initial lateral spread of 0.14 μm will be produced for an arsenic implant of 40 keV, 1 x 10^{16} cm^{-2}.

The diffusion from this initial two-dimensional distribution on heating has been calculated following the method suggested by Tielot (1982). The computer model developed includes the effects of concentration-dependent diffusion of arsenic using the values in SUPREM II (Antoniadis and Dutton 1979). Figure 7 shows a two-dimensional iso-concentration contour plot obtained from the model for a 40 keV 1 x 10^{16} cm^{-2} arsenic implant annealed at 950°C for one hour. It can be seen that the contours between 10^{19} cm^{-3} and 10^{15} cm^{-3} are closely packed (~40 nm) due to the concentration-dependent diffusivity of arsenic. Superimposed on the figure is the junction delineated in the TEM etched cross-section. The agreement between the measured junction depth and the concentration contour for

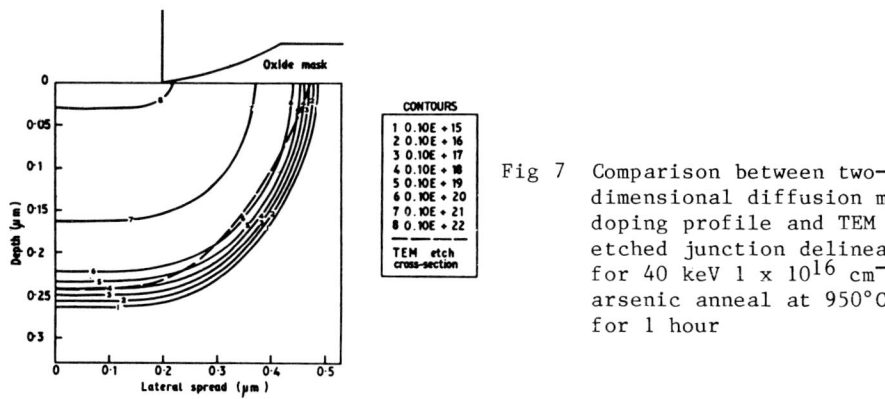

Fig 7   Comparison between two-dimensional diffusion model doping profile and TEM etched junction delineation for 40 keV 1 x 10^{16} cm^{-2} arsenic anneal at 950°C for 1 hour

10^{17} cm^{-3} is encouraging. However, further work is required both in the TEM cross-section preparation and in the diffusion modelling before a more quantitative interpretation is possible.

The TEM etched cross-section results can also be compared with the diffusion model as a function of time at 950°C. Figure 8 shows the measured and

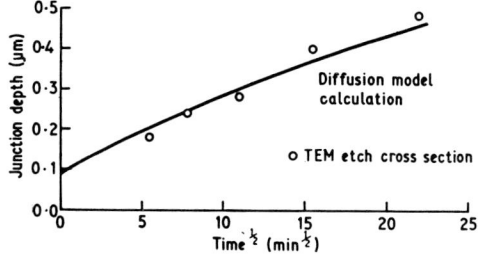

Fig 8   Measured junction depth (X_j) as a function of (anneal time)$^{1/2}$. Model calculation is shown as solid line.

calculated junction depth for a 40 keV, 1×10^{16} cm^{-2} arsenic implant. The agreement between the two techniques is excellent. Junction depths measured using a bevel and stain technique also confirm the reliability of the TEM cross-section method. Figure 9 shows the measured and calculated lateral diffusion (Y_S) as a function of time. Again the agreement between the two methods is encouraging.

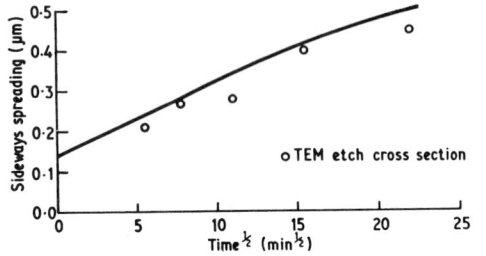

Fig 9   Measured sideway spread (Y_S) as a function of (anneal time)$^{1/2}$. Model calculation is shown as solid line.

It has been shown that the TEM cross-section technique can be used to obtain reliable measurements of the geometry of fine dimension MOS transistors. The thickness of the gate oxide can be determined to an accuracy of ±1 nm. The shape of the polysilicon gate electrode may also be monitored. Using the approach of Sheng and Marcus (1981), the two-dimensional profiles of n^+-p arsenic source/drain junctions have been delineated to an accuracy of ±10 nm. These measurements have been compared with two-dimensional diffusion modelling of heat treatments at 950°C for a range of times. The agreement in all cases is good. Due to the increased diffusivity of arsenic at high doping concentrations, the arsenic concentration profile falls off very rapidly near the junction. Although this makes absolute calibration of the TEM cross-section technique difficult, the delineated depth is expected to be within 40 nm of the metallurgical junction. Further experiments are underway to enable characterisation of the concentration level corresponding to the TEM measurements to be established.

Acknowledgements

We would like to thank SERC for financial support of the work at Oxford.

References

Antoniadis D A and Dutton R W 1979 IEEE Trans Electron Devices ED-26 490
Ashburn P, Bull C, Nicholas K H and Booker G R 1977 Solid State Electronics 20, p 731
Bull C, Ashburn P, Booker G R and Nicholas K H 1979 Solid State Electronics 22, p 95
Mader S and Michel A E 1976 Phys Stat Sol (a) 33 p 793
Runge H 1977 Phys Stat Sol (a) 39 p 595
Sheng T T and Marcus R B 1981 J Electrochem Soc 128 p 881
Tielot R 1982 Nato advanced study institute on processing and device simulation for MOS-VLSI circuits July 12-23

Inst. Phys. Conf. Ser. No. 67: Section 10
Paper presented at Microsc. Semicond. Mater. Conf., Oxford, 21–23 March 1983

473

Transmission electron microscope studies of MOS silicon device structures

N G Chew, A G Cullis, J C White and T I Cox

Royal Signals and Radar Establishment, Malvern, Worcs. WR14 3PS

Abstract   Cross-sectional transmission electron microscopy has been used to study three types of MOS device structure. The epitaxial recrystallization of deposited polycrystalline Si following laser induced melting has been demonstrated and the effects of this regrowth on devices fabricated in these layers investigated. Processing features encountered in the early stages of development of a self-aligned gate MOS device have been identified. Stepped-gate device structures have been studied and characteristics of the small-scale topography arising from the non-planar geometry revealed.

1. Introduction

As device geometries decrease in size and new potential device structures emerge, the need for more detailed analysis is increasing. Cross-sectional transmission electron microscopy (TEM) offers a unique capability for detailed study of the small scale topography and microstructure present in these devices at all stages of fabrication (see for example Sheng and Marcus (1980), Oppolzer and Huber (1983)).

This work focuses on studies of three types of metal-oxide-semiconductor (MOS) device structure. Field effect transistors (FETs) fabricated in laser recrystallized polycrystalline Si (Cox et al 1983) have been investigated both for their inherent importance and as a diagnostic probe for detailed characterization of the recrystallized layers. Self-aligned gate FETs, of interest for their potential in very short channel length devices, (White et al 1983) have been examined in the early stages of development to identify possible processing defects. Finally the stepped-gate FET, a novel device possessing unique threshold-voltage characteristics (White 1982) has been studied to characterize in detail the non-planar topography and microstructure present.

2. Experimental

2.1 FETs fabricated in laser recrystallized polycrystalline Si.

Polycrystalline layers used in this work were deposited on (001), single crystal, high resistivity substrates by the thermal decomposition of silane at 625°C. Doping of these layers was carried out by the implantation of As^+ or B^+ ions to give n- or p-type material, respectively. Recrystallization of the implanted layers was achieved either by laser annealing using spatially uniform 30ns pulses of 694nm radiation obtained from a Q-switched ruby laser system or by annealing in a furnace for 30mins at 1100°C. MOS devices were fabricated in these layers using standard processing techniques.

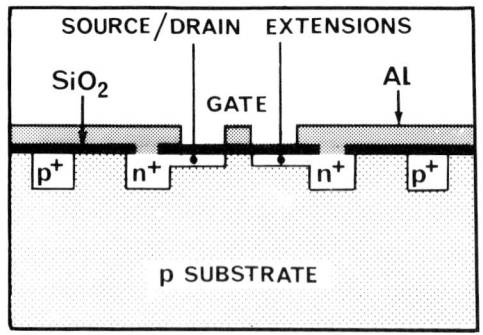

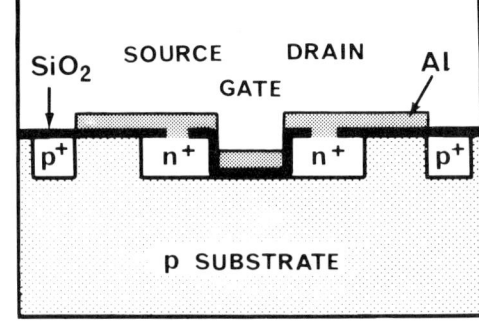

Fig. 1   Diagram of self-aligned gate device.

Fig. 2   Diagram of stepped-gate device.

2.2 Self-aligned gate FETs

These devices, shown schematically in Fig. 1, were fabricated in (001) single crystal Si. The Al gate metallization was defined using a "lift-off" lithography process. Self-aligned source/drain extensions were then formed by the implantation of 40keV As$^+$ at a dose of 5×10^{15} ions/cm^2 and a subsequent low temperature anneal (Scovell and Young 1980).

2.3 Stepped-gate FETs

The stepped-gate devices, shown schematically in Fig. 2, were fabricated in (110) single crystal Si, the stepped geometry being obtained by the use of a KOH anisotropic wet etch (etch ratio <110> to <111> $\approx$600:1). The bottom of the n-type source/drain region provided a positive end-stop during etching, the side walls being defined by the {111} planes perpendicular to the slice surface. After thermal growth of the gate oxide, source, drain and gate metallizations were deposited simultaneously by Al evaporation.

All the TEM specimens employed in these studies were thinned to electron transparency by sequential mechanical polishing and low voltage Ar$^+$ ionmilling. Thinned specimens were examined in a TEM operated at 120keV.

3. Results

3.1 FETs fabricated in laser recrystallized polycrystalline Si

The as-deposited polycrystalline layer is shown in cross-section in Fig. 3a. It consisted of a mass of fine grains, typically $\sim$200Å in width and extending through the thickness of the film ($\sim$3000Å). After the implantation of dopant atoms, layers of this type were further implanted with 150keV ^{29}Si$^+$ at a dose of 10^{15} ions/cm^2 to enhance absorption of laser radiation by the formation of a damaged surface layer. The resulting ion damage, illustrated in Fig. 3b, had saturated to give a continuous amorphous band extending $\sim$1000Å into the layer. The topography of the amorphous/crystal boundary closely followed the microscopically rough surface of the original sample, which was preserved during the implantation process. When these implanted layers were annealed with laser energy densities of $\sim$2.0J/cm^2, transient surface melting and subsequent epitaxial recrystallization from the substrate resulted in the formation of single crystal layers, as illustrated in Fig. 3c. Residual defects remaining in the layers were mainly V-shaped

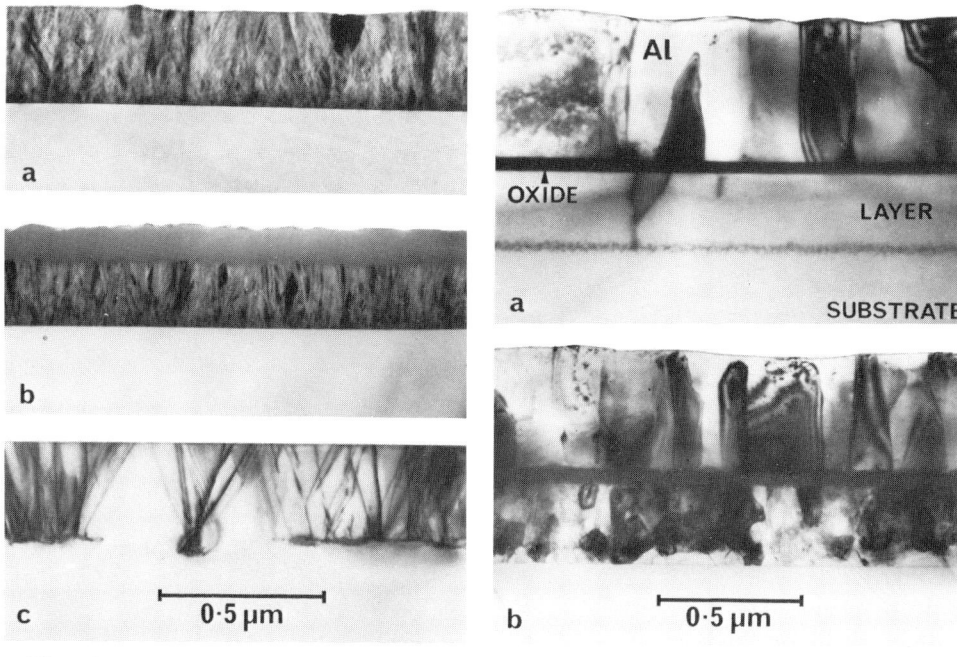

Fig. 3 Cross-sectional TEM images of deposited polycrystalline layers: a) as deposited;   b) after $^{29}Si^+$ ion implantation;   c) after implantation and laser annealing.

Fig. 4 Cross-sectional TEM images of gate regions of finished devices (~ 500Å gate oxide): a) laser annealed layer; b) furnace annealed layer.

dislocation pairs nucleated at the position of the original layer/substrate interface. These were attributed to contamination present at the interface not being dispersed during laser melting and giving rise to dislocation nucleation during resolidification. Indeed it was found that improved substrate cleaning prior to deposition significantly reduced the defect density in the annealed layer - see Fig. 4a which shows the gate region of a device fabricated in one of these laser recrystallized layers. The Si region is single crystal with few remaining defects. These defects consisted of a band of fine disorder at the position of the original layer/substrate interface with occasional dislocation pairs propagating to the surface. The Si surface was microscopically smooth and planar. This resulted in the formation of a uniform gate oxide layer, which would be expected to have good dielectric properties. In comparison a cross-section from the gate region of a device fabricated in a polycrystalline layer after furnace annealing is shown in Fig. 4b. Significant grain growth had taken place in the Si but the surface remained microscopically roughened. This roughness was reproduced during growth of the gate oxide resulting in an undulating oxide layer with the possibility of many high field points and degradation of the dielectric properties of the oxide.

The device surface mobility (holes) is of particular significance (Cox et al 1983), this parameter being critically dependent upon the near surface morphology of the Si layer. Very poor results ($\sim 2cm^2/s$-V) obtained for the thermally annealed devices contrast with the relatively good performance ($\sim 90cm^2/s$-V) of the laser recrystallized devices - single crystal control

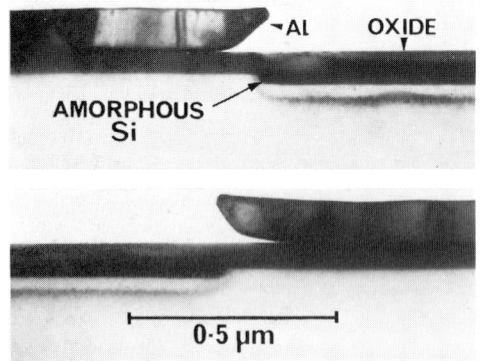

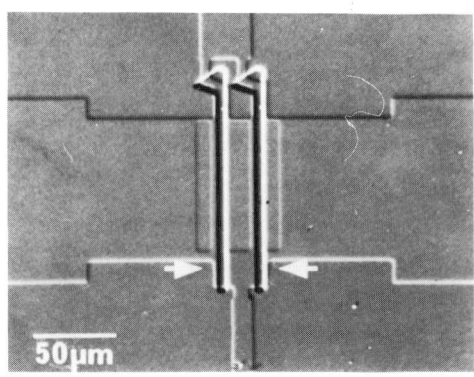

Fig. 5 Cross-sectional TEM images of self-aligned gate devices showing source/drain extensions and edges of gate regions.

Fig. 6 Scanning electron micrograph of a double-gate stepped-gate device. (Arrows indicate the position of TEM cross-section shown in Fig. 7).

samples exhibited a surface mobility of $\sim 145 \text{cm}^2/\text{s-V}$. These electrical measurements correlate well with the near surface topography and microstructure of the devices, shown to be present by the TEM observations.

3.2 Self-aligned gate FETs

The edge of the gate channel is of particular importance in determining the electrical characteristics of these devices, hence cross-sectional specimens were prepared to investigate the structure present in these regions. Two examples which illustrate the results obtained from the first batch of prototype devices are illustrated in Fig. 5. The implant used for the source/drain extensions had been only partly regrown by the low temperature annealing process and $\sim 300 \text{Å}$ of amorphous material remained at the Si surface. In addition there was a band of residual fine damage at the position of the original implant tail. This damage band (expected to be present after the low temperature anneal) was assumed to be electrically inactive and therefore could be used to determine accurately the delineation of the source/drain junctions in these devices. These observations enabled annealing conditions to be modified during processing of subsequent batches to enable complete regrowth of the amorphized Si. The edge shape of the Al gate metallization was revealed and seen to have been undercut during the lithography process. The degree of undercutting occurring in the gate metallization is of crucial importance to the device characteristics, the uncontacted gate region greatly affecting the current flow in the channel region (Sugino et al 1982).

3.3 Stepped-gate FETs

An example of the stepped-gate structures studied is illustrated in Fig. 6 which shows a double gate device. To investigate in detail the topography present in the gate regions cross-section specimens were prepared, particular care being taken to achieve uniformly thinned material in the area of the gates. A cross-section from the position indicated in Fig. 6 is illustrated in Fig. 7a where both gates are visible. It can be seen that the two gates were of very similar depth and width. One of these gates is shown of higher

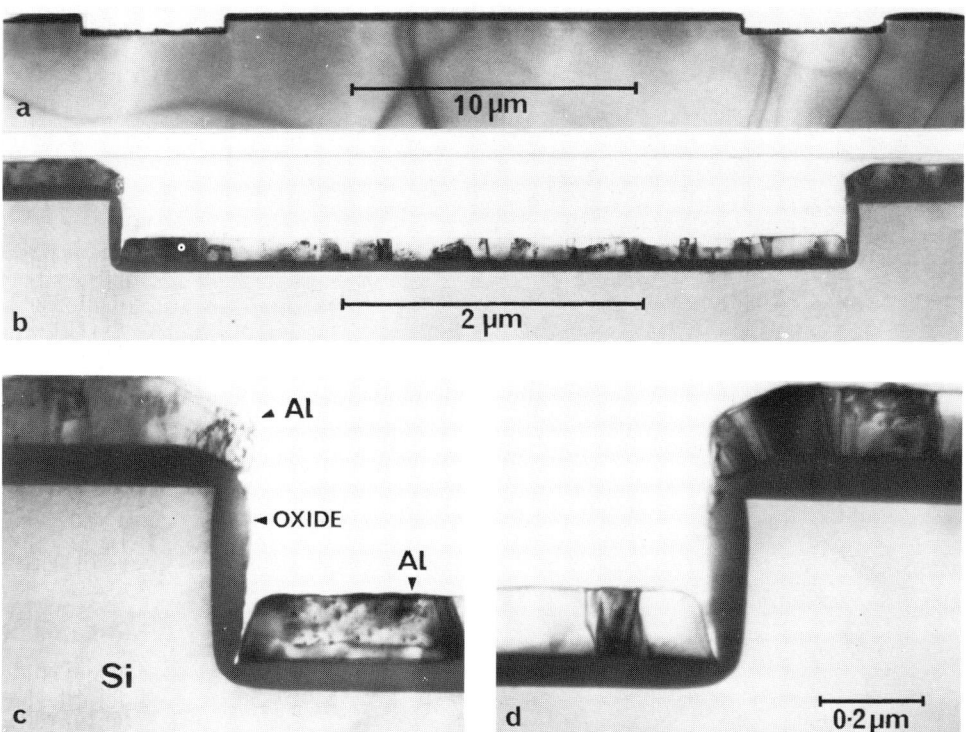

Fig. 7 Cross-sectional TEM images of stepped-gate device structure showing: a) both gates; b) single gate at higher magnification; c) & d) details of the step corners.

magnification in Fig. 7b where the excellent depth uniformity of the stepped region is revealed. The geometry of the step corners is of particular significance for device performance. Electrical isolation of the source, drain and gate regions depends on achieving a complete break in the metallization layer at the edge of the stepped groove during the Al evaporation. (Contact to the gate region is obtained by "ramping" the Al out of the end of the groove on the slope defined by the set of {111} planes making an acute angle with the surface - see Fig. 6.) These edges are shown in Fig. 7c and d. The side walls are vertical and well-defined, the thermally grown gate oxide following this topography. Some reduction in

Fig. 8 Cross-sectional TEM image of stepped-gate structure with metallization defect.

oxide thickness is apparent at the corners of the steps, this behaviour being consistent with that observed previously (Shankoff et al 1980). The rounding of the Si at the bottom corners of the steps is significant, any angular features in this region would be expected to have a large effect on the field distribution at the edge of the channel region and also therefore on the device threshold characteristics. The metallization process was extremely effective in these devices, the gate region being completely covered and the source/drain metallization terminating cleanly at the top corner of each step. In devices from one processing batch, however, occasional metallization defects were observed, an example being shown in Fig. 8. In this instance the Al had been deposited partly down the side wall of the step groove and although this example would be expected to be adequate electrically, more severe problems of this nature would lead to a short circuit between the gate and source/drain regions. This may be a limiting factor in determining the minimum useful step height in this device structure.

4. Conclusions

This work has identified a number of processing characteristics and defects in MOS device structures. These include residual defects remaining in the silicon after laser and thermal annealing, gate oxide thickness variations and non-optimum metallization edge profiles. Most of these features were related to specific processing steps. TEM particularly in cross-sectional configuration provides a unique capability for studying the small scale topography and microstructure of modern electronic devices and is particularly useful when used in the early stages of device development.

Acknowledgement

The authors would like to acknowledge collaboration with H C Webber (RSRE) in some of the work described.

References

Cox T I, Deshmukh V G I, Hill J R, Webber H C, Chew N G and Cullis A G 1983 IEEE Trans. Electron Devices ED30
Oppolzer H and Huber V 1983 This proceedings volume
Scovell P D and Young J M 1980 Electron. Lett. 16 614
Shankoff T A, Sheng T T, Haszko S E, Marcus R B and Smith T E 1980 J. Electrochem. Soc. 127 216
Sheng T T and Marcus R B 1980 J. Electrochem. Soc. 127 737
Sugino M, Akers L and Rebeschini M 1982 Proc. IEDM (Washington DC) p 462
White J C 1982 British Patent Application No. 8229,071
White J C, Wilkinson S, Janes T and Binnie C 1982 Proc. Microcircuit Engineering '82 (Grenoble)

Inst. Phys. Conf. Ser. No. 67: Section 10
Paper presented at Microsc. Semicond. Mater. Conf., Oxford, 21–23 March 1983

TEM, microdiffraction and electrical studies of buried SiO$_2$ layers formed by high dose oxygen implantation

D Fathy, O L Krivanek, R W Carpenter and S R Wilson*

Center for Solid State Science, Arizona State University, Tempe, Arizona 85257
*Semiconductor Research and Development Laboratory, Motorola Inc., Phoenix, Arizona 85008

Abstract   A study has been made of the SiO$_2$ insulating layer in SOI (Silicon on Insulator) structures produced by high energy oxygen implantation into Si.   The SiO$_2$ layer and especially its two bounding Si-SiO$_2$ interfaces have been examined by high resolution transmission electron microscopy (HREM) before and after annealing for two different implantation dose rates. The results indicate that the interface structure and oxide layer quality are 'dose-rate' dependent. For higher dose rates, less Si-SiO$_2$ intermixing is observed, the Si-SiO$_2$ interfaces are sharper, and there is less residual damage in the top Si overlayer.

1. Introduction.   During the past decade, there has been an increasing interest in finding possible insulators on which thin layers of defect-free silicon can be grown. Semiconductor on Insulator (SOI) structures offer a number of advantages over standard processing in bulk silicon (Malhi et al, 1982), thus making them one of the best prospective candidates for fabricating large scale integrated devices. Silicon on Sapphire (SOS) has received the most attention. The microstructural details and the quality of silicon have been studied using various techniques, including HREM (Hutchison et al, 1981). The results indicate that the structure suffers from crystal defects in silicon, degradation in mobility and an increase in the leakage current with time (Duffy et al, 1982).
      Even more recently SOI structures formed by high energy implantation of large doses of oxygen and consisting of a nearly damage-free Si layer, which can be used as a substrate for epitaxial growth of Si, on top of a buried SiO$_2$ layer, have attracted attention. Preliminary results show a likely substantial improvement over SOS structures (Ohwada et al, 1981). In contrast to SOS structure preparation, in the implanted oxygen case, the implantation energy, direction, dose and dose rate can be adjusted to yield structure with different properties.
      Systematic high resolution electron microscopy (HREM) studies of SOI structures prepared under different conditions have so far not been carried out.   EM observations such as have been made have been limited to low magnification and low resolution. For a review, see Pinizotto, 1983. In the present study, we concentrate on the dose-rate dependence of SOI structure and electrical properties, and use HREM to elucidate the improved structures observed at higher dose rates.

2. Experimental.   Phosphorous-doped 10 Ω-cm silicon (100) wafers were implanted with molecular oxygen (O$_2$+) at an energy of 400 keV with a dose of $\approx 2 \times 10^{18}$ oxygen atoms/cm^2 at two different dose rates of 18µA/cm^2 and 25µA/cm^2, both along the (100) channelling direction. The specimens were

maintained thermally insulated from the rest of the system, leading to higher substrate temperatures for higher dose rates.

Following the implantation, some of the specimens were annealed at 1050°C, followed by the deposition of 1μm of 1Ω-cm p type Si, grown epitaxially by CVD methods utilizing $SiCl_4$. Cross-sectionally thinned specimens were produced by a combination of mechanical polishing to a thickness of 50 μm and subsequent argon beam polishing (Krivanek et al, 1978). Thin regions of around 10 nm thickness suitable for high resolution electron microscopy were obtained.

HREM and micro-analysis were performed using a JEOL 200 CX and a Philips 400 ST (FEG) electron microscope respectively. The structure of the interface was investigated by the lattice imaging of (111) and (200) planes of 0.31 nm and 0.27 nm spacing with the electron beam parallel to (011). The Si overlayer resistivity and the oxide leakage current were measured using a spreading resistance probe and current versus voltage methods on samples prepared identically to the ones used for the HREM studies.

3. <u>Observations</u>. The structure of an as-implanted sample for the low dose rate implantation is shown in Fig. 1. The buried oxide layer is at a depth of 0.3-0.6μm from the marked surface. The overlying silicon consists of two separate regions, as shown also in Fig. 2a taken at a higher magnification of the same region. The area close to the Si-SiO_2 interface on the Si side consists mainly of polysilicon and low density planar faults, whereas the area close to the surface of the silicon is relatively damage-free, except for isolated faults. At intermediate magnifications, the interface looks fairly abrupt. However, at high resolution, poly-silicon of thickness ≈ 50 nm can be seen at the interface, with characteristic twins and stacking faults in this region. The oxide layer itself consists of a large number of small islands of silicon of the order of 10 nm in size.

The structure of low dose rate specimens after annealing is shown in Fig. 2b. As can be observed, the thickness of the oxide layer remains the same under annealing temperatures. The damage in the overlying silicon is reduced and limited to a region of the order of 0.1μm in the vicinity of the interface. This region consists of about 10 nm of polysilicon, plus the rest consisting of dislocated single crystal Si.

Fig. 1 Cross-sectional bright field image taken at 200 keV, showing the depth geometry of the 400 keV/O_2^+ as-implanted oxide region relative to the overlying Si and the substrate.

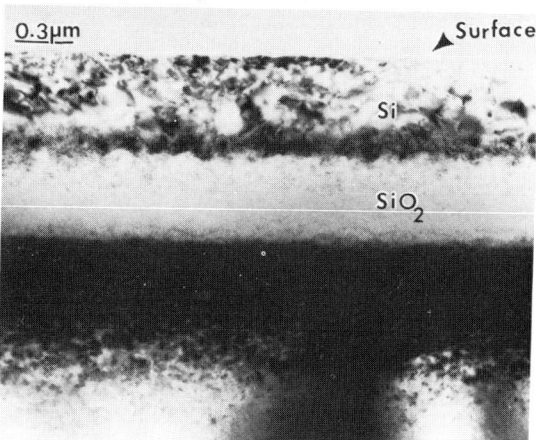

The silicon islands, existing previously in the amorphous layer, have now grown in size. They mainly have preferential orientation to that of the silicon overlay. In some cases, the appearance of Moire's fringes of

1 nm spacing indicates that different grains with different orientation exist in the same island. See also Fig. 3a. The nature of the islands, which have been checked both by measuring the lattice spacing and micro-diffraction, appears to be the same as that of Si. The density of the islands particularly increases as we approach the top $Si-SiO_2$ interface and only a relatively island-free region exists at the depth of 0.2-0.3μm from the interface.

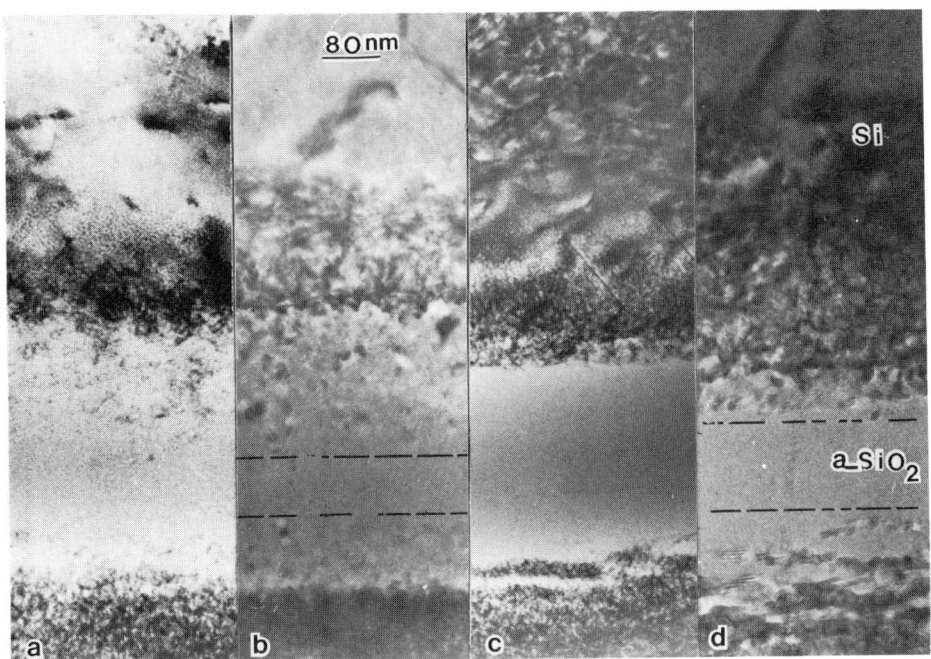

Fig. 2 Bright field images, taken at 200 keV, showing (a) Low dose rate before annealing: a region extending from 0.15-0.25μm below the top interface is relatively island-free. Except for isolated defects, the top silicon layer is damage-free. (b) Image of low dose rate after annealing, showing the increase in the size of the Si islands. (c) Image of a high dose rate implanted specimen before annealing, showing the position of the island-free oxide layer. (d) Image of high dose rate after annealing, showing a strong B-W diffraction contrast from the interlying oxide regions close to the interface.

As for higher dose rates before and after annealing, the oxide layer has formed at the same depth but slightly narrower in total width; see Fig. 2c&d. The amorphous oxide contains fewer silicon islands, as observed in these figures and the high resolution image of Fig. 3b, relative to that formed at low dose rates. The overlying silicon has few defects close to the oxide interface. The interface, as shown in Fig. 3b, consists of a number of large silicon areas connected to each other, with the spacing between the areas filled with amorphous SiO_2.

As can be seen, the oxide layer is amorphous with no visible Si islands. From this image and that of Fig. 2d, it can also be observed that the B-W contrast is mainly caused by localized absorption and diffraction contrast effects from the amorphous and crystalline regions.

The Si-SiO$_2$ interface is atomically sharp and only small isolated defects exist in this region. Silicon itself is in single crystal form with amorphous oxide inclusions decreasing in density towards the free surface of silicon, causing the regions of light contrast, as also observed in Fig. 2c&d.

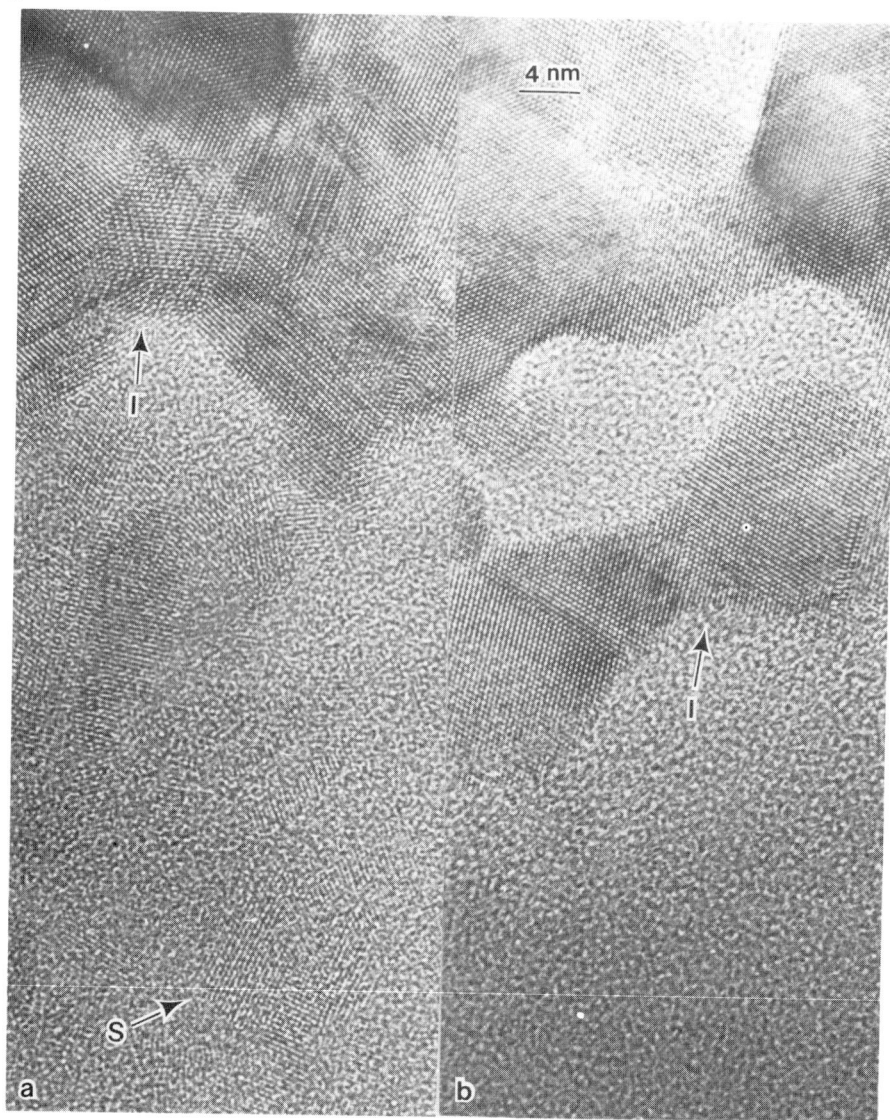

Fig. 3(a) High resolution image of the interface region of Fig. 2b, showing the presence of a very thin polysilicon region (I). Note also the lattice fringes and the Moire fringes in the silicon islands. (b) High resolution image of the interface region of Fig. 2d, showing the presence of closely connected Si islands (I).

The images of the low and high dose rate implanted regions and the

overlying Si after annealing and epitaxial growth are shown in Fig. 4a&b.
These images also show that for high dose rates, after annealing, the
overall thickness of the oxide layer is greatly reduced. Further, the
area with the strong B-W contrast, close to the interface, is also twice
as wide (see Fig. 4b).

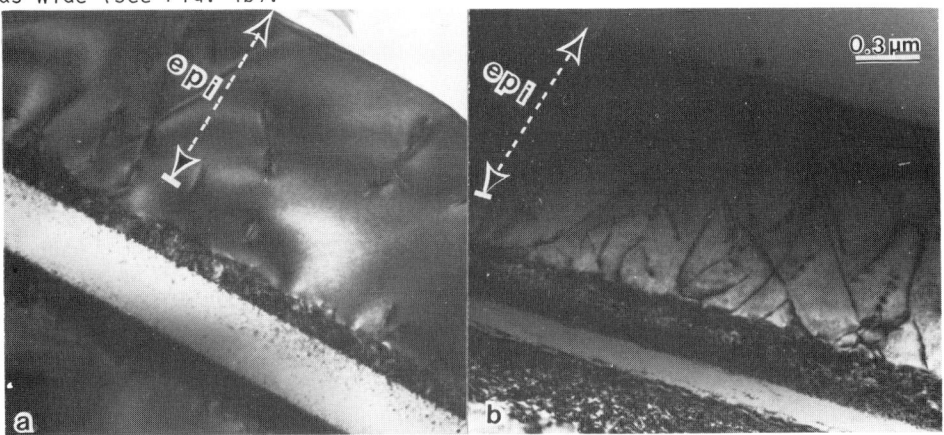

Fig. 4 Cross-sectional bright field image taken at 200 keV. (a)
Image of low rate implanted region after annealing and epi-layer growth,
showing the presence of line dislocations in the epi-layer. (b) Image of
the high dose rate specimen; the overall thickness of the oxide layer is
less than that shown in Fig. 4a. A higher dislocation density originating
from the region of high B-W diffraction contrast near the interface is
also observable.

Electron energy loss spectroscopy (ELS) measurements were done on the
above specimens. The results obtained from the concentration measurements,
and the observation of bond change induced shape changes in the Si L edge
from the oxide, indicate that the ratio of Si to O in the amorphous
regions does not exceed 1:2, in agreement with AES measurements (Maeyama
et al, 1982) which indicate that oxygen does not exceed stoichiometric
values in Si.
The results obtained, after annealing and epitaxial growth, from
mobility and conduction through the oxide measurements, show a very good
crystalline and electronic quality epitaxial layer. The results of the
mobility measurements yield values similar to those of epitaxial layers
grown on bulk silicon (Wilson et al, 1983). The results of voltage versus
current measurements performed on the implanted oxide layers for two
different dose rates show a much higher breakdown voltage and lower
leakage current (an improvement by a factor of 2) for the high dose rate
implanted oxide layers.
4. Interpretation of the results. When oxygen is implanted into silicon
at doses below the stoichiometric value, the distribution of oxygen into
silicon can be approximated to a Gaussian shape and calculated by the
L.S.S. (Lindhard-Scharff-Schiott) method. The regions of Si under the
peak of the Gaussian distribution of oxygen, are heavily displaced by
knock-on damage, thus forming an oxide amorphous layer. However, at
higher doses, where the peak atomic ratio is greater than the stoichiometric ratio, the profile changes to a trapezoid form, i.e. the oxygen
in silicon does not stay above the stoichiometric value of 2, as indicated
earlier, and the excess oxygen diffuses into nearby Si material causing

the shape of the profile to change. As a result, sharp distribution gradients form across which the oxygen falls from the value of $\approx 2 \times 10^{22}$ O_2/cm^3 required to produce SiO_2 to values of less than $10^{19} O_2/cm_3$ in a distance less than 100 nm from the interface or otherwise the end of the plateau.

The results obtained from Rutherford back-scattering (Maeyama et al, 1982) also indicate that in fact the plateau height itself is dose-rate dependent, and only at high dose rates approaches that of stoichiometric values. These results also indicate that the quality of the oxide and the sharpness of the crystalline to amorphous transition improves for higher dose rates, where the plateau height approaches more closely the stoichiometric value.

These results are confirmed by the high resolution images. For cases where stoichiometric SiO_2 is formed, the high resolution images show a continuous amorphous layer. In regions where the oxygen starts to fall below the stoichiometric value, both Si islands and amorphous regions exist. For low dose rate implantation, only a very narrow region of the order of $0.1 \mu m$ of stoichiometric SiO_2 must have been formed. However, in this case a low rate of fall in the oxygen concentration is evident by the gradual change in the increase in the number of silicon islands. Similarly it can be argued that for a high dose rate implantation, a much wider layer of stoichiometric SiO_2, $\approx 0.2 \mu m$ thick, with a much sharper transition region is formed; see Fig.2d and 3b. The effect of annealing however does not change the oxygen distribution so much as it affects the crystallinity of the Si overlayer. As observed earlier, it only affects the size of the Si islands and consequently the regrowth of Si into the oxide layer in the high dose rate case.

5. <u>Conclusion</u>. These initial observations indicate that a combination of high resolution electron microscopy together with elemental and electrical measurements can valuably assist in solving the problems which presently exist with this structure.

References

Duffy M T, Corboy J F, Cullen G W, Smith R T, Soltis R A, Harbeke G, Sanderock J R and Blumenfeld M 1982 J. Crystal Growth 58, 10
Hutchison J L, Booker G R, Abrahams M S 1981 Inst. of Physics Conf. Ser. No.60, 3, 139
Krivanek O L, Tsui D C, Sheng T T and Kamgar A 1978 Proc. of Int. Topical Conf. on Physics of SiO_2 and its Interfaces, 357
Maeyama S and Kajiyama K 1982 Japanese Journal of Applied Physics, Vol 21, 5, 744
Malhi S D S, Lam H W, Pinizotto R F, Hamdi A H and McDaniel F D 1981 Technical Digest, IEDM, pp1-4
Ohwada K, Omura Y and Sana E 1981 IEEE Trans Elect. Dev. ED-28, 1084
Pinizotto R F 1983 Material Research Conference at Fort Collins (to be published in J. of Crystal Growth)
Wilson S R and Fathy D 1983 (to be published)

Acknowledgement

We are thankful to Dr. W.M. Paulson of Motorola Inc. for valuable suggestions. This work was supported at A.S.U. by a grant from N.S.F. (Contract #DMR 8117052).

Inst. Phys. Conf. Ser. No. 67: Section 10
Paper presented at Microsc. Semicond. Mater. Conf., Oxford, 21–23 March 1983

485

TEM and Auger electron spectroscopy observations of C-MOS and N-channel silicon devices incorporating buried oxide implanted isolating regions

M R Taylor, C G Tuppen, R P Arrowsmith, R M Dobson, A E Glaccum, M C Wilson*, G R Booker*, and P Hemment**

British Telecom Research Laboratories, Martlesham Heath, Ipswich, IP5 7RE
*Department of Metallurgy and Science of Materials, University of Oxford
**Department of Electronic and Electrical Engineering, University of Surrey

Abstract   Cross sectional TEM and Auger electron spectroscopy have demonstrated that devices fabricated on substrates incorporating a buried oxide layer have electrical characteristics related to the oxide structure. Excellent electrical isolation was found when the buried oxide was homogeneous stoichiometric SiO_2. Observations of the quality of silicon epitaxial layers grown on oxygen implanted substrates, and on the structure of the buried oxide and surrounding material are reported.

1. Introduction

There is increasing interest in buried oxide structures as substrates for fast integrated 'silicon on insulator' circuits. An electrically isolating silica layer is formed by oxygen implantation of sufficiently high dose to generate SiO_2, and of sufficiently high energy for the oxide layer to be buried. Such structures have been studied by a number of workers, including Das et al (1981), Homma et al (1982), Maeyama et al (1982), and Pinizzotto et al (1982). Devices have been fabricated and their electrical characteristics reported (eg Izumi et al, 1978).

This paper presents the results of examination by cross sectional TEM and Auger electron spectroscopy (AES) of some devices, and relates them to measurements of their electrical characteristics. Some observations of the influence of implantation dose, thermal annealing treatment, and the quality of epitaxial layer growth are included. The cross sectional TEM photos selected for inclusion, (BTRL results), were chosen because of their direct relationship to the devices. However the project as a whole has evolved following cross sectional TEM examinations of many such slices during the last two years, and some of these results, (Oxford work), will be reported later (M C Wilson, G R Booker and P L F Hemment).

2. Experimental

Buried oxide structures were prepared by ion implantation of O_2^+ ions at 400 keV, which is equivalent to an implantation of twice the dose of O^+ at 200 keV. All doses quoted below are 'equivalent doses' at 200 keV. The irradiations were carried out on the heavy ion accelerator at Surrey University, using an electrostatically scanned beam. Specimens were

implanted through a 2 cm x 2 cm silicon mask, and were supported   on
silicon points to reduce heat losses by conduction from the back of the
wafer, thus allowing a high implantation temperature to be attained.
Specimen BT8, (a 3 cm x 3 cm slice), received a dose of 1.8×10^{18} 0
cm^{-2} at a flux of 50 - 55 μA, and implantation temperature of
approximately 550 °C.   Devices BT3 and BT6, (5 cm_8 diameter
wafers),   received doses of 1.4×10^{18} and 2.0×10^{18} 0 cm^{-2}
respectively at a flux of 40 - 45 μA and estimated implantation
temperature of 490 °C. Implantation temperature calibration curves were
measured using IR pyrometry on test samples (Hemment et al, 1983).

17 - 23 ohm-cm, boron doped, p-type silicon substrates of (100)
orientation (to within 1°) were used. Devices were prepared at British
Telecom using normal 3 μm N-MOS fabrication procedures; the essential
difference from standard polysilicon gate devices was simply the presence
of the electrically isolating buried oxide layer. A relevant feature was a
high temperature anneal before fabrication, as described below. A silica
capping layer was deposited on the surface by low temperature Chemical
Vapour Deposition (CVD), (SiH_4/H_2), and the slice was annealed in a
nitrogen ambient at 1150 °C for 2 hours. (Slice BT8, which was not
fabricated into devices, was also capped and annealed under dry nitrogen,
either for 30 mins at 1000 °C, or at 1150 °C for 2 hours). A thin
epitaxial layer was deposited at 1170 °C for approximately 1 min 15 secs
by CVD of $SiCl_4/H_2$. The layer only reached this temperature for about
4 mins. Other heat treatments during device fabrication amounted to a
total of approximately 35 hours at temperatures between 825 and 875 °C,
mostly under steam or oxygen. There was also a 15 min argon anneal at 1000
°C, and other lower temperature heat treatments. TEM examination was
carried out on a JEM 200 CX TEMSCAN system. Cross sectional specimens were
prepared by ion thinning essentially as described by Fletcher et al
(1980). An Ion-Tech system was used with argon ions at approximately 5 KV,
60 μA, and 30 deg glancing angle. AES was carried out on a VG Scientific
MA500 scanning Auger microscope, with 5 kV argon ion sputtering for depth
profiling.

3. Results

In section 3.2, n-channel devices will be shown to have electrical
characteristics which bear an obvious relationship to the structure of the
buried oxide. Whilst elucidating a number of very important features,
earlier reported structural examinations of buried oxides have either been
carried out at lower implantation energies, and without the benefits of
cross sectional TEM (eg Homma et al, 1982), or have not covered the dose
range most appropriate for the fabrication of devices (eg Das et al,
1981). Section 3.1 attempts to provide some of the missing information.

3.1 Unannealed and annealed material : Specimen BT 8

Unannealed material, (1.8×10^{18} 0 cm^{-2}, 200KeV), displays a sharply
demarcated buried oxide layer surrounded by highly defective regions (Fig
1). Careful examination by microdiffraction demonstrated that the buried
oxide was both amorphous and homogeneous. No evidence was found for
microcrystallites of silicon, or any other crystalline phase in this
region. It should be noted that the defective zone adjacent to the buried
oxide extends to the top surface. Microdiffraction and selected area
diffraction showed this zone, and also the region below the oxide, to be
single crystal of the same orientation as the substrate. A very slight

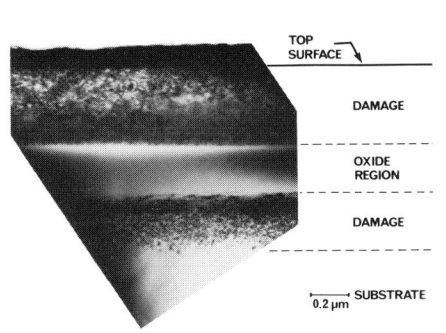

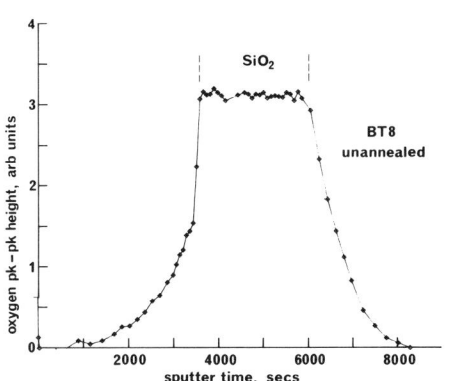

Fig 1. Unannealed oxygen implanted silicon, specimen BT8. 90° cross section, TEM bright field, near a <110> type pole, 220 reflection.

Fig 2. Auger electron spectroscopy plot of oxygen Pk to Pk height vs. sputter time (a measure of depth below the top surface layer).

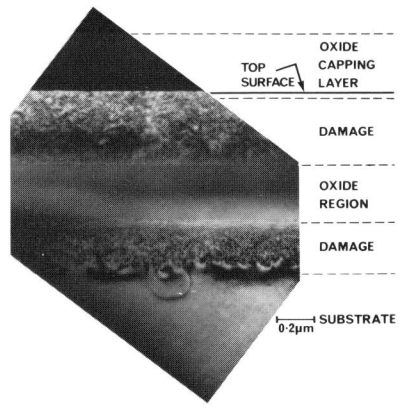

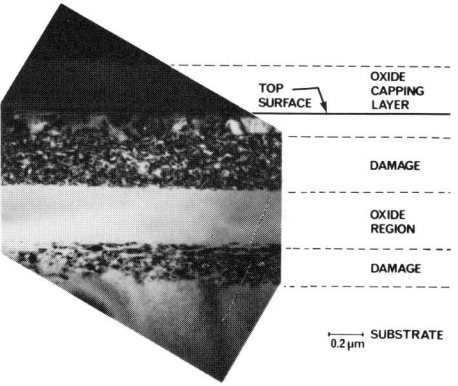

Fig 3. As Fig 1, but annealed to 1000°C for 30 mins. TEM bright field, near a <110> type pole, 400 reflection.

Fig 4. As Fig 1, but annealed to 1150°C for 2 hours. TEM bright field, near a <110> type pole, 220 reflection.

tendency to 'banding' at the lower buried oxide interface should be noted; this effect develops on annealing. Measurements of cross sectional TEM negatives of specimen BT8 (unannealed) revealed that the buried oxide was 0.28 μm thick, and that the top silicon layer was 0.40 μm thick (approximate errors ± 5%). Catalayse crystals were used for calibration (Wrigley, 1968).

An AES depth profile of unannealed BT8 is given in Fig 2. Relative elemental sensitivity factors measured for silicon and oxygen in a sample of SiO_2 prepared by CVD were used to show the buried oxide to be stoichiometric SiO_2. This was confirmed by monitoring the energy of the KLL Si Auger transition, which was found to range from 1618.5 eV in the silicon substrate to 1611.0 eV in the buried oxide layer. The total shift was therefore -7.5 eV, in good agreement with Fellenberg et al (1982), who measured a chemical shift of -7.3 eV (Si KLL transition) between elemental silicon and stoichiometric SiO_2. The defective regions surrounding the buried oxide were shown to be material with appreciable oxygen concentration (Fig 2), the highest levels being in the regions closest to the oxide.

After annealing at 1000 $^{\circ}$C for 1/2 hour, a region of relatively good crystalline quality approximately 25 nm thick develops at the top surface of the silicon adjacent to the silica capping layer (Fig 3). Annealing at 1150 $^{\circ}$C for 2 hours yields a wider region approximately 110 nm thick (Fig 4). These improved quality regions nevertheless contain approximately 10^9 to 10^{10} threading dislocations cm^{-2}. The AES depth profile was similar to that found for unannealed material (Fig 2). Thus, the buried oxide layer was shown to be amorphous stoichiometric SiO_2, and there was no detectable change in oxygen concentration in the damaged material adjacent to the buried layer. The silicon forming the inner buried oxide/silicon interface has developed a clear 'banded' structure. Crystalline material bent away from the substrate orientation by a few degrees, and containing a few low angle boundaries, was found in this region. There has also been annealing of the lower defective region of silicon at its boundary with the underlying substrate, leading to the formation of some extended defects.

3.2 Devices

Two n-channel device structures were prepared and examined by cross-sectional TEM. The first specimen, BT6, (Fig 5), was prepared on material implanted with 2.0 x 10^{18} O cm^{-2}, and shows a clearly delineated amorphous buried oxide layer with no internal structure. In contrast, the second specimen, BT3, (Fig 6), was fabricated on material implanted with 1.4 x 10^{18} O cm^{-2}. The buried layer is no longer homogeneous, consisting of a mixture of amorphous oxide and silicon crystallites with sizes ranging up to 40 or 50 nm, and of many orientations. The leakage current from device to substrate in sample BT3 was 10^{-6} A cm^{-2} at only 0.9 V, substantially higher than the 10^{-13} A cm^{-2} at 20 V measured on device BT6.

The growth of an epitaxial layer of silicon on the implanted material can be distinguished by the presence of a number of dark dots at and below the interface. Their nature is at present uncertain. The epitaxial layers themselves were thinner than intended, being only 25-50 nm thick (see Figs 5 and 6). Threading dislocations propogate through the epitaxial layer/substrate interface without apparent disturbance. (In the case of specimen BT 3, the gate oxide between the epitaxial layer and the polycrystalline silicon layer above it is not missing, but does not appear in Fig 6 because the cross section has been prepared in a region where it is absent).

4. Discussion

Observations of the implanted device structures, BT3 and BT6, strongly suggest that the high leakage currents found in specimen BT3 are a consequence of the structure of the buried oxide. Silicon crystallites presumably provide electrically conducting paths through the isolating

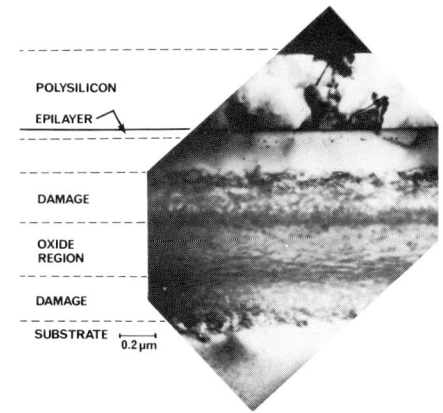

POLYSILICON
GATE · OXIDE
EPILAYER
DAMAGE
OXIDE REGION
DAMAGE
SUBSTRATE
0.2 μm

POLYSILICON
EPILAYER
DAMAGE
OXIDE REGION
DAMAGE
SUBSTRATE
0.2 μm

Fig 5. Device BT6, 90° cross section, TEM bright field, near a <110> type pole, 220 reflection.

Fig 6. Device BT3, 90° cross section, TEM bright field, near a <110> type pole, 111 reflection.

buried oxide layer; when crystallites are absent, (BT6), leakage is low. It is thus seen to be desirable that devices should be fabricated on material which has received a sufficiently high dose to form homogeneous oxide. At 200 KeV, under the implantation conditions described, our results suggest that a dose of 1.8×10^{18} O cm^{-2} is acceptable, but that a dose of 1.4×10^{18} O cm^{-2} is not (see also Hemment et al, 1983).

As mentioned by other workers, eg Das et al (1981), annealing is necessary if a zone of improved crystalline quality is to be formed at the top surface of the implanted silicon structure. The early stages of annealing are revealed in Fig 3, where it is clearly demonstrated that further heat treatment is necessary if devices are to be fabricated directly on the substrate without an intervening epitaxial layer.

Epitaxial layer deposition did not achieve the ideal of providing a perfect crystalline quality, dislocation free, region of silicon isolated by the buried oxide from the substrate. Nevertheless the $SiCl_4/H_2$ CVD system was shown to produce single crystal material which allowed successful device fabrication. The threading dislocation density was at least not higher than in the immediately underlying implanted and annealed substrate region.

A batch of C-MOS buried oxide devices has now been prepared, and results are encouraging, with good electrical isolation having been found in initial measurements.

The combination of cross sectional TEM (microscopic structural information) and AES (chemical information) has provided a useful input for modelling of the implantation process in buried oxide structures, Dobson et al (1983). Particularly important is the accurate depth scale which can be associated with chemical and structural effects.

5. Conclusions

1. N-channel devices fabricated on oxygen implanted silicon possessed good electrical isolation when the buried oxide layer was homogeneous, amorphous stoichiometric SiO_2, but leakage was high when the buried layer was a mixture of amorphous oxide and crystalline silicon.

2.   A dose equivalent to 1.8×10^{18} O cm^{-2} at 200keV, and at an implantation temperature of approximately $550^{\circ}C$, was sufficient to form a homogeneous amorphous stoichiometric buried oxide suitable for device fabrication.

3. Annealing at 1000 $^{\circ}C$ for 30 mins under dry nitrogen with a silicon capping layer revealed the early stages in the formation of a zone of reasonable crystalline quality at the substrate surface. Annealing at 1150 $^{\circ}C$ for 2 hours provided a region suitable for device fabrication.

4. Epitaxial layers suitable for device fabrication were successfully deposited by the $SiCl_4/H_2$ CVD system, though threading dislocation densities were not reduced below the levels in the underlying substrate.

6. Acknowledgements

Acknowledgement is made to the Director of Research, British Telecom, for permission to publish this paper. It is a pleasure to thank Mr Ancell, British Telecom, for the mechanical polishing of specimens.

References

Das K, Butcher J B, Wilson M C, Booker G R, Wellby D W, Hemment P L F and Anand K V 1981 Inst. of Phys. Conf. Ser. No. 60 307
Dobson R M and Arrowsmith R P 1983 To be published
Fellenberg R, Streubel P and Heisel A 1982 Phys. Status Solidi b 112 55
Fletcher J, Titchmarsh J M and Booker G R 1980 Inst. of Phys. Conf. Ser. No. 52 153
Hemment P L F, Maydell-Ondrusz, Stephens K G, Butcher J, Ioannou J and Alderman J To be published
Homma Y, Oshima M and Hayashi T 1982 Jap. J. Appl. Phys. 21 890
Izumi K, Doken M and Ariyoshi H 1978 Electronics Lett. 14 593
Maeyama S and Kajiyama K 1982 Jap. J. Appl. Phys. 21 744
Pinizzotto R F, Vaandrager B L and Lam H W 1982 Proc. Mater. Res. Soc. 7 401
Wrigley N G 1968 J. Ultrastructure Research 24 454

Inst. Phys. Conf. Ser. No. 67: Section 10
Paper presented at Microsc. Semicond. Mater. Conf., Oxford, 21–23 March 1983

TEM investigation of thin oxides in MOS structures

P Pongratz, H Oppolzer[*], D Schmitt-Landsiedel[*], K Hofmann[*], G Dorda[*] and
P Skalicky

Inst. f. Applied Physics, Karlsplatz 13, TU-Vienna, Austria
*) Siemens AG, Research Laboratories, Munich, Fed. Rep. Germany

Abstract High resolution lattice imaging of cross-sectional specimens
of thin MOS structures was used to determine the thickness of the gate
oxide. This geometrical thickness is compared with thickness values obtained from C-V measurements, measurements of the Shubnikov-de Haas oscillations at 4.2 K, Fowler-Nordheim tunneling and ellipsometry. To get
most probable thickness values for the SiO_2 layer, for the transition
region SiO_x and for the average distance of the electrons from the Si/
SiO_2 interface, all five methods have been applied on the same sample.

1. Introduction

The design of VLSI circuits is based on MOS transistors with small channel
lengths. The scaling principle (Dennard at al., 1974) implies that for
small lateral dimensions also thin gate oxides have to be used to reduce
short channel effects. The thickness t_{ox} of the amorphous SiO_2 layer between the Si substrate and the polysilicon of an MOS structure can be determined accurately by lattice imaging TEM of thin cross-sections (Krivanek,
1980). This geometrical thickness can be compared with t_{ox} values obtained
by measuring electrical properties as capacitance (C-V), Shubnikov-de Haas
(SdH) oscillations at 4.2 K and Fowler-Nordheim (FN) tunneling current.
Furhtermore t_{ox} can be determined optically by ellipsometry. The film
thicknesses obtained from C-V and FN measurements and by evaluation
of SdH oscillations at 4.2 K, include the average distance d_{el} of the
electrons from the Si/SiO_2 interface. To get the most probable thickness
values for the SiO_2 layer, for the transition region SiO_x and for d_{el}, all
measurements have been performed on the same sample.

2. Experimental

Thin layers of SiO_2 were grown on (100)-Si (p-type, 1 $\Omega \cdot cm$) by thermal
oxidation in dry O_2 + 2 % HCl at 900°C for approximately 2 min. Polysilicon was deposited by LPCVD at 625°C and highly P-diffused at 900°C.
Cross-sectional samples for TEM investigations with the foil normal of the
substrate Si oriented along [110] were prepared in the usual way (Henghuber et al. 1980) and thinned by ion beam milling to get transparent
foils with a thickness of less than 40 nm in the area of interest. A JEOL
200 CX electron microscope with a side entry goniometer and a high resolution objective lens (point resolution 0.3 nm at 200 kV) was used in the
lattice imaging mode. The Si crystal was oriented in the eucentric position with its [110] zone parallel to the beam direction (axial illumination) and an aperture was chosen to include all type {111} and {200} dif-

fracted beams symmetrically. A multi-beam image can thus be made with the Si/SiO$_2$ interface lying just perpendicular to the image plane. Structural detail at the interface between the regular lattice planes of the Si substrate and the amorphous gate oxide is visible and the interface roughness can be estimated (Fig. 1). In favourable cases a low-index lattice plane or even a zone axis of a polysilicon grain is parallel to the substrate zone axis and lattice fringes appear on both sides of the gate oxide (Fig. 1 and 2). Imaging of both interfaces by the transition of crystalline to amorphous material allows the determination of the oxide thickness independent of black and white contrast from Fresnel fringes. Specimen thickness and defocus must be chosen very carefully to obtain good contrast of the lattice fringes. Typical specimen thickness was 30 nm and defocus was about 200 nm, although due to local bending of the crystal, it is not possible to define imaging conditions very exactly. Accurate magnification calibration (0.5 %) is achieved from the spacing of the lattice fringes (e.g. (111)-type, which are in strong contrast). It is mainly the interface roughness which is more pronounced on the polysilicon side, where variations with a periodicity of 5 to 10 nm occur, that limits the accuracy of the thickness measurements. Image processing of the micrographs could improve the delineation of the interfaces. Figures 1 and 2 although taken with different defocus values show the same oxide thickness of t_{ox} = 5.8 nm. The error, as determined from the foregoing consideration is $\pm$ 0.3 nm.

The oxide capacitance was measured in two ways: 1) with a 1 MHz capacitance bridge, correcting also the errors caused by the substrate and contact series resistances, and 2) with a slow ramp method. The oxide capacitance is related to the oxide thickness by the plate condenser formula. It must be emphasized, however, that the mobile carriers inside the semiconductor

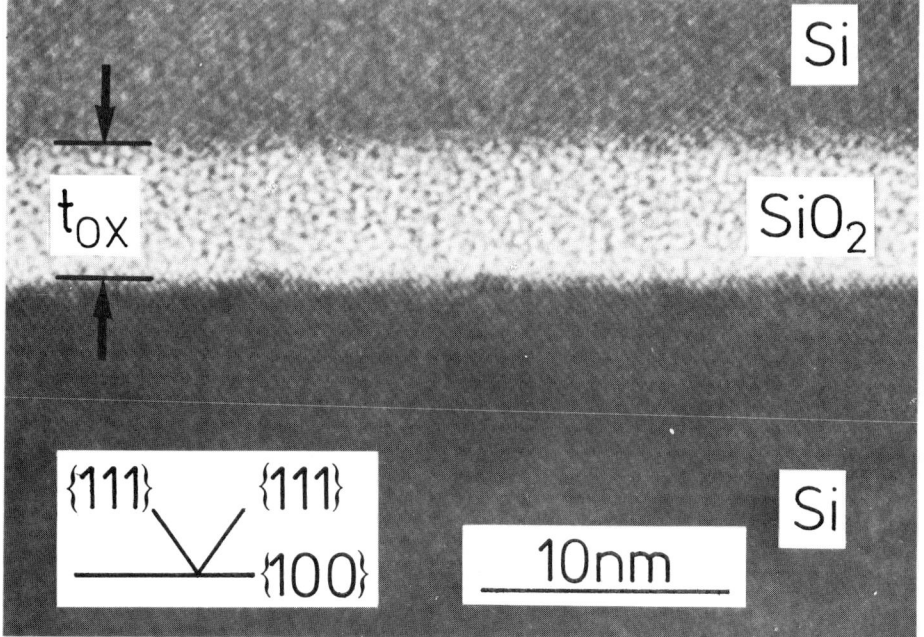

Fig. 1: Multi-beam image of cross section through thin gate oxide between Si (bottom) and polysilicon (top) crystals. SiO$_2$ thickness t_{ox}= 5.8±0.3 nm.

inversion or accumulation layer have a finite average distance d_{el} from the oxide. This distance has been calculated quantum mechanically in dependence of the electron concentration (Stern, 1972). Since the capacitance is de-termined by the average distance between the charge layers on the condenser plates, the average electron distance d_{el} - corrected for the different di-electric moduli in oxide and semiconductor - is included in the value of the "effective oxide thickness" $t_{ox,eff}$ which is obtained from C-V measurement.

The Fowler-Nordheim tunneling method for oxide thickness determination (Hofmann, 1983) uses the current of electrons tunneling through the surface barrier from the Si substrate into the SiO_2 conduction band at high applied electric fields (6-8 MV/cm). The current density depends only on the elec-trical field in the oxide outside the tunnel distance and is determined by the applied voltage and the effective capacitance of the MOS structure. The elctrical field E_o to produce a given current density $j_o = j(E_o)$ is obtained with high accuracy from the voltage and from the ellipsometric thickness of a thick oxide layer. Thus an effective thickness $t_{ox,FN}$ of thinner oxides can be determined from the applied voltage V which is necessary to achieve the same current density j_o by setting $V = E_o \cdot t_{ox,FN}$. The FN - current depends very sensitively (exponentially) on the field E and thus also on small oxide thickness variations. For thin oxides with such thickness variations this could result in a smaller effective value than the geometrical average thickness.

<u>Fig. 2:</u> Multi-beam image of cross section through MOS structure. A micro-twin in the polysilicon is out of contrast. The thickness of the amorphous SiO_2 is not uniform throughhout the film as can be seen on the right side of the polysilicon interface. The gate oxide thickness is t_{ox} = 5.8 nm.

Transmission
electron microscopy

Ellipsometry

Fowler-Nordheim
tunneling

Shubnikov-de Haas
oszillations at 4.2 K

Capacitance voltage
measurement

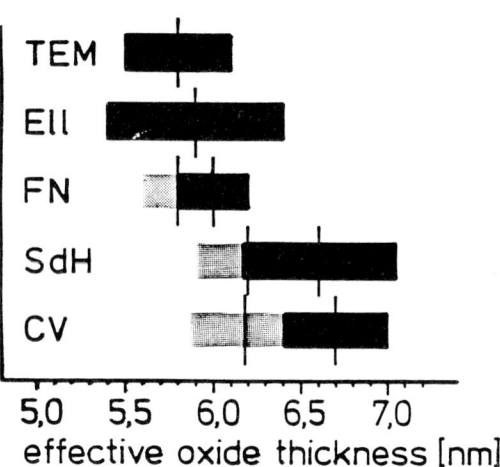

Fig. 3: Comparison of effective oxide thickness for all methods employed. Shaded error bars indicate values after correction for d_{el}.

The energy band of electrons in an inversion layer are quantized into Landau levels if there is a magnetic field at low temperature (Dorda et al. 1978). For a varying electron concentration or varying gate voltage the Fermi level crosses these Landau levels and the resistivity shows an oscillatory behaviour. The difference in electron concentration between two minima of these Shubnikov-de Haas oscillations can be calculated and compared to the difference in gate voltage, yielding a further possibility to evaluate the oxide capacitance and oxide thickness. Here again the average electron distance from the oxide - at 4.2 K - is included in the obtained "effective oxide thickness". As only a DC voltage and the magnetic field have to be measured the SdH oscillations provide an accurate, alternative method to determine the electrically effective oxide thickness. The ellipsometric measurement of t_{ox} was performed by means of an automatic ellipsometer Auto EL II from Rudopf Research.

3. Discussion

Figure 3 shows a comparison of t_{ox} for the various methods. Error bars and the corrections for the average electron distance d_{el} from the oxide (CV, SdH) and the potential drop inside the semiconductor (FN) (shaded error bars) are indicated. A value of t_{ox} = 6.0 nm lies within the error bars of any method. It is possible that corrections for surface roughness and/or an atomic layer of nonstoichiometric SiO_x must be included to explain the slightly larger values for t_{ox} from CV and SdH measurements. It is, however, very satisfactory that all values of t_{ox} lie within an interval of 0.5 nm. The transition region SiO_x between Si and SiO_2 extends only over one or two atomic layers.

References

Dennard R H et al. 1974 J. Solid-State Circuits 9 256
Dorda G 1973 Festkörperprobleme XIII Ed. Sauter (Vieweg, Braunschweig) 215
Dorda G, Eisele I and Gesch H 1978 Phys. Rev. B 17 1785
Henghuber G, Oppolzer H and Schild S 1980 Siemens F.-u.E.-Ber. 9 363
Hofmann K R 1983 to be published
Krivanek O L and Mazur J H 1980 Appl. Phys. Letters 37 392
Stern F 1972 Phys. Rev. B 5 4891

Inst. Phys. Conf. Ser. No. 67: Section 10
Paper presented at Microsc. Semicond. Mater. Conf., Oxford, 21–23 March 1983

495

Transmission electron microscopy of Si₃N₄/SiO₂ structures for local oxidation processes

J Vanhellemont★, C Claeys☆, J Van Landuyt★, G Declerck☆, S Amelinckx★
and R Van Overstraeten☆

★Universiteit Antwerpen, R U C A ,Groenenborgerlaan 171, B-2020 Antwerpen,
 Belgium
☆K U Leuven, E S A T ,Kardinaal Mercierlaan 94, B-3030 Heverlee, Belgium

Abstract High voltage electron microscopy and high resolution electron
microscopy analysis are performed to investigate the properties of
Si₃N₄/SiO₂ structures used in local oxidation processes (LOCOS). Special
attention is given to "bird's beak" formation, gate oxide thinning and
defect generation in the silicon substrate. Several models to explain
the dislocation generation are discussed.

1. Introduction

The local oxidation technique is widely used for the isolation of active
regions of LSI and VLSI (Very Large Scale Integrated) circuits. A Si₃N₄
mask protects the gate oxide areas during the isolation oxidation. The
generation of dislocations by the high intrinsic stresses in the nitride
is prevented by a thin SiO₂ layer (pad oxide) that separates the mask from
the Si-substrate.
The breakthrough of several NMOS and CMOS processes resulted in a huge
amount of investigations in order to characterize the different aspects of
Si₃N₄/SiO₂ interface structures. During local oxidation the isolation oxide
penetrates under the nitride mask and lifts it up. The resulting configuration is called a "bird's beak". The length of this beak limits the packing density of the integrated circuits. After removing nitride layer
and pad oxide, the gate oxide is grown. Often one observes a thinning of
the gate oxide near the edge of the LOCOS pattern, the so-called "white
ribbon" effect (E Kooi et al 1976, T A Shankoff et al 1980). This thinning degrades the electrical performance of the transistors by lowering
the breakdown voltage. This paper discusses some of the phenomena occurring at the nitride edge during local oxidation.

2. Experimental Techniques

In this study n-type, 8-12 Ωcm, 5 cm ∅, 250 µm thick, (001) oriented
Czochralski silicon wafers are used. After growing a 60 nm pad oxide in
dry oxygen at 1000°C, 120 nm LPCVD (Low Pressure Chemical Vapour Deposited)
Si₃N₄ is deposited.
The nitride pattern, defined by wet etching in hot H₃PO₄ acid with an undoped silox layer as a mask consists of stripes parallel to a <110> direction. The width of the stripes is 12 µm, the spacing between them 10µm.
The field oxidation is carried out at 975°C in wet oxygen for variable
times. After etching the 'Si₃N₄ sandwich', the 60 nm gate oxide is grown in
a chlorine containing ambient at 1025°C. In some experiments an additio-

nal wet oxidation step at 975°C is carried out before etching the pad oxide.
To investigate the bird's beak parameters it is necessary to observe the
structures in a plane perpendicular to the nitride edge. A vertical cross-
sectioning technique was optimized to prepare the required samples (Vanhellemont et al 1983a). The specimens are investigated by both high voltage
(1000kV) and high resolution electron microscopy. The length of the bird's
beak is also determined from electrical measurements.

The generation of defects at the nitride edge is studied on wafers with the
Si_3N_4 directly on the silicon substrate (no pad oxide). After field oxidation only part of the nitride layer is removed thus preventing the silicon
from being attacked by the H_3PO_4 etch. The thick oxide is completely removed with buffered HF. Samples are thinned chemically from the backside in
order to obtain "surface" specimens. Due to the remaining nitride layer
the edge of the structure can easily be located. The distribution of the
defects as function of the depth under the surface is studied in cross-section by both optical microscopy after Wright etching and TEM.
Dislocations are generated at the nitride edge by heating the samples in
the high voltage microscope.

3. Results and Discussion

Figure 1 shows a bird's beak configuration after local oxidation for 10 h.
The length is 1,1 µm, in fair agreement with the electrical measurement
which gives a value of 0 ,75 µm.
Thick oxide, pad oxide, nitride layer and the thin oxide on the nitride can
easily be distinguished from absorption contrast. The thick layer covering
the field oxide and the bird's beak is a 450 nm LPCVD polysilicon layer
used to protect the structure during specimen preparation and to improve
the contrast at the oxide surface. Higher magnifications allow accurate
measurement of the thickness of each layer, which is not possible by other

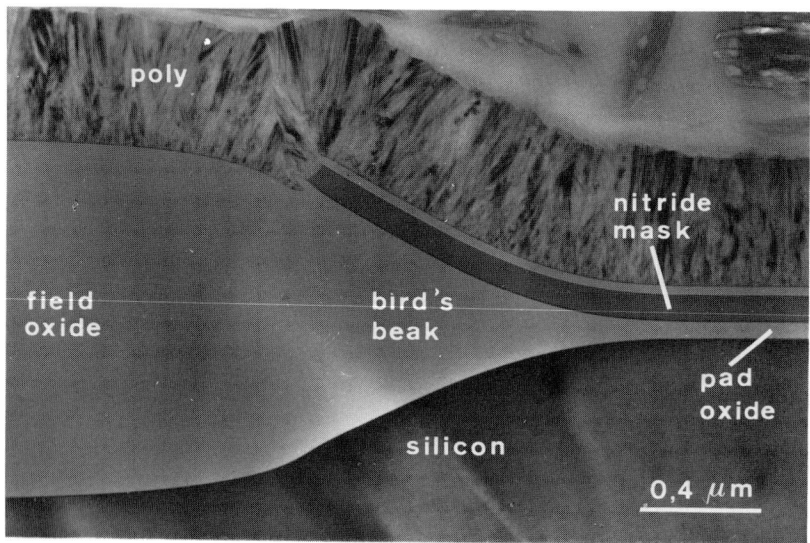

Fig.1 Bird's beak configuration after a 10 h wet oxidation at 975°C

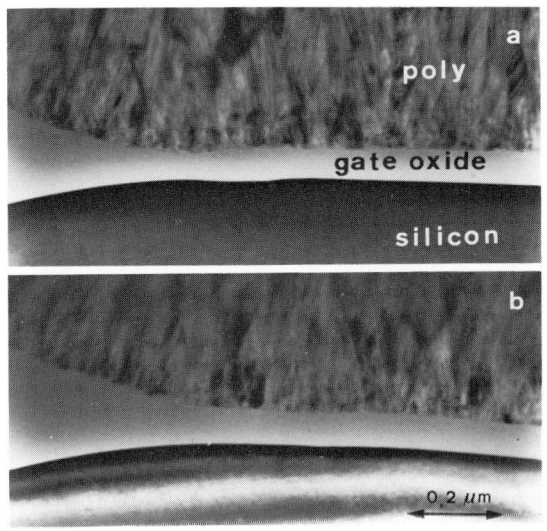

Fig. 2
a) Gate oxide thinning
   of 20%
b) No gate oxide thin-
   ning after a sacri-
   ficial wet oxidation
   step

means. This points out that after 10 h local oxidation about 10 nm of nitride is converted in SiO_2. Figure 2a shows the gate oxide at the LOCOS
edge. A gate oxide thinning of about 20% is observed. Kooi et al (1976)
assume that this effect is caused by an oxidation barrier consisting of
either Si_3N_4 or an oxynitride at the Si/SiO_2 interface. An additional wet
oxidation step for 15 min at 975°C before etching the pad oxide removes the
oxidation barrier. After subsequent gate oxidation no thinning is observed
(fig.2b). By increasing the local oxidation time to 30 h a gate thinning
of up to 100 %, depending on the overetching with BHF while removing the
pad oxide, is observed when no sacrificial wet oxidation step is used. When
a stoichiometric Si_3N_4 is assumed to form the barrier this would require a
minimum nitride thickness of 3 nm. However, as we were not able to observe
this by HREM presumably very little nitridation of the silicon surface occurs but rather some kind of oxynitride layer is formed in the pad oxide
near the interface (Habraken et al 1982). High resolution electron microscopy shows the SiO_2/Si interface of both the gate oxide and the thick oxide

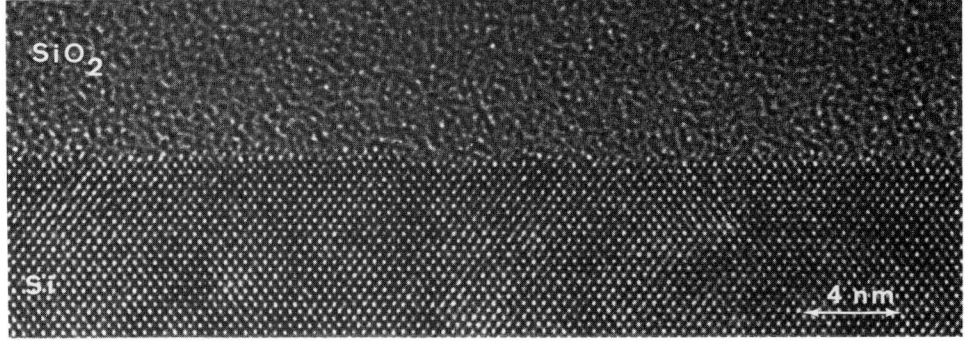

Fig.3 High resolution image of the field oxide/silicon interface

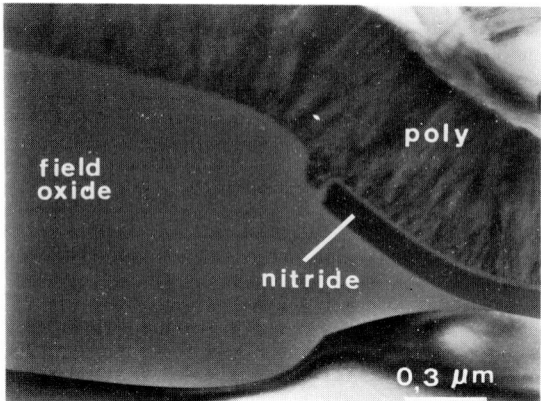

Fig.4 Bird's beak con-
figuration after 10 h
field oxidation (with-
out pad oxide)

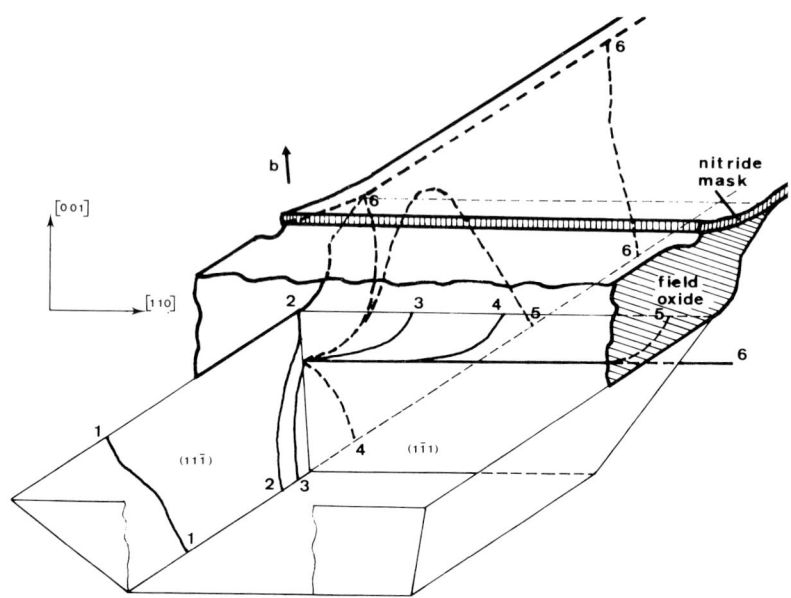

Fig. 5 Cross-glide mechanism at the nitride edge. Dislocation (1) glides
inwards from the rim of the wafer in a (11$\bar{1}$) plane. The stresses exerted
by the nitride mask and the field oxide reach a maximum near the bird's
beak and force the upper part of the dislocation to cross-glide in the
(1$\bar{1}$1) plane parallel to the <110> edge (Kolbesen et al 1981), while the
deeper part continues to glide in the original (11$\bar{1}$) plane and forms a
half loop (1-6) (Hu et al 1976).

to be very flat and without a visible transition zone.Small steps (typical
0,3 to 0,6 nm) are formed on the silicon surface (fig.3).

Wafers with the nitride mask directly on the Si substrate are oxidized for
respectively 1h and 10h. After 1h no dislocations are found. This indicates
that even without a pad oxide the intrinsic stresses in the nitride layer
are not large enough to generate dislocations. After 10h essentially two
types of dislocations are generated : dislocation half loops perpendicular
to the nitride edge and straight 60° dislocations parallel to the <110>
edge. Cross-section TEM inspection reveals that a short bird's beak with a
sharp bend in the SiO_2/Si interface under the nitride edge is formed(fig.4).
A large notch is created in the thick oxide at the edge. The dislocation
half loops have their endpoints at the beak's point and at the bend in the
interface. They are of the type described by Hu et al (1976), who suggest
that they result from interaction of screw dislocations, gliding inwards
from the rim of the wafer, with the nitride edges. Often cross-glide of
the dislocation part under the thick oxide is observed, resulting in a
straight 60° dislocation line. This can be explained by a mechanism similar to that proposed by Kolbesen (1981). The combination of the two mechanisms is schematically shown in figure 5. In regions with a low dislocation density, the majority of the 60° dislocations seems to originate in
this manner. They all have the same Burgers vector, of the $\frac{a}{2}[011]$ type inclined to the surface, and lie in close neighbouring $(1\bar{1}1)$ planes under the
thick oxide. Cross-section investigation reveals that they extend to a
depth of about 10 μm.
Where the dislocation density is very high often interactions of another
type of dislocation half loop with the 60° dislocations occur forming dense
networks (fig.6). These half loops, perpendicular to the edge but with a
Burgers vector parallel to the surface of the $\frac{a}{2}[1\bar{1}0]$ type are never found
isolated. Their formation can be explained by the same mechanism (Hu et al
1976) but in this case 60° dislocations traversing the wafer thickness
glide inwards and interact with the straight dislocations at the edge.
The reactions observed are of the type :

$$\frac{1}{2}[1\bar{1}0] + \frac{1}{2}[011] \rightarrow \frac{1}{2}[101]$$

$$\frac{1}{2}[1\bar{1}0] + \frac{1}{2}[10\bar{1}] \rightarrow \frac{1}{2}[2\bar{1}\bar{1}]$$

In each $(1\bar{1}1)$ plane there are two possible Burgers vectors for 60° dislocations parallel to the <110> direction : $\frac{1}{2}[011]$ and $\frac{1}{2}[10\bar{1}]$.

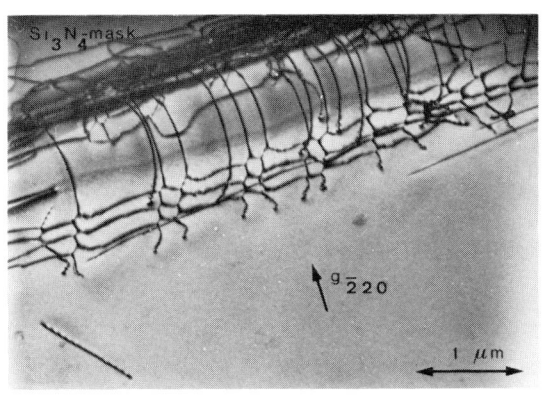

Fig.6 Dislocation net-
work formed at the
nitride edge

In regions of high dislocation density we sometimes observed pairs of 60°
dislocations lying in the same (1$\bar{1}$1) plane, with alternating Burgers vector
(Magdo and Bohg,1978).They interacted close to the surface as follows :

$$\tfrac{1}{2}[011] + \tfrac{1}{2}[10\bar{1}] \rightarrow \tfrac{1}{2}[110]$$

Not only in the surface region but also in the bulk, below 5µm under the
surface, a high density of defects (small stacking faults, precipitates,
prismatic punching, perfect dislocation loops) are formed. When a pad oxide
is used only very few bulk and surface defects are observed.
One of the possible generation mechanisms of the 60° dislocations is studied
by heating samples without pad oxide in the electron microscope (at 500kV
to prevent radiation damage)(Vanhellemont et al 1983b).Specimen preparation
locally causes small scratches on the silicon surface. At a temperature of
about 600°C one observes the generation of small half loops at the surface
damage. These half loops lie in (1$\bar{1}$1) planes parallel to the <110> direction of the mask. Further heating causes the half loops to grow and when
one end reaches the undamaged surface it starts to glide very quickly, depending on the temperature, parallel to the <110> direction and a 60° dislocation is formed. Dislocation lenghts up to 800µm can easily be generated when the surface damage lies very close to the edge. Often more than one
60° dislocation is generated at the same scratch. They sometimes have alternating Burgers vectors and may interact with each other. It is clear
that this source of dislocations is very effective and is activated in the
first minutes of the local oxidation.

4. Conclusion

It has been demonstrated that HVEM and HREM are very useful tools to characterize the different aspects of Si_3N_4/SiO_2 structures. The use of a pad
oxide has a very large influence on both the characteristics of the bird's
beak and the defect generation in the silicon substrate.

Acknowledgement

G. Declerck is indebted to the Nationaal Fonds voor Wetenschappelijk Onder-
zoek (N.F.W.O.) for his fellowship as Onderzoeksleider. This work is performed with financial support of the Belgian Science Foundation (IIKW).

References

Habraken F H P M, Kuiper A E T, Tamminga Y and Theeten J B 1982 J. Appl.
    Phys. 53 6996
Hu S M, Klepner S P, Schwenker R O and Seto D K 1976 J. Appl.Phys.47 4098
Kolbesen B O and Strunk H 1981 Inst.Phys. Conf. Ser. N°57 21
Kooi E, van Lierop J G and Appels J A 1976 J. Electrochem.Soc.123 1117
Magdo I and Bohg A 1978 J. Electrochem. Soc. 125 932
Shankoff T A, Sheng T T, Haszko S E, Marcus R B and Smith T E 1980 J.
    Electrochem. Soc. 127 216
Vanhellemont J, Bender H, Claeys C, Van Landuyt J, Declerck G, Amelinckx S
    and Van Overstraeten R 1983a to be published in Ultramicroscopy
Vanhellemont J, Claeys C, Van Landuyt J, Declerck G, Amelinckx S and Van
    Overstraeten R 1983b to be published Proc. 7th International Conference
    on High Voltage Electron Microscopy, Berkeley

Inst. Phys. Conf. Ser. No. 67: Section 10
Paper presented at Microsc. Semicond. Mater. Conf., Oxford, 21–23 March 1983

Characterization of TiN films obtained by ion implantation

A Armigliato, A Garulli, D Govoni and P Ostoja

C.N.R. - Istituto LAMEL, Via Castagnoli, 1 - 40126 Bologna (Italy)

Abstract   Titanium nitride films have been prepared by implanting N_2^+ ions in titanium layers deposited on silicon single crystals. The electrical resistivity and the optical properties of such films are better, even in the as-implanted form, than the ones obtained by evaporation and sputtering techniques. Thermal treatments performed in vacuum or in H_2, as well as by electron beam, improved the electrical resistivity, whereas transient annealing by incoherent light resulted in better optical properties. The applicability of these films as a conductive and/or transparent material is now under investigation.

1. Introduction

The properties of titanium nitride films are at present extensively investigated by a number of researchers (Mäenpää et al 1980-1982, Wittmer et al 1981, Rivory et al 1981, Armigliato et al 1982a,b). In particular the electrical performances of such layers when deposited onto silicon wafers have proved to be very promising in the metallization schemes of both the electronic devices and the solar cells technology. Instead, the optical properties of the films obtained by the conventional reactive sputtering process have not been found satisfactory, in view of an application of TiN as a transparent conductive film for silicon solar cells. To this end, we have considered the possibility of obtaining TiN films with good electrical and optical properties by an alternative preparation method, i.e. by implantation of nitrogen ions into silicon wafers covered with a layer of metallic titanium.

2. Experimental

The films of titanium were deposited by electron gun evaporation onto (100) silicon wafers of high nominal resistivity (100 ohm.cm). Two films thicknesses have been chosen, namely 57 and 76 nm; they should result, after implantation, in TiN layers approximately 61 and 82 nm thick, respectively (Armigliato et al 1982a), and hence give rise, for a refractive index n=2, to a minimum in the reflectance curve at $\lambda = 530 \div 680$ nm, according to the quarter wavelength criterion. The specimens were implanted with a dose of 2×10^{17} N_2 ions/cm^2 by means of a LINTOTT III ion implanter, operating at an energy of 40 KeV and a maximum beam current of 4 mA. This leads to an estimated projected range R_p of about 38 nm with a standard deviation $\Delta R_p = 23$ nm; these figures were not affected by any sputtering of Ti, which might occur during the implantation process, as proved by backscattering analysis. The TiN films so obtained were annealed at temperatures of 600°C and 700°C in vacuum and in H_2 for

15 min. Subsequently, additional annealing experiments were performed by electron beam and incoherent light annealing. These two techniques have been developed in our Institute and are described elsewhere (Lulli & Merli 1983, Pedulli & Correra 1982).

The film structure was studied by TEM on plan and cross sectional specimens; the stoichiometry was determined by Rutherford backscattering of 1.8 MeV of He$^+$ ions (Ti analysis) and (d,p) nuclear reactions (N and O analysis). The electrical resistivity was determined by the four point probe method and, in some specimens, with the Van der Pauw geometry, as realized by a photolithographic process; the optical reflectance as a function of wavelength was measured by means of a Jobin Yvon monochromator, equipped with a Barnes Engineering specular reflectance apparatus, whereas the complex refractive index (n-ik) at a wavelength of 546 nm was evaluated with a Gaertner L-119 ellipsometer.

3. Results and Discussion

The as-implanted films were observed by TEM and found to consist of a polycrystalline layer having the structure of cubic titanium nitride, covered with an amorphous film. This latter gave rise to an electron diffraction pattern which was previously attributed to amorphous TiN (Armigliato et al 1982a); actually, nuclear (d,p) reaction analysis evidenced the presence in this surfacial layer of a high carbon content, due to a contamination effect, occurring during the ion-implantation experiments. After removal of this surfacial layer by RF etch-sputtering, the thickness of the TiN films were found to be 66 and 85 nm, respectively. Henceforth these films will be referred to as A and B films, respectively. The size of the TiN crystallites ranged from about 20 up to 80 nm in both the films (Fig. 1a) and a pronounced (111) orientation was evident from the diffraction patterns (see insert in Fig. 1a). The polycrystalline structure of the implanted films is in agreement with the criterion reported by Rauschenbach and Hohmuth (1982) which states that a film produced by implantation has to be crystalline when the ratio R between the atomic radii of the implanted and of the target species is lower than 0.59. In our case R=r(N)/r(Ti)=0.5, and hence a polycrystal is expected. The in-depth structure of the film is visible in the cross section in Fig. 1b, which also evidences the good quality of the film/substrate interface.

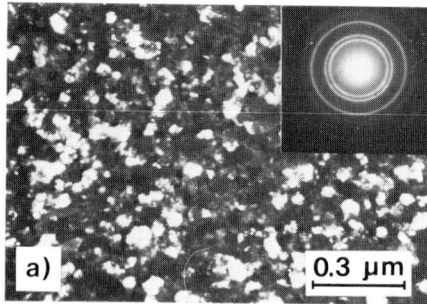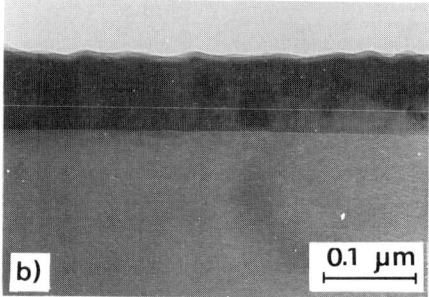

Fig. 1 TEM micrographs of an as-implanted B-TiN film. a) (220) dark-field plan view and SAD pattern; b) cross-sectional image

The nitrogen content was found by N^{14}(d,p)N^{15} reaction to be 3x10^{17}

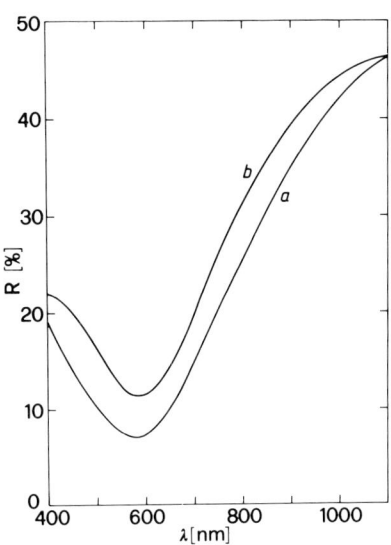

Fig. 2 Curves of reflectance vs
wavelength for an as-implanted
B film (a) and our best evaporated TiN (b) (see text)

at/cm^2, whereas the Ti concentrations
were 2.9×10^{17} and 4.1×10^{17} at/cm^2
in the A and B film, respectively.
There is an excess of nitrogen atoms
in the A film, which has to be located beyond the TiN/Si interface, due
to the tail of the implanted profile;
this is in agreement with the previously reported values of $R_p \pm \Delta R_p$,
which indicates that about 15% of the
total implanted dose lies in the Si
substrate. The electrical sheet resistivity of the A and B as-implanted
films was found to be as low as 27 $\Omega/\square$
and 34 $\Omega/\square$. These values compare
favourably with the best TiN films
obtained previously by e-gun evaporation, which presented a sheet resistivity of 40 $\Omega/\square$. From the optical
point of view, the curves of reflectance exhibited a minimum value
of 7%, for the as-implanted B film,
whereas for the best evaporated one
it was as high as 12%. The lower
reflectance of the implanted film all
throughout the visible range is quite
evident in Fig. 2, although in the
near-infrared range it is still
rather high to be useful for photovoltaic applications. It is worthwhile
to remark that the films obtained by the more usual reactive sputtering
technique presented in general a worse trade-off between sheet resistivity and optical reflectance, as compared with the above quoted evaporated one. The complex refractive index of the A and B films was evaluated
by ellipsometry and turned out to be $n-ik=2.16-i0.4$ and $1.85-i0.6$, respectively. The extinction coefficient k was rather high for a good transparency. It is known to depend critically on the oxygen content, which
amounted to about 4×10^{16} at/cm^2, in both A and B films; this value is
very near to the minimum value we measured in our evaporated or sputtered TiN films (Armigliato et al 1982b). The electrical resistivity and
the optical reflectance were found to improve after annealing in vacuum
and in H$_2$ atmosphere (see also Armigliato et al 1982a) in the range 600
to 700°C for 15 min. The values of sheet resistance in the A and B films
for the different annealing treatments are reported in Table 1.

Table 1 Sheet resistances ($\Omega/\square$) of TiN films annealed at
600 and 700°C in vacuum and in hydrogen

	vacuum		hydrogen	
Film	600°C	700°C	600°C	700°C
A	21.8	20.4	19.4	18.0
B	19.5	7.3	17.5	6.9

It is evident the marked decrease in sheet resistance in film B annealed at 700°C in both the ambients. This is probably related to the formation of titanium silicide at the TiN/Si interface, as can be deduced from the RBS spectra reported in Fig. 3; in fact, titanium silicides are known to be effective as conductive materials in silicon device technology (Murarka 1980, Mohammadi 1981). From the analysis of the spectrum in Fig.3b it comes out that a $TiSi_2$ phase of roughly 10 µg/cm^2 mass thickness takes place after an annealing at 700°C in vacuum. This result is con-

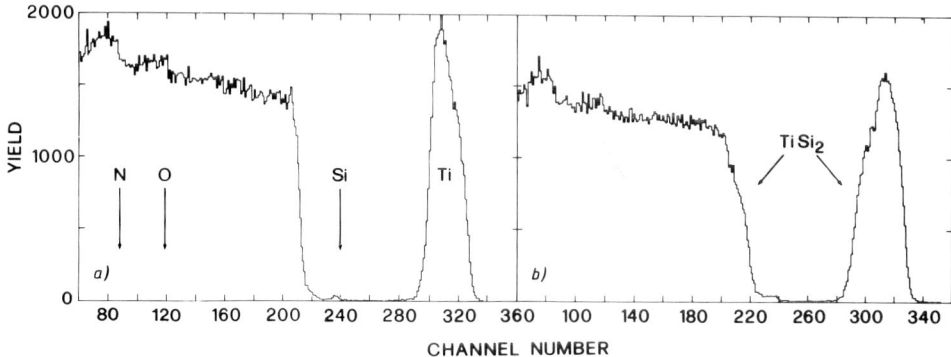

Fig. 3 RBS spectra of B-TiN. a) as implanted and b) after an annealing at 700°C for 15 min in vacuum. The formation of $TiSi_2$ is evident ·in b). A small segregation effect of Si at the surface is also visible

firmed by TEM observations, which revealed the presence of additional rings in the TiN diffraction patterns, whose corresponding interplanar spacings are consistent with the ones of the orthorhombic $TiSi_2$ (ASTM card No.10-225). In Fig. 4 is reported a dark field micrograph taken with the (200) reflection of $TiSi_2$, which exhibits the presence of crystallites some tens nm in diameter, also in form of agglomerates. The formation of a silicide was noticed also in films annealed at 600°C in vacuum, but in this case the RBS analysis assigned a TiSi stoichiometry; this is not surprising, because it is known that $TiSi_2$ is stable at temperatures higher than TiSi. It has not been so far possible to confirm the presence of TiSi by electron microscopy. The formation of silicides only in B films can be explained by taking into account the energy of formation of TiN, TiSi and $TiSi_2$, which are $\Delta G=-84$, -31 and -32 kcal/mole, respectively. When a high concentration of N is present at the TiN/Si interface, like in A films, the silicide formation is prevented, whereas it becomes possible in B films, where the interface is located far beyond the $R_p + \Delta R_p$ position of the depth profile of the implanted nitrogen. On the other hand, the thickness of the silicide layer, as determined from RBS spectra, resulted smaller in films annealed in H_2 atmosphere as compared to the vacuum-annealed ones. This depends probably on the higher oxygen content which was detected by the $O^{16}(d,p)O^{17}$ reaction in H_2-annealed films ($C_o=10^{17}$ at/cm^2) and due to oxygen contamination, unavoidable even in the purest hydrogen gas. As it is known, the presence of oxygen favours the formation of TiO_2 instead of Ti silicides, as the energy of formation of oxide is $\Delta G=-229$ kcal/mole (Goldschmidt 1967, Mäenpää et al 1982). The minimum value of the optical reflectance curve of the annealed specimens was found to decrease, as compared to the as-implanted ones, but with a corresponding in-

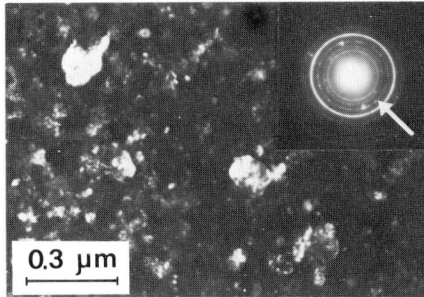

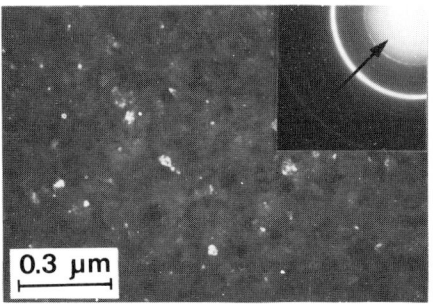

Fig. 4 Dark-field TEM micrograph
of the film in Fig. 3b, taken with
the $TiSi_2$ reflection arrowed in the
SAD pattern in the insert

Fig. 5 Dark-field TEM micrograph
of an A-TiN incoherent light annealed at 800°C, obtained with a
TiO_2 reflection (see insert)

crease in the near-infrared range. Moreover, a moderate decrease in the
value of the extinction coefficient was noticed, as previously reported
(Armigliato et al 1982a), but the ultimate value was still higher than
0.1. We employed also two additional annealing techniques, namely the
electron beam rapid isothermal and the incoherent light annealing, with
the aim of improving the films characteristics. In the first case, we
used a beam with power densities between 3.25 and 7.5 W/cm^2, which
correspond to stationary temperatures of 600, 700 and 800°C, respectively, impinging for 30 sec onto specimens 1.5x1.5 cm^2 wide. After the
annealing, the sheet resistivity was measured and found to decrease with
temperature, particularly for the B films, where an ultimate value of
3 $\Omega/\square$ was attained (see Table 2). To our knowledge, this is the lowest
value obtained in TiN films, although it is related to the formation of
silicides, which occurred also in this case.

Table 2 Sheet resistances ($\Omega/\square$) of
TiN films annealed at 600, 700 and
800°C by electron beam

Film	600°C	700°C	800°C
A	22.2	17.7	12.3
B	17.7	11.7	3.2

The optical constants of these films did not result to be better than
the ones obtained in thermally annealed specimens, in particular as
regards the near-infrared range of the reflectance curve and the value
of the extinction coefficient. The transient annealing by incoherent
light was carried out in air with a xenon arc lamp as a light source.
The current in the lamp was set in our preliminary experiment, so that a
temperature of 800°C was maintained for about 5 sec. The specimens were
kept rotating during the annealing experiment, in order to ensure the
thermal homogeneity. As a result, the following values were obtained:
R_s=25 $\Omega/\square$, n-ik =2.15-i0.07, R_{min} <1% at λ=600 nm. The marked improve-

ment in the optical characteristics of the films are due to the increase in the oxygen content ($C_o = 1.6 \times 10^{17}$ at/cm^2), which gives rise to the formation of small TiO_2 crystallites dispersed in the TiN film, as can be seen from the diffraction pattern reported in the insert in Fig. 5. The TiO_2 grains are visible in the dark field image of this figure.

4.Conclusions

The TiN films prepared by ion implantation exhibit in general a better trade-off between electrical and optical properties as compared to the ones obtained by evaporation or sputtering techniques. Thermal treatments performed in vacuum or in hydrogen atmosphere, as well as electron beam treatments, improved the electrical resistivity, particularly when the formation of silicides occurs, whereas the optical properties are markedly dependent on the oxygen content, as in the case of the transient annealing by incoherent light, performed in air. It is worthwhile to notice that the formation of crystallites of TiO_2 has not a significant adverse effect on the film resistivity, as usually occurs in evaporated or sputtered films. Work is now in progress to gather more information on the applicability of these films in metallization schemes or as a transparent conductive material.

Acknowledgments

The authors are indebted to M Berti and A V Drigo for the RBS and nuclear analysis with the Van de Graaff accelerator of the Laboratori Nazionali di Legnaro; to G Lulli, L Correra and G Angelucci for electron beam, incoherent light and thermal annealings, respectively. The skillful technical assistance of S Guerri, C Boccafogli, P Deluca, E Gabilli and R Lotti is also gratefully acknowledged.

References

Armigliato A, Celotti G, Garulli A, Guerri S, Lotti R, Ostoja P and Summonte C 1982a Appl. Phys. Lett. 41 446
Armigliato A, Celotti G, Garulli A, Guerri S, Martinelli G, Ostoja P and Rosa R 1982b Thin Solid Films 92 341
Goldschmidt H J 1967 Interstitial Alloys (New York: Plenum Press) p 219
Lulli G and Merli P G 1983 Materials Chem. in press
Mäenpää M, von Seefeld H, Cheung N, Nicolet M-A and Cullis A G 1980 Proc. Symp. Thin Film Interfaces and Interactions, Vol. 80-2 eds J E E Baglin and J M Poate (Princeton: The Electrochem. Soc.) p 199
Mäenpää M, Nicolet M-A, Suni I and Colgan E G 1981 Solar Energy 27 283
Mäenpää M, Suni I, Sigurd D, Finetti M and Nicolet M-A 1982 Phys. Stat. Sol.(a) 72 763
Mohammadi F 1981 Solid State Technol. Jan/81 p 65
Murarka S P 1980 J. Vac. Sci. Technol. 17 775
Pedulli L and Correra L 1982 in Laser and Electron Beam Interactions with Solids eds B R Appleton and G K Celler (New York: Elsevier) p 777
Rauschenbach B and Hohmuth K 1982 Phys. Stat. Sol. (a) 72 667
Rivory J, Behaghel JM, Bertier S and Lafait J 1981 Thin Sol Films 78 161
Wittmer M, Studer B and Melchior H 1981 J. Appl. Phys. 52 5722

Inst. Phys. Conf. Ser. No. 67: Section 10
Paper presented at Microsc. Semicond. Mater. Conf., Oxford, 21–23 March 1983   507

Microstructural investigation of large area and device size AuGeNi contacts on InP

R J Graham and J W Steeds

H.H. Wills Physics Laboratory, University of Bristol, Tyndall Avenue, Bristol BS8 1TL.

Abstract   The results of investigations by electron microscopy of the AuGeNi contact on (100) InP are presented. Various contacts have been investigated including large area, device sized and contact resistance measurement structures. A correlation between measured contact resistance and observed interface microstructure is discussed.

1. Introduction

AuGeNi contacts on InP are prepared by evaporating Ge, Au and Ni to thicknesses of 500 Å, 5000 Å and 50 Å respectively on (100) InP. These contacts are then heat treated at temperatures between 300°C and 400°C. The structures investigated include large area contacts (several mm²), grid structure metallisation contacts (grid dimension 1 μm), dot metallisation contacts (200 μm diameter) and contact resistance measurement structures. Several contacts were subsequently heat treated for up to several hundred hours to simulate the elevated temperatures encountered in devices during operation.

Extraction replicas of the metal semiconductor interface were prepared after removal of the metallisation layer with Hg and the InP with HCl. These were examined by transmission electron microscopy (TEM), scanning electron microscopy (SEM) and energy dispersive X-ray analysis (EDX). Measurements of contact resistance were made on samples given standard heat treatments and longer heat treatments simulating device operation temperatures.

2. Large area AuGeNi contacts on InP

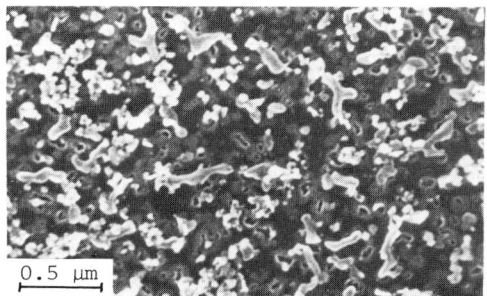

0.5 μm

Fig. 1. InP face of interface (375°C 1 min).

Examination of large area contacts comprised the early part of our investigations. The contacts studied were heat treated at temperatures between 350°C and 400°C and exhibited widely varying microstructural features including a previously unknown phase, nickel germanium phosphide (Graham et al 1981). This phase often occurs as strand or 'Y' shape material as seen by TEM on the extraction

replica and SEM of the InP face of the interface (Fig. 1). It is seen
to lie on a continuous fine grained sheet of particles approximately
0.1 µm thick of a further ternary phase containing Au, Ge and P.
Convergent beam diffraction of isolated single crystals of this phase
approximately 0.1 µm in size, found on 400°C fired contacts, has shown this
phase to be monoclinic (a = 6.45 Å, b = 5.84 Å, c = 6.35 Å, β = 127°,C2/c)
and isostructural with the phase AuGeAs observed at the interface of
AuGeIn contacts to GaAs Gunn diodes (Rackham and Steeds 1981). Subsequent
examination of selected area diffraction patterns of the fine grained
interfacial sheet showed that this consisted predominantly of the ternary
AuGeP phase.

3. Grid structure AuGeNi contacts on InP

The grid structure metallisation contacts investigated consisted of Ge, Au
and Ni deposited as 1 µm wide bars 5 µm apart to thicknesses of 55 Å,
1000 Å and 45 Å respectively, with a heat treatment of 390°C for 2 minutes.
Extraction replicas of the interface revealed small particles of nickel

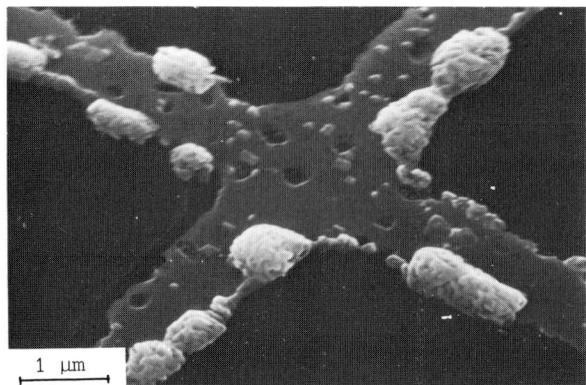

phosphides along approx-
imately 5% of the grid bar
edges, with an interfacial
layer entirely absent. SEM
and EDX of the Au face of
the interface revealed
polycrystalline phases
0.5 µm in size containing
Au and In along the grid
bar edges growing into the
InP to a depth of 0.5 µm
(Fig. 2). The other light
feature visible, particularly at the edge of the
grid bars, may be embryonic
stages of similar Au - In
features.

1 µm

Fig. 2. Au face of interface of grid node.

4. Dot AuGeNi contacts on InP

The dot metallisation contacts consisted of Ge, Au and Ni (500 Å, 5000 Å,
50 Å) dots, 200 µm in diameter, on mesa etched InP. These contacts had
a similar structure to real devices, and were fired at a range of temperatures from 340°C to 390°C. The gold face of the interface of these contacts was examined in SEM after removal of the InP with HCl. A continuous
sheet of small particles was observed but larger protruding features
around the edge of the discs corresponding to similar features seen on the
grid metallisation structures were entirely absent on all samples.

5. Contact resistance measurement structures

The contact resistance measurement structures examined were based on the
transmission line model (TLM) for the horizontal type of planar contact as
described and modelled by Berger (1972), and consisted of a series of
300 µm square GeAuNi (500 Å, 5000 Å, 50 Å) contacts on 5 µm thick
$1.2 \times 10^{18} cm^3 S$ doped (100) InP with spacing between adjacent pads from
10 - 20 µm in 2 µm steps. By injecting known currents between two adjacent
pads and measuring the voltages across the gap between them, the resistance
may be found and plotted against pad separation, yielding the contact re-

sistance via the intercepts and gradient using

$$\rho_c = \frac{R_c^2 \, W}{g} \quad \text{(Berger, 1972)}$$

where ρ_c is contact resistivity, R_c = half the intercept, g = gradient and W = contact width.

The results of contact resistance measurements are summarised in Table 1.

Initial firing condition	Mean contact resistance x10^{-10} Ωm^2	Semiconductor sheet resistance Ω/square
250°C 15 mins (14 samples)	7.46 ± 2.46	3.85 ± 0.71
300°C 15 mins (9 samples)	2.71 ± 1.42	2.39 ± 0.41
340°C 1 min (7 samples)	1.36 ± 0.24	3.05 ± 0.31
360°C 1 min (11 samples)	2.23 ± 0.53	2.33 ± 0.13
375°C 1 min (16 samples)	1.69 ± 0.36	2.16 ± 0.14
390°C 1 min (13 samples)	1.79 ± 0.37	2.70 ± 0.19

Table 1

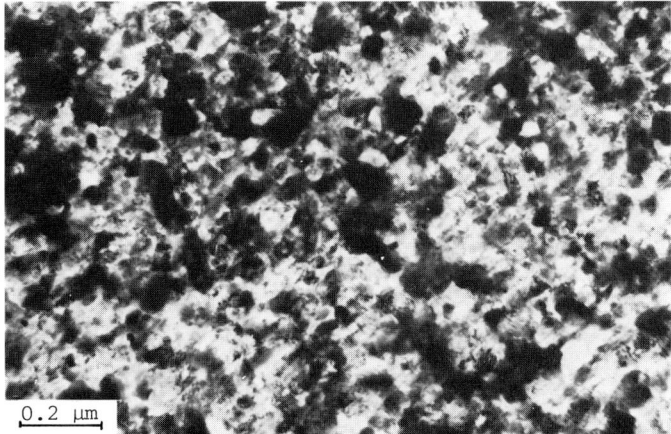

0.2 μm

Fig. 3. Extraction replica of contact interface (390°C for 1 minute)

Extraction replicas of some of these contacts were sub- sequently examined in TEM. All those examined exhibited a continuous sheet of small grain AuGeP - an example is shown in Fig. 3. Several contacts given an initial heat treatment of 300°C/15 mins and 375°C/1 min were then given an ex- tended heat treat- ment at 300°C for up to several hund- red hours to simulate the temperatures encountered in devices during their operation and to investigate what effect this might have on the resistance and microstructure of the contact. Extraction replicas of the contact interfaces were examined in TEM and SEM and are compared with contact resistance measurements in Fig. 4.

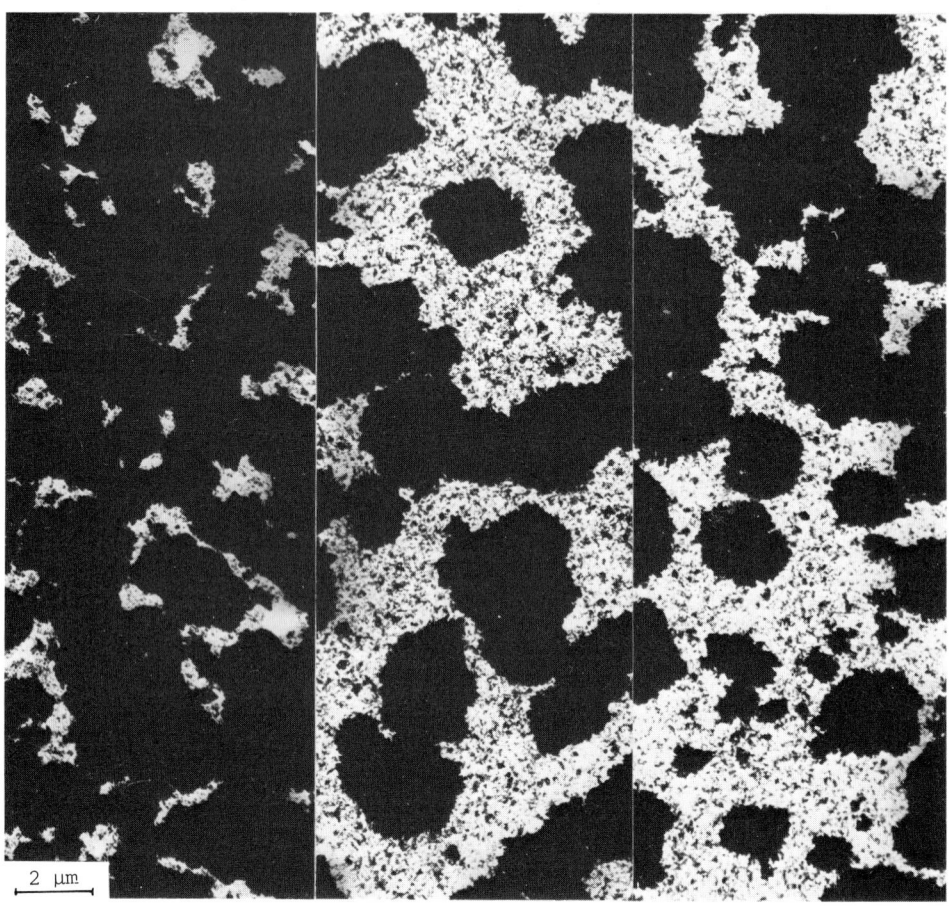

2 μm

Initial firing 300°C
for 15 minutes
    T = 19 hours
CRes = 0.31 x $10^{-10} \Omega m^2$

R_s = 3.69 Ω/square

Initial firing 375°C
for 1 minute.
    T = 72 hours
CRes= 1.12±0.21x$10^{-1} \Omega m^2$

R_s = 2.32±0.16 Ω/square

Initial firing 375°C
for 1 minute
    T = 195 hours
CRes=1.30±0.23x$10^{-10} \Omega m^2$

R_s = 2.12±0.13 Ω/square

Fig. 4. Contacts after extended heat treatment, CRes = mean contact resistance, R_s = mean semiconductor sheet resistance, T = number of hours at 300°C in subsequent heat treatment.

EDX shows the large dark features to contain Au and P which are probably
Au_2P_3. SEM of the extraction replicas shows these Au_2P_3 lumps to be
polycrystalline and extending up to 2 μm into the InP as shown in Fig: 5.

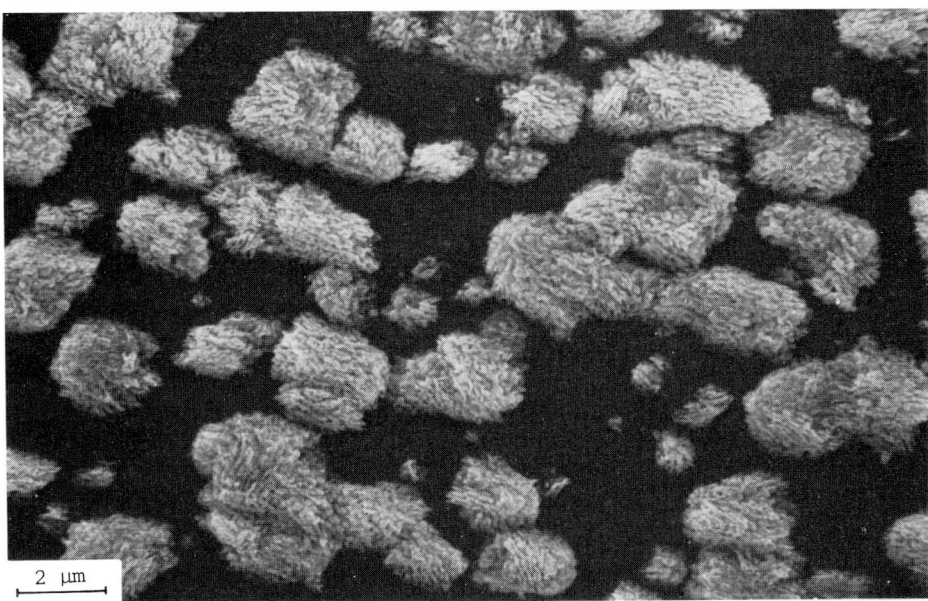

2 μm

Fig. 5. Surface of extraction replica corresponding to Au face of interface of contact (initial heat treatment 375°C for 1 minute, then 300°C
for 72 hours.)

6. Discussion

The shape of the nickel germanium phosphide phase suggests that Ni diffuses
down the Au grain boundaries to the interface where it reacts with Ge and
P to produce $Ni_6Ge_2P_5$. The reason for putting the nickel on the contact
is to prevent the gold 'balling up' during heat treatment and it is
possible that the presence of the $Ni_6Ge_2P_5$ at the gold grain boundaries
inhibits this movement of the gold.

The presence of nickel phosphides along the edges of the grid bars of the
grid metallisation sample suggests that nickel diffuses down these edges
to the InP where reaction and formation of Ni_2P and NiP_2 occurs. A similar
diffusion route can account for the presence of the Au-In features extending into the InP. The absence of a continuous interfacial layer of AuGeP
may be due to thinner metallisation layers of the grid structure contacts
providing insufficient material to form AuGeP. Alternatively, InP surface
contamination prior to metallisation can prevent phase formation as in
'white spot' features as seen on InP (Graham et al 1981). The dot metallisation contacts did not show features attributable to surface diffusion
down contact edges, however a thin interfacial layer of AuGeP was present.
It is therefore possible that this layer somehow limits surface diffusion
along contact edges and initially prevents larger phase growth into the
InP.

The contact resistance and semiconductor sheet resistances can be seen to be generally similar for the range of firing temperatures investigated. The large spread in values is attributed to some dependence of voltage reading on probe position on the contact pads because the metallisation is thin. Relative probe positions on the contacts were kept as consistent as possible. The similar contact resistances for the various heat treatments correlates with the similar general appearance of the contact interface extraction replicas as seen in TEM.

After prolonged heat treatment at 300°C however, obvious differences in microstructure and contact resistance exist between contacts given different initial heat treatments. Contacts initially fired at 375°C for 1 min exhibit similar coverages of Au_2P_3 features growing into the InP for heat treatments at 300°C lasting 72 hours and 195 hours. This correlates with similar contact resistances measured for both samples. For contacts initially fired at 300°C for 15 minutes, however, the coverage of Au_2P_3 is much greater for a heat treatment at 300°C for only 19 hours. This appears to correlate with highly variable but reduced contact resistances and increased semiconductor sheet resistance measurements. There is evidence that the transmission line model may no longer be valid for such samples. Without further data we do not yet wish to interpret and model fully the evolution of such contacts, but it is clear that the growth of Au_2P_3 into the InP and active region of any device is likely to impair device performance.

7. Conclusion

Examination of large area and other contacts has revealed the existence of a continuous fine grained (0.1 μm) sheet of AuGeP at nearly all contact interfaces. The presence of this layer may affect the formation of features due to surface diffusion down contact edges. In view of the large differences in microstructure and contact resistances observed for contacts given prolonged heat treatments with different initial heat treatments, it seems likely that the microstructural nature of the AuGeP layer is significant in controlling diffusion of Au into the InP and subsequent reaction. Investigations in this field continue and will be complemented by a high resolution CL study of the underlying InP.

Acknowledgements

We wish to acknowledge the support of RSRE for supplying samples for our investigations, in particular Mr. M. Slater for fabricating the contact resistance measurement structures and Drs. D. Anderson and J. Staromlynska. The Plessey Co. supplied the grid structure contacts. One of us (R.J.G) holds a CASE studentship with RSRE Malvern.

References

Berger H H, 1972 Solid State Electronics 15 pp 145-158.

Graham R J, Mansfield J F and Rackham G M 1981 Electron Microscopy and
    Analysis 1981 (Bristol: Inst. Phys.) pp 545-546.
Rackham G M and Steeds J W 1981 Microscopy of Semiconducting materials
    1981 (Bristol: Inst. Phys.) pp 397-402.

Author Index

Subject Index†

†Page numbers refer to the first pages of the papers in which the citations occur.